Ironworking

Level 1

Trainee Guide

Second Edition

Prentice Hall

Boston Columbus Indianapolis New York San Francisco Upper Saddle River
Amsterdam Cape Town Dubai London Madrid Milan Munich Paris Montreal Toronto
Delhi Mexico City Sao Paulo Sydney Hong Kong Seoul Singapore Taipei Tokyo

NCCER

President: Don Whyte
Director of Product Development: Daniele Stacey
Ironworking Project Managers: Jennifer Wilkerson/Rob Richardson
Production Manager: Tim Davis
Quality Assurance Coordinator: Debie Ness
Editor: Chris Wilson
Desktop Publishing Coordinator: James McKay
Production Assistant: Laura Wright

NCCER would like to acknowledge the contract service provider for this curriculum: Topaz Publications, Liverpool, New York.
Lead Writer/Project Manager: Rob Hopkins
Desktop Publisher: Joanne Hart:
Art Director: Megan Paye
Permissions Editors: Andrea LaBarge, Alison Richmond
Writers: Roy Parker, Troy Staton, Gerald Shannon

PEARSON

www.pearsonhighered.com

ISBN 13: 978-0-13-213714-0

Preface

To the Trainee

Congratulations! If you are training under an NCCER Accredited Training Sponsor, you have successfully completed *Core Curriculum*. Now you are well on your way to more specific skills training in ironworking.

As the first of three levels of *Ironworking*, this curriculum provides a basic foundation in ironworking safety, tools, rigging, and welding. This training is a rich source of knowledge and experience that will give you a competitive edge as you grow in your profession.

We wish you success as you progress through this training program. Should you have any comments on how NCCER might improve upon this textbook, please complete the User Update form located at the back of each module and send it to us. We will always consider and respond to input from our customers.

We invite you to visit the NCCER website at www.nccer.org for information on the latest product releases and training, as well as online versions of the *Cornerstone* newsletter and Pearson's product catalog.

Your feedback is welcome. You may email your comments to curriculum@nccer.org or send general comments and inquiries to info@nccer.org.

Opportunities for Dual Credentials

Opportunities for dual credentials now exist if your training program is being administered through an NCCER Accredited Training Program Sponsor. Successful completion of this training program could lead to a completion certificate for *Ironworking Level One* as well as module completion credit for the *Oxyfuel Cutting* module. This is because the *Oxyfuel Cutting* module is presented as it originally appeared in the *Welding* curriculum.

If you are not training through an NCCER Accredited Training Program Sponsor, find out where one exists near you. Call NCCER at 1.888.622.3720 or visit us at www.nccer.org.

NCCER Standardized Curricula

NCCER is a not-for-profit 501(c)(3) education foundation established in 1995 by the world's largest and most progressive construction companies and national construction associations. It was founded to address the severe workforce shortage facing the industry and to develop a standardized training process and curricula. Today, NCCER is supported by hundreds of leading construction and maintenance companies, manufacturers, and national associations. The NCCER Standardized Curricula was developed by NCCER in partnership with Pearson Education, Inc., the world's largest educational publisher.

Some features of NCCER's Standardized Curricula are as follows:

- An industry-proven record of success
- Curricula developed by the industry for the industry
- National standardization providing portability of learned job skills and educational credits
- Compliance with Office of Apprenticeship requirements for related classroom training (CFR 29:29)
- Well-illustrated, up-to-date, and practical information

NCCER also maintains a National Registry that provides transcripts, certificates, and wallet cards to individuals who have successfully completed modules of NCCER's Standardized Curricula. *Training programs must be delivered by an NCCER Accredited Training Sponsor in order to receive these credentials.*

NCCER Standardized Curricula

NCCER's training programs comprise more than 80 construction, maintenance, and pipeline areas and include skills assessments, safety training, and management education.

Boilermaking
Cabinetmaking
Carpentry
Concrete Finishing
Construction Craft Laborer
Construction Technology
Core Curriculum: Introductory Craft Skills
Drywall
Electrical
Electronic Systems Technician
Heating, Ventilating, and Air Conditioning
Heavy Equipment Operations
Highway/Heavy Construction
Hydroblasting
Industrial Coating and Lining Application Specialist
Industrial Maintenance Electrical and Instrumentation Technician
Industrial Maintenance Mechanic
Instrumentation
Insulating
Ironworking
Masonry
Millwright
Mobile Crane Operations
Painting
Painting, Industrial
Pipefitting
Pipelayer
Plumbing
Reinforcing Ironwork
Basic Rigger
Intermediate Rigger
Advanced Rigger
Scaffolding
Sheet Metal
Signal Person
Site Layout
Sprinkler Fitting
Tower Crane Operator
Welding

Green/Sustainable Construction

Building Auditor
Fundamentals of Weatherization
Introduction to Weatherization
Sustainable Construction Supervisor
Weatherization Crew Chief
Weatherization Technician
Your Role in the Green Environment

Energy

Introduction to the Power Industry
Introduction to Solar Photovoltaics
Introduction to Wind Energy
Power Industry Fundamentals
Power Generation Maintenance Electrician
Power Generation I&C Maintenance Technician
Power Generation Maintenance Mechanic

Pipeline

Control Center Operations, Liquid
Corrosion Control
Electrical and Instrumentation
Field Operations, Liquid
Field Operations, Gas
Maintenance
Mechanical

Safety

Field Safety
Safety Orientation
Safety Technology

Management

Fundamentals of Crew Leadership
Project Management
Project Supervision

Supplemental Titles

Applied Construction Math
Careers in Construction
Tools for Success

Spanish Translations

Rigging Fundamentals (Principios Básicos de Maniobras)
Carpentry Fundamentals (Introducción a la Carpintería, Nivel Uno)
Carpentry Forms (Formas para Carpintería, Nivel Trés)
Concete Finishing, Level One (Acabado de Concreto, Nivel Uno)
Core Curriculum: Introductory Craft Skills (Currículo Básico: Habilidades Introductorias del Oficio)
Drywall, Level One (Paneles de Yeso, Nivel Uno)
Electrical, Level One (Electricidad, Nivel Uno)
Field Safety (Seguridad de Campo)
Insulating, Level One (Aislamiento, Nivel Uno)
Ironworking, Level One (Herrería, Nivel Uno)
Masonry, Level One (Albañilería, Nivel Uno)
Pipefitting, Level One (Instalación de Tubería Industrial, Nivel Uno)
Reinforcing Ironwork, Level One (Herreria de Refuerzo, Nivel Uno)
Safety Orientation (Orientación de Seguridad)
Scaffolding (Andamios)
Sprinkler Fitting, Level One (Instalación de Rociadores, Nivel Uno)

Acknowledgments

This curriculum was revised as a result of the farsightedness and leadership of the following sponsors:

Buckner Companies
Crossland Construction Company, Inc.
KBR, Inc.
Miller Safety Consulting, Inc.
LPR Construction Company
Miller and Long Concrete Construction
Steel Erectors Association of America

This curriculum would not exist were it not for the dedication and unselfish energy of those volunteers who served on the Authoring Team. A sincere thanks is extended to the following:

John Barton
Stacey Cobb
Ernest Garcia
Bryan McClure
Steve Miller
Nelson Plumb
Chip Pocock
Rudy Runko
Eddie Williams

NCCER Partners

American Fire Sprinkler Association
Associated Builders and Contractors, Inc.
Associated General Contractors of America
Association for Career and Technical Education
Association for Skilled and Technical Sciences
Carolinas AGC, Inc.
Carolinas Electrical Contractors Association
Center for the Improvement of Construction Management and Processes
Construction Industry Institute
Construction Users Roundtable
Design Build Institute of America
Merit Contractors Association of Canada
Metal Building Manufacturers Association
NACE International
National Association of Manufacturers
National Association of Minority Contractors
National Association of Women in Construction
National Insulation Association
National Ready Mixed Concrete Association
National Technical Honor Society
National Utility Contractors Association
NAWIC Education Foundation
North American Technician Excellence
Painting & Decorating Contractors of America
Portland Cement Association
SkillsUSA®
Steel Erectors Association of America
U.S. Army Corps of Engineers
University of Florida, M.E. Rinker School of Building Construction
Women Construction Owners & Executives, USA

NCCER Business Partners

ACT
Prov
JUDGMENT INDEX
PEARSON
NACB
CONSTRUCTNET INTERNATIONAL

Contents

30109-11
Structural Ironworking One

Identifies the types of construction that utilize structural steel, the components of the structures, and the process involved in erecting a steel structure. Explains the principles of structural stresses and the requirements of bolted connections. (7.5 Hours)

30110-10
Plumbing, Aligning, and Guying

Describes the purpose and function of aligning and plumbing steel structures, the tools that are used, and the procedures for performing the plumbing and aligning. Identifies and explains column base and baseplate components and foundation failures. (5 Hours)

29102-09
Oxyfuel Cutting

Explains the safety requirements for oxyfuel cutting. Identifies oxyfuel cutting equipment and setup requirements. Explains how to light, adjust, and shut down oxyfuel equipment. Provides instruction on cutting techniques that include straight line, piercing, bevels, washing, and gouging. (17.5 Hours)

(*continued*)

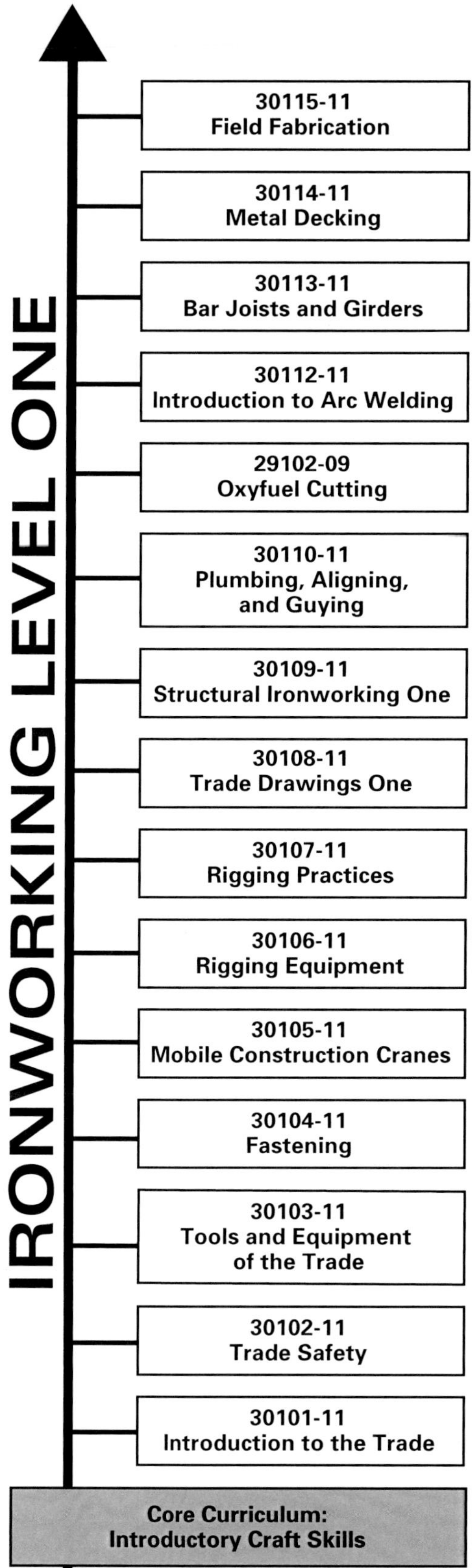

This course map shows all of the modules in *Ironworking Level One*. The suggested training order begins at the bottom and proceeds up. Skill levels increase as you advance on the course map. The local Training Program Sponsor may adjust the training order.

30112-11
Introduction to Arc Welding

Identifies different welding equipment and processes. Describes the safety precautions associated with arc welding. Explains how to identify weld joints, their dimensions, and applications from welding symbols and drawings. Instructs the trainee on how to set up and use SMAW equipment and explains the governing welding codes. (22.5 Hours)

30113-11
Bar Joists and Girders

Explains how to recognize the various types of bar joists and how they are designated. Describes the proper procedures for rigging and storing steel joists. Explains the use of joist girders in steel joist construction systems and the proper erection procedures for bar joists. Includes *OSHA Subpart R.* (5 Hours)

30114-11
Metal Decking

Identifies and explains decking types and profiles and how decking is packaged, shipped, and stored. Describes erecting decking and job-site safety. Discusses the effects of deck penetrations and damage. Includes *OSHA Subpart R.* (10 Hours)

30115-11
Field Fabrication

Identifies the safety hazards associated with field fabrication. Describes how to use common layout tools. Explains how to fabricate angle iron, channel, T-shapes, and W-shapes to given dimensions. (15 Hours)

nccer

Introduction to the Trade

30101-11

Trainees with successful module completions may be eligible for credentialing through NCCER's National Registry. To learn more, go to **www.nccer.org** or contact us at **1.888.622.3720**. Our website has information on the latest product releases and training, as well as online versions of our *Cornerstone* newsletter and Pearson's Contren® product catalog.

Your feedback is welcome. You may email your comments to **curriculum@nccer.org,** send general comments and inquiries to **info@nccer.org**, or use the User Update form at the back of this module.

V.1 1/11

30101-11
INTRODUCTION TO THE TRADE

Objectives

When you have completed this module, you will be able to do the following:

1. Identify the personal qualities that contribute to successful employment.
2. Describe the historical development of the trade.
3. Identify the principles and purpose of apprenticeship training.
4. Identify employer and employee safety obligations.

Performance Tasks

This is a knowledge-based module; there are no performance tasks.

Trade Terms

Apprentice
National Center for Construction Education and Research (NCCER)
Occupational Safety and Health Administration (OSHA)
Related technical instruction

Contents

Topics to be presented in this module include:

Figures and Tables

1.0.0 INTRODUCTION

Opportunity is driven by knowledge and ability, which are in turn driven by education and training. This program of the **National Center for Construction Education and Research (NCCER)** was designed and developed by the construction industry for the construction industry. It is the only nationally accredited, competency-based, construction training program in the United States. A competency-based program requires that the trainee demonstrate the ability to safely perform specific job-related tasks in order to receive credit. This approach is unlike other **apprentice** programs, which merely require a trainee to put in the required number of hours in the classroom and on the job.

The primary goal of the NCCER is to standardize construction craft training throughout the country so that both employers and employees will benefit from the training, no matter where they are located. A trainee in an NCCER program will become part of a national registry. The trainee will receive a certificate for each level of training completed. If the trainee applies for a job with any participating contractor in the country, a transcript of their training will be available. If that training is incomplete when a job transfer is made, the trainee can pick up where they left off, because every participating contractor is using the same training program. In addition, many technical schools and colleges are using the program.

2.0.0 IRONWORKING TRADE

The use of iron and steel in buildings is the most important development in construction since the times of the ancient Greeks and Romans. Steel framing makes it possible to build stronger and taller structures with less use of material than stone, brick, or wood. It can also allow for greater unsupported spans over openings and more interior space.

2.1.0 History of Structural Steel Building Materials

The growth and change of the ironworking trade has been a natural consequence of the production of steel, the development of steel alloys, and the technological advances in the construction trade. The history of steel-framed buildings and steel bridges represents the history of the people involved in their construction. Ironworkers can consider such buildings as the Eiffel Tower, the Empire State Building, the Astrodome, the Vehicle Assembly Building at Cape Canaveral, Florida, and such bridges as the 1,200-foot Port Mann arch bridge in Vancouver, British Columbia, as monuments to their efforts.

Although structural steel framing has always offered great flexibility to accommodate a variety of architectural and structural challenges, several recent developments have advanced the technology of steel-framed construction. Among them are production methods, fabricating capabilities, and technological advances.

Today's production methods use recycled scrap to produce steels that are nearly 40-percent stronger, yet 30-percent less expensive than the steels of just 20 years ago. Modern fabricating capabilities have held the labor cost involved in creating the components of a structural steel frame in check while steadily increasing the potential to efficiently produce even the most complex structural form.

The new standards developed to make buildings more environmentally beneficial call for the reuse of materials from other buildings. Those building materials may be from the ones torn down to make room for the new structures, or materials from other buildings being demolished elsewhere. The result is major savings in energy use and raw materials for new construction (*Figure 1*).

Finally, advances in design and construction technologies and concepts have improved the function of steel solutions in the majority of applications. New designs for buildings take advantage of the capabilities of steel construction to produce fanciful building shapes (*Figure 2*).

2.2.0 Ironworking

Ironworkers perform a variety of tasks that are all centered around the erection of large steel or concrete structures. They erect or repair the steel frame and other metal parts of buildings, bridges, and structures, such as towers and storage tanks. They also rig heavy construction machinery and deliver it to new work sites. Ironworking is divided into three related trades:

- Structural ironworkers erect, align, and fasten the steel framework of a structure (*Figure 3*).
- Riggers and machine movers set up and rig the hoisting equipment used to erect and dismantle structural steel frames. They also move heavy construction equipment and machinery (*Figure 4*). Only a designated and trained competent person should be allowed to direct crane operators.

101F01.EPS

Figure 1 A building built with recycled structural materials.

101F02.EPS

Figure 2 An example of a modern building structure.

- Ornamental ironworkers install metal ladders, stairways, catwalks, floor gratings, cabinets, window frames, lamp posts, fences, and decorative ironwork. They also bolt or weld prefabricated frames and panels to buildings.

Most ironworkers spend time working at various heights off the ground. The safety procedures involved in such work must be followed at all times (*Figure 5*). Ironworkers are the industrial workers most likely to be injured, most commonly either by falling or by objects falling on them. If a worker carefully watches overhead objects, and avoids standing under objects, that worker probably will not be struck by falling objects. An ironworker who always uses fall protection equipment correctly when off the ground probably will not fall. There are hazards of design or situations involving other workers that cannot always be controlled, but every worker has the personal responsibility of being safety conscious at all times.

3.0.0 Opportunities in the Construction Industry

The construction industry employs more people and contributes more to the nation's economy than any other industry. Our society will always need new homes, roads, airports, hospitals, schools, factories, and office buildings. This means that there will always be a source of well-paying jobs and

101F03.EPS

Figure 3 Fastening beams together.

101F04.EPS

Figure 4 Crane in use.

career opportunities for ironworkers and other construction trade professionals. As shown in *Figure 6*, the opportunities are not limited to work on construction projects. A skilled, knowledgeable ironworker can work in a number of areas.

As a construction worker, an ironworker can progress from apprentice through several levels:

- *Journeyman ironworker* – After successfully completing an apprenticeship, a trainee becomes a journeyman. The term *journeyman* originally meant to journey away from the master and work alone. A person can remain a journeyman or advance in the trade. Journeymen may have additional duties such as supervisor or estimator. With larger companies and on larger jobs, journeymen often become specialists.
- *Master ironworker* – A master craftsperson is one who has achieved and continuously demonstrates the highest skill levels in the trade. The master is a mentor and teacher of those who follow. Master ironworkers often start their own businesses and become contractors/owners.
- *Foreman/lead ironworker* – This individual is a front-line leader who directs the work of a crew of craftworkers and laborers.
- *Supervisor* – Large construction projects require supervisors who oversee the work of crews made up of foremen, apprentices, and journeymen. They are responsible for assigning, directing, and inspecting the work of construction crew members. The supervisor is most likely to handle the day-to-day operations on the job site.

101F05.EPS

Figure 5 Fall protection equipment in use.

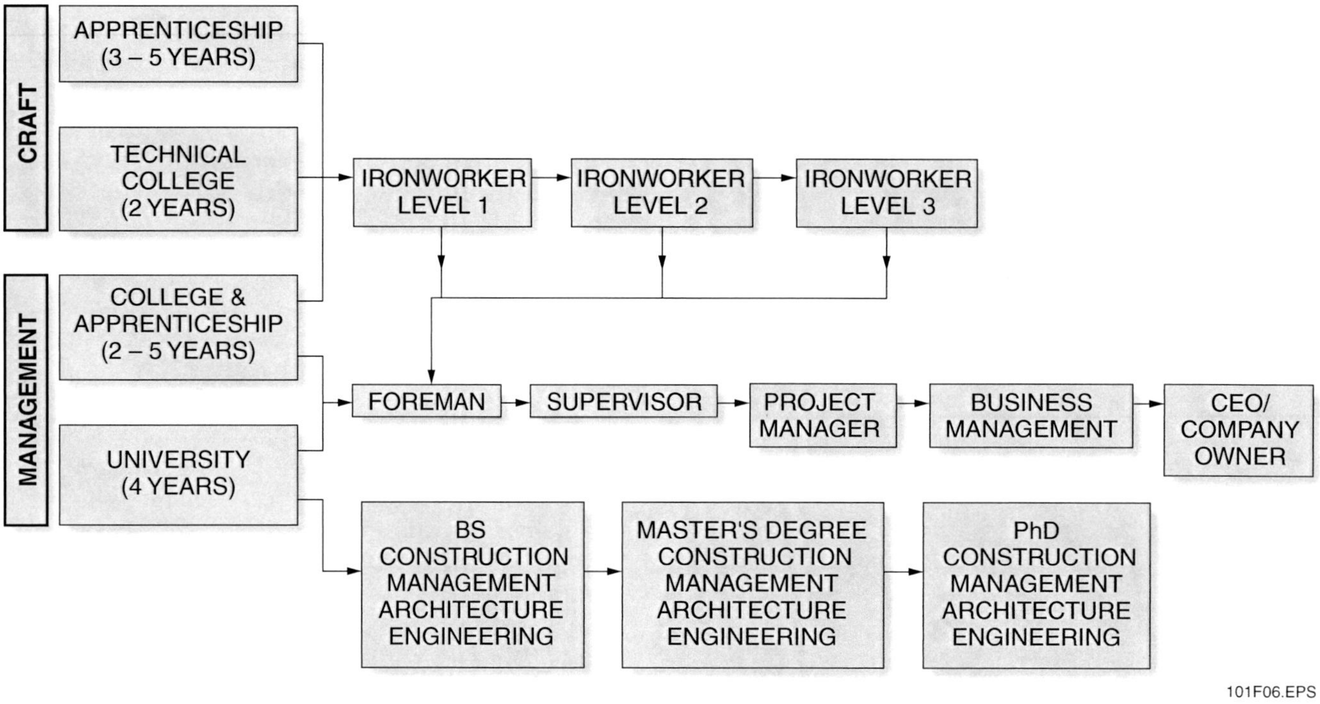

Figure 6 Opportunities in the construction industry.

- *Field engineer* – The field engineer is responsible for resolving engineering problems that arise on the site. This might include designing field constructions necessary to achieve the design requirements, or clarifying questions about the design.
- *Project engineer* – The project engineer is the design engineer for a particular project, and is responsible for the project being designed correctly.
- *Surveyor* – A surveyor is the person who performs the site layout work and places all the initial pieces where they belong. He or she works with optical and electronic equipment to ensure the accuracy of both the initial layout and the location of structural points, such as floor levels, columns, stairways, etc.
- *Safety manager* – The safety manager is responsible for project safety and health-related issues, including development of the safety plan and procedures, safety training for workers, and regulatory compliance.
- *Project manager/administrator* – Business management and administration deal with the scope and direction of the business and such concerns as payroll, taxes, and employee benefits. Larger contracting firms may have one or several managers/administrators. A project manager is responsible for worker output and must determine the best methods to use and the best way to apply workers to accomplish the job. A project administrator is responsible for a contractor's support operations, such as accounting, finance, and secretarial work.
- *Estimator* – Estimators work for contractors and building supply companies. They make careful estimates of the materials and labor required for a job. Based on these estimates, the contractor submits bids for jobs. Estimating requires a complete understanding of construction methods as well as the materials and supplies required. Only experienced ironworkers who possess good math skills and the patience to prepare detailed, accurate estimates are employed to do this work. This is a highly responsible position since errors in estimates can result in financial losses to the contractor. Depending on the size and type of the business, the job of estimating may be done by the owner, manager, administrator, or an estimating specialist. Today's estimators need solid computer skills because advances in computer software have revolutionized the field of estimating.
- *Architect* – An architect is a person who is licensed to design buildings and oversee their construction. A person normally needs a specialized degree in architecture to qualify as an architect.
- *General contractor* – A general contractor is an individual or company that manages an entire construction project. The general contractor plans and schedules the project, buys the materials, and usually contracts with carpentry,

plumbing, electrical, and other trade contractors to perform the work. The general contractor usually works with architects, engineers, and clients and/or the client's construction manager in planning and implementing a project. General contracting is a natural career path for a master ironworker because, of the many trades involved in a construction project, the ironworker generally plays the largest role and is more likely to have knowledge of the other trades. The general (prime) contractor is also responsible for safety on site.

- *Construction manager* – The role of the construction manager (CM) is different from that of the general contractor. The CM is usually hired by the building owner to represent the owner's interests on the project. The CM is the individual who works with the general contractor and architect to ensure that the building meets the owner's requirements.
- *Contractor/owner* – Construction contractors/owners are those who have established a contracting business. Generally, they hire apprentices, journeymen, and master ironworkers to work for them. Depending upon the size of the business, contractors may work with the crew or they may manage the business full-time. Very small contractors may have only one or two people do everything, including managing the business, preparing estimates, obtaining supplies, and doing the work on the job. This group includes specialty subcontractors who perform specialized tasks such as decking, joist hanging, and truss work.

The important thing to learn is that, regardless of the path chosen, a career is a lifelong learning process. An effective ironworker needs to keep up to date with new tools, materials, and methods. If a worker chooses to work in management, or to someday start their own construction business, they need to learn management and administrative skills on top of keeping their ironworking skills honed. Almost every successful manager and business owner started the same way, and they all have one thing in common: a desire and willingness to keep on learning. The learning process begins with apprentice training.

As the trainee develops ironworking skills and gains experience, he or she will have the opportunity to earn greater pay for services. There is great financial incentive for learning and growing within the trade. No one can get to the top, however, without learning the basics.

4.0.0 YOUR TRAINING PROGRAM

The Department of Labor's Office of Apprenticeship sets the minimum standards for training programs across the country. These programs rely on mandatory classroom instruction and on-the-job learning (OJL). They require at least 144 hours of classroom instruction per year and 2,000 hours of OJL per year. In a typical apprenticeship program, trainees spend a total of at least 576 hours in classroom instruction and 6,000 hours in OJL before receiving journeyman certificates issued by registered apprenticeship programs.

4.1.0 Standardized Training by NCCER

NCCER is a not-for-profit education foundation established by the nation's leading construction companies. NCCER was created to provide the industry with standardized construction education materials, the Contren® Learning Series, and a system for tracking and recognizing students' training accomplishments—NCCER's National Registry. Refer to the *Appendix* for examples of NCCER credentials.

NCCER also offers accreditation, instructor certification, and skills assessments. NCCER is committed to developing and maintaining a training process that is internationally recognized, standardized, portable, and competency-based.

Working in partnership with industry and academia, NCCER has developed a system for program accreditation that is similar to those found in institutions of higher learning. NCCER's accreditation process ensures that students receive quality training based on uniform standards and criteria. These standards are outlined in NCCER's Accreditation Guidelines and must be adhered to by NCCER Accredited Training Sponsors.

More than 518 training and assessment centers across the U.S. and eight other countries are proud to be NCCER Accredited Training Sponsors. Millions of craft professionals and construction managers have received quality construction education through NCCER's network of Accredited Training Sponsors and the thousands of Training Units associated with the Sponsors. Every year the number of NCCER Accredited Training Sponsors increases significantly.

A craft instructor is a journeyman craft professional or career and technical educator trained and certified to teach NCCER's Contren® Learning Series. This network of certified instructors ensures that NCCER training programs will meet the standards of instruction set by the industry. There are more than 4,544 master trainers and 47,374 craft instructors within the NCCER instructor network. More information is available at www.nccer.org.

4.2.0 Apprenticeship Program

Apprenticeship as a means of transferring practical knowledge from one generation to another, goes far back in human history. Provisions for the teaching of apprentices are found in the Babylonian Code of Hammurabi written 4,000 years ago. In ancient Rome, the high level of craftsmanship that produced aqueducts, great public buildings like the Colosseum, and a large network of roads was the result of craftsmen and artisans whose skills were handed down from father to son.

One existing apprentice contract provided that a man named Heraclas be taught weaving over a period of five years. He was given food, 20 holidays a year, and a yearly tunic. After two and a half years he was paid 12 drachmas; in his fifth year he was to make double that.

During the Middle Ages (300 AD to about 1500 AD), merchants and skilled artisans organized themselves into guilds, which attempted to control their respective fields in order to keep down competition and provide training for apprentices. The guilds established the indenture system. Sons of journeymen were indentured to a master who agreed to train the apprentice in the skills of the craft as well as to provide food, clothing, and shelter. The length of apprenticeship varied from two to ten years, depending on the trade.

The basic principles of apprenticeship training have not changed over time. First, it is a means for individuals entering the craft to learn from those who have mastered the craft. Second, it focuses on learning by doing; real skills versus theory. Although some theory is presented in the classroom, it is always presented in a way that helps the trainee understand the purpose behind the skill that is to be learned.

4.2.1 Apprenticeship Standards

All apprenticeship standards prescribe certain work-related or on-the-job learning. This on-the-job learning is broken down into specific tasks in which the apprentice receives hands-on training during the period of the apprenticeship. In addition, a specified number of hours is required in each task. The total number of hours for the ironworking apprenticeship program is traditionally 8,000, which amounts to about four years of training. In a competency-based program, it may be possible to shorten this time by testing out of specific tasks through a series of performance exams.

The traditional apprenticeship program required OJL of 2,000 hours per year and a minimum of 144 hours of classroom-related training. Apprenticeship programs therefore varied in length from 2,000 hours to 10,000 hours. However, on October 29, 2008, the U.S. Department of Labor (DOL) published new regulations to modernize the National Apprenticeship System. These regulations provide for more flexibility in how related technical instruction (RTI) can be delivered. They also provide registration agencies with the option of issuing interim credentials to offer active apprentices official recognition of their accomplishments and to equip them with a portfolio of skills and incentives to complete their programs and continue their career preparation. Finally, these regulations allow program sponsors to offer three different ways for apprentices to complete a registered apprenticeship program:

- The traditional, time-based approach, which requires the apprentice to complete a specific number of OJL and RTI hours.
- A competency-based approach, which requires the apprentice to demonstrate competency in the defined subject areas and requires OJL and RTI.
- A hybrid approach, which requires the apprentice to complete a minimum number of OJL and RTI hours and demonstrate competency in the defined subject areas.

The apprentice must log all work time and turn it in to the apprenticeship committee so that accurate time control can be maintained. Another important aspect of keeping work records up to date is that after each 1,000 hours of related work, the apprentice will receive a pay increase as prescribed by the apprenticeship standards.

The classroom-**related technical instruction** and work-related training will not always run concurrently due to such reasons as layoffs, type of work needed to be done in the field, etc. Furthermore, apprentices with special job experience or coursework may obtain credit toward their classroom requirements. This reduces the total time required in the classroom while maintaining the total on-the-job learning requirement. These special cases will depend on the type of program and the regulations and standards under which it operates.

Informal on-the-job learning provided by employers is usually less thorough than that provided through a formal apprenticeship program. The degree of training and supervision in this type of program often depends on the size of the employing firm. A small contractor may provide training in only one area, while a large company may be able to provide training in several areas.

For those entering an apprenticeship program, a high school or technical school education is desirable, as are courses in shop, mechanical drawing, and general mathematics. Manual dexterity, good physical condition, and quick reflexes are important. The ability to solve problems quickly and accurately and to work closely with others is essential, along with having a high concern for safety.

The prospective apprentice must submit certain information to the apprenticeship committee. This may include the following:

- Aptitude test (General Aptitude Test Battery or GATB Form Test) results (usually administered by the local Employment Security Commission)
- Proof of educational background (candidate should have school transcripts sent to the committee)
- Letters of reference from past employers and friends
- Results of a physical examination
- Proof of age
- If the candidate is a veteran, a copy of DD Form 214
- A record of technical training received that relates to the construction industry and/or a record of any pre-apprenticeship training
- High school diploma or General Equivalency Diploma (GED)

The apprentice must do the following:

- Wear proper safety equipment on the job
- Purchase and maintain tools of the trade as needed and required by the contractor
- Submit a monthly on-the-job learning report to the committee
- Report to the committee if a change in employment status occurs
- Attend classroom-related instruction and adhere to all classroom regulations such as attendance requirements

5.0.0 Responsibilities of the Employee

In order to be successful, the professional must have the skills to use current trade materials, tools, and equipment to finish the task in a minimum period of time. An ironworker must be adept at adjusting methods to meet each situation and continuously train to remain knowledgeable about technical advancements and gain the skills to use them. A professional never takes chances with regard to personal safety or the safety of others.

5.1.0 Professionalism

The word *professionalism* is a broad term that describes the desired overall behavior and attitude expected in the workplace. Professionalism is too often absent from the construction site and the various trades. Most people would argue that professionalism must start at the top in order to be successful. It is true that management support of professionalism is important to its success in the workplace, but it is more important that individuals recognize their own responsibility for professionalism.

Professionalism includes honesty, productivity, safety, civility, cooperation, teamwork, clear and concise communication, being on time and prepared for work, and regard for one's impact on co-workers. It can be demonstrated in a variety of ways in the workplace. Most important is that the workers do not tolerate the unprofessional behavior of co-workers. This is not to say that anyone shuns the unprofessional worker; instead, everyone should work to demonstrate the benefits of professional behavior.

Professionalism is both a benefit to the employer and the employee. It is a personal responsibility. Our industry is what each individual chooses to make of it; choose professionalism and the industry image will follow.

5.2.0 Honesty

Honesty and personal integrity are important traits of the successful professional. Professionals pride themselves in performing a job well, and in being punctual and dependable. Each job is completed in a professional way, never by cutting corners or reducing materials. A valued professional maintains work attitudes and ethics that protect property such as tools and materials belonging to employers, customers, and other trades from damage or theft at the shop or job site.

Honesty and success go hand in hand. It is not simply a choice between good and bad, but a choice between success and failure. Dishonesty will always catch up with people. Whether someone is stealing materials, tools, or equipment from the job site or simply lying about their work, it will not take long for employers to find out. While a dishonest worker may find another employer, this option will ultimately run out.

Honesty means more than just not taking things that you do not own, however. It means giving a fair day's work for a fair day's pay. It means carrying out your side of a bargain. It means that your words convey true meanings and actual facts. Your thoughts as well as your actions should be honest. Employers place a high value on an employee who is strictly honest.

To be successful, enjoy continuous employment and consistency of earnings, and to be sought after as opposed to seeking employment, start out with a standard of honesty in the workplace and you will reap the benefits.

5.3.0 Loyalty

Employees expect employers to look out for their interests, to provide them with steady employment, and to promote them to better jobs as openings occur. Employers feel that they, too, have a right to expect their employees to be loyal to them—to keep their interests in mind, to speak well of them to others, to keep any minor troubles strictly within the plant or office, and to keep absolutely confidential all matters that pertain to the business. Both employers and employees should keep in mind that loyalty is not something to be demanded; rather, it is something to be earned. If a fellow worker complains about the company, or about policies, the worker should listen carefully to the complaints. Certainly, a worker should neither argue nor join in the complaints. However, it is a good idea to add some comments, in a tactful manner, on the good policies and actions of the company, to help deflect the other employee's anger.

5.4.0 Willingness to Learn

Every office and plant has its own way of doing things. Employers expect their workers to be willing to learn these ways. Adapting to change and a willingness to learn new methods and procedures as quickly as possible are key. Sometimes changes in safety regulations or the purchase of new equipment makes it necessary for even experienced employees to learn new methods and operations. It is often the case that employees resent having to accept improvements because of the retraining that is involved. However, employers will no doubt think they have a right to expect employees to put forth the necessary effort. Methods must be kept up to date in order to meet competition and show a profit. It is this profit that enables the owner to continue in business and provide jobs for the employees.

5.5.0 Willingness to Take Responsibility

Most employers expect their employees to see what needs to be done, then go ahead and do it. It is very tiresome to have to ask again and again that a certain job be done. It is obvious that having been asked once, an employee should assume the responsibility from then on. It is true that, in general, responsibility should be delegated and not assumed. Once the responsibility has been delegated, however, the employee should continue to perform the duties without further direction. Every employee has the responsibility for working safely.

5.6.0 Willingness to Cooperate

To cooperate means to work together. In our modern business world, cooperation is the key to getting things done. Each worker must learn to work as a member of a team with employers, supervisors, and fellow workers in a common effort to get the work done efficiently, safely, and on time.

5.7.0 Rules and Regulations

People can work together well only if there is some understanding about what work is to be done, when and how it will be done, and who will do it. Rules and regulations are a necessity in any work situation and should be so considered by all employees.

5.8.0 Tardiness and Absenteeism

Tardiness means being late for work, and absenteeism means being off the job for one reason or another. Consistent tardiness and frequent absences are an indication of poor work habits, unprofessional conduct, and a lack of commitment.

Everyone is, to some extent, a creature of habit. What people do once they will tend to do again, unless the results are too unpleasant. The habit of always being late may have begun in a person's school days, when they found it hard to get up in the morning. This habit can get students into trouble at school, and it can continue to get them into trouble when they finish with school and go to work.

Work life is governed by the clock. Everyone is required to be at work at a definite time. Failure to get to work on time results in confusion, lost time, and resentment on the part of those who do come on time. In addition, it may lead to penalties, including dismissal. Although it may seem that a few minutes out of a day are not very important, remember that a principle is involved. It is the worker's obligation to be at work at the time indicated. Workers agree to the terms of work when they accept the job.

Look at the matter from the point of view of the supervisor. Supervisors cannot keep track of people if they come in any time they please. It is not fair to others to ignore tardiness. Failure to be on time may hold up the work of fellow workers. Better planning of the worker's morning routine will often keep them from being delayed and so prevent a late arrival. In fact, arriving a little early indicates interest and enthusiasm for the work, which is appreciated by employers. The habit of being late is one thing that stands in the way of promotion.

It is sometimes necessary to take time off from work. No one should be expected to work when sick or when there is serious trouble at home. However, it is possible to get into the habit of letting unimportant and unnecessary matters keep us from the job. This results in lost production and hardship on those who try to carry on the work with less help. Again, there is a principle involved. The person who hires a worker has a right to expect them to be on the job unless there is some very good reason for staying away. Workers should certainly not let some trivial reason keep them at home. They should not stay up nights until they are too tired to go to work the next day. If workers are ill, they should use the time at home to recover as quickly as possible. This, after all, is what anyone would expect of a person they had hired to work for them, and on whom they depended to do a certain job.

If it is necessary to stay home, then the worker should at least phone the office early in the morning so that the supervisor can find another worker for the day. Time and again, employees remain home without sending any word to their employer. This is the worst possible way to handle the matter. It leaves those at work uncertain about what to expect. They have no way of knowing whether the employee has merely been held up and will be in later, or whether immediate steps should be taken to assign their work to someone else. Simple courtesy demands that workers let their supervisor know if they cannot come to work.

The most frequent causes of absenteeism are illness, death in the family, accidents, personal business, and dissatisfaction with the job. Some of the causes are legitimate and unavoidable, while others can be controlled. People can usually plan to conduct personal business affairs after working hours. Frequent absences will reflect unfavorably on a worker when promotions are being considered.

Employers sometimes resort to docking pay, demotion, and even dismissal in an effort to control tardiness and absenteeism. No employer likes to impose restrictions of this kind. However, in fairness to those workers who do come on time and who do not stay away from the job, an employer is sometimes forced to discipline those who will not follow the rules.

6.0.0 Human Relations

Most people underestimate the importance of working well with others. There is a tendency to pass off human relations as nothing more than common sense. What exactly is involved in human relations? One response would be to say that part of human relations is being friendly, pleasant, courteous, cooperative, adaptable, and sociable.

6.1.0 Making Human Relations Work

As important as the previously noted characteristics are for personal success, they are not enough. Human relations is much more than just getting people to like each other. It is also knowing how to handle difficult situations as they arise.

Human relations has to do with knowing how to work with supervisors who are often demanding and sometimes unfair. It involves understanding one's own personality traits and those of the people one works with. Building sound working relationships in various situations is important. If working relationships have deteriorated for one reason or another, restoring them is essential. Learning how to handle frustrations without hurting others is part of human relations.

6.2.0 Human Relations and Productivity

Effective human relations is directly related to productivity. Safe productivity is the key to business success. Every employee is expected to produce at a certain level. Employers quickly lose interest in an employee who has a great attitude but is able to produce very little. There are work schedules to be met and jobs that must be completed.

All employees, both new and experienced, are measured by the amount of quality work they can safely turn out. The employer expects every employee to do his or her share of the workload.

However, doing one's share in itself is not enough. If a worker is to be productive, he or she must do their share (or more than their share) without antagonizing fellow workers. Workers must perform their duties in a manner that encourages others to follow that example. It makes little difference how ambitious a worker is or how capably they perform. Nobody can become the kind of employee they want to be or the type of worker management wants them to be without learning how to work with their peers. Employees must do everything possible to build strong, professional working relationships with fellow employees, supervisors, and clients.

6.3.0 Attitude

A positive attitude is essential to a successful career. First, being positive means being energetic, highly motivated, attentive, and alert. A positive attitude is essential to safety on the job. Second, a positive employee contributes to the productivity of others. Both negative and positive attitudes are transmitted to others on the job. A persistent negative attitude can spoil the positive attitudes of others. It is very difficult to maintain a high level of productivity while working next to a person with a negative attitude. Third, people favor a person who is positive. Being positive makes a person's job more interesting and exciting. Finally, the kind of attitude transmitted to management has a great deal to do with future success in the company. Supervisors can determine someone's attitude by their approach to the job, reactions to orders, and the way they handle problems.

6.4.0 Maintaining a Positive Attitude

A positive attitude is far more than a smile, which is only one example of an inner positive attitude. As a matter of fact, some people transmit a positive attitude even though they seldom smile. They do this by the way they treat others, the way they look at their responsibilities, and the approach they take when faced with problems. Here are a few suggestions that will help to maintain a positive attitude:

- Remember that a positive attitude helps in all areas of life. If someone makes a greater effort to be a more positive person in their social and personal lives, it will automatically help them on the job. The reverse is also true. One effort will complement the other.
- Negative comments are seldom welcome on the job. Stay positive and be complimentary. Constant complainers do not build good relationships.
- Look for the good things in people on the job, especially your supervisor. Nobody is perfect and almost everyone has a few worthwhile qualities. If a worker dwells on people's good features, it will be easier to work with them.
- Look for the good things about work. What are the factors that make it a good place to work? Is it the hours, the physical environment, the people, the actual work being done? Or is it the atmosphere? Keep in mind that nobody can expect to like everything. No work assignment is perfect, but if the employee concentrates on the good things, the negative factors will seem less important and bothersome.
- Look for the good things in the company. Just as there are no perfect assignments, there are no perfect companies. Nevertheless, almost all organizations have good features. Is the company progressive? What about promotional opportunities? Are there chances for self-improvement? What about the wage and benefit package? Is there a good training program? No one can expect to have everything they would like, but there should be enough to keep them positive. In fact, if a worker decides to stay with a company for a long period of time, it is wise to look at the good features and think about them. If the employee thinks positively, he or she will act the same way.
- One may not be able to change the negative attitude of another employee, but one can protect their own attitude from becoming negative.

7.0.0 Employer and Employee Safety Obligations

The U.S. Congress passed the *Occupational Safety and Health Act* in 1970. This act also created the **Occupational Safety and Health Administration (OSHA)**. It is part of the U.S. Department of Labor. The job of OSHA is to set occupational safety and health standards for all places of employment, enforce these standards, ensure that employers provide and maintain a safe workplace for all employees, and provide research and educational programs to support safe working practices.

OSHA requires each employer to provide a safe and hazard-free working environment. OSHA also requires that employees comply with OSHA rules and regulations that relate to their conduct on the job. To gain compliance, OSHA can perform spot inspections of job sites, impose fines for violations, and even stop any more work from proceeding until the job site is safe. Remember, safety is everyone's responsibility (*Figure 7*).

According to OSHA standards, all employees are entitled to on-the-job safety learning. Each employer must do the following:

101F07.EPS

Figure 7 Safety is everyone's responsibility.

- Show workers how to do the job safely
- Provide the required PPE
- Warn workers about specific hazards
- Supervise workers for safety while performing the work

OSHA was adopted with the stated purpose to ensure as far as possible every working man and woman in the nation safe and healthful working conditions and to preserve our human resources.

The enforcement of this act of Congress is provided by the federal and state safety inspectors who have the legal authority to make employers pay fines for safety violations. The law allows states to have their own safety regulations and agencies to enforce them, but they must first be approved by the U.S. Secretary of Labor. For states that do not develop such regulations and agencies, federal OSHA standards must be obeyed.

These standards are listed in *OSHA Safety and Health Standards for the Construction Industry (29 CFR, Part 1926)*, sometimes called *OSHA Standards 1926* (*Figure 8*). Other safety standards that apply to construction are published in *OSHA Safety and Health Standards for General Industry (29 CFR, Parts 1900 to 1910)*.

The most important general requirements that OSHA places on employers in the construction industry are as follows:

- The employer must perform frequent and regular job site inspections of equipment.
- The employer must instruct all employees to recognize and avoid unsafe conditions, and to know the regulations that pertain to the job so they may control or eliminate any hazards.
- No one may use any tools, equipment, machines, or materials that do not comply with *OSHA Standards 1926*.
- The employer must ensure that only qualified individuals operate tools, equipment, and machines.

Employees also have safety obligations. In exchange for the benefits of employment and wellbeing, he or she agrees to work safely. In other words, employees are obligated to work safely. Employees are also obligated to make sure anyone they supervise or work with is working safely.

Some employers have safety committees. If an employee works for such an employer, he or she is then obligated to that committee to maintain a safe working environment. This means they must do two things:

- Follow the safety committee's rules for proper working procedures and practices.
- Report any unsafe equipment and conditions directly to the committee or the supervisor.

On the job, if something is not safe, report it! Do not ignore it. It will not correct itself. Each employee has an obligation to report safety issues to the immediate supervisor.

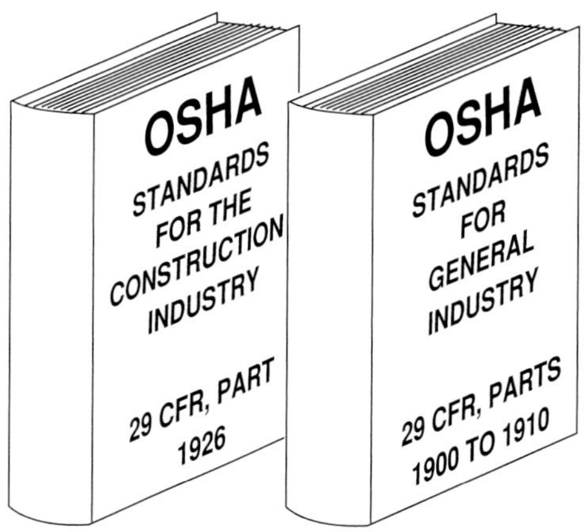

101F08.EPS

Figure 8 OSHA standards govern workplace safety.

In the long run, an unsafe condition affects everyone. Always report unsafe conditions. Do not think an employer will be angry because productivity suffers while the condition is corrected. On the contrary, an employer would be more likely to criticize a worker for not reporting a problem.

Employers know that the short time lost in making conditions safe again is nothing compared with shutting down the whole job because of a major disaster. If that happens, employees are out of work anyway. So do not ignore an unsafe condition. In fact, OSHA regulations require workers to report hazardous conditions.

This applies to every part of the construction industry. Whether employed by a large contractor or a small subcontractor, everyone is obligated to report unsafe conditions. The easiest way to do this is to tell the supervisor. If that person ignores the unsafe condition, report it to the safety director. If it is the owner who is being unsafe, the worker should let that person know their concerns. Think about it in terms of someone's life being at risk.

SUMMARY

There are many job and career opportunities for skilled ironworkers. An apprenticeship program that combines competency-based, hands-on training with classroom instruction has proven to be the most effective means for a person to learn and advance in the ironworking craft. Developing job skills is only part of the solution; it is just as important for the worker to learn good work habits, convey a positive, cooperative attitude to those around them, and practice good safety habits every day.

Review Questions

1. The most important change in building construction since the times of the Greeks and Romans has been the use of _____.
   a. cement
   b. iron and steel
   c. masonry
   d. precast

2. The growth and change in the ironworking trade has been a natural consequence of the production of steel, the technological advances in the construction trade, and the _____.
   a. rising cost of lumber
   b. shortage of synthetic fibers
   c. development of steel alloys
   d. widespread use of plastics

3. The position that follows an apprenticeship is _____.
   a. master ironworker
   b. journeyman
   c. general contractor
   d. estimator

4. The day-to-day operations on the job site are most likely to be handled by the _____.
   a. supervisor
   b. project manager
   c. architect
   d. owner

5. The typical total of on-the-job learning needed for an ironworking apprentice to advance to journeyman is approximately _____.
   a. 2,000 hours
   b. 4,000 hours
   c. 6,000 hours
   d. 8,000 hours

6. A competency-based training program is one that requires the student to _____.
   a. receive at least four years of classroom training
   b. receive on-the-job training for at least two years
   c. demonstrate the ability to perform specific job-related tasks
   d. pass a series of written tests

7. Which of the following is true with respect to honesty?
   a. It is okay to borrow tools from the job site as long as you return them before anyone notices.
   b. You are doing your company a favor by using lower-grade materials than those listed in the specifications.
   c. It is okay to take materials or tools from your employer if you feel the company owes you for past efforts.
   d. Not giving a fair day's work for a fair day's pay is the same as stealing from your employer.

8. If one of your co-workers complains about your company, you should _____.
   a. contribute your own complaints to the conversation
   b. find some good things to say about the company
   c. suggest that the person look for another job
   d. agree with the person to avoid conflict

9. If you see an unsafe condition on the job, the best action you should take is to _____.
   a. ignore it because it is not your job
   b. tell a co-worker
   c. contact the owner
   d. report it to the immediate supervisor

10. The purpose of OSHA is to _____.
    a. catch people breaking safety regulations
    b. make rules and regulations governing all aspects of construction projects
    c. ensure that the employer provides and maintains a safe workplace
    d. assign a safety inspector to every project

Trade Terms Introduced in This Module

Apprentice: A worker who is learning a craft.

National Center for Construction Education and Research (NCCER): An organization whose primary goal is to standardize construction craft training throughout the country so both employers and employees will benefit from the training, no matter their location.

Occupational Safety and Health Administration (OSHA): A division of the U.S. Department of Labor that is responsible for establishing and enforcing safety and health standards for all places of employment.

Related technical instruction: Classroom instruction that is designed to support on-the-job learning.

Appendix A

Samples of NCCER Training Credentials

NATIONAL CENTER FOR CONSTRUCTION EDUCATION AND RESEARCH

August 27, 2009

Sample Student
National Center for Construction Education and Research
3600 NW 43rd St Bldg G
Gainesville, FL 32614-1104

Dear Sample,

On behalf of the National Center for Construction Education and Research, I congratulate you for successfully completing NCCER's Contren® Learning Series program. I also congratulate you for choosing construction as a career.

You are now a valuable member of one of our nation's largest industries. The skills you have acquired will not only enhance your career opportunities, but will help build America.

Enclosed are your credentials from the National Registry. These industry-recognized credentials give you flexibility in planning your career and ensure your achievements follow you wherever you go.

To access your training accomplishments through the Automated National Registry, follow these instructions:

1. Go to www.nccer-anr.org.
2. Click the "Individuals" button.
3. Enter the NCCER card number, located on front of your wallet card or transcript, and your PIN.
   Note: The default PIN is the last four digits of your SSN. You may change your PIN after you login.
4. First-time users will be directed to answer a few security questions upon initial login.
5. Contact the registry department with any questions.

NCCER applauds your dedication and wishes you the best in your future endeavors.

Sincerely,

Donald E. Whyte

Donald E. Whyte
President, NCCER

Enc.

3600 NW 43rd St, Bldg G ◦ Gainesville, FL 32606 ◦ P 352.334.0911 ◦ F 352.334.0932 ◦ www.nccer.org

National Center for Construction Education and Research

This is to certify that

Sample Student

has fulfilled the requirements for

Ironworking

in NCCER's standardized training curriculum
this Tenth day of December, 2010

NATIONAL CENTER FOR CONSTRUCTION EDUCATION AND RESEARCH • FOUNDED 1996

Donald E. Whyte
President

NATIONAL CENTER FOR CONSTRUCTION EDUCATION AND RESEARCH
BUILDING TOMORROW'S WORKFORCE
3600 NW 43rd St, Bldg G ◦ Gainesville, FL 32606
P 352.334.0911 ◦ F 352.334.0932 ◦ www.nccer.org
Affiliated with the University of Florida

August 27, 2009

Official Transcript

Sample Student
National Center for Construction Education and Research
3600 NW 43rd St Bldg G
Gainesville. FL 32614-1104

Current Employer/School:

Card #: 2781481

Course / Description		Instructor	Training Location	Date Compl.
00101	Basic Safety	Don E Whyte		1/1/2001
00102	Basic Math	Don E Whyte		1/1/2001
00103	Introduction to Hand Tools	Don E Whyte		1/1/2001
00104	Introduction to Power Tools	Don E Whyte		1/1/2001
00105	Introduction to Blueprints	Don E Whyte		1/1/2001
00106	Basic Rigging	Don E Whyte		1/1/2001
26101-02	Electrical Safety	Don E Whyte		5/1/2001
26102-02	Hand Bending	Don E Whyte		5/1/2001
26103-02	Fasteners and Anchors	Don E Whyte		5/1/2001
26104-02	Electrical Theory One	Don E Whyte		5/1/2001
26105-02	Electrical Theory Two	Don E Whyte		5/1/2001
26106-02	Electrical Test Equipment	Don E Whyte		5/1/2001
26107-02	Introduction to the National Electrical Code	Don E Whyte		5/1/2001
26108-02	Raceways, Boxes, and Fittings	Don E Whyte		5/1/2001
26109-02	Conductors	Don E Whyte		5/1/2001
26110-02	Introduction to Electrical Blueprints	Don E Whyte		5/1/2001

Page 1

Donald E. Whyte
President, NCCER

To access the Automated National Registry (ANR), visit
http://www.nccer-anr.org
and enter the number on the front of this card.
If found, drop in mailbox. Postage paid by
NCCER
3600 NW 43rd Street Building G
Gainesville FL 32606
NATIONAL CENTER
FOR CONSTRUCTION
EDUCATION AND RESEARCH
Sample Student
2781481

Additional Resources

This module presents thorough resources for task training. The following resource material is suggested for further study.

Occupational Safety and Health Standards for the Construction Industry, 29 CFR, Part 1926, Latest Edition. Washington, DC: OSHA Department of Labor, U.S. Government Printing Office.

Figure Credits

Topaz Publications, Inc., Module opener
LPR Construction, Figures 1 – 5

NCCER CURRICULA — USER UPDATE

NCCER makes every effort to keep its textbooks up-to-date and free of technical errors. We appreciate your help in this process. If you find an error, a typographical mistake, or an inaccuracy in NCCER's curricula, please fill out this form (or a photocopy), or complete the on-line form at **www.nccer.org/olf**. Be sure to include the exact module ID number, page number, a detailed description, and your recommended correction. Your input will be brought to the attention of the Authoring Team. Thank you for your assistance.

Instructors – If you have an idea for improving this textbook, or have found that additional materials were necessary to teach this module effectively, please let us know so that we may present your suggestions to the Authoring Team.

NCCER Product Development and Revision

13614 Progress Blvd., Alachua, FL 32615

Email: curriculum@nccer.org

Online: www.nccer.org/olf

❑ Trainee Guide ❑ Lesson Plans ❑ Exam ❑ PowerPoints Other ______________________

Craft / Level: Copyright Date:

Module ID Number / Title:

Section Number(s):

Description:

Recommended Correction:

Your Name:

Address:

Email: Phone:

Trade Safety

30102-11

Trainees with successful module completions may be eligible for credentialing through NCCER's National Registry. To learn more, go to **www.nccer.org** or contact us at **1.888.622.3720**. Our website has information on the latest product releases and training, as well as online versions of our *Cornerstone* newsletter and Pearson's Contren® product catalog.

Your feedback is welcome. You may email your comments to **curriculum@nccer.org,** send general comments and inquiries to **info@nccer.org**, or use the User Update form at the back of this module.

V.1 1/11

30102-11
TRADE SAFETY

Objectives

When you have completed this module, you will be able to do the following:

1. Describe the consequences of on-the-job accidents.
2. Explain the special responsibilities of OSHA.
3. List potential hazards for ironworkers.
4. Describe safe work practices when near cranes.
5. List major health hazards classified by OSHA.
6. Identify and explain the safe operation of various pieces of light equipment, including:
   - Aerial lifts
   - Generators
   - Compressors
   - Forklifts

Performance Tasks

Under the supervision of the instructor, the trainee should be able to do the following:

1. Position and use an aerial work platform.
2. Demonstrate the proper use of a fall protection system.
3. Identify points of inspection on specified pieces of equipment.

Trade Terms

Aerial lift
Capacity
Circuit breaker
Compliance
Compressor
Forklift
Fuse holder
Generator
Ground fault circuit interrupter (GFCI)
Interlocking systems
Powered industrial trucks
Proportional control
Straight blade duplex connector
Twist-lock connector

Contents

Topics to be presented in this module include:

Contents (continued)

Contents (continued)

Figures and Tables

1.0.0 INTRODUCTION

Most accidents can be prevented through safe work habits and a thorough understanding of the kinds of behavior and conditions that cause accidents. According to the National Safety Council, work accidents destroy over 90,000 lives each year in spite of increased concern from management and the federal government for human welfare on the job. These accidents include losses such as the following:

- Death or disabling injury to the person or persons involved
- Loss of job time and wages
- Medical fees and/or funeral expenses
- Loss of skills, experience, or training
- Increased insurance rates

This module discusses the effects that on-the-job accidents have on employees, employers, and the public. It also covers OSHA regulations and other safety standards that apply to the ironworker trade. A description of job-site safety concerns is included, as well as ways to develop safety consciousness. This module also discusses the safe and effective use of light equipment, such as personnel lifts and other means of access to the work.

2.0.0 ON-THE-JOB ACCIDENTS

On-the-job accidents not only affect the involved workers, but also have tremendous consequences for their employers. A fall from a beam, a hoisting line failure, or the improper handling of welding electrodes or torches may result in personal injury, death, and/or the loss of equipment and materials.

An employer may be forced to spend thousands of dollars to repair or replace equipment and materials. Additional losses that can affect an employer are as follows:

- Production slowdowns
- Fees for legal actions
- Increased insurance coverage for workers
- Loss of public confidence

If the employer loses money, so do the employees. The funds available for raises and bonuses are greatly depleted by a serious on-the-job accident. A good safety program keeps everyone in business.

Work habits and work attitudes are closely related. The employee who resists authority is easily distracted from the task at hand. Employees who horse around on the job site are demonstrating a poor work attitude and can disrupt or interrupt work. This can also cause a serious accident.

2.1.0 Examples of Hazardous Conditions

A wide range of conditions can fall under the classification of a job-site hazard. Examples of hazardous conditions on the job site include the following:

- Poor housekeeping
- Defective or unsafe tools and equipment
- Improper ventilation
- Inadequate lighting
- High noise levels
- Unsafe storage of flammable fluids
- Slippery floors
- Poorly constructed and rigged scaffolding
- Improperly stacked construction materials
- Unmarked low ceilings

2.2.0 Ironworking Accident Statistics

Statistics kept by the United States Bureau of Labor Statistics, as well as by OSHA and other state and federal agencies, show fatalities and injuries for the industry. Such statistics are listed on various websites and in printed publications. The facts show that most accidents that involve fatalities and major injuries are caused by not using common sense. The accidents are made worse by not following safety procedures, such as wearing and properly hooking up fall protection equipment. In 2001, 751.8 out of every 10,000 workers were injured in accidents. In 2007, 75.6 out of every 100,000 workers were killed in accidents, in most cases by falls. In almost every case, the accident or at least the injuries could have been prevented by following proper safety procedures.

2.3.0 Causes of Construction Accidents

According to the Bureau of Labor Statistics, the largest percentage of the fatalities in the construction industry from 2004 to 2007 involved transportation accidents of various types. Other major causes were homicides and suicides. However, under the umbrella of accidents in the workplace, 2,679 fatalities broke down into 1,711 falls, 557 workers struck by an object, 229 crushed by collapsing equipment or structures, and 182 caught or compressed in moving equipment. Most accidents, but not all, can be prevented by workers following safety procedures.

The proper use of fall protection equipment at all times can certainly help prevent most fatalities and injuries from falls. The most common circumstance of falling is from a roof. This is probably because many roofs are sloped and do not provide good anchorages for fall protection, especially in the case of residential roofs. Ladders and scaffoldings or staging locations present other common circumstances of falls. The relative instability of both supports makes them places where great caution should be employed. Some risk certainly can be avoided by tying off ladders and properly securing fall protection systems.

In the case of material and equipment collapses, avoiding the areas under crane booms and under unsecured material or material held on cranes is the best approach. When the worker has to be in a risky position, constant vigilance is an absolute necessity. Pieces being moved by any machine can escape control, but if the worker is constantly alert, it is often possible to dodge threats.

The same rules apply to all forms of threats. Keep a watch on all directions, especially overhead. If a crane brings an object close, keep an eye on it. If a machine is operating nearby, be aware of the hazards that could occur. A spinning wheel or drum that catches a worker's sleeve or pants leg can pull the worker into it. An object being moved can strike a column or other obstruction and shift suddenly. Always be ready to stay out of the way of danger. Follow all safety procedures, and remain alert at all times.

3.0.0 Occupational Safety and Health Administration

The Occupational Safety and Health Act of 1970 authorized OSHA in the Department of Labor to do the following: set occupational safety and health standards for all places of employment; enforce the standards; and provide research and educational programs to support safe work practices. OSHA requires each employer to provide a safe and hazard-free place of employment for each employee. Employees are required to comply with OSHA rules and regulations that relate to their conduct on the job. OSHA regulations that relate to ironworkers involved in steel erection are included in *Section 1926 Subpart R*. These regulations are available at www.osha.gov. According to OSHA standards, employees are entitled to on-the-job instruction.

Specifically, as a new employee you must be properly informed and equipped as follows:

- Told what to do on the job
- Shown how to do what you have been assigned to do
- Shown how to do that job safely
- Provided with the proper tools and equipment, including appropriate PPE
- Warned about specific hazards in the work area and in the surrounding area
- Supervised for safety while performing work

3.1.0 Fall Protection

While OSHA's general fall protection requirements are 6 feet, *OSHA Section 1926.760* describes the fall protection requirement during steel erection activities. This regulation is intended to provide each worker with 100-percent fall protection. This protection includes establishing perimeter safety cable systems (*Figure 1*), guardrail systems, safety nets, and personal fall arrest and fall restraint systems for every ironworker who is working or walking on a surface that is 15 feet or more above the next lower level.

102F01.EPS

Figure 1 Horizontal safety cable.

3.2.0 Controlled Decking Zones

Controlled decking zones (CDZ) may be established for areas where metal decking is being installed more than 15 feet but less than 30 feet above the next lower level. Access to the CDZ is restricted to those employees who are performing leading-edge work and who have been trained for CDZ work.

A CDZ is a substitute for fall protection in areas where decking is initially being installed. Only workers trained for CDZ work are permitted in the areas. The CDZ must be clearly marked. The zone can be no wider than 90 feet wide and 90 feet deep from any leading edge.

3.3.0 Opening Covers

All openings in floors or roofs must be covered with materials capable of supporting twice the weight of workers and equipment that may be placed on the cover. These covers must be painted with high-visibility paint or marked with the word HOLE or COVER to provide warning of the hazard.

3.4.0 OSHA-Required Training

OSHA requires training to be provided on the proper use of each of the fall protection devices used at the work site. The employer must also provide special training for all employees engaged in installing and maintaining fall protection equipment, establishing CDZs, and connecting structural members.

4.0.0 Developing Safety Awareness

As a trainee, it is important to have an awareness of safety. Always understand the hazards that exist and the reasons for the safety rules. Practice following safety rules until they become second nature. The safe way is always the right way to perform any activity. The following is a partial list of hazards that ironworkers may encounter:

- Falling from heights
- Being caught between a stationary object and a moving truck or railway car when unloading steel
- Being struck by falling materials, tools, and equipment or swinging hooks, chains, and cables
- Hearing damage from continuous exposure to high noise levels
- Burns and eye damage from welding, cutting, and chipping
- Sprains and strains from improper lifting and carrying methods
- Electrical shock from poorly grounded or defective tools and equipment
- Injuries due to defective tools and equipment
- Injuries due to improper tool or equipment use

Stay focused on the job being performed. If tired or distracted, immediately stop what you are doing. Just a moment's lapse in attentiveness when working above the ground can cause a serious accident. The same guidelines should be observed when approaching co-workers. Wait until they have finished a particular operation before attracting their attention.

4.1.0 Maintain a Safety Conscious Attitude

Observe all warning signs. Use tools and equipment as prescribed, and maintain good housekeeping. Safety conscious workers can greatly reduce risks by anticipating hazards and observing general safety precautions such as the following:

- Comply with OSHA requirements for head and foot protection in hard hat zones.
- Locate and observe all warning and accident prevention signs, and confine movements only to the work area.
- Only enter restricted areas with permission from the superintendent or foreman.
- Anticipate the movement paths of all motorized equipment.
- Keep out from under raised loads, unless duties require it.
- Never ride on a load or crane hook.
- Walk around scaffolding areas. If it is necessary to work under raised platforms, be sure that the work site is covered and protected.

4.1.1 Contaminated and Dangerous Areas

Check the immediate work area for major health hazards. If work must be done in contaminated and dangerous areas, always use protective clothing and equipment. Keep track of exposure time and spend as little time as possible in these areas. OSHA has identified the following as major hazards:

- Air contaminants
- Asbestos
- Coal tar pitch volatiles
- Cancer-causing agents
- Occupational noise
- Ionizing radiation
- Nonionizing radiation

4.1.2 Hazardous Materials

Check the job site for the location of highly flammable, volatile, corrosive, and explosive materials. The hazardous materials ironworkers may come into contact with are as follows:

- Compressed gases (acetylene, oxygen, propane)
- Hydrogen
- Nitrous oxide
- Gasoline
- Diesel

Always check the MSDS and Hazcom for the material being used.

4.2.0 Dress Safely for the Job

The work environment determines the kinds of clothing needed for ironworking. Ironworkers are usually working with metals. They often perform activities such as wire-brush stripping, grinding, cutting, and welding. Because of this, they must wear clothing that will protect them from flying debris. Proper clothing includes long-sleeved shirts, full-length pants with no cuffs, and high-topped leather work boots (*Figure 2*).

Most job sites and plant operations require workers to wear additional safety clothing. Among these items are hard hats or helmets, aprons, goggles or safety glasses, and regulation ear plugs or muffs for excessive noise levels. The proper use of safety equipment is determined by the specific hazards of the job. OSHA has specific requirements for the use of safety equipment. For example, proper eye protection should be worn when cutting, chipping, or welding.

The following items should be considered before starting any ironworking activities:

- *Loose clothing and accessories* – Do not wear loose or torn clothing, or rings or other jewelry on the job. Clothing and accessories that do not fit tightly to the body can become caught in moving machinery.
- *Clothing materials* – Because of the flying sparks and intense heat created by grinding, cutting, and welding activities, work clothing cannot be made of polyester or other synthetic fibers. Sparks or intense heat will melt these materials, possibly causing severe burns. Because wool and cotton are more resistant to sparks, choose work clothing made from these materials.
- *Cuffs and flaps* – A shirt with long sleeves protects the upper body, and a pair of full-length pants (with no cuffs) shields the lower half of the body. The cuffless pants should hang straight through the leg and over the top of the work boots. Pants with cuffs must be avoided. Cuffs can catch sparks, which could start a fire. Pockets on shirts should be covered with flaps that can be buttoned. Always keep the pocket flaps closed and buttoned and the shirt buttoned all the way to the top collar button. This will keep flying debris or sparks from getting down inside a shirt or shirt pockets. Long sleeves and a closed shirt collar, along with long pants, also provide protection from exposure to the flash and ultraviolet rays resulting from cutting and welding.
- *Worn clothing* – Clothing must not be worn-out, frayed, or have material hanging down from it. Clothing cannot protect the body if it is worn to the point of being ragged. Clothing that is frayed or ragged could become snagged on machinery or could catch a stray spark and ignite.
- *Footwear* – Footwear should be high-top leather work boots, with safety (reinforced) toes recommended. The shanks of the work boots should be at least eight inches tall to protect against sparks and any arc flash generated by

102F02.EPS

Figure 2 PPE.

welding. Low-top shoes should never be worn when welding or grinding.

- *Headgear* – The type of headgear worn by ironworkers depends on where they are working. Hard hats are required in areas where falling or moving objects could cause head injuries. Ironworkers often work and move under structures. In these situations, hard hats also protect them from injuries resulting from impact with stationary objects. When welding, some ironworkers wear a solid material (non-mesh) cap with the bill pointing toward the back. Wear this cap under a hard hat and welding helmet. This protects the back of the neck and will help prevent sparks from going down your collar. Safety glasses are worn in most situations.
- *Gloves* – Ironworkers are almost constantly handling metal objects. Because of this, they must wear good leather work gloves to protect themselves from cuts and burns. Ironworkers performing or assisting in welding activities must wear welding gloves instead of basic work gloves.

Weather conditions also affect the type of clothing needed. In the summer heat, it is tempting to open shirt collars and roll up sleeves. For proper protection, keep shirt collars closed and sleeves rolled down and buttoned. Winter weather usually results in extra clothing, such as coats and coveralls. These are worn over normal work clothes. For safety reasons, additional clothing items, such as outerwear, should also be kept buttoned in the same manner as normal work clothing.

When ironworkers are working above the ground, they may be required to wear a safety harness (*Figure 3*). Remember that a safety harness is useless unless it is used correctly. It should be buckled and worn tightly enough so that there is no danger of slipping out if a fall occurs.

4.3.0 Ventilation Needs and Respirator Devices

Ironworkers sometimes work in areas that are dusty or where there is a heavy concentration of fumes from welding equipment. The gases, dust, and fumes produced by shielded metal arc welding can be hazardous if the appropriate safety precautions are not taken. Safety precautions include wearing the proper ventilation or respirators.

Respirators may be needed if welding is to be performed in a confined space (less than 10,000 cubic feet of air per worker). Additional precautions are necessary when welding metals that are galvanized or are painted with epoxy paint. Hot

102F03.EPS

Figure 3 Safety harness in use on test dummies.

work permits must be obtained before any welding or cutting is performed, if this is required on the job site. Specific safety precautions will be covered when the ironworker receives training on particular welding procedures.

4.3.1 Ventilation

The electrodes used for welding give off fumes and smoke. These are not considered dangerous as long as there is adequate ventilation. Adequate ventilation can be a problem in tight or cramped work areas. The following general rules can be used to determine if the welding workplace ventilation is adequate:

- The air must be circulating.
- The air circulation must not be blocked by partitions, structural barriers, or equipment.

Even when there is adequate ventilation, personnel in the welding area must try to avoid inhaling the fumes and smoke. The heated fumes and smoke generally rise straight up from the welding arc. Observe the column and position of smoke so as to avoid it. A small fan may also be used to divert the smoke, but care must be taken to keep the fan from blowing directly on the welding arc. The welding fumes and smoke must be present at the arc in order to protect the molten metal from the air.

> **NOTE**
> The removal of the protective smoke shield at the welding arc will cause weld defects. Never point a fan or compressed air stream directly at the arc.

4.3.2 Respirators

If adequate ventilation cannot be ensured, or if the person welding cannot avoid the welding fumes and smoke, a respirator (*Figure 4*) should be used. For most welding applications, a filter respirator offers adequate protection. The filter respirator is a mask that covers the nose and mouth. The air intake contains a cloth or paper filter to remove impurities.

Certain welding electrodes give off fumes that are extremely toxic. When welding with these electrodes, an air-line respirator must be used. An air-line respirator is similar to a filtered respirator, except that it provides breathing air through a hose connected between an external source and the respirator mask. The breathing air can come from a cylinder of compressed breathing air, from a special compressor that furnishes breathing air, or from breathing-air filters attached to compressed air lines.

WARNING

Air-line respirators must only be used with pure breathing air sources. Standard compressed air (often called plant air) contains oil, which is toxic. Standard compressed air can only be used for breathing if it is cleaned by a special breathing-air filter installed in the line just before the line connects to the respirator. Always install this filter when using standard compressed air. Failure to use this filter when using standard compressed air can cause personal injury or death.

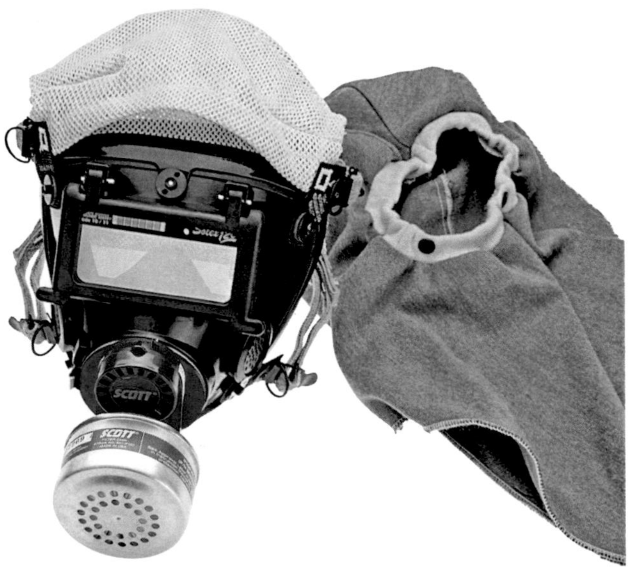

102F04.EPS

Figure 4 Respirator.

4.3.3 Air-Purifying Respirators

Air-purifying respirators provide the lowest level of protection. They should only be used in atmospheres that have enough oxygen to sustain life (at least 19.5 percent). Air-purifying respirators use special filters and cartridges to remove specific gases, vapors, and particles from the air. The respirator cartridges contain charcoal, which absorbs certain toxic vapors and gases. When the person wearing the respirator detects any taste or smell, the charcoal's absorption **capacity** has been reached and the cartridge can no longer remove the contaminant. The respirator filters remove particles, such as dust, mists, and metal fumes, by trapping them within the filter material. Filters should be changed when the wearer begins to notice difficulty in breathing. Depending on the contaminants, cartridges can be used alone or in combination with a filter/prefilter and filter cover. Air-purifying respirators should only be used for protection against the types of contaminants listed on the filters and cartridges and on the NIOSH approval label found on each respirator carton and replacement filter/cartridge carton.

4.4.0 Tool Safety

Specific safety rules for using hand and power tools are discussed in the modules covering tools, but there are some general rules that apply to all tools and processes. These rules should become part of a positive attitude toward safety. Always use the right tool for the job. This requires knowing the capabilities of each tool and understanding how each tool is used.

4.4.1 Tool Selection

Selecting the wrong tool for the job may result in injury. A wrench that is the wrong size can slip or break; a file used as a pry bar may break; and a poorly selected cable for lifting a heavy load could snap.

4.4.2 Power Tools

Portable power tools present special hazards to their users. Because they are handheld and can be easily moved, the cutting, shaping, or penetrating part of the tool may contact the operator's body. This can cause burns, cuts, scrapes, or even more serious injuries. Another source of danger is the possibility of dropping a handheld power tool. Because the source of the tool's power is close to the operator, this can create electrical shock, explosion, and fire hazards. Be aware that the size and mobility of these tools make them difficult to guard properly.

4.4.3 Tool Care

Proper care of tools prolongs their use and may prolong a worker's life. Caring for tools properly requires that a number of checks are carried out. Always be sure of the following:

- Tools are free of grease and oil before using them.
- Tools are properly dressed, free of defects, and sharpened where appropriate.
- Electrical hand tools are equipped with an automatic shut-off control.
- Portable electric saws and drills are properly guarded and grounded.
- Damaged tools are repaired before use.
- Never perform any welding on tools.

4.5.0 Report Damaged Equipment

All damaged equipment must be immediately reported to the job supervisor. Take steps to see that the equipment is not used until it has been checked out. Always disconnect the power source, and post a sign to prevent others from using the equipment until it can be repaired.

4.6.0 Welding Safety

Ironworkers are often required to perform welding and cutting operations. Some basic safety precautions must be observed at all times when welding or cutting. Eye protection (safety glasses) should be worn at all times, even under a welding helmet.

Additional appropriate personal protective equipment that must be worn while welding or cutting includes welding leathers, such as aprons, sleeves or jackets; welding gloves; and welding helmets or shields. The gases, dust, and fumes produced during welding can be hazardous if the appropriate precautions are not observed. Welding should only be done in well-ventilated areas. Ironworkers must avoid inhaling excessive welding fumes and smoke. Respirators may be needed if welding is to be performed in a confined space (less than 10,000 cubic feet of air per worker). Additional precautions are necessary when welding metals that are galvanized or are painted with epoxy paint. Hot work permits must be obtained before any welding or cutting is performed, if this is required on the job site. Specific safety precautions will be covered when the ironworker receives training on particular welding procedures.

5.0.0 JOB SITE SAFETY

Hazards at the job site vary with the size and complexity of the job. However, there are certain fundamental principles for job site safety that should be observed by every worker. Most accidents in the field result from inadequate maintenance of equipment, insufficient instruction in safe practices, or failure to follow safety regulations. The ironworker trainee needs to know and pay special attention to the hazards involved when working above the ground. The worker must also be aware of situations where work operations and the moving of equipment and materials pose an increased risk.

5.1.0 Cranes

Cranes are a potential danger to the ironworker. This danger can be minimized by taking preventive action. The first step in prevention is knowing how and where serious accidents occur. The area close to the crane is especially dangerous. Over 40 percent of all accidents occur to workers in this area.

5.1.1 Crane Accident Statistics

Crane operations involve lifting and moving large and frequently heavy objects. Because of this, accidents occur with some frequency. Crane operations have been studied to show the sources of accidents. The following are crane accident statistics:

- 25 percent of those injured were electrocuted.
- 21 percent of the injuries resulted from being struck by loads.
- 20 percent of the injuries were caused by being struck by a crane.
- 14 percent were caused by crane collapses.
- 9 percent of injuries were caused by falls.
- 5 percent resulted from workers getting caught in/between the crane.

5.1.2 Observation of Safety Practices

As is the case with all accidents, most crane accidents can be prevented. The trainee working in the crane area can prevent a serious accident by observing the following safety procedures:

- Do not walk under raised loads unless specifically instructed to do so.
- Always be aware of the position of the load when hooking on, arranging loads, or moving across the construction site. Treat all lifting operations as if equipment failures could occur.

- Keep hands out of pinch points when assisting with hooking or unhooking a load. Use hand hooks whenever possible. *Figure 5* shows typical pinch points to avoid.
- Do not use a crane to pull a sling out from under a load.
- Always stand clear of the rear of the rotating super structure (*Figure 6*).
- Keep clear of power lines. Contact with power lines is the largest single cause of fatalities associated with cranes. *Figure 7* shows the hazardous locations created when a crane contacts a power line.
- Learn the crane signals, especially the emergency stop signal. Use it only when it may prevent an accident from happening. The crane signals should always be posted on the job site and/or crane.

NOTE

Crane operations should stop when the crane operator cannot see or hear the signal person.

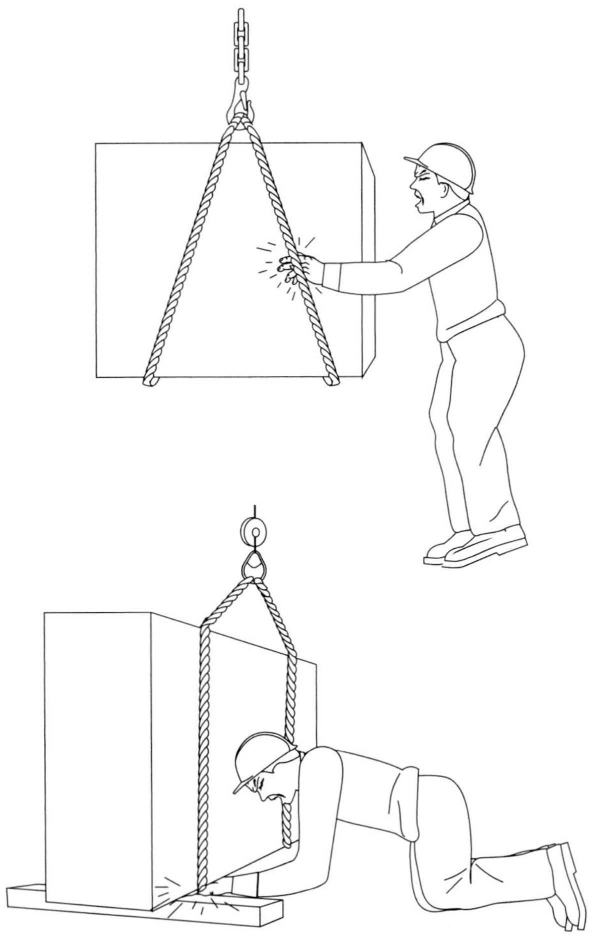

Figure 5 Pinch points.

5.2.0 Material Handling and Storage

Approximately 25 percent of all occupational injuries occur when manually handling or moving construction materials. Much of the nonmechanized moving, stacking, loading, and unloading tasks are performed by the less experienced members of the trade. In many cases, these are trainees. Strains, sprains, fractures, and crushing injuries can be minimized by having the proper knowledge of safe lifting and handling procedures and by using good body mechanics. General points to keep in mind when handling or moving construction materials include the following:

- Inspect materials for slivers and for rough or sharp edges.
- Determine the weight of the load before attempting to move it.
- Know your own capacity for lifting objects.
- Be sure that the pathway is free from obstacles.
- Keep hands free of oil and grease.
- Take a firm grip on the object before it is moved, being careful to keep fingers from being pinched.
- Use good body mechanics.
- Use gloves when handling sharp-edged materials.
- Get help when the load is too large or heavy.

6.0.0 Light Equipment

In addition to cranes, a number of types of light equipment are used in ironwork applications. These include personnel lifts, **compressors**, and

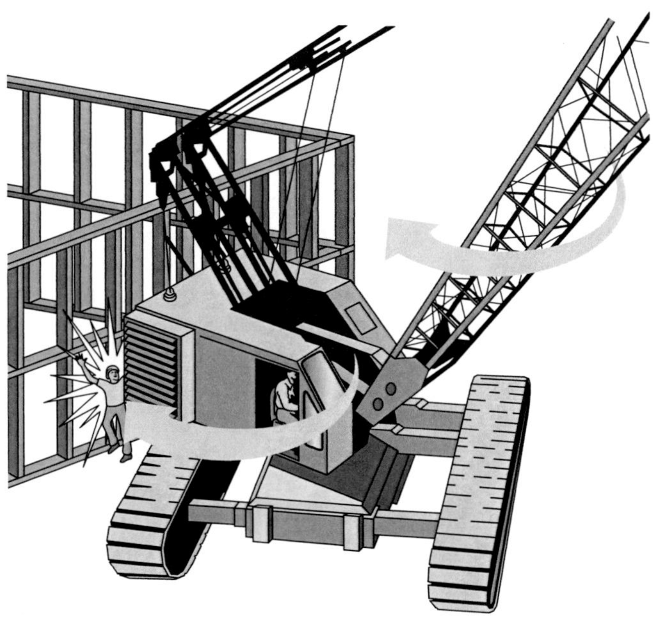

Figure 6 Cab swing.

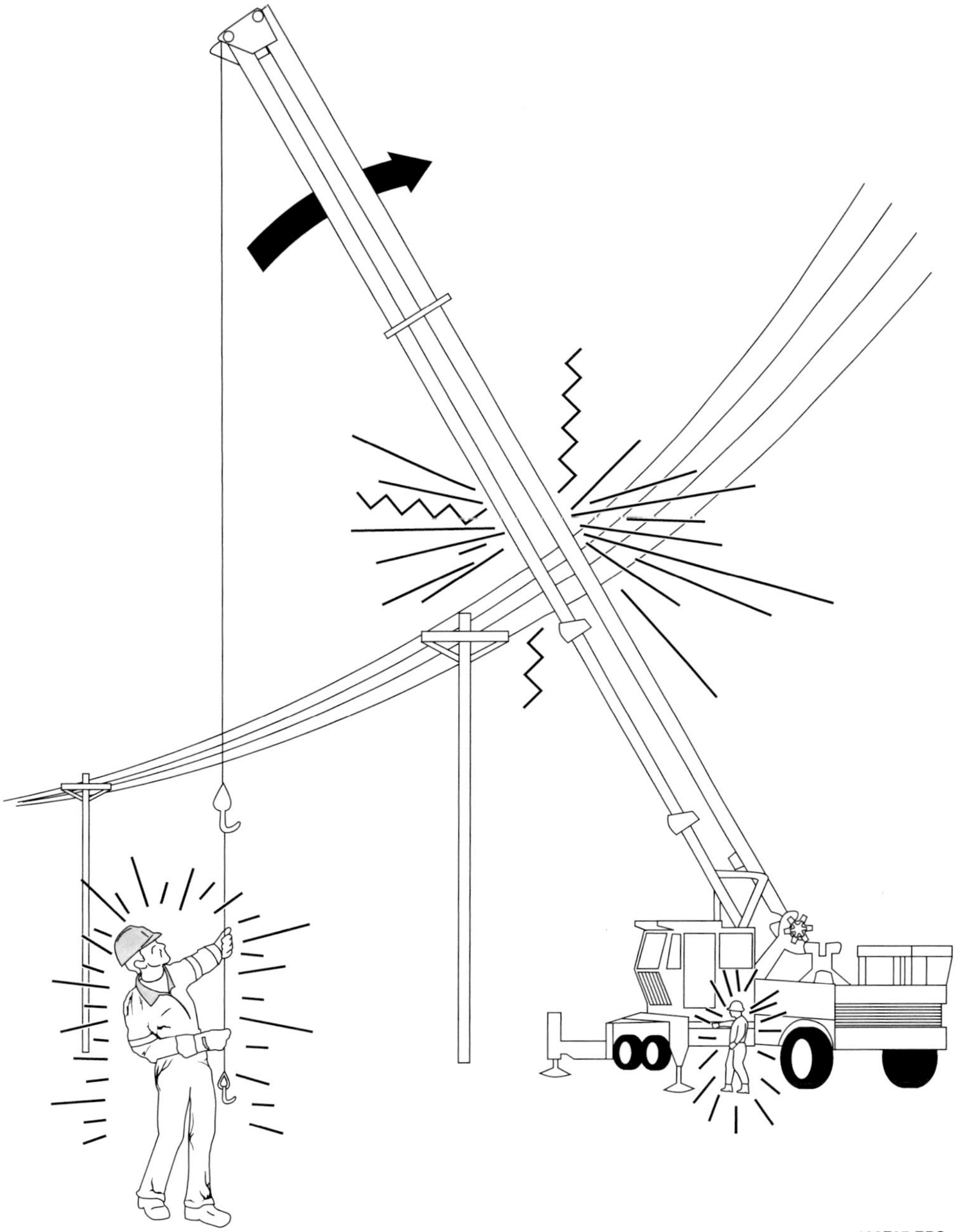

Figure 7 Power line hazards.

generators. Safety is the primary concern no matter what type of equipment is being operated. Most of the equipment mentioned in this module share common safety precautions. These safety precautions are discussed in this section. As each piece of light equipment is covered, remember this safety discussion and think about how these precautions apply to the equipment. Additional safety precautions that are unique to a particular type of equipment are discussed later. Do not leave any machine running unattended. Also, make sure that any internal combustion engines, such as gas, propane, or diesel-powered machines, have a fire extinguisher close at hand. Locate the fire extinguisher before starting the machine.

The manufacturer provides safety precautions for each piece of light equipment. These safety precautions should be observed whenever any equipment is being used. All of the equipment discussed can cause serious injury or death if the safety precautions are not followed.

> **WARNING**
>
> Anyone operating power equipment must be properly trained in its use. Part of this training includes understanding the associated safety precautions. Some localities may require the operator to be trained and qualified in the use of certain equipment. The primary points to remember are as follows:
>
> - Read and fully understand the operator's manual and follow the instructions and safety precautions.
> - Gasoline and diesel engines produce deadly carbon monoxide. If they are used in a closed environment, the exhaust must be vented to the outdoors.
> - Know the capacity and operating characteristics of the equipment being operated.
> - Inspect the equipment before each use to make sure everything is in proper working order. Have any defects repaired before using the equipment. Never modify or remove any part of the equipment unless authorized to do so by the manufacturer.
> - Check for hazards above, below, and all around the job site. Be sure to maintain a safe distance from electrical power lines and other electrical hazards.
> - Learn as much about the work area as possible before beginning work.
> - Fasten seat belts or operator restraints before starting.
> - Set up warning barriers, and keep others away from the equipment and job site.
> - Know where to get assistance.
> - Know how to use a first aid kit and fire extinguisher.

6.1.0 Transporting Equipment

Transporting equipment between jobs involves the general risks of roads and machinery. When transporting equipment between jobs, remember these safety guidelines:

- Park, unload, and load the equipment from a trailer on level ground.
- To prevent tipping, connect the trailer to the tow vehicle before loading or unloading.
- Follow the manufacturer's requirements for towing and transporting the equipment.
- Always maintain the correct tire pressure.
- Inspect the tires and wheels before towing. Do not tow with high or low tire pressures, cuts, excessive wear, bubbles, damaged rims, or missing lug bolts or nuts.
- Do not inflate tires with flammable gases or from systems using an alcohol injector.

6.2.0 Hydraulic Systems

Hydraulic systems involve high pressures and the attendant risks. When operating equipment that uses hydraulics, always remember the following:

- Before disconnecting any hydraulic lines, relieve system pressure by cycling controls.
- Before pressurizing the system, be sure all connections are tight and the lines are undamaged.

> **WARNING**
>
> Do not perform any work on any equipment unless authorized and qualified to do so.

- Check for leaks using a piece of cardboard or wood; never use bare hands.

> **WARNING**
>
> Hydraulic fluid escaping through pinholes in hoses and hose fittings can be almost invisible. Escaping hydraulic fluid under pressure can penetrate the skin or eyes, causing serious injury. When inspecting light construction equipment, hydraulic hoses, and hose couplings for fluid leaks, always wear protective clothing and eye protection. Do not trace for leaks along hydraulic hoses and hydraulic system components with bare hands; instead, use a piece of cardboard or wood. In the event that hydraulic fluid penetrates the skin, seek immediate medical attention from a doctor familiar with this type of injury.

6.3.0 Fueling Safety

Fueling safety precautions are applicable to all light equipment using gasoline. Make sure to follow all the manufacturer's instructions before fueling the equipment. The following are general safety precautions associated with fueling light equipment:

- Never fill the fuel tank with the motor running, while smoking, or when near an open flame.
- Be careful not to overfill the tank or spill fuel. If fuel is spilled, clean it up immediately.
- Use clean fuel only.
- Do not operate the equipment if fuel has been spilled inside or near the unit.
- Ground the fuel funnel or nozzle against the filler neck to prevent sparks.
- Replace the fuel tank cap after refueling.

6.4.0 Battery Safety

Battery safety precautions are applicable to all light equipment that has a battery for starting or operating electric components. Follow these guidelines when using light equipment supplied with a battery:

- Chargers can ignite flammable materials and vapors. They should not be used near fuels, grain dust, solvents, or other flammables.
- To reduce the possibility of electric shock, a charger should only be connected to a properly grounded single-phase outlet.
- If the battery is frozen, do not charge or attempt to jump-start the equipment as the battery may explode.
- Battery and fuel fumes could ignite and cause explosions and burns. Keep batteries away from flames or sparks. Do not smoke near batteries.

7.0.0 AERIAL LIFTS

Aerial lifts are used to raise and lower workers to and from elevated job sites. There are two main types of lifts: boom lifts and scissor lifts. Both types are available in various models. Some are transported on a vehicle to a job site, where they are unloaded. Others are trailer-mounted and towed to the job site by a vehicle. Still others are permanently mounted on a vehicle. Depending on their design, aerial lifts can be used for indoor work, outdoor work, or both. *Figure 8* shows two types of commonly used aerial lifts.

Boom lifts are designed for both indoor and outdoor use. Boom lifts have a single arm that extends a work platform/enclosure capable of holding one or two workers. Some models have a jointed (articulated) arm that allows the work platform to be positioned both horizontally and vertically. Scissor lifts raise a work enclosure vertically by means of crisscrossed supports.

Most models of aerial lifts are self-propelled, allowing workers to move the platform as work is performed. The power to move these lifts is provided by several means, including electric motors, gasoline, propane, or diesel engines, and hydraulic motors.

All startup and shutdown procedures must follow manufacturers' instructions.

7.1.0 Aerial Lift Assemblies

Aerial lifts normally consist of three major assemblies: the platform, a lifting mechanism, and the base. *Figure 9* shows the components of a scissor lift.

The platform of an aerial lift is constructed of a tubular steel frame with a skid-resistant deck surface, railings, toe board, and midrails. Entry to the platform is normally from the rear. The entry opening is closed either with a chain or a spring-returned gate with a latch. The work platform may also be equipped with a retractable extension platform. The lifting mechanism is raised and lowered either by electric motors and gears or by one or more single-acting hydraulic lift cylinders. A pump driven by either an AC or DC motor provides hydraulic power to the cylinders. The base provides a housing for the electrical and hydraulic components of the lift. These components are normally mounted in swing-out access trays. This allows easy access when performing maintenance or repairs on the unit. The base also contains the axles and wheels for moving the assembly. In the case of a self-propelled platform, electrical or hydraulic motors drive two or more of the wheels to allow movement of the lift from one location to another. Brakes are incorporated on one or more of the wheels to prevent accidental movement of the lift.

7.2.0 Aerial Lift Operator Qualifications

Only trained and authorized workers may use an aerial lift. Safe operation requires the operator to understand all limitations and warnings, operating procedures, and operator requirements for maintenance of the aerial lift. The following is a list of requirements that the operators must meet:

- Understand and be familiar with the associated operator's manual for the lift being used.
- Understand all procedures and warnings in the operator's manual and those posted on decals on the aerial lift.
- Be familiar with the employer's work rules and all related government (OSHA) safety regulations.
- Demonstrate their understanding, and operate the associated model of aerial lift during training in the presence of a qualified trainer.

Some equipment comes with **interlocking systems**. Most, if not all lifts have interlocking control systems to prevent an operator on the ground or in the air from causing an accident. The controls can only perform a particular action from one station or the other. Such interlocks prevent an operator on the ground from unknowingly placing an operator on the lift at risk, or the operator in the lift from injuring the operator on the ground. Never defeat or modify such interlocks, or any other safety device.

Figure 8 Aerial lifts.

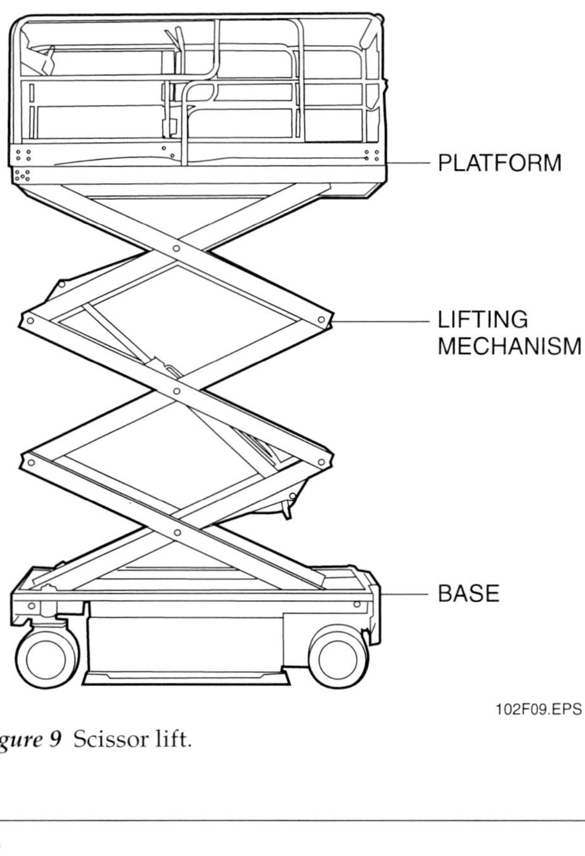

Figure 9 Scissor lift.

7.3.0 Typical Aerial Lift Controls

An example of an electrical panel and its associated controls is shown in *Figure 10*. This is only an example. Other models of lifts may have different control configurations. This electrical panel contains the following switches and controls:

- *Up/down toggle switch* – By holding this switch in the Up position, the platform can be raised to the desired level. The platform stops moving when the switch is returned to the center position. By holding this switch in the Down position, the platform can be lowered.
- *Buzzer alarm* – An audible alarm sounds when the platform is being lowered. On some models, this alarm may sound when any control function is being performed.
- *Hour meter* – This meter records the number of hours the platform has been operating. The meter will only register when the electric or hydraulic motor associated with operating the aerial lift is running.

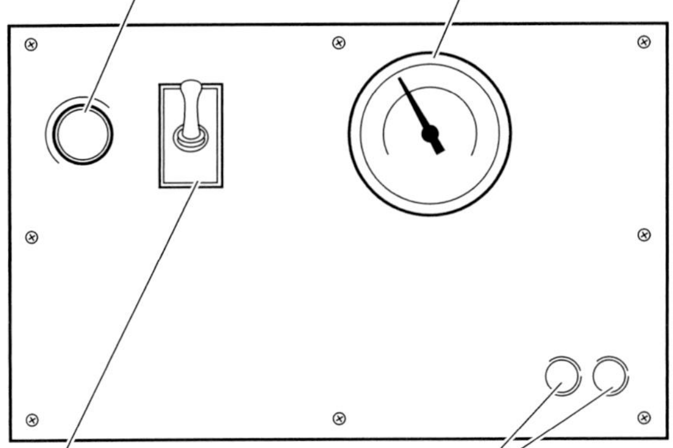

Figure 10 Aerial lift electrical panel.

One other control that may be available to the operator is an emergency battery disconnect switch. When this switch is placed in the Off position, it will disconnect power to all control circuits.

The basic controls for an aerial lift with a hydraulic system generally include the following:

- *Emergency lowering valve* – This valve allows the platform to be lowered in the event of an electrical/hydraulic system failure.
- *Free wheeling valve* – Opening this valve allows hydraulic fluid to flow through the wheel motors. This allows the aerial lift to be pushed by hand and prevents damage to the motors when the aerial lift is moved between job site locations. There are usually strict limits to how fast the aerial lift may be moved without causing damage to hydraulic system components.

The following are other types of controls that may be available on aerial lifts:

- *Parking brake manual release* – This control allows manual release of the parking brake. It should only be used when the aerial lift is on a level surface.
- *Safety bar* – The safety bar is used to support the platform-lifting hardware in a raised position during maintenance or repair.
- *Up/down selector switch* – This switch is used to raise and lower the platform.
- *Emergency stop button* – This button is used to cut off power to both the platform and base control boxes.

Stabilizer hardware is required on some aerial lifts. At a minimum, this hardware includes a stabilizer leg, stabilizer lock pin and cotter key, and a stabilizer jack at each corner of the aerial lift.

Most aerial lifts are equipped with a base control box. The base control box includes a switch to select whether operation will be controlled from the platform or the base control box. The base control overrides the platform control.

The worker on the platform, who is designated as the operator, uses the platform controls. In some aerial lifts, operation from the platform is limited to raising and lowering the platform. On other models, the aerial platform is designed to be driven by the operator using controls provided on the platform. In either case, the operator has an operator control box located on the platform assembly. *Figure 11* shows a typical platform control box for a self-propelled aerial lift.

Controls available for the operator include (but are not limited to) the following:

- *Drive/steer controller* – This is a one-handed toggle-type (joystick) lever control for controlling speed and for steering the aerial platform. This is usually a deadman switch that returns to neutral and locks when released. The handle is moved forward to drive the aerial platform forward. The platform speed is determined by how far forward the handle is moved. The handle is moved backward to drive the aerial platform backward, and speed is selected as before. Releasing the stick stops the motion of the aerial lift. Steering is performed by depressing a rocker switch on the top of the stick in the desired direction of travel, either right or left.
- *Up/down selector switch* – Placing the switch in the Up position raises the platform, and placing the switch in the Down position lowers it. When released, the switch returns to the middle position, which stops the movement of the platform.
- *Lift/off/drive selector switch* – The Lift position energizes the lift circuit, and the Off position removes power to the control box. The Drive position energizes the drive and steering controls.
- *Emergency stop push button* – When pushed, this button disconnects power to the platform control circuit. In an emergency, push the button in. To restore power, pull the button out.
- *Lift enable button* – When pushed and held down, this button enables the lift circuit. The button must be held down when raising or lowering the platform. Releasing the button stops the motion of the platform.

Different aerial lifts have different controls. It is important that the operator fully understand all

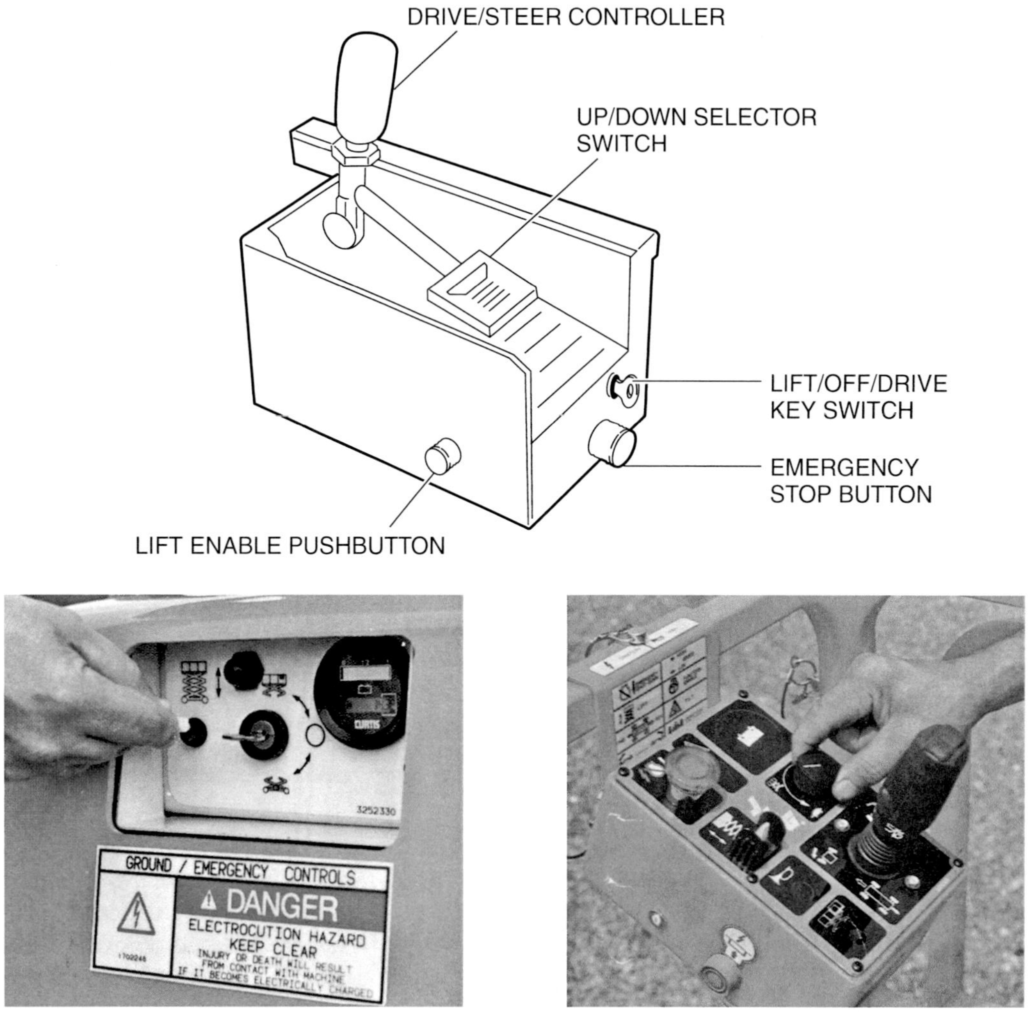

102F11.EPS

Figure 11 Aerial lift platform control box.

controls and their functions before operating the aerial lift. The operator must read and fully understand all control and safety information provided by the manufacturer before attempting to operate the equipment.

7.4.0 Aerial Lift Safety Precautions

Safety precautions unique to aerial lifts must also be followed. Remember that the other safety precautions already discussed in this module also apply to aerial lifts. Each manufacturer provides specific safety precautions in the operator's manual that comes with the equipment. Specifically, *OSHA Standard 1926.453* defines and governs the use of aerial lifts as expressed in the following precautions:

- Avoid using the lift outdoors in stormy weather or in strong winds.
- Prevent people from walking beneath the work area of the platform.
- Use personal fall arrest equipment (body harness and lanyard) as required for the type of lift being used. Use approved anchorage points.
- Do not use an aerial lift on uneven ground.
- Lower the lift and lock it into place before moving the equipment. Also, lower the lift, shut off the engine, set the parking brake, and remove the key before leaving it unattended.
- Stand firmly on the floor of the basket or platform. Do not lean over the guardrails of the platform, and never stand on the guardrails. Do not sit or climb on the edge of the basket or use planks, ladders, or other devices to gain additional height.

7.5.0 Aerial Lift Operating Procedure

The following procedures are for a scissor-type, self-propelled aerial lift. Remember that operating procedures vary from model to model. The operator must become familiar with all operating procedures, controls, safety features, and associated safety precautions before operating any type of aerial lift.

A typical **proportional control** procedure involves the operation of a lever or foot pedal to cause the aerial lift to move. The farther the control is moved, the more power is applied to the motor, and the faster the aerial lift moves.

The following tasks must be performed before operating the aerial lift:

- Carefully read and fully understand the operating procedures in the operator's manual and all warnings and instruction decals on the work platform.
- Check for obstacles around the work platform and in the path of travel, such as holes, drop-offs, debris, ditches, and soft fill. Operate the aerial lift only on firm surfaces.
- Check overhead clearances. Make sure to stay at least 10 feet away from overhead power lines.
- Make sure batteries are fully charged (if applicable).
- Make sure all guardrails are in place and locked in position.
- Perform an operator's checklist.
- Never make unauthorized modifications to the components of an aerial lift.

Figure 12 shows an example of a typical operator's checklist.

The following is an example of the operator controls used to drive the aerial lift forward. The operator selects the drive position with the lift drive select switch. The operator then lifts the handle lock ring and moves the drive/steer controller forward. The speed can be adjusted by continuing to move the controller forward until the desired speed is reached. By releasing the controller, the forward motion of the lift is stopped. To drive in reverse, the lever is moved in the opposite direction.

7.6.0 Aerial Lift Operator's Maintenance Responsibility

The operator must be assured that the work platform has been properly maintained before using it. The operator should perform daily checks on the equipment.

WARNING

Death or injury to workers and damage to equipment can result if the aerial platform is not maintained in good working condition. Inspection and maintenance should be performed by personnel who are authorized for such procedures.

As a minimum, if the operator is not responsible for the maintenance of the aerial lift, the operator should perform the daily checks. *Figure 13* shows an example of a maintenance and inspection schedule for a typical aerial lift. The daily checks should be performed at the beginning of each shift or at the beginning of the day if only one shift is worked. The aerial lift must never be used until these checks have been completed satisfactorily. Any deficiencies found during the daily check must be corrected before using the aerial lift.

Figure 13 is broken down into four major categories. Each category is further divided into the components that should be checked daily, weekly, monthly, every three months, every six months, and yearly. Footnotes (the numbers in parentheses) following each component tell the operator what type of inspection is required. For example, under the Electrical category, battery fluid level should be checked daily or at the beginning of

OPERATOR'S CHECKLIST

INSPECT AND/OR TEST THE FOLLOWING DAILY OR AT BEGINNING OF EACH SHIFT

___ 1. OPERATING AND EMERGENCY CONTROLS

___ 2. SAFETY DEVICES

___ 3. PERSONNEL PROTECTIVE DEVICES

___ 4. TIRES AND WHEELS

___ 5. OUTRIGGERS (IF SO EQUIPPED) AND OTHER STRUCTURES

___ 6. AIR, HYDRAULIC, AND FUEL SYSTEM(S) FOR LEAKS

___ 7. LOOSE OR MISSING PARTS

___ 8. CABLES AND WIRING HARNESSES

___ 9. DECALS, WARNINGS, CONTROL MARKINGS, AND OPERATING MANUALS

___ 10. GUARDRAIL SYSTEM

___ 11. ENGINE OIL LEVEL (IF SO EQUIPPED)

___ 12. BATTERY FLUID LEVEL

___ 13. HYDRAULIC RESERVOIR LEVEL

___ 14. COOLANT LEVEL (IF SO EQUIPPED)

___ 15. CHECK SCISSORS FOR CRACKS, DEFORMATIIONS

___ 16. CHECK FOR OPERATOR'S MANUAL

102F12.EPS

Figure 12 Operator's checklist.

each shift. The (*1*) refers the operator to the Notes portion of the schedule. Note 1 tells the operator to perform a visual inspection of the battery fluid level.

Different models of aerial lifts have different maintenance requirements. Always refer to the operator's manual to determine exactly what checks are required before operation.

	Daily	Weekly	Monthly	3 Months	6 Months	*Annually
Mechanical						
Structural damage/welds (1)	✓					✓
Parking brakes (2)	✓					✓
Tires and wheels (1)(2)(3)	✓					✓
Guides/rollers/slides (1)	✓					✓
Railings/entry chain/gate (2)(3)	✓					✓
Bolts and fasteners (3)	✓					✓
Rust (1)			✓			✓
Wheel bearings (2) King pins (1)(8)	✓					✓
Steer cylinder ends (8)				✓		✓
Electrical						
Battery fluid level (1)	✓					✓
Control switches (1)(2)	✓					✓
Cords and wiring (1)	✓					✓
Battery terminals (1)(3)	✓					✓
Terminals and plugs (3)	✓					✓
Generator and receptacle (2)	✓					✓
Limit switches (2)	✓					✓
Hydraulic						
Hydraulic oil level (1)	✓					✓
Hydraulic leaks (1)	✓					✓
Lift/lowering time (10)				✓		✓
Hydraulic cylinders (1)(2)		✓				✓
Emergency lowering (2)	✓					
Lift capacity (7)			✓			✓
Hydraulic oil/filter (9)					✓	✓
Miscellaneous						
Labels (1)(11) Manual (12)	✓					✓

Notes:

(1) Visually inspect. (3) Check tightness. (5)(6) N/A. (8) Lubricate.

(2) Check operation. (4) Check oil level. (7) Check relief valve setting. Refer to serial number nameplate. (9) Replace.

(10) General specifications.

(11) Replace if missing or illegible.

(12) Proper Operating Manual *must* be in the manual tube.

*Record Inspection Date

102F13.EPS

Figure 13 Example lift maintenance and inspection schedule.

General maintenance rules applicable to any aerial lift include the following:

- Disconnect the battery ground negative (–) lead before performing any maintenance.
- Properly position safety devices before performing maintenance with the work platform in the raised position.

Preventive maintenance is easier and less expensive than corrective maintenance.

7.7.0 Scissor Lifts

Scissor lifts (*Figure 14*) are powered personnel lifts. They use hydraulic rams to raise and lower the platform. Power is supplied to the hydraulic system by a battery, a gasoline engine, or a propane-powered (LPG) engine. The controls for the lift are located at an operator's station on the platform itself. Some models also include an operator's station at the ground and the platform level. In addition to controlling the height of the platform, the operator is also able to move the unit to the front or rear and to steer it as it moves.

Lift heights and capacities vary by model and manufacturer. Lift heights may be as high as 41 feet and capacities may reach 2,000 pounds. Both the height and **capacity** of the lift can usually be found on a decal conspicuously placed on the lift. Some manufacturers offer optional manual outriggers for added stability.

102F14.EPS

Figure 14 Self-propelled elevating work platform (scissor lift).

7.7.1 Operating Precautions

No one is permitted to operate a scissor lift without training in the proper operating procedures for the particular model. This section of the module discusses general operating procedures, but is not intended to serve as an all-inclusive guide to scissor lift operation. Keep the following safety guidelines in mind when operating a scissor lift:

- Read and observe the warnings and cautions on the unit.
- Keep the vehicle and work platform at least 15 feet from electrical power lines. Scissor lifts are not insulated to protect against arcing or electrical contact.
- Anything that is not working correctly must be repaired before using the lift.
- Do not remove any parts or safety devices.
- Make certain that the immediate area is clear of all personnel and obstructions before raising or lowering the lift.
- Keep the guardrails and chains in place when operating the lift.
- Do not bridge between two scissor lifts or between a scissor lift and a building.
- Never crawl into the scissors at any point.
- Operate the vehicle only on firm, solid ground. Do not operate the vehicle near a drop-off.
- Do not impose side loads on the work platform.
- Never exceed the rated capacity of the lift.
- Do not operate the vehicle in an explosive atmosphere.
- Never climb on railings.
- Only operate per manufacturer's instructions.

7.7.2 Controls and Indicators

Each manufacturer uses different control systems. Always get familiar with the particular controls of the unit to be used beforehand. The following controls are typically found on scissor lifts:

- *Power switch* – This may be key-operated and may include three positions: Off, On, and Start.
- *Steering* – This may be a toggle switch that is used to turn the vehicle to the right or left.
- *Platform* – This may be a toggle switch that moves the platform up and down.
- *Drive* – This switch may control both the speed and direction of the vehicle. Some models have

dual speeds for both forward and reverse directions.

- *Emergency stop* – This switch may have two positions: Run and Emergency Stop. It is usually a protected toggle switch. In order for the power switch to work, the Emergency Stop switch must be in the Run position.

7.7.3 Other Controls and Indicators

The following controls and indicators may also be found on a scissor lift. Make sure to become familiar with the controls of the unit prior to use.

- *Manual override* – This may be a push-button switch to lower the platform manually during an emergency. It is usually labeled with a decal and may be found at an opening on the hydraulic panel cover.
- *Emergency release for brake* – This is used if the vehicle stalls during operation. It may be a lever located behind the hydraulic tank and labeled with a decal. Once the vehicle has been restarted, the switch should be returned to the normal position.
- *Fuel gauges* – These indicate the levels of propane or gasoline. Propane-powered models also have a supply valve to open or shut the LPG supply.
- *Oil dipstick and hydraulic oil indicators* – These allow the operator to check the supply of oil in the crankcase of engine-powered models and the amount of hydraulic oil in the reservoir. Learn to check oil levels before operating the vehicle.

7.7.4 Operation

The following steps cover the general operating procedures for scissor lifts. This is not intended to serve as a substitute for the operator's manual for a particular unit.

Startup and shutdown procedures are different depending upon how the vehicle is powered. For battery-powered models, turn the key to the Power position (or the equivalent). Then activate all functions before applying load to warm up the hydraulic oil. To shut down a battery-powered model, first lower the platform, and then turn the key to the Off position. Remove and store the key.

Follow these steps to operate engine-powered models:

Step 1 Turn the key slowly to the Start position. Release it quickly when the engine starts. If the unit does not start, turn the key to the Off position, and wait about five seconds.

CAUTION

To prevent damage to the starter, never hold the key in the Start position for more than 10 seconds.

NOTE

If the vehicle does not start, turn the switch to Off and check the gasoline or LPG supply gauge. Check the LPG supply valve to make sure it is open. Then check the condition of the batteries.

Step 2 Let the engine warm up for about two minutes. This allows the hydraulic oil to warm up.

To shut down an engine-powered model, lower the platform, turn the key to the Off position, and close the LPG valve.

7.7.5 Maneuvering

Before traveling with a scissor lift, check the immediate route. Avoid personnel, obstructions, other traffic, rocks, holes, debris, soft surfaces, and steep grades.

On most models, the switches that control direction and steering must be held manually. When pressure is released, these switches usually return to the center position. It may be necessary to use both hands when traveling and steering. Follow these steps to move the scissor lift to another location:

Step 1 Choose the speed range, if the vehicle is equipped with this control.

Step 2 Push and hold the Forward-Reverse switch with one hand in the desired direction of travel. At the same time, push the Steering switch to the Right or Left with the other hand.

NOTE

On most models, when the Forward-Reverse switch is released, the vehicle decelerates and stops. When the Steering switch is released, the wheels remain in the same position until moved again.

On most models, the Up and Down switches must be held in position to raise or lower the platform. The platform stops automatically when the switch is released.

CAUTION

Before operating the platform, be sure that the areas above and below it are clear of personnel, obstructions, and power lines, and that the vehicle is on firm, level ground. These precautions are necessary to prevent personal injury.

7.7.6 Maintenance

Always follow the manufacturer's recommendations for daily and periodic maintenance. General safety checks for loose or missing parts and damage must be conducted before operating the vehicle. Always check the oil before starting the engine, and never operate a vehicle that is low on hydraulic oil.

At the end of the day, the batteries on electrically powered models should be recharged. Periodically check the batteries for proper electrolyte level, the tightness and condition of the cables, and any signs of corrosion or visible damage.

Check the hydraulic oil reservoir when the unit is shut down. The oil level should be up to the neck strainer, unless otherwise specified by the manufacturer. Always check the hydraulic connections for signs of leaks.

7.8.0 Boom Lifts

Boom lifts (*Figure 15*) are work platforms mounted on the end of an arm that can be raised and lowered and may be able to extend beyond the mobile base. The boom section may consist of two or more telescoping sections. The machines usually can be driven by the operator from the platform. Boom lifts are generally restricted to supporting one or two workers at a time with their tools.

7.8.1 Application

Boom lifts come in two basic categories: those designed as off-slab units and those designed for use on paved/slab surfaces. The off-slab units have truck-type tires or even deep tread mud-type tires. Some units have oscillating axles and four-wheel drive to maneuver more easily on rough terrain. These units have a wide variety of uses in construction and outdoor maintenance. *Figure 16* shows the major parts of a boom lift.

The paved/slab units are designed for use on improved surfaces. Some are intended for indoor use and are small enough to fit through interior doorways. The units designed for indoor use are LPG or battery powered to avoid any exhaust problems. These units have non-marking tires, and are primarily used by maintenance personnel.

7.8.2 Operation

Boom lifts have two sets of controls. The first is on the work platform and the second is on the base of the unit. The work platform controls are the primary control station. The base-mounted controls are mainly used in emergency situations. If a worker is injured, the base controls can be used to lower the person to the ground to obtain the necessary assistance. As an additional safety feature, all boom lifts can even be lowered when there is no power to the unit.

NOTE

Be sure to read and understand the vendor-provided operator's manual before attempting to use any boom lift.

The following information is provided as a general guideline for operating a boom lift, but is not specific to any individual lift. *Figure 17* shows a typical boom lift control panel.

Notice that the control panel switches are guarded. This serves two purposes: it protects the switch handles, and it allows the switch to move in a logical direction. For example, the switch to steer right and left moves to the right and left. The control panel shown is for an engine-powered lift similar to the lift shown in *Figure 16*. The engine could be either gasoline- or LP-fueled. The engine is started by setting the Ignition switch to the On position and pressing the Start push button. If necessary, the Choke switch may be used during starting. Be sure to release the button as soon as the engine starts to avoid damaging the starter and engine flywheel. Once the engine has been started and allowed to warm up briefly, the lift can be maneuvered into position. The engine provides power to move the unit along the ground, as well as to operate the boom lifting, telescoping, and swing. The High Engine and High Drive switches are set to the On position while moving the unit from place to place with the boom retracted and in the down position. The hydraulic pumps used to raise and maneuver the work platform are not powered while the High Engine and High Drive switches are in the On position. Once the unit has been situated, these switches are returned to the Off position.

ARTICULATING BOOM LIFT

TELESCOPIC BOOM LIFT

102F15.EPS

Figure 15 Articulating and straight boom lifts.

Movement of the boom is controlled using levers. The left lever, Swing, controls the boom movement to the right and left. The first lever on the right of the main control panel, Raise, is used to raise and lower the work platform. Slow movements of the unit can be made while the boom is raised using the right hand lever, Drive, and the Steer switch on the central panel. Operators must be very careful to avoid overhead hazards while moving the unit with the boom raised.

7.8.3 Maintenance

Maintenance of boom lifts must be done in accordance with the manufacturer's recommendations. Lubrication of the boom is a large part of the maintenance requirements. Normal engine maintenance is required on the LP- or gasoline-powered units, and recharging is required on battery-powered units. Servicing the hydraulic-powered units of boom lifts requires checking fluid levels, cleaning or replacing filters regularly, and keeping the units clean for ease in detecting any leaks.

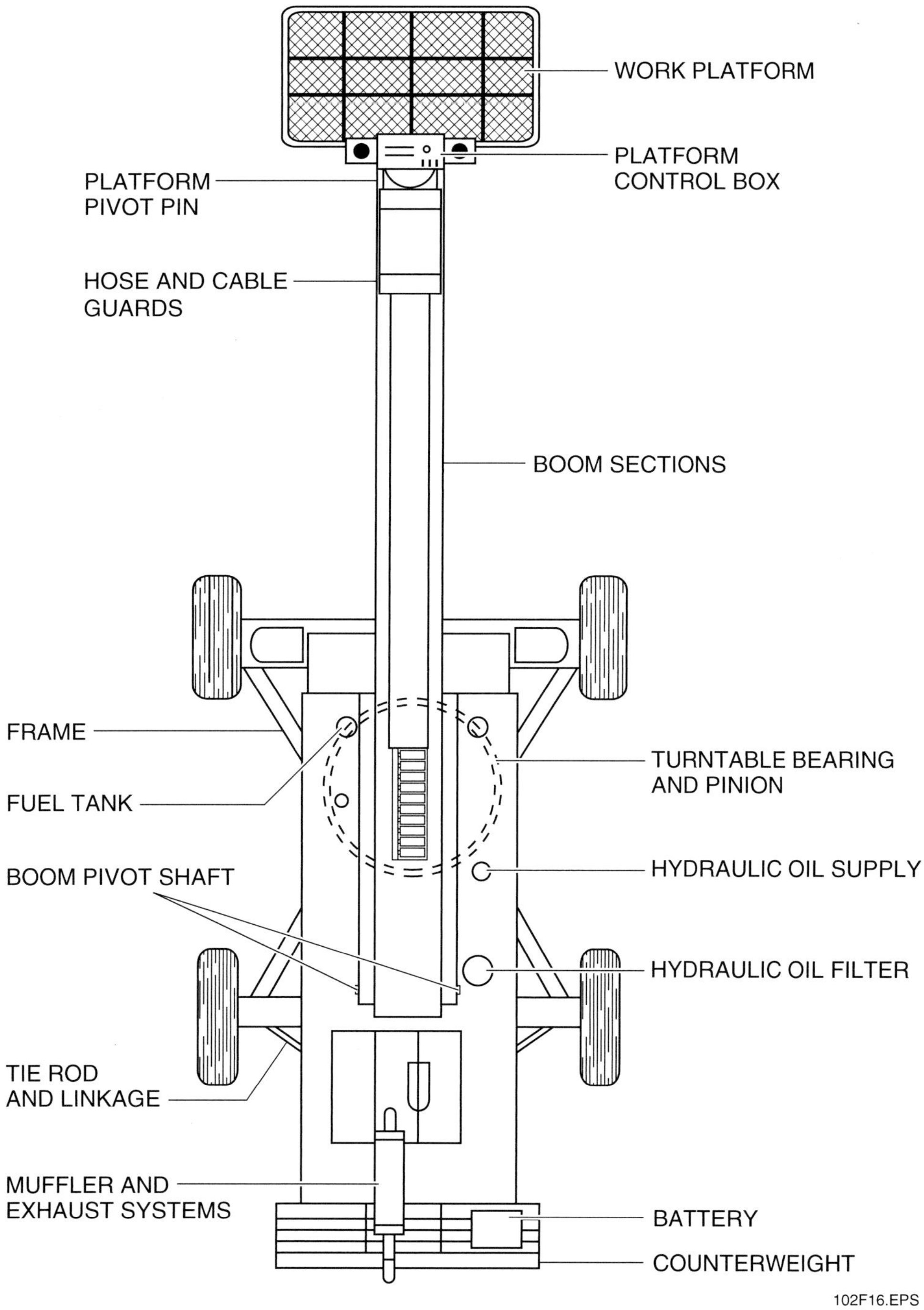

Figure 16 Parts of a boom lift.

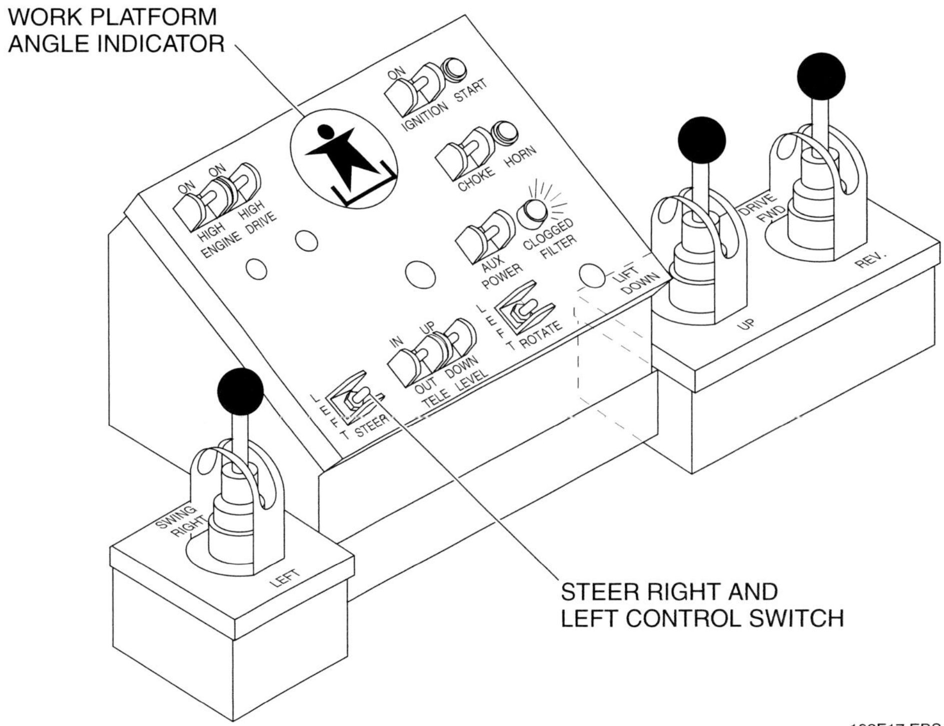

Figure 17 Boom lift control panel.

8.0.0 GENERATORS

Generators are used to provide electrical power and/or lighting at the job site. Generators are available in many different sizes and configurations (*Figures 18* and *19*). They range from small portable machines to large installed backup power systems for emergency power generation. Construction site requirements for power vary depending on the scale of the work being performed. This section describes a typical tow-behind generator used to provide electrical power to a job site.

Figure 18 Generators.

Figure 19 Tow-behind generator for lighting.

8.1.0 Generator Assemblies

Like any piece of equipment, a generator has several parts, each performing a different aspect of the machine's work. A tow-behind generator is comprised of the following main components:

- Towing assembly and frame
- Protective covers and doors
- Engine and generator assemblies
- Operator control panels

8.2.0 Typical Generator Controls

Tow-behind generators have operator controls for the engine and an electrical control panel to control and monitor the power produced by the generator. An engine control panel (*Figure 20*) provides the operator with the controls required to start and stop the motor and monitor critical motor functions.

Controls and indicators include the following:

- *Engine over-speed indicator* – Monitors engine speed. If speed exceeds the manufacturer's set limit, the engine will shut down automatically.
- *High engine temperature indicator* – Indicates when the engine coolant temperature has exceeded the manufacturer's set limit. The engine will shut down automatically.
- *Low engine oil pressure indicator* – Shows when the engine oil pressure has fallen below the manufacturer's preset limit. The engine will shut down automatically.
- *Alternator not charging* – Indicates that the engine alternator is not outputting enough energy to charge the unit's battery.
- *Ignition switch* – Place this switch in the On position to run; place it to the Off position to stop.
- *Start switch* – Pressing this switch activates the engine starting motor.
- *Safety circuit bypass* – Pressing this switch bypasses automatic shutdowns when starting the engine.
- *Emergency stop* – Pressing this switch causes the engine to shut down with no other operator action required.
- *Engine tachometer gauge* – Indicates engine rpm.
- *Engine oil pressure gauge* – Displays the engine oil pressure.
- *Engine coolant temperature* – Shows the temperature of the engine coolant.
- *Ammeter gauge* – Indicates the charge rate of the engine alternator.
- *Fuel level gauge* – Indicates the level of fuel in the fuel tank(s).
- *Hour meter* – Records total engine operating hours. It is used in determining the maintenance schedule on the unit.

The second control panel in a tow-behind generator is the generator control panel. This panel contains meters, monitor switches, a voltage regulator, **circuit breakers**, and receptacles to control and monitor the output of the generator. *Figure 21* shows a typical generator control panel with the following controls and indicators:

- *AC voltmeter* – Indicates the generator output voltage level.
- *AC amperes meter* – Indicates the generator output load in amperes.

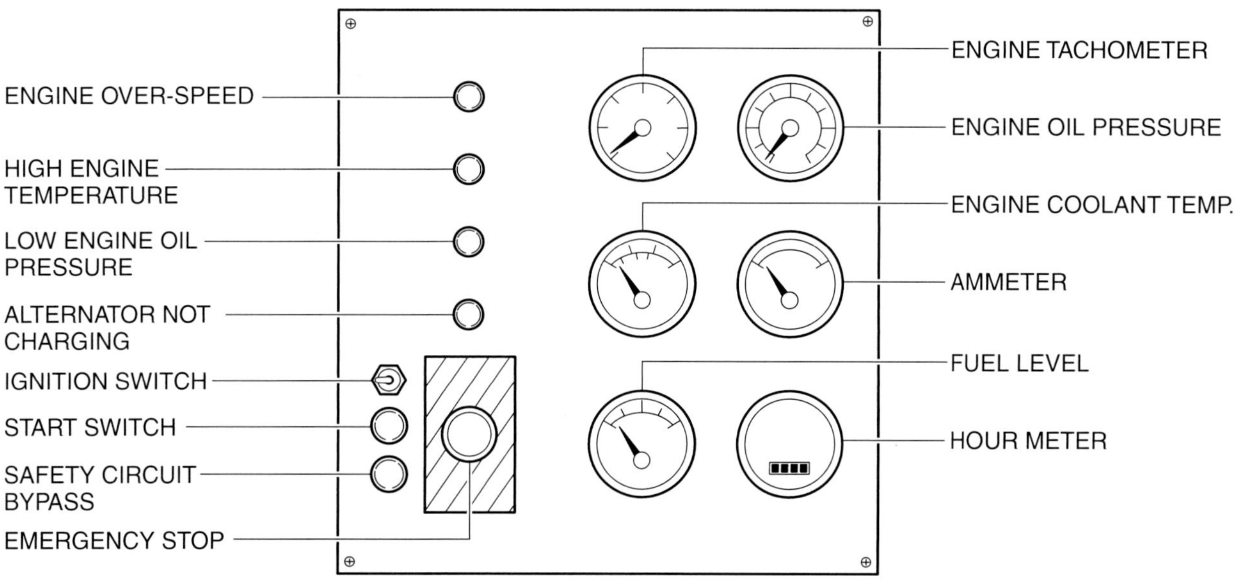

Figure 20 Generator engine control panel.

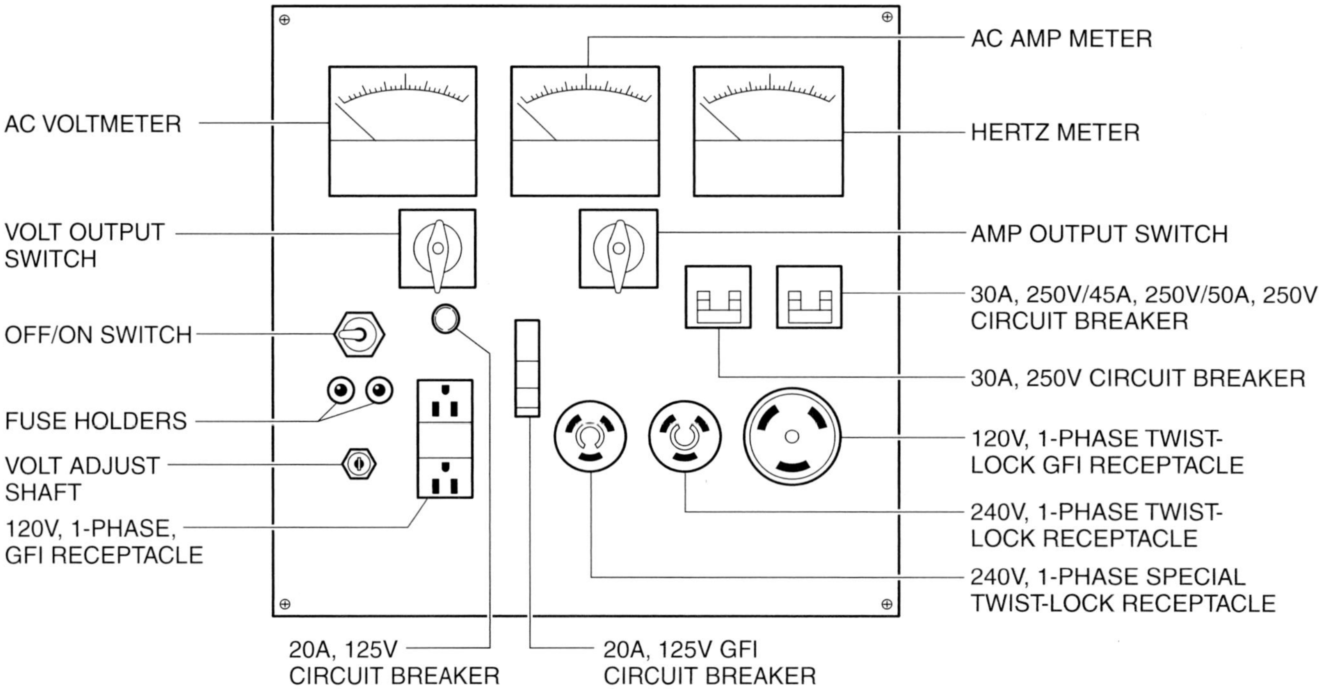

Figure 21 Generator control panel.

- *Hertz meter* – Indicates the frequency of the generator output.
- *Voltage output monitor switch* – Selects the reference for the generator voltage displayed on the AC voltmeter.
- *Amperage output monitor switch* – Selects the line-to-line (phase) amperage displayed on the AC ammeter.
- *Voltage regulator On/Off switch* – In the Off position, this switch removes excitation to the generator field, stopping the generation of power.
- *Voltage regulator fuse holders* – The **fuse holders** house the fuses for the voltage regulator.
- *Voltage regulator voltage adjust shaft* – This shaft is turned to adjust generator output voltage.
- *Circuit breakers* – Various styles of circuit breakers are used, including push-to-reset, **ground fault circuit interrupter (GFCI)**, and flip-to-reset styles.
- *Receptacles* – Various types of receptacles are provided, depending on the size and style of generator used. They include GFCI **straight blade duplex connectors** and **twist-lock connectors**.

8.3.0 Generator Safety Precautions

The safe operation of a generator is the operator's responsibility. Keep in mind that the other safety precautions already discussed in this module also apply.

WARNING

High voltages are present when the generator is operating. Exercise caution to avoid electric shock.

The following are general safety precautions that must be observed when operating a generator:

- Do not operate electrical equipment when standing in water, on wet ground, or when hands or shoes are wet.
- Grounding should be performed in **compliance** with local electrical codes and in accordance with the manufacturer's manual.

NOTE

If the generator is supplying only equipment that is mounted on the generator, or equipment plugged into the receptacles on the generator, and the parts of the generator are grounded to the frame of the generator, the frame can act as the ground. If the generator is connected by way of a transfer switch to a structure, then there must be a grounding rod or electrode. For example, if a shop is being supplied with power by the generator, then the ground rod is necessary, but if a deck saw is powered by the generator, the rod is not needed.

- Use a generator as an alternate power supply only after the main feed at the service entrance panel has been opened and locked.
- Do not change voltage selection while the engine is running. Voltage selection, adjustment, and electrical connections may only be performed by qualified personnel.
- Do not exceed the generator power (wattage) rating during operation.
- Never make electrical connections with the unit running.
- If welding is required on the unit, follow the manufacturer's instructions to prevent damage to the circuitry.

8.4.0 Generator Operating Procedure

Before operating the generator, local requirements for grounding must be investigated and followed. The generator set can produce high voltages, which can cause severe injury or death to personnel and damage to equipment. The generator should have proper internal and external grounds when this is required by the *National Electrical Code*® (*NEC*®). A qualified, licensed electrical contractor, knowledgeable in local codes, should be consulted. Before operation, follow all manufacturer's requirements to ensure that the generator is connected properly.

The following section covers operation of a typical generator. An ironworker should observe these operations to ensure a safe work area.

Each manufacturer provides detailed operating instructions with the equipment. Refer to the operator's manual provided with the generator for specific instructions. The general guidelines given here are intended to help in understanding the operator controls. The operator should perform a complete inspection of the generator and correct any deficiencies before beginning operation.

8.4.1 Generator Setup and Preoperational Checks

The following is a typical procedure for setting up a generator for use at the job site. Follow these steps when performing generator setup and preoperational checks:

Step 1 Place the unit as level as possible. Follow the manufacturer's directions for equipment placement and any special considerations for equipment location.

Step 2 Chock the wheel of the generator.

Step 3 Disconnect the generator from the towing vehicle.

Step 4 Unlock the jack and lower it to the service position.

Step 5 Lock the jack in the service position.

Step 6 Disconnect the safety chains, and crank the jack to raise the coupling off the hitch.

Before putting the generator into operation, check it for evidence of arcing on or around the control panel. If any arcing is noted, contact your supervisor because the problem must be located and repaired before beginning operation. Check for loose wiring or loose routing clamps within the housing.

Perform the following additional checks before operating:

- Check for any engine fluid leaks on or under the engine.
- Check the engine oil and coolant levels.
- Check for frayed or loose fan belts, hoses, or wiring insulation.
- Check for obstructions in air vents.
- Check the fuel tank level, and fill the tank at the end of each day to prevent condensation.
- Check the engine air cleaner.

Maintain the engine according to the manufacturer's recommendations.

8.4.2 Generator Setup, Startup, and Shutdown

The operator should study the correct startup and shutdown procedure for the equipment. Start and shut down the generator according to the manufacturer's detailed instructions.

8.5.0 Generator Operator's Maintenance Responsibility

In a typical preventive maintenance schedule for a generator (*Figure 22*), the schedule is laid out with the performance period across the top and the items to be checked listed down the side of the page. For example, under the Daily column, the following items would need to be checked:

- Check for any fluid leaks on or under the engine
- Evidence of arcing around electrical terminals
- Loose wire routing clamps
- Engine oil and coolant levels
- Grounding circuit
- Instruments
- Fan belts, hoses, and wiring insulation
- Air vents
- Fuel/water separator
- Service air indicator

PREVENTIVE MAINTENANCE SCHEDULE	DAILY	WEEKLY	MONTHLY/ 150 HRS.	3 MONTHS/ 250 HRS.	6 MONTHS/ 500 HRS.	YEARLY 1000 HRS.
Check for Any Fluid Leaks on or Under the Engine	✓					
Evidence of Arcing Around Electrical Terminals	✓					
Loose Wire Routing Clamps	✓					
Engine Oil and Coolant Levels	✓					
Proper Grounding Circuit	✓					
Instruments	✓					
Frayed/Loose Fan Belts, Hoses, and Wiring Insulation	✓					
Obstructions in Air Vents	✓					
Fuel/Water Separator (drain)	✓					
Service Air Indicator	✓					
Precleaner Dumps		✓				
Tires		✓				
Battery Connections		✓				
Engine Radiator (exterior)			✓			
Air Intake Hoses and Flexible Hoses			✓			
Fasteners (tighten)			✓			
Emergency Stop Switch Operation			✓			
Engine Protection Shutdown System			✓			
Diagnostic Lamps			✓			
Voltage Selector and Direct Hook-up Interlock Switches				✓		
Air Cleaner Housing				✓		
Control Compartment (interior)					✓	
Fuel Tank (fill at end of each day)					Drain	
Fuel/Water Separator Element					Replace	
Wheel Bearings and Grease Seals					Repack	
Engine Shutdown System Switches (settings)						✓
Exterior Finish				(as needed)		
Engine		Refer to Engine Operator				
Decals		Replace decals if removed, damaged, or missing.				
✓ = Check or Clean (and Adjust or Replace, if necessary)						

102F22.EPS

Figure 22 Example of a generator preventive maintenance schedule.

Proper maintenance will extend the life and performance of the equipment.

9.0.0 Air Compressors

Compressors are used to provide compressed air for tools at the job site. Compressors are available in many different sizes and configurations (*Figure 23*). They range from portable home-use machines to large installed units for industrial applications. Construction site requirements for compressed air vary depending on the scale of the work being performed. This section describes a typical tow-behind compressor.

9.1.0 Compressor Assemblies

A tow-behind compressor is shown in *Figure 24*. Its major components include the following:

- Towing assembly and frame
- Protective cover and doors
- Engine and compressor assembly
- Operator control panel located behind the access door

9.2.0 Typical Compressor Controls

Compressors normally have operator controls for the engine and compressor (*Figure 25*). The controls discussed here are associated with a tow-behind compressor. The panel shown provides the operator with the controls required to start and stop the motor and to monitor critical motor and compressor functions.

Controls and indicators include the following:

- *Engine over-speed indicator* – Monitors engine speed. If the speed exceeds the manufacturer's preset limit, the engine shuts down automatically.
- *High engine temperature indicator* – Shows that the engine coolant temperature has exceeded the manufacturer's preset limit; the engine shuts down automatically.
- *Low engine oil pressure indicator* – Points out that the engine oil pressure has fallen below the manufacturer's preset limit; the engine shuts down automatically.
- *Alternator not charging* – Indicates that the engine alternator is not outputting enough energy to charge the unit's battery.
- *Ignition switch* – Place this switch in the On position to run and in the Off position to stop.
- *Start switch* – Pressing the Start switch activates the engine starting motor.
- *Safety circuit bypass* – Pressing this switch bypasses automatic shutdowns when starting the engine.
- *Emergency stop* – Pressing this switch causes the engine to shut down with no other operator action required.
- *Air pressure gauge* – Indicates the output air pressure level.
- *Engine tachometer gauge* – Indicates engine rpm.
- *Engine oil pressure gauge* – Displays the engine oil pressure.
- *Engine coolant temperature* – Indicates the temperature of the engine coolant.
- *Ammeter gauge* – Indicates the charge rate of the engine alternator.
- *Fuel level gauge* – Indicates the level of fuel in the fuel tank(s).
- *Hour meter* – Records total engine operating hours. It is used for scheduling maintenance on the unit.
- *Temperature gauge* – Indicates the temperature of the compressed air.

Figure 23 Compressors.

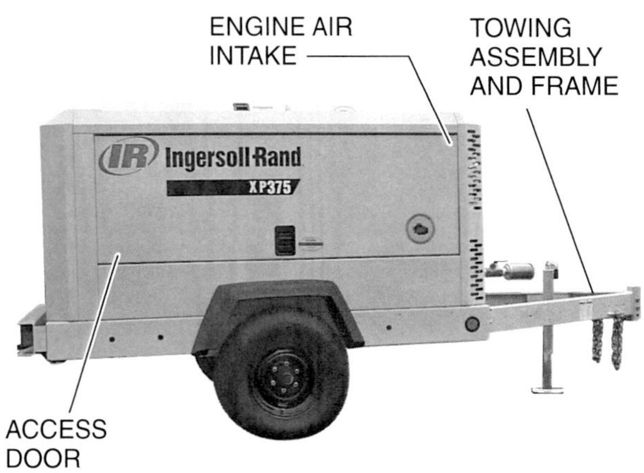

Figure 24 Large tow-behind compressor.

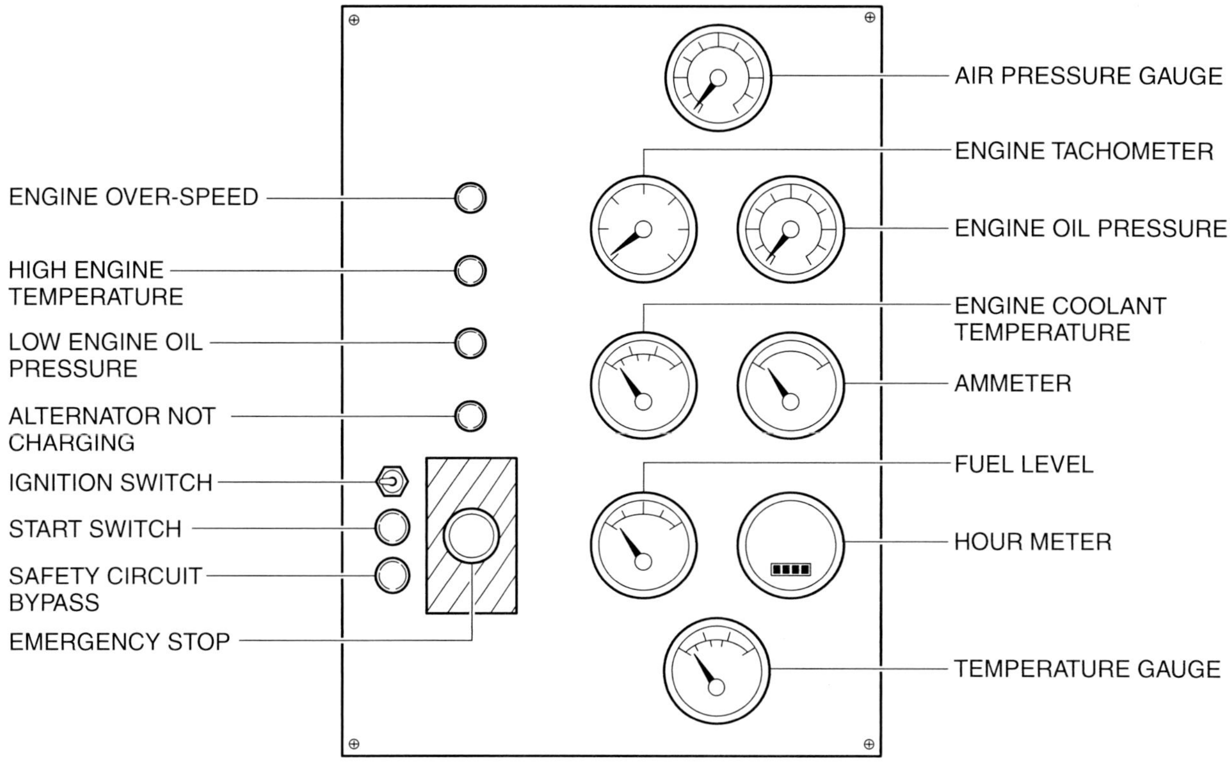

Figure 25 Compressor control panel.

9.3.0 Compressor Safety Precautions

The safe operation of an air compressor is the operator's responsibility. The following are general safety precautions that are applicable to air compressors. Keep in mind that the other safety precautions already discussed in this module also apply. Refer to the operator's manual for specific safety precautions that apply to the machine being used.

- Wear eye protection.
- Compressed air venting to the atmosphere can cause hearing damage. Make sure that proper hearing protection is worn by everyone in the area.
- Ether is highly flammable. Do not inject ether into a hot engine or an engine equipped with a glow-plug type of preheater.
- Do not inject ether into the compressor air filter or into a common air filter for the engine and compressor.
- Do not attempt to move the compressor or lift the drawbar without adequate personnel or equipment to handle the weight.
- Do not exceed the machine's air pressure rating.
- Do not use compressed air for breathing.
- Never direct compressed air at anyone.
- Make sure that there is no pressure in the system before removing the compressor filler cap.
- Do not use tools that are rated for a pressure lower than that provided by the compressor.
- Make sure that all connections are securely made and hoses under pressure are secured to prevent whipping. Use appropriate safety devices to secure hoses. Only use the appropriate manufacturer suggested clamps on all hoses.
- Do not weld or perform any modifications on the air compressor receiver tank.
- Use a safe, nonflammable solvent when cleaning parts.

9.4.0 Compressor Operating Procedure

Each manufacturer provides detailed operating instructions with the equipment. Refer to the operator's manual provided with the compressor for specific instructions. The general outline provided here is intended to help in understanding the operator controls. The operator should perform a complete inspection of the compressor and correct any deficiencies before beginning operation.

The following sections cover the type of inspection the operator performs before beginning work. Each manufacturer provides specific inspection procedures in the operator's manual.

9.4.1 Compressor Setup and Preoperational Checks

Before beginning operation, the operator must perform a setup and preoperational checklist. Each manufacturer provides a checklist with the equipment. This section covers the setup and preoperational checks for compressors.

Place the unit in as level a position as possible. Follow the manufacturer's directions for equipment placement and any special considerations for equipment location. Disconnect the compressor from the towing vehicle. Chock the wheels of the compressor. Unlock the jack, and lower it to the service position. Lock the jack in the service position. Disconnect the safety chains, and crank the jack to raise the coupling off the hitch.

Vent all internal pressure before opening any air line, fitting, hose, valve, drain plug, connector, oil filler, filler caps, and before refilling anti-ice systems. Make sure that all hoses, shutoff valves, flow-limiting valves, and other attachments are connected according to the manufacturer's instructions. If the air compressor will be used along with other sources of air, make sure that there is a check valve installed at the service valve.

Be sure to also check for the following on the compressor:

- Loose wiring and routing clamps
- Any engine fluid leaks
- Engine oil and coolant levels
- Frayed or loose fan belts, hoses, or wiring insulation
- Obstructions in the air vents
- Fuel tank level (the tank should be filled at the end of each day to prevent condensation)
- Engine and compressor air cleaner(s)

Maintain the engine according to the manufacturer's recommendations.

9.4.2 Compressor Startup and Shutdown

The operator should study the correct startup and shutdown procedure for the equipment. Start and shut down the compressor according to the manufacturer's detailed instructions.

9.5.0 Air Compressor Operator's Maintenance Responsibility

The manufacturer's manual provides a schedule and procedures to be followed when performing maintenance on the air compressor. Before performing any maintenance, the operator must be trained, authorized, and have the proper tools to perform any procedures. The following are examples of the types of maintenance that could be performed on an air compressor:

- Changing the engine oil and filter
- Changing the engine and compressor air filters
- Changing the engine coolant
- Performing battery maintenance
- Lubricating parts
- Inspecting all guards and safety devices

10.0.0 FORKLIFTS

Forklifts are used to move, unload, and place materials, such as pallets of shingles or stacks of plywood, at required locations on the job site. Forklifts are also sometimes referred to as **powered industrial trucks**. In this module, the term *forklift* is used. This module introduces safe forklift operation. Forklifts are often misused. A firm foundation in forklift safety ensures greater productivity and a safer working environment.

10.1.0 Forklift Assemblies

Many different types of forklifts are available to meet different needs. Most forklifts used in construction are diesel-powered. They fall into one or two broad categories: fixed-mast and telescoping-boom. Within each of these categories, forklifts are further differentiated by their drive train, steering, and capacity.

There are two types of fixed-mast forklifts: rough terrain and warehouse. Rough terrain forklifts are by far the most common type of fixed-mast forklift used in construction (*Figure 26*). These forklifts are made for outside use. They have higher ground clearances than warehouse forklifts, larger tires, and leveling devices. Warehouse forklifts are designed for inside use. They have hard rubber tires that are usually all the same size.

When the term *fixed-mast forklift* is used in this module, it may be assumed that it refers to the rough terrain forklift.

10.2.0 Forklift Operator Qualifications

Only trained, authorized persons are permitted to operate a powered forklift. To operate the equipment safely, operators of powered forklifts must be qualified as to visual, auditory, physical, and mental ability. Training must be conducted under the direct supervision of a qualified trainer.

102F26.EPS

Figure 26 Fixed-mast rough terrain forklifts.

10.3.0 Typical Forklift Controls

Forklifts have either two-wheel drive or four-wheel drive. The distinction is the same as that used with other vehicles.

A two-wheel drive forklift has a drive train that transmits power to the front wheels. Often the front wheels are larger than the rear wheels, and they do not steer the forklift. These forklifts are good for general use, but may lack power in rough terrain.

A four-wheel drive forklift has a drive train that transmits power to all four wheels. These forklifts are well suited to rough terrain and other conditions that might require additional traction.

The rear wheels of a forklift with two-wheel steering move when the steering wheel is turned. This feature is standard on most forklifts and provides increased maneuverability. The rear wheels are generally smaller than the front wheels.

When the four-wheel steer is engaged on a forklift with this feature, all the wheels move. However, some forklifts offer the following three options:

- The rear wheels may be locked to allow the front wheels to move.
- All wheels may move in the same direction. This is called crab steering or oblique steering.
- The front and rear wheels may move in opposite directions. This is called articulated steering or four-wheel steering.

10.3.1 Fixed-Mast Forklifts

The upright member along which the forks travel is called the mast. A typical fixed-mast forklift has a mast that may tilt as much as 19 degrees forward and 10 degrees rearward. This type of forklift is suitable for placing loads vertically and traveling with loads, but its horizontal reach is limited to how close the machine can be driven to the pick-up or landing point. For example, a fixed-mast forklift cannot place a load of purlins beyond the edge of the roof because of its limited reach.

The mast can be either two-stage or three-stage. This designation refers to the number of telescoping channels built into the mast. A two-stage mast has one telescoping channel, and a three-stage mast has two telescoping channels. The purpose of the telescoping channels is to provide greater lift height.

10.3.2 Telescoping-Boom Forklifts

Telescoping-boom forklifts (*Figure 27*) provide more versatility in horizontal and vertical placement than a fixed-mast forklift. A telescoping-boom forklift, sometimes called a shooting-boom forklift, is really a combination of a telescoping-boom crane and a forklift.

Some models of telescoping-boom forklifts have a level reach fork carriage, sometimes called a squirt boom. A squirt boom allows the fork carriage to be moved horizontally while the boom remains in a stationary position.

10.4.0 Forklift Safety Precautions

Safe operation is the responsibility of the operator. Operators must develop safe working habits and recognize hazardous conditions to protect themselves and others from death or injury. Always be aware of unsafe conditions to protect personnel, as well as the load and the forklift, from damage. Be familiar with the operation and function of all controls and instruments before operating a fork-

102F27.EPS

Figure 27 Telescoping-boom forklift.

lift. Before operating any forklift, read and fully understand the operator's manual.

The following safety rules are specific to forklift operation. In addition to these requirements, keep in mind the other safety precautions already discussed in this module.

- Always wear a seat belt.
- Never put any part of the body into the mast structure or between the mast and the forklift.
- Never put any part of the body within the reach mechanism.
- Understand the limitations of the forklift.
- Do not permit passengers to ride unless a safe place has been provided by the manufacturer.
- Never leave the forklift running unattended.
- The operator must be certified to operate the forklift.

Safeguard pedestrians at all times by observing the following rules:

- Always look in the direction of travel.
- Do not drive the forklift up to anyone standing in front of an object or load.
- Make sure that personnel stand clear of the rear swing area before turning.
- Exercise particular care at blind spots, cross aisles, doorways, and other locations where pedestrians may step into the travel path.
- The use of a spotter is recommended when landing an elevated load with a telescoping-boom forklift.

10.5.0 Forklift Operation

The following are general guidelines for operation of a forklift. To ensure a safe work area, an ironworker should be present to observe that these guidelines are being followed.

The most important factor to consider when using any forklift is its capacity. Each forklift is designed with an intended capacity, and this capacity must never be exceeded. Exceeding the capacity jeopardizes not only the equipment but also the safety of everyone on or near it. Each manufacturer supplies a capacity chart for each forklift model. Be sure to read and follow the capacity chart.

10.5.1 Picking Up a Load

Some forklifts are equipped with a side shift device that allows the operator to horizontally shift the load several inches in either direction with respect to the mast. A side shift enables more precise placing of loads, but it changes the center of gravity and must be used with caution. If the forklift being used is equipped with a side shift, be sure to return the fork carriage to the center position before attempting to pick up a load.

Step 1 Check the position of the forks with respect to each other. They should be centered on the carriage. If the forks have to be moved, check the operator's manual for the proper procedure. There is usually a pin at the top of each fork. When the pin is lifted, it allows each fork to be slid along the upper backing plate until the fork centers over the desired notch.

Step 2 Travel to the area at a safe rate of speed. Always keep the forks lowered when traveling with a forklift.

Step 3 Before picking up a load with a forklift, make sure the load is stable. If it looks like the load might shift when it is picked up, secure the load. Knowing the load's center of gravity is crucial, especially when picking up tapered sections. If necessary, make a trial lift to determine and adjust its center of gravity.

Step 4 Approach the load so that the forks straddle the load evenly. It is important that the weight of all loads be distributed evenly on the forks. Overloading one fork at the expense of the other can damage the forks. In some cases, it may be advisable to measure the load and mark its center of gravity.

Step 5 Drive up to the load with the forks straight and level. If the load being picked up is on a pallet, be sure the forks are low enough to clear the pallet boards.

Step 6 Move forward until the leading edge of the load rests squarely against the back of both forks. If unable to see the forks engage the load, ask someone to signal. This prevents costly injuries and equipment damage.

Step 7 Raise the carriage, then tilt the mast rearward until the forks contact the load. Raise the carriage until the load safely clears the ground. Then tilt the mast fully rearward to cradle the load. This minimizes the chance that the load may slip during travel.

10.5.2 Traveling With a Load

Always travel at a safe rate of speed with a load. Never travel with a raised load. Keep the load as low as possible, and be sure the mast is tilted rearward to cradle the load.

When traveling, keep eyes open, and stay alert. Watch the load and the conditions ahead, and alert others. Avoid sudden stops and abrupt changes in direction.

Be careful when downshifting because sudden deceleration can cause the load to shift or topple. Be aware of front and rear swing when turning.

If traveling with a telescoping-boom forklift, be sure the boom is fully retracted.

To drive on a slope, keep the load as low as possible. Do not drive across steep slopes. To turn on an incline, make the turn wide and slow.

10.5.3 Placing a Load

Position the forklift at the landing point so that the load can be placed where it is needed. Be sure everyone is clear of the load.

The area under the load must be clear of obstructions and must be able to support the weight of the load. If placement cannot be seen, use a signaler.

With the forks in the unloading position, lower the load and tilt the forks to the horizontal position. When the load has been placed and the forks are clear from the underside of the load, back away carefully to disengage the forks or retract the boom on variable-reach units.

10.5.4 Placing Elevated Loads

Special care must be taken when placing elevated loads. Some forklifts are equipped with a leveling device that allows the operator to rotate the fork carriage to keep the load level during travel. When placing elevated loads, it is extremely important to level the machine before lifting the load.

One of the biggest potential safety hazards during elevated load placement is poor visibility for the operator. There may be workers in the immediate area who cannot be seen. The landing point itself may not be visible. Depth perception decreases as the height of the lift increases. To be safe, use a signaler to help spot the load.

Drive the forklift as close as possible to the landing point with the load kept low. Set the parking brake. Raise the load slowly and carefully while maintaining a slight rearward tilt to keep the load cradled. Under no circumstances should the load be tilted forward until it is over the landing point and ready to be set down.

If the forklift's rear wheels start to lift off the ground, stop immediately, but not abruptly. Lower the load slowly and reposition it, or break it down into smaller components, if necessary. If surface conditions are poor at the unloading site, it may be necessary to provide reinforcement in order to gain more stability.

10.5.5 Traveling With Long Loads

Traveling with long loads presents special problems, particularly if the load is flexible and subject to damage. Traveling multiplies the effect of bumps over the length of the load. A stiffener may be added to the load to give it extra rigidity.

The forklift may be used to carry pieces of rigging equipment. This method requires the use of slings and a spreader bar.

In some cases, long loads may be snaked through openings that are narrower than the load itself. This is done by approaching the opening at an angle and carefully maneuvering one end of the load through the opening first. Avoid making quick turns because abrupt maneuvers cause the load or its center of gravity to shift.

10.6.0 Forklift Operator's Maintenance Responsibility

It is important to be familiar with some of the preventive maintenance procedures required of forklift operators. These procedures include the following activities and checks:

- Check the operator's manual for lubrication points and suggested lubrication periods. These are given in terms of the number of hours of operation.
- Check the engine and the ground for leaking fluids. Always check all fluid levels, and use the recommended fluids when refilling.
- Because forklifts often operate in dusty and dirty environments, it is important to keep the moving parts of the machine well greased. This reduces wear, prolongs the life of the machine, and ensures safe operation. Follow the manufacturer's recommendations for lubricants.

Forklifts are also provided with a daily checklist from the manufacturer. The checklist covers all the items that must be checked on the forklift every day. *Figure 28* shows an example of a daily checklist, including the items usually checked. These may differ from unit to unit.

11.0.0 Fall Protection Equipment

Ironworkers spend a large part of their time working in elevated work areas. Falls from such high places can cause serious injuries or death when improper or faulty fall protection equipment is used or when the equipment is used improperly.

11.1.0 Standards for Fall Protection Equipment

The three common types of fall protection equipment are guardrails, personal fall arrest systems, and safety nets. The following sections focus on personal fall arrest systems and safety nets. These devices and their use are governed by *OSHA Standard 1926 Subpart M.*

OSHA Standard 1926 Subpart M also covers guardrails, but the rules provided in this subpart do not apply to guardrails on scaffolds. The rules covering guardrails on scaffolds are contained in *OSHA Standard 1926 Subpart L.* OSHA basically requires that all workers use guardrail systems, safety net systems, or personal fall arrest systems generally designed to protect themselves from falling more than 6 feet (1.8 meters) to the ground or a lower work level.

11.2.0 Personal Fall Arrest Systems

Personal fall arrest systems catch workers after they have fallen. These systems are designed to activate only if a fall occurs. They are rigged to prevent a worker from falling a distance of more than 6 feet (1.8 meters) to the ground or a lower work area. The following terms must be understood when describing personal fall arrest systems:

- *Free fall distance* – The vertical distance of a fall before a deceleration device is activated.
- *Deceleration device* – A device, such as a shock-absorbing lanyard or self-retracting lifeline, that brings a falling person to a stop without injury.
- *Deceleration distance* – The distance required before a person comes to a stop. The required deceleration distance for a fall arrest system is a maximum of 3½ feet (1.07 meters).
- *Arresting force* – The force needed to stop a person from falling. The greater the free fall distance, the more force is needed to stop or arrest the fall.

Equipment used in personal fall arrest systems includes the following:

- Body harnesses and belts
- Lanyards
- Deceleration devices
- Lifelines
- Anchoring devices and equipment connectors

11.2.1 Body Harnesses

Full-body harnesses with back D-rings (*Figure 29*) are used in personal fall arrest systems. They are made of straps, which are designed to be worn securely around the user's body. This allows the arresting force to be distributed through the harness straps and throughout the body, including the shoulders, legs, torso, and buttocks. This distribution decreases the chance of injury. When a fall occurs, this keeps the worker in an upright position and helps to distribute the arresting force. The worker is suspended in a relatively comfortable position while awaiting rescue.

Selecting the right full-body harness depends on a combination of job requirements and personal preference. Harness manufacturers normally provide selection guidelines in their product literature. Some types of full-body harnesses can be equipped with front chest D-rings, side D-rings, or shoulder D-rings. Harnesses with front chest D-rings are typically used in ladder climbing and personal positioning systems. Those with side D-rings are also used in personal positioning sys-

OPERATOR'S DAILY CHECKLIST

Check Each Item Before Start Of Each Shift — Date:___________

Check One: ☐ Gas/LGP/Diesel Truck ☐ Electric Sit-down ☐ Electric Stand-up ☐ Electric Pallet

Truck Serial Number:_______________ Operator:_________________ Supervisor's OK: _________________

Hour Meter Reading: _______________

Check each of the following items before the start of each shift. Let your supervisor and/or maintenance department know of any problem. DO NOT OPERATE A FAULTY TRUCK. Your safety is at risk.

After checking, mark each item accordingly. Explain below as necessary.

Check boxes as follows: ☐ OK ☐ NG, needs attention, or repair. Circle problem and explain below.

OK	NG	Visual Checks
		Tires/Wheels: wear, damage, nuts tight
		Head/Tail/Working Lights: damage, mounting, operation
		Gauges/Instruments: damage, operation
		Operator Restraint: damage, mounting, operation, oily, dirty
		Warning Decals/Operators' Manual: missing, not readable
		Data Plate: not readable, missing adjustment
		Overhead Guard: bent, cracked, loose, missing
		Load Back Rest: bent, cracked, loose, missing
		Forks: bent, worn, stops OK
		Engine Oil: level, dirty, leaks
		Hydraulic Oil: level, dirty, leaks
		Radiator: level, dirty, leaks
		Fuel: level, leaks
		Battery: connections loose, charge, electrolyte low
		Covers/Sheet Metal: damaged, missing
		Brakes: linkage, reservoir fluid level, leaks
		Engine: runs rough, noisy, leaks

OK	NG	Visual Checks
		Steering: loose/binding, leaks, operation
		Service Brake: linkage loose/binding, stops OK, grab
		Parking Brake: loose/binding, operational, adjustment
		Seat Brake (if equipped): loose/binding, operational, adjustment
		Horn: operation
		Backup Alarm (if equipped): mounting, operation
		Warning Lights (if equipped): mounting, operation
		Lift/Lower: loose/binding, excessive drift, leaks
		Tilt: loose/binding, excessive drift, "chatters," leaks
		Attachments: mounting, damaged, operation, leaks
		Battery Test (electric trucks only): indicator in green
		Battery: connections loose, charge, electrolyte low while holding full forward tilt
		Control Levers: loose/binding, freely return to neutral
		Directional Control: loose/binding, find neutral OK

Explanation of problems marked above: ____

102F28.EPS

Figure 28 Example of a forklift operator's daily checklist.

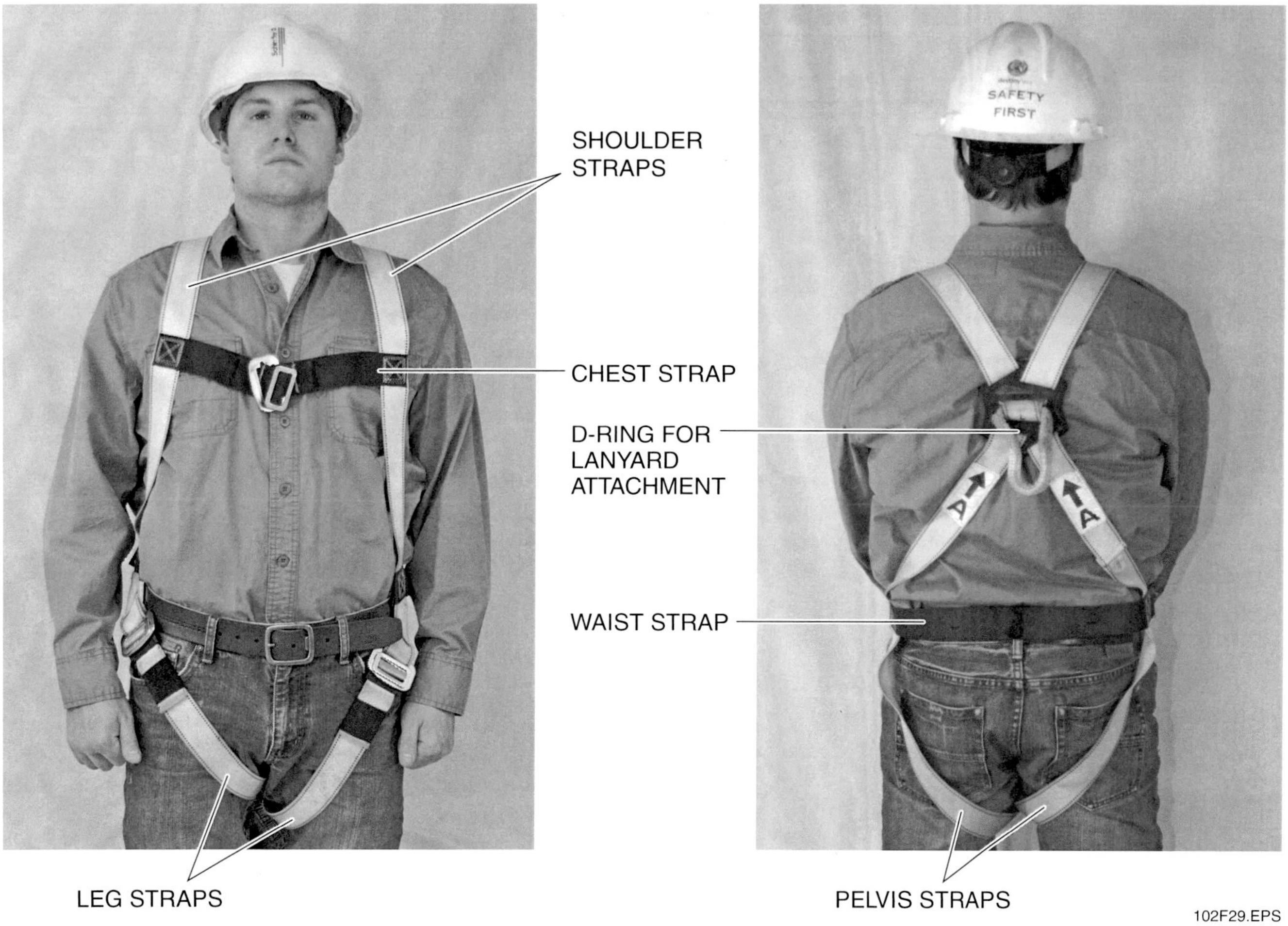

Figure 29 Full-body harness.

tems. Personal positioning systems are systems that allow workers to hold themselves in place, keeping their hands free to accomplish a task. According to OSHA regulations, a personal positioning system should not allow a worker to free fall more than 2 feet (0.61 meter). The anchorage to which the system is attached should be able to support at least twice the impact load of a worker's fall or 3,000 pounds, whichever is greater.

Harnesses equipped with shoulder D-rings are typically used with a spreader bar or rope yoke for entry into and retrieval from confined spaces.

In the past, body belts (*Figure 30*) were frequently used instead of a full-body harness as part of a fall arrest system. As of January 1, 1998, body belts have been banned from use as fall arrest equipment. This is because they concentrate all of the arresting force in the abdominal area. If a fall occurs, the worker also hangs in an uncomfortable and potentially dangerous position while awaiting rescue. Today, the use of body belts is limited to personal positioning systems.

11.2.2 Lanyards

Lanyards are short, flexible lines with connectors on each end. They are used to connect a body harness or body belt to a lifeline, deceleration de-

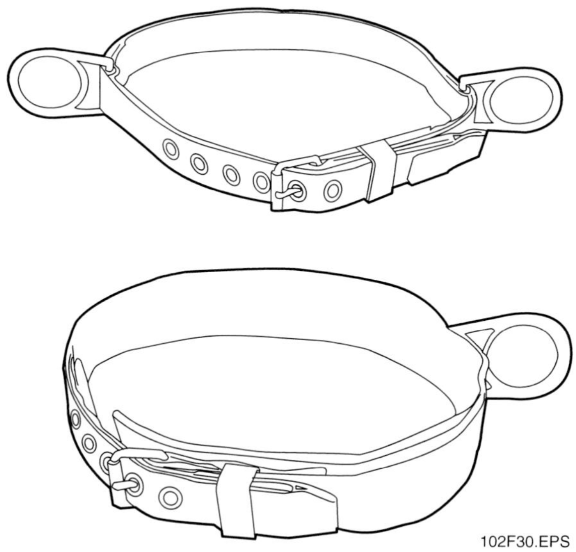

Figure 30 Body belt.

vice, or anchorage point. There are many types of lanyards designed for different uses and climbing situations. All must have a minimum breaking strength of 5,000 pounds. Lanyards come in both fixed and adjustable lengths. They are made from steel, rope, or nylon webbing. Some have a shock absorber that absorbs up to 80 percent of the arresting force when a fall is being stopped. When choosing a lanyard for a particular job, it is always a good idea to follow the manufacturer's recommendations. Shock absorbers are required in harnesses and lanyard systems (*Figure 31*) used for fall arrest.

CAUTION

When activated during the fall arresting process, a shock-absorbing lanyard stretches as it acts to reduce the fall arresting force. This increase in length must always be taken into consideration when determining the total free-fall distance from an anchor point.

11.2.3 Deceleration Devices

Deceleration devices (*Figure 32*) limit the arresting force that a worker is subjected to when the fall is suddenly stopped. Rope grabs and self-retracting lifelines are two common deceleration devices. A rope grab connects to a lanyard and attaches to a lifeline. In the event of a fall, the rope grab is pulled down by the attached lanyard, causing it to grip the lifeline and lock in place.

Self-retracting lifelines provide workers with unrestricted movement and fall protection when climbing and descending ladders or when working on multiple levels. Self-retracting lifelines usually have an 8-foot to 100-foot wire rope or synthetic web cable that automatically takes up the slack in the attached lanyard and keeps the lanyard out of the worker's way. In the event of a fall, a centrifugal braking mechanism engages to limit the worker's free fall. According to OSHA requirements, self-retracting lifelines and lanyards that limit the free fall distance to 2 feet (0.61 meter) or less must be able to support a minimum tensile load of 3,000 pounds. Those types that do not limit the free fall to 2 feet (0.61 meter) or less must be able to hold a tensile load of at least 5,000 pounds.

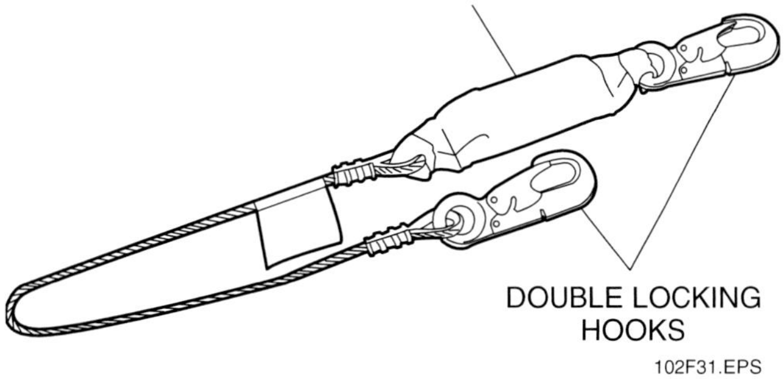

Figure 31 Typical shock-absorbing lanyard.

11.2.4 Lifelines

Vertical lifelines are usually flexible wire or fiber ropes that are attached to an anchorage. They provide a means for tying off personal fall protection equipment. Vertical lifelines are suspended from a fixed anchorage at the upper end. A fall arrest device, such as a rope grab or wire cable, is attached to the lifeline. Vertical lifelines must have a minimum breaking strength of 5,000 pounds. Workers must use separate vertical lines. A vertical lifeline must be terminated so that it will keep the worker from moving past the end, or it must extend to the ground or the next lower working level.

Horizontal lifelines are connected between two fixed anchorages. A fall arrest device is attached to the lifeline. Horizontal lifelines must be designed, installed, and used under the supervision of a qualified, competent person. The more workers tied off to a single horizontal line, the stronger the line and anchors must be.

11.2.5 Anchoring Devices and Equipment Connectors

Anchoring devices, commonly called tie-off points, must support the entire weight of the fall arrest system. The anchorage must be capable of supporting 5,000 pounds or twice the intended load or the safety factor of at least two for each worker attached. Types of anchorage points include eye bolts, overhead beams, and integral parts of building structures.

The D-rings, buckles, and snap hooks that fasten and/or connect the parts of a personal fall arrest system are called connectors. OSHA regulations specify how connectors are to be made and require D-rings and snap hooks to have a minimum tensile strength of 5,000 pounds. All components should be designed for use with the attached hardware. As of January 1, 1998, only locking-type snap hooks are permitted for use in personal fall arrest systems.

11.3.0 Using Personal Fall Arrest Equipment

Before using fall protection equipment on the job, employers should provide employees with training in the basics of fall protection and the

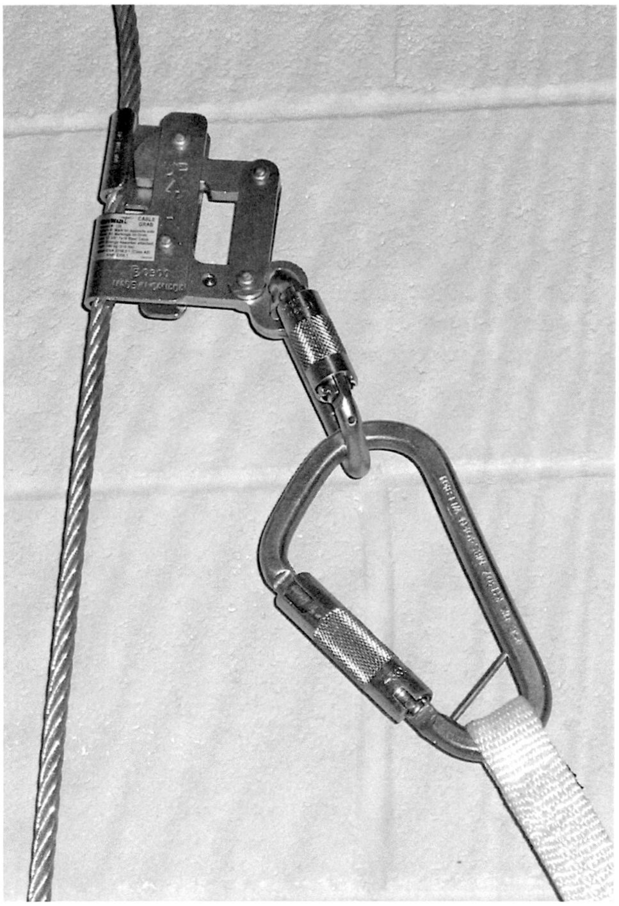

ROPE GRAB

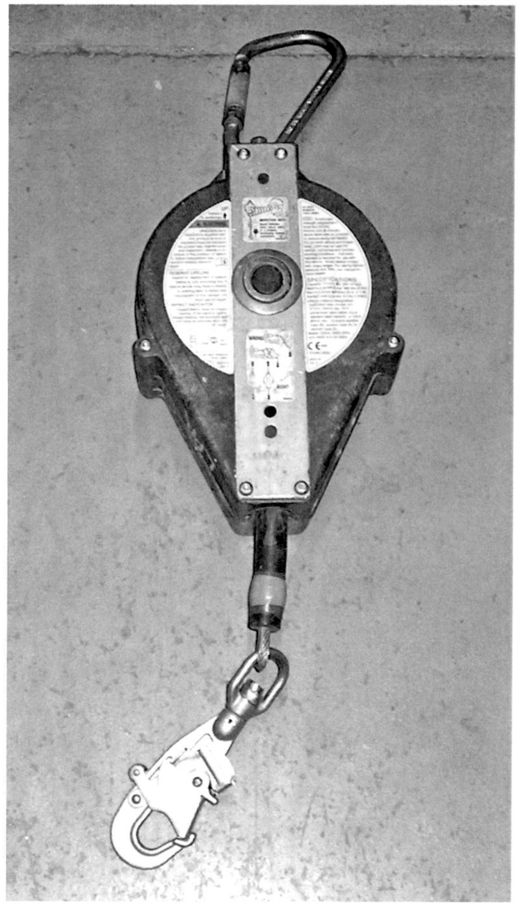

RETRACTABLE LIFELINE

102F32.EPS

Figure 32 Deceleration devices.

proper use of the equipment. All equipment supplied by employers must meet OSHA standards for strength. Before each use, always read the instructions and warnings on any fall protection equipment. Inspect the equipment using the following guidelines:

- Examine harnesses and lanyards for mildew, wear, damage, and deterioration.
- Make sure no straps are cut, broken, torn, or scraped.
- Check for damage due to fire, chemicals, or corrosives.
- Hardware should be free of cracks, sharp edges, and burns.
- Snap hooks should close and lock tightly, and buckles should work properly.
- Check ropes for wear, broken fibers, pulled stitches, and discoloration.
- Make sure lifeline anchors and mountings are not loose or damaged.

A supervisor must approve all equipment substitutions. All damaged or defective parts must be immediately taken out of service and tagged as unusable or be destroyed. If the equipment has been subjected to impact from a fall, remove it from service until it can be inspected by a qualified person.

11.3.1 Wearing a Full-Body Harness

A harness is only useful if it is put on and used correctly. Follow these steps to put on a sliding back D-ring full-body harness:

Step 1 Hold the harness by the back D-ring, and shake the harness, allowing all the straps to fall into place.

Step 2 Unbuckle and release the waist and/or leg straps.

Step 3 Slip the straps over the shoulders so that the D-ring is located in the middle of the back.

Step 4 Fasten the waist strap. It should be tight, but not binding.

Step 5 Pull the straps between each leg and buckle.

Step 6 Tighten all friction buckles so that the harness fits snugly but allows full range of movement.

Step 7 Pull the chest strap around the shoulder straps and fasten in the mid-chest area. Tighten enough to pull the shoulder straps taut.

11.3.2 Selecting an Anchor Point and Tying Off

After the full-body harness has been put on, the next step is to connect it either directly or indirectly to a secure anchor point. This is called tying off. Tying off is always done before getting into a position where a fall could occur. Always follow the manufacturer's instructions for the best tie-off methods for the equipment. Use the following general instructions and guidelines when selecting an anchorage and tying off.

When selecting an anchorage point, ensure that it is as follows:

- Directly above the worker (when this is possible)
- Easily accessible
- Damage-free and capable of supporting 5,000 pounds per worker or twice the intended load or the safety factor of at least two for each worker attached.
- High enough so that no lower level will be struck should a fall occur

When tying off, consider the following:

- Tie-offs that use knots are weaker than other methods of attachment. Knots can reduce the lifeline or lanyard strength by 50 percent or more. A stronger lifeline or lanyard should be used to compensate for this.
- A shorter free fall can reduce the chances of falling into obstacles, being injured by the arresting force, or damaging the equipment. To limit the free fall, a shorter lanyard can be used between the lifeline and harness. Raising the tie-off point on the lifeline can reduce the amount of slack in the lanyard. The tie-off to the lifeline or anchor must always be higher than the connection to the harness. Tie-off to the highest point practical.
- When using retractable systems, always be aware of the pendulum effect.

11.4.0 Safety Net Systems

Safety nets are used for fall protection on bridges and similar types of construction. They must be installed as close as is practical below the work area, but not more than 30 feet (9.14 meters) below the work area. OSHA requires enough clearance under a safety net to prevent a worker who falls into it from hitting the surface below. There must be no obstruction between the work area and the net. Depending on the actual vertical distance between the net and the work area, the net must extend at least 8 to 13 feet (2.44 to 3.96 meters) beyond the edge of the work area. Mesh openings in the net are limited to 36 square inches (91.44 square centimeters) and 6 inches (15.24 centimeters) on the side. The border rope must have a 5,000 pound (22.2 kilo Newton) minimum breaking strength, and connections between net panels must be as strong as the nets. Safety nets must be inspected at least once a week and after any event that might have damaged or weakened them. Worn or damaged nets must be removed from service.

11.5.0 Rescue After a Fall

Every elevated job site should have a rescue and retrieval plan in case it is necessary to rescue a fallen worker. Planning is especially important in remote areas that may not have easy telephone access. Make sure to know the requirements of each employer's rescue plan in order to be prepared if a fall should occur. Find out what rescue equipment is available and where it is located. Learn how to use the equipment for self-rescue and the rescue of others.

If a fall occurs, any employee hanging from the fall arrest system must be rescued safely and quickly. The employer should have a previously determined method of rescue for fall victims. This may include equipment that lets the victim rescue themselves, a system of rescue for co-workers, or a way to alert a trained rescue squad. If a fall rescue depends on calling for outside help, such as from the fire department or rescue squad, all related phone numbers must be posted in plain view at the work site. Always follow the employer's rescue plan. Call any special rescue services needed. Communicate with the victim, and monitor them constantly during the rescue.

Summary

Safety is, and should be the primary concern at every level of responsibility in ironwork. Employers are responsible for providing the means, support, and training for safe work. Employees are responsible for maintaining a safety awareness at all times. *OSHA Standard 1926 Subpart R* provides safety guidelines for construction work in general. All forms of ironwork require close attention to the appropriate safety procedures for that aspect of the work. Working around cranes requires awareness of the location and movement of loads and cranes.

Working with power tools demands that the worker know the hazards associated with those tools. Working at heights demands constant attention to the environment and to the activities going on around the worker. Even the clothing a worker wears is a safety factor. Take all possible practical precautions.

Aerial work platforms and scissors lifts are used to raise and lower workers to and from elevated work areas. *OSHA Standard 1926 Subpart N* defines and governs the use of aerial lifts.

Falls from ladders, scaffolds, lifts, or unprotected work areas can cause serious injuries or death when no fall protection equipment or the wrong kind of equipment is used. Similar mishaps can occur when the right type of equipment is improperly used. The three types of fall protection commonly used when working at elevated levels are guardrails, personal fall arrest systems, and safety nets. Personal fall arrest system and safety net types and uses are governed by *OSHA Standard 1926 Subpart M.* This OSHA standard requires that all workers use guardrail systems, safety net systems, or personal fall arrest systems to protect themselves from free falling more than 6 feet (1.8 meters) to the ground or a lower work area.

Review Questions

1. The primary consideration when using any equipment is _____.
   a. safety
   b. stability
   c. height-lifting ability
   d. weight-lifting ability

2. To prevent the trailer from tipping when loading a piece of equipment onto a trailer, _____.
   a. load the back of the trailer
   b. lash the dunnage in place first
   c. weight down the front of the trailer
   d. hitch the trailer to the tow vehicle before loading

3. The operator should check for leaks in a pressurized hydraulic system by _____.
   a. checking the hydraulic oil level in the supply tank
   b. feeling all the hoses and connectors
   c. checking the lines using a piece of wood or cardboard
   d. using a flashlight and an inspection mirror

4. When refueling a piece of equipment, keep the nozzle or fuel funnel in contact with the filler neck to _____.
   a. prevent the possibility of sparks
   b. lower the possibility of a fuel spill
   c. protect you from breathing the fumes
   d. make it easier to see when the fuel tank is full

5. When a battery is frozen, _____.
   a. disconnect the negative lead before attaching the charger
   b. do not charge it or attempt to jump-start it
   c. jump-start the equipment following the manufacturer's instructions
   d. replace the battery and charge the old battery

6. The components of an aerial lift include a _____.
   a. base, platform, and lifting mechanism
   b. base, auger, and bucket
   c. platform, boom, and compactor
   d. compactor, control panel, and forks

7. Which of the following allows the hydraulic fluid to flow and the aerial lift to be pushed by hand?
   a. Emergency lowering valve
   b. Free wheeling valve
   c. Stabilizer hardware
   d. Steer controller

8. If a problem with an aerial lift is found while performing daily checks, _____.
   a. note the problem and continue working
   b. contact the manufacturer immediately
   c. disconnect the battery and continue the daily checks
   d. correct the problem before beginning work

9. When trying to start the engine (motor) of a scissor lift, *never* hold the key in the Start position for more than _____.
   a. 5 seconds
   b. 10 seconds
   c. 15 seconds
   d. 20 seconds

10. On a boom lift, the primary control station is the _____.
    a. remote control box
    b. work platform controls
    c. base-mounted control box
    d. steering wheel-mounted controls

11. A generator will automatically be shut down by a(n) _____.
    a. high engine temperature indicator
    b. ammeter gauge
    c. engine coolant temperature gauge
    d. engine tachometer

12. When evidence is found of arcing on the control panel of a tow-behind generator, _____.
    a. continue with operation of the generator
    b. the control panel must be replaced
    c. contact the supervisor about the problem before operating
    d. call the manufacturer for additional information

13. When a portable air compressor is used along with another air source, make sure that _____.
   a. there is a check valve installed at the service valve
   b. the engine and air compressor air filters are clean
   c. the compressor is on level ground
   d. all of the air vents are clear

14. The most common type of fixed-mast forklift used in construction is the _____.
   a. telescoping boom forklift
   b. warehouse forklift
   c. rough terrain forklift
   d. squirt boom forklift

15. Personal fall arrest systems are designed to prevent a worker from falling more than _____.
   a. 4 feet
   b. 6 feet
   c. 10 feet
   d. 12 feet

Trade Terms Introduced in This Module

Aerial lift: A mobile work platform designed to transport and raise personnel, tools, and materials to overhead work areas.

Capacity: The total amount of weight the lift is capable of lifting. This includes personnel, tools and materials, and/or equipment.

Circuit breaker: A device designed to protect circuits from overloads, which is able to be reset after tripping.

Compliance: Acting in accordance with regulations.

Compressor: A mechanical device that generates compressed air to power a variety of tools.

Forklift: A machine designed to move bulk items around the job site; also known as a powered industrial truck.

Fuse holder: A device used to hold a fuse that may be part of a circuit or that may hold a spare fuse.

Generator: A machine designed to generate electricity.

Ground fault circuit interrupter (GFCI): A device that detects grounds on a power line and interrupts power flow to the attached outlet(s).

Interlocking systems: Control systems that are set to prevent accidents by keeping one system from overriding the other.

Powered industrial trucks: See *forklift*.

Proportional control: A control that increases speed in proportion to the movement of the control.

Straight blade duplex connector: An electrical connector or style of outlet that pushes in to connect.

Twist-lock connector: A type of electrical connector that twists to connect.

Additional Resources

This module presents thorough resources for task training. The following resource material is suggested for further study.

The complete OSHA Safety and Health Regulations for construction are located at www.osha.gov.

Figure Credits

Topaz Publications, Inc., Module opener, Figures 15 (telescopic boom lift), 18 (portable generator), 19, 23, 29, and 32

LPR Construction, Figures 1-3

Scott Health and Safety, Figure 4

JLG Industries, Inc., Figures 8, 11 (photos), 14, and 15 (articulating boom lift)

Ingersoll Rand Construction Technologies, Figures 18 (tow-behind generator) and 24

Sellick Equipment Ltd., Figure 26

Manitou North America, Inc., Figure 27

NCCER CURRICULA — USER UPDATE

NCCER makes every effort to keep its textbooks up-to-date and free of technical errors. We appreciate your help in this process. If you find an error, a typographical mistake, or an inaccuracy in NCCER's curricula, please fill out this form (or a photocopy), or complete the online form at **www.nccer.org/olf**. Be sure to include the exact module ID number, page number, a detailed description, and your recommended correction. Your input will be brought to the attention of the Authoring Team. Thank you for your assistance.

Instructors – If you have an idea for improving this textbook, or have found that additional materials were necessary to teach this module effectively, please let us know so that we may present your suggestions to the Authoring Team.

NCCER Product Development and Revision
13614 Progress Blvd., Alachua, FL 32615
Email: curriculum@nccer.org
Online: www.nccer.org/olf

❑ Trainee Guide ❑ Lesson Plans ❑ Exam ❑ PowerPoints Other ________________________

Craft / Level: Copyright Date:

Module ID Number / Title:

Section Number(s):

Description:

Recommended Correction:

Your Name:

Address:

Email: Phone:

Tools and Equipment of the Trade

30103-11

Trainees with successful module completions may be eligible for credentialing through NCCER's National Registry. To learn more, go to **www.nccer.org** or contact us at **1.888.622.3720**. Our website has information on the latest product releases and training, as well as online versions of the *Cornerstone* newsletter and Pearson's Contren® product catalog.

Your feedback is welcome. You may email your comments to **curriculum@nccer.org,** send general comments and inquiries to **info@nccer.org**, or send us the User Update form at the back of this module.

V.1 1/11

30103-11
TOOLS AND EQUIPMENT OF THE TRADE

Objectives

When you have completed this module, you will be able to do the following:

1. Identify and explain commonly used safety tools and equipment.
2. Identify and describe the proper use of common ironworking hand tools.
3. Identify the power sources for common ironworking tools.
4. Identify and describe the proper use of common ironworking power tools.

Performance Tasks

Under the supervision of the instructor, the trainee should be able to do the following:

1. Demonstrate the proper use of appropriate personal protective equipment.
2. Demonstrate the safe and effective use of available ironworking hand tools.
3. Demonstrate the safe and effective application of available power sources for ironworking power tools.
4. Demonstrate the safe and effective use of available ironworking power tools.

Trade Terms

Drift pin
Frog
Mushrooming
Reciprocating saw
Sleever bar
Spud wrench
Thread chaser
Torque

Contents

Topics to be presented in this module include:

Contents (continued)

Figures and Tables

1.0.0 INTRODUCTION

Most work activities performed by ironworkers require the use of some type of tool. This module describes the hand and power tools used specifically by the average ironworker, and which were not covered in the *Core Curriculum*.

2.0.0 IRONWORKING PPE

Most ironwork has specific hazards associated with cutting, shaping, and welding steel. These are activities that produce high-impact bits of debris, so face and hand protection is required. There is risk of hearing damage from the noise level of much of the work. There must also be protection against the harmful arcs of welding and torch work. Clothing covers the skin, but a worker's face is usually closest to the work.

2.1.0 Face Shields

Ironworkers sometimes work in situations where basic safety glasses or goggles are not enough to protect the face and eyes. Grinding and cutting activities often generate small pieces of flying debris. Wires from the wire brushes used on powered strippers often fly several feet away from the brush and can easily penetrate the skin of anyone nearby. To protect against flying debris, wear a face shield that covers all of the face. A basic face shield consists of an adjustable headband with either a flat or rounded face shield attached to it. The headband is normally positioned on the head so that the shield will extend down in front of the face.

Although there are many types of face shields, they can be divided into two basic groups: clear safety face shields and tinted face shields for welding helmets (*Figure 1*).

2.1.1 Clear Face Shields

Most clear face shields are made of lightweight molded plastic. The headband portion supports a clear visor that extends down over the face. The headband is size-adjustable. The visor can be either rounded or flat. Clear face shields are often used for protection against splashes, sprays, or flying particles encountered when performing light machining work, chipping, grinding, stripping, or sanding. If safety glasses are required in a work area, a face shield can be worn over the safety glasses to provide additional protection.

2.1.2 Face Shields for Welding Helmets

Because of the infrared and ultraviolet rays created during oxyfuel cutting and welding, a special face shield with a darkened visor is made to be worn over a welding helmet. These face shields are not a substitute for a welding helmet. They are intended to provide additional protection when workers cut or weld for prolonged periods.

2.2.0 Protective Leathers

For additional protection, ironworkers and welders often wear special clothing made from leather. The protective leathers are worn over the usual work clothing. Leather aprons, split-leg aprons, sleeves, capes, and jackets are flexible enough to allow ease of movement while providing extra protection from sparks and heat. These protective leathers can also be useful to ironworkers handling steel parts. *Figure 2* shows different types of protective leather clothing.

Figure 1 Welding/cutting goggles and face shield combinations.

Figure 2 Leather and fire-resistant protective coverings.

2.3.0 Welding Gloves

Gauntlet-type leather gloves (*Figure 3*) are specifically designed for welding activities. The longer sleeves of the gloves extend up the arm to protect from hot metal splatters and from the ultraviolet and infrared arc rays associated with cutting and welding. These gloves must be worn when performing any type of arc welding.

> **CAUTION**
>
> Picking up hot metal with leather welding gloves burns the leather and makes it shrivel and become hard. To prevent damage to the gloves, never wear them when handling hot metal. Use machinery or tongs to handle hot metal.

2.4.0 Welding Shields

Ironworkers sometimes perform welding as part of their daily activities. Welding shields (also called welding helmets or hoods) provide eye and face protection. Some shields are equipped with handles and are held up to shield the face and eyes. Most types are designed to be worn on the head. The shields either connect to helmet-like headgear or attach to a hard hat. Shields worn on the head have pivot points where they attach to the sides of the headgear. These shields can be raised or removed when they are not needed. *Figure 4* shows typical welding shields.

Welding shields are made of dark, nonflammable materials. Near the center of the shield is an observation window that allows the person wearing it to observe the welding arc. The window sizes are 2½" × 4¼" and 4½" × 5¼". The shield window contains a shaded glass filter plate and a clear glass or plastic safety lens. The shaded filter plate provides eye protection. The safety lens on the outside of the window protects the more costly filter plate from damage caused by splattering metal or debris. For additional filter plate protection, a second clear safety lens may be installed on the inside of the shield's window and behind the filter plate. On most welding shields, the window is fixed in the shield itself. Since the window contains the shaded filter plate, the person wearing the shield must remove it to see clearly when not welding.

2.4.1 Hinged Window

Some welding shields are equipped with a hinged window containing the shaded filter plate. This hinged window allows the person doing the welding to see or to clean welding components during the welding process without having to repeatedly remove the welding shield. Clear safety glass installed inside the shield and behind the window protects the eyes when the window is raised. A hinged window can be flipped up so the person wearing the shield can clearly

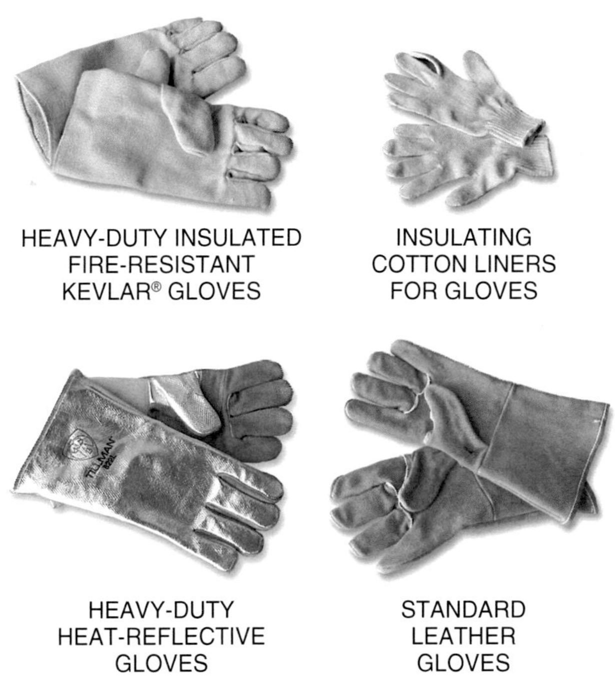

Figure 3 Gauntlet-type welding gloves.

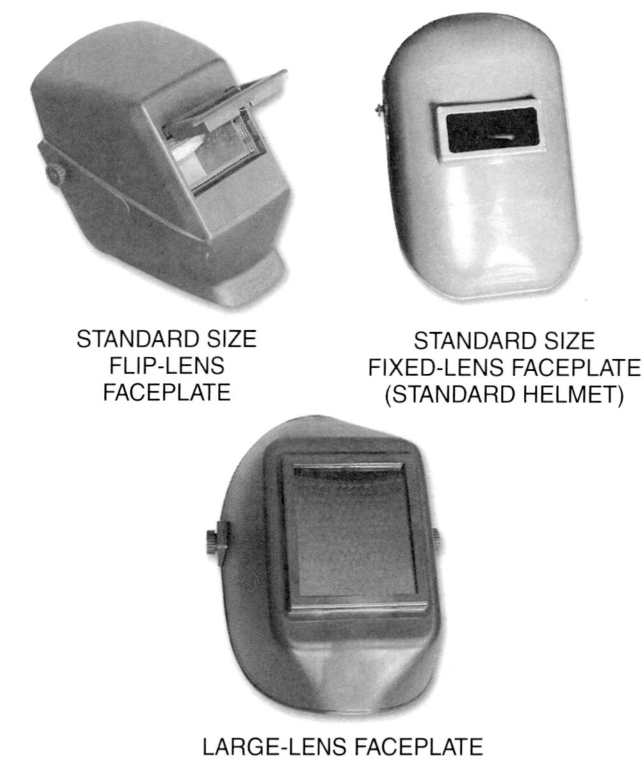

Figure 4 Electric arc welding shields.

see without removing the shield. Eye protection is provided by the safety glass mounted inside the shield. The hinged window should never be raised while welding.

> **WARNING**
>
> Never raise the window of a welding shield during a welding process. Serious eye damage can result from observing a welding arc without the protection of a shaded filter plate.

2.4.2 Filter Plates

Filter plates come in various shades. The shade required depends on the maximum amount of welding amperage being used. The higher the amperage, the darker the filter plate must be. Filter plates are graded by number. The larger the number, the darker the filter plate. Filter plate recommendations for various welding amperage ranges are as follows:

- *Shade 10* – 75 to 200 amperes
- *Shade 12* – 200 to 400 amperes
- *Shade 14* – 400 amperes and above

2.4.3 Protection of the Lens

During normal welding operations, the window in the welding shield becomes dirty from welding smoke and splatter. To ensure proper visibility, the window must be cleaned on a regular basis. The safety lens and filter plate can be removed and can be cleaned in the same manner that safety glasses are cleaned, by wiping them with a clean, soft cloth. With use, the outer safety lens can become scratched, pitted, or saturated with welding slag. When this occurs, replace the outer safety lens.

2.5.0 Fall Protection Devices

The commonly used beamer (*Figure 5A*) is used to attach the lanyard to beams. The other method of providing protection between vertical beams is a horizontal lifeline, most frequently attached to a set of uprights mounted on a beam before the beam is installed and connected, as shown in *Figure 5B*. This allows the workers to work on connections to the beam as soon as it is adequately connected.

Beamers may be attached to the beam on which the worker is standing or sitting, or on the beam above. It should be installed on the highest point practical under the circumstances.

3.0.0 Hand Tools

When working with heavy metal, hand tools may seem old-fashioned. However, many of these tools are still in widespread use today. The following types of hand tools are identified and explained in this module:

- Measuring tools
- Clamping or securing tools
- Wrenches
- Cleaning tools
- Cutting tools
- Threading tools
- Alignment tools
- Belts and bags
- Welding tools
- **Frogs**
- **Sleever bars**
- **Spud wrench**
- Hammers

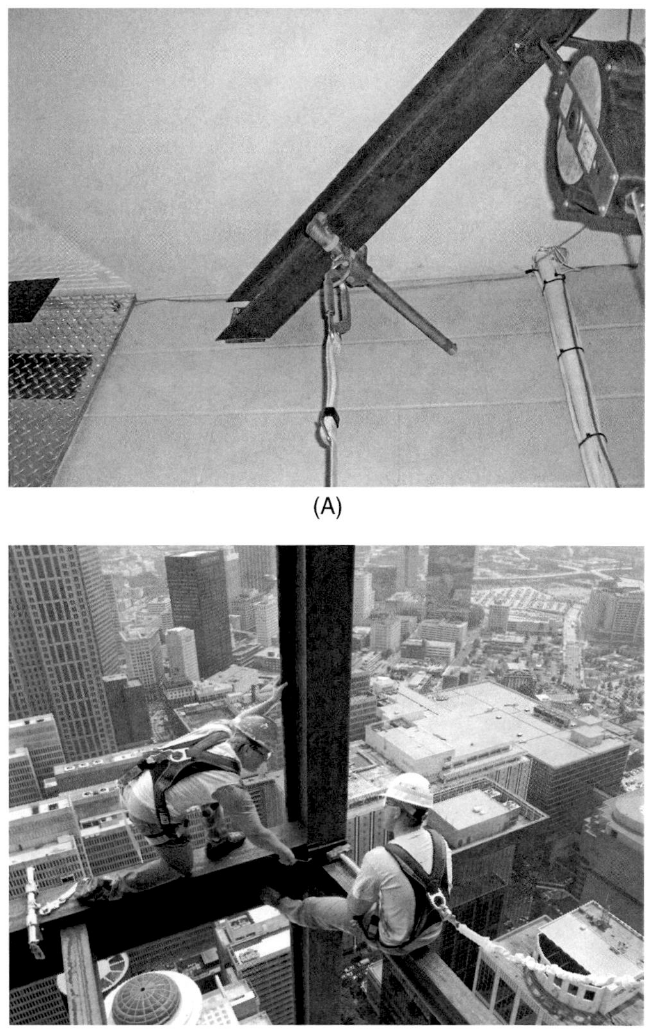

(A)

(B)

103F05.EPS

Figure 5 Variations of fall protection equipment installations and uses.

3.1.0 Commonly Used Measuring Tools

In addition to common measuring tools, such as rulers, tape measures, squares, and levels, which were covered in *Core Curriculum*, ironworkers also use marking devices and fillet gauges.

3.1.1 Marking Devices

Because of the building materials used by ironworkers and the environments in which they work, special marking devices must be used to indicate their measurement points on building materials. For welding activities, special temperature-sensitive marks are used to show when a metal object has reached the desired temperature. The marking devices commonly used by ironworkers are as follows (*Figure 6*):

- *Pencils* – Common carpenter's pencils are often used for routine marking on materials that allow the pencil marks to be visible.
- *Soapstone* – The most common marking device used by ironworkers is soapstone. Soapstone is actually a type of rock called steatite. The soapstone material is a hard, slightly greasy chalk that is cut into sticks about 6 inches long, ½ inch wide, and ⅛ inch thick. The color of soapstone is grayish white. When rubbed against metal objects, soapstone leaves a mark of the same color. Special holders are available that hold and protect the soapstone sticks during use.
- *Paint sticks* – Special colored markers, shaped like small sticks, are often used by ironworkers to mark on a variety of surfaces, including oily, rough, smooth, wet, dry, icy, and rusty surfaces. Paint sticks may be used on glass, plastic, lumber, cement, cardboard, and rubber. The markers are made to work in varying temperature ranges. All-purpose paint sticks are made to work in temperatures ranging from –50°F to 150°F. Heat-treating and annealing paint sticks can withstand temperatures up to 1,600°F. Some markers are designed to bleed through the primer coating of metal materials.
- Lumber crayons are used to mark concrete block, lumber, metal, plastic, rubber, paper, cardboard, and ceramics.

3.1.2 Fillet Gauges

Fillet gauges are used to measure the size and shape of fillet welds. The gauge is available in a pocket size, and allows the worker to determine whether a weld meets specifications. While the weld will be examined by an inspector and will be checked for conformance at that time, it is better to find out before the inspector comes and correct the weld than to find out after the weld has been rejected.

3.2.0 Commonly Used Clamping or Securing Tools

When assembling, some type of device is frequently required to maintain the relationship between the pieces of metal. Other equipment is used to pull or move materials. The devices used by ironworkers to hold materials together prior to welding or pulling include the following:

- Bar clamps
- Finger clamps
- Button punch
- C-clamp
- Vise grip

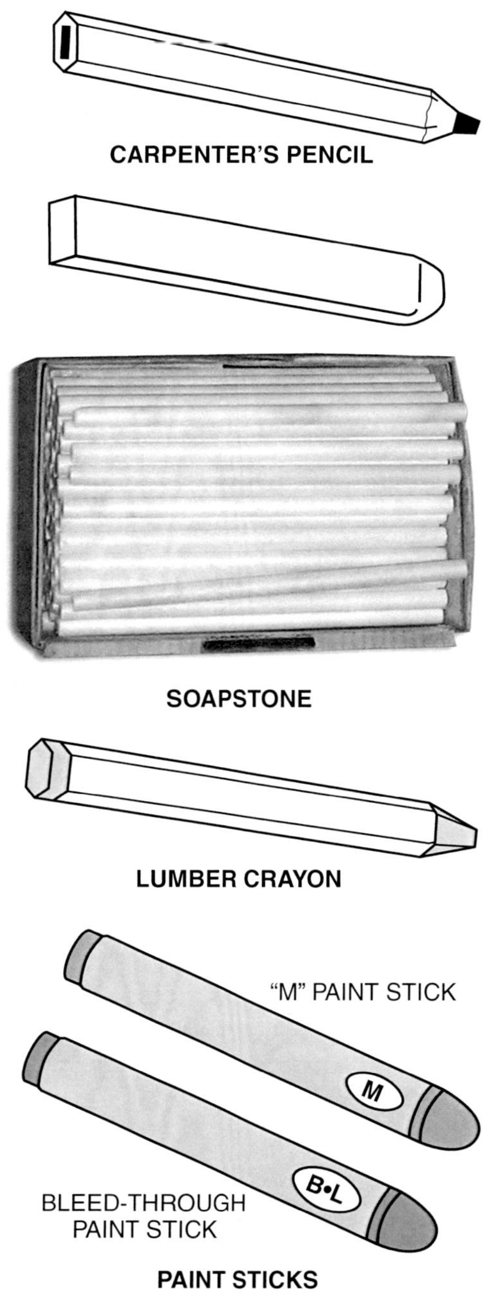

Figure 6 Marking devices.

3.2.1 Bar Clamps

The jaws, or pinchers, of most clamping devices are usually pulled together by either springs or by mechanical methods, such as a screw jack. Several variations of the screw-style clamp exist, but ironworkers are more likely to use bar clamps (*Figure 7*). The two styles of bar clamps commonly used by ironworkers are the basic bar clamp and the I-beam bar clamp.

Both the basic bar clamp and the I-beam bar clamp have similar components.

- *Basic bar clamp* – A basic bar clamp is a light-duty device designed for general clamping applications. Basic bar clamps are easy to handle and may be used for simple jobs, such as holding together two pieces of lightweight metal that have been aligned for welding or bolting.
- *I-beam bar clamp* – An I-beam bar clamp is a heavy-duty clamping device designed for heavy everyday use. Typical uses include holding pieces of heavyweight metal together for welding or bolting and pulling two small I-beams together.

3.2.2 Finger Clamps

Finger clamps, or locking clamps, are plier-like devices that are used to hold lightweight objects together. The shape of the jaws on these devices shows why they are called finger clamps. Although several variations of these clamps are available, ironworkers most often use the locking C-clamps and the locking sheet metal clamps. *Figure 8* shows finger clamps often used in ironworking activities.

Finger clamps have an adjustment screw in the handle that allows the opening of the jaws to be adjusted to securely clamp any object. After the jaws have been positioned and clamped onto an object, the handles lock closed when squeezed together. The locking handles of these clamps are similar to the pliers called vise grips. A trigger device inside the handle permits the jaws of the clamp to be quickly and easily released. This allows the jaws to be repositioned or the clamp to be removed from the object(s) being clamped.

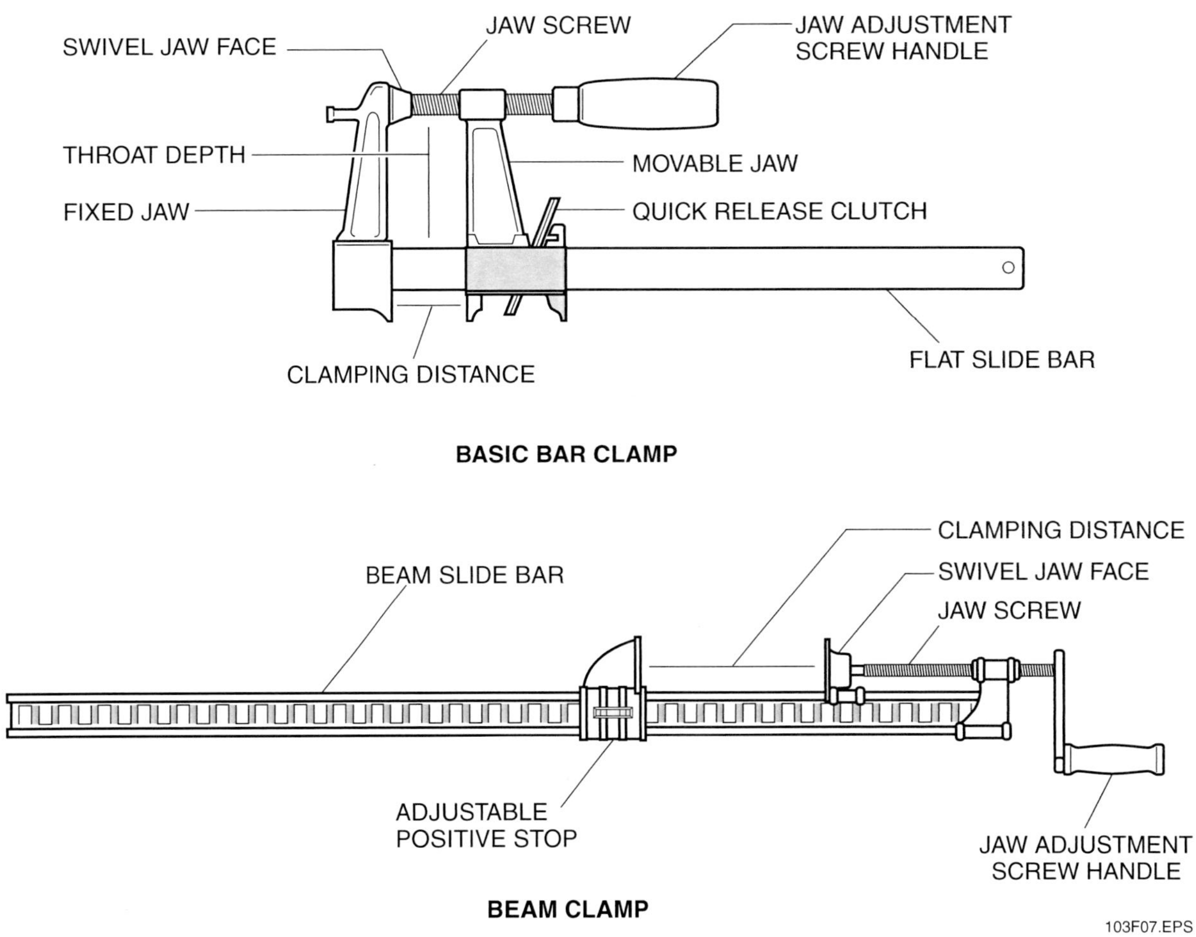

Figure 7 Bar clamps.

The following are some common types of finger clamps:

- *Spring clamps* – These are lightweight tools used for holding very lightweight materials together for measuring or other purposes.
- *Locking C-clamps* – Locking C-clamps are typically used when welding lightweight pieces of metal.
- *Locking sheet metal clamps* – Personnel fabricating with sheet metal often use these locking sheet metal clamps because of the wide lip on each of the clamp's jaws. The jaws/lips on these clamps make folding sheet metal edges into a 90-degree shape much easier. The wide jaws/lips also make it much easier to hold two pieces of sheet metal together for welding.

3.2.3 Button Punch

The button punch (*Figure 9*) is a specialized tool used to join standing seams on metal decking. The tool's jaws are made of aircraft-grade stainless steel. The button punch produces a raised button on the two sheets of decking to be joined. This button is often enough to hold the decking, or it may be used to crimp the decking prior to welding.

3.2.4 C-Clamps

The old-fashioned screw-based C-clamp (*Figure 10*) is very powerful and easy to use. These clamps have a great deal of strength, and are available in many different sizes. In addition to the screw-based version, a powerful hydraulic C-clamp is manufactured as part of the Porta Power system.

3.3.0 Common Wrenches (Standard or Metric)

Many of the tasks performed by an ironworker require installing and removing screws or bolts and nuts. The ironworker has to use some kind of a specialized tool to install or remove some of these screws, bolts, and nuts. The following sections cover those wrenches that were not described in *Core Curriculum*.

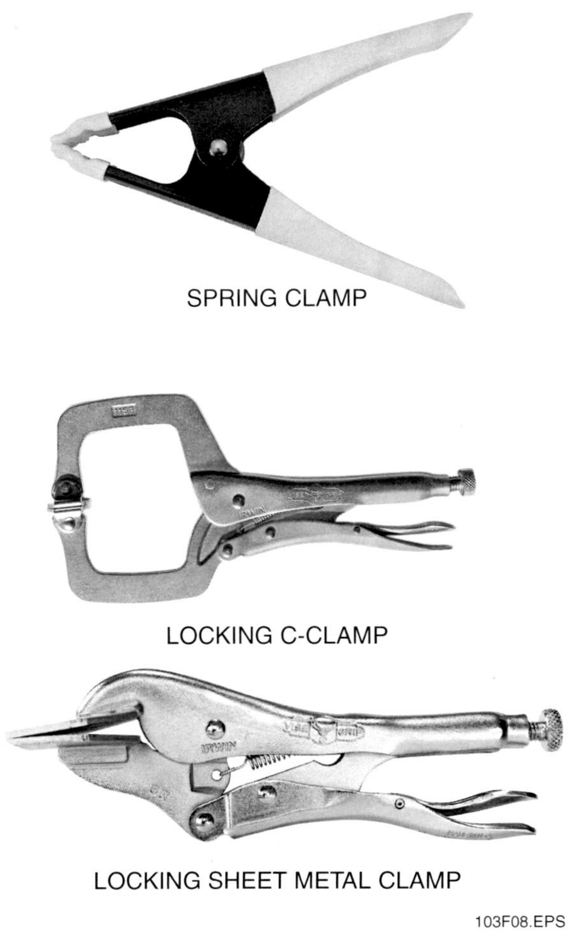

Figure 8 Common finger clamps.

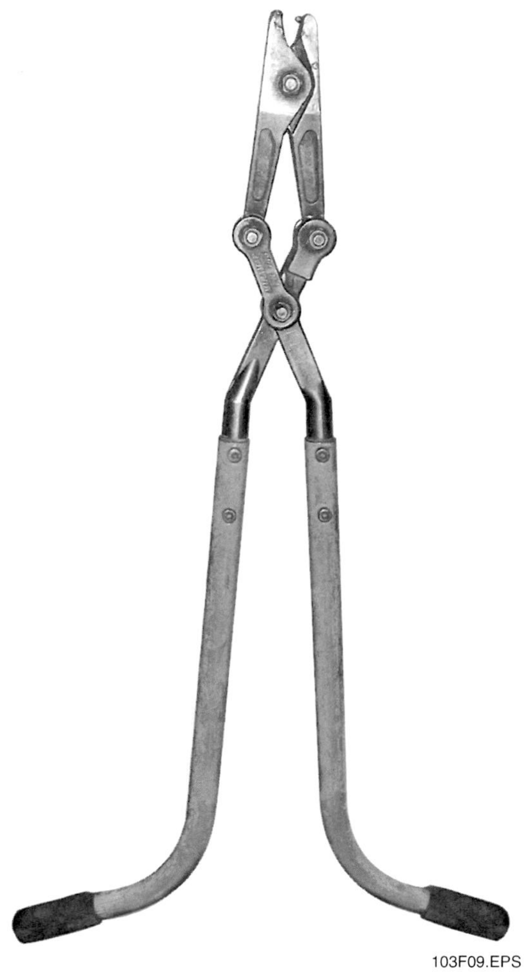

Figure 9 Button punch.

Figure 10 Screw-based C-clamp.

3.3.1 Allen Wrenches

Setscrews (or bolts) are special fastening devices that have round heads. These screws, or bolts, are often used to hold parts together when conventional bolts and nuts or screws cannot be used. Because of the round heads, regular wrenches cannot be used to install or remove these screws or bolts.

In the center of the head on these fasteners is a six-sided hole. A six-sided wrench is used to install and remove them. These fasteners are often called hex screws or bolts. The term *hex* comes from the word *hexagon*. A hexagon is a shape having six sides. Hex wrenches are also called Allen wrenches (*Figure 11*).

An Allen, hex, or setscrew wrench is an L-shaped, six-sided wrench made from tool steel. These wrenches are usually bought in sets with sizes varying up to ¾ of an inch for standard measurements. These wrenches also are made for metric measurements.

To correctly use these wrenches, the user must first select a wrench size that fits the hex hole inside the screw or bolt head. The hole must be cleaned out so that the end of the wrench can be inserted into it. Always make sure that the selected wrench fits the hole perfectly. If a wrench is too small, the corner shoulder of the hole or of the wrench will be rounded off when pressure is applied to remove or tighten the screw or bolt. A wrench that is too large should not be forced into the hex hole. This will damage both the hole and the wrench.

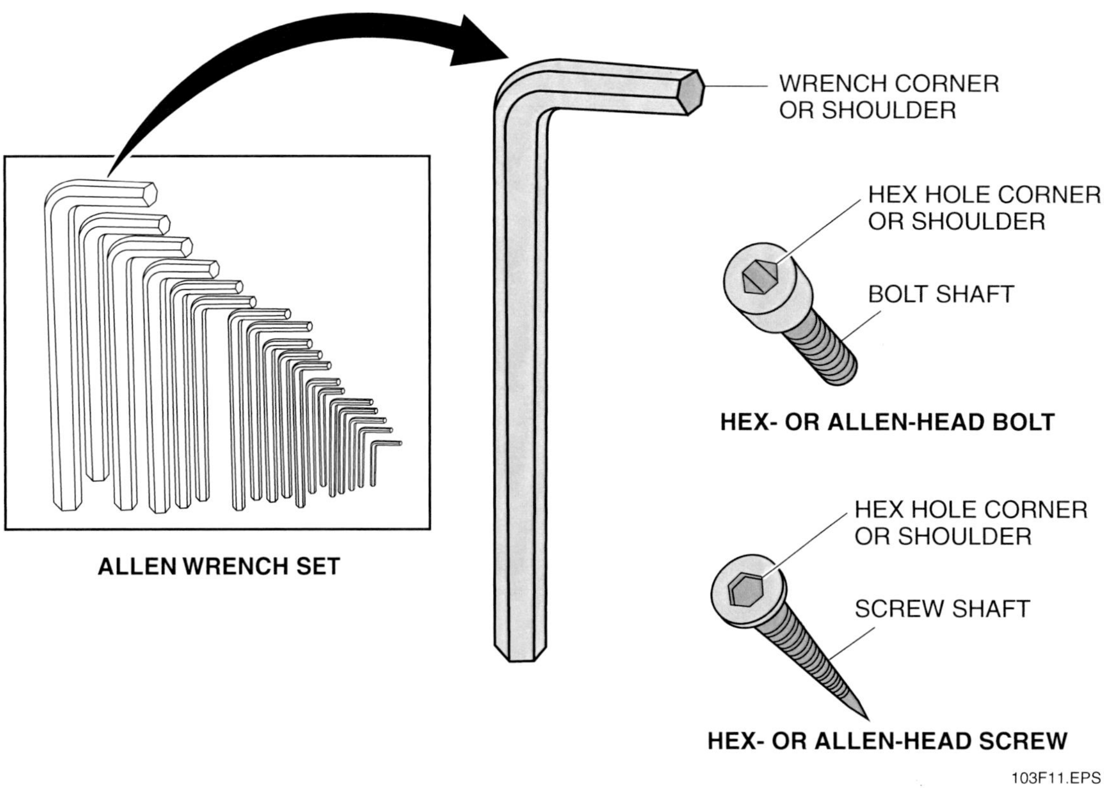

Figure 11 Allen wrenches.

CAUTION

When tightening or loosening hex screws or bolts, make sure that the correct size wrench is selected. This will help to prevent stripping or rounding off the corners of the hole or the wrench.

L-shaped wrenches are designed for specific functions. The tip of the wrench's long leg is inserted into the hole of a hex screw or bolt and turned to loosely install (or remove) the screw or bolt. The tip of the wrench's short leg is used to tighten (or loosen) the screw or bolt. When tightening or loosening hex screws or bolts, the long leg of the wrench provides extra leverage.

3.3.2 Knocker (Striking) Wrenches

Ironworking often requires installing or removing large bolts and nuts that are used to fasten steel beams and devices. When nuts are too tightly fastened (frozen), normal wrenches may not be able to break them loose for removal. Knocker or striking wrenches (also called slugging wrenches) are special heavyweight box wrenches used to help break loose nuts that are frozen to bolts. Although knocker wrenches are usually used to loosen nuts from bolts, they can also be used to tighten a nut onto a bolt.

Knocker or striking wrenches are available in a variety of sizes, both standard and metric, with either 6-point or 12-point wrench heads (*Figure 12*). The handles of these wrenches are designed to be struck with a hammer.

CAUTION

Never add a length of pipe to a knocker wrench in order to gain more leverage. This could damage the tool or result in injury to the operator.

After a knocker wrench has been installed onto a nut, deliver a hammer blow to the striking face on the wrench handle. The hammer blow creates a shock effect that travels through the wrench to the nut. This helps to loosen the nut from the bolt. Striking the wrench also creates a twisting effect that helps to loosen the nut.

WARNING

To prevent personal injury, always wear appropriate PPE when hitting a knocker wrench with a hammer. Metal chips can break away from either the hammer or the wrench. Take the necessary precautions to prevent the wrench from flying off the nut when the wrench is struck. Always secure the wrench when working at heights to prevent it from falling.

Knocker wrenches must be inspected for damage prior to being used. The inner and exterior surfaces of the wrench walls must not be cracked, worn, or rounded off. The striking face and handle must not be damaged to the point of being twisted or deformed. If the handle is twisted or the striking face is damaged to the point that the hammer's striking force will not move the wrench in the desired direction, the wrench must be replaced.

3.4.0 Cleaning Tools

The metals used in ironworking are often exposed to the weather and other elements that may cause their surfaces to rust or become contaminated. Dirt, paint, and other debris may also accumulate on the metal. Ironworkers often use hand tools to

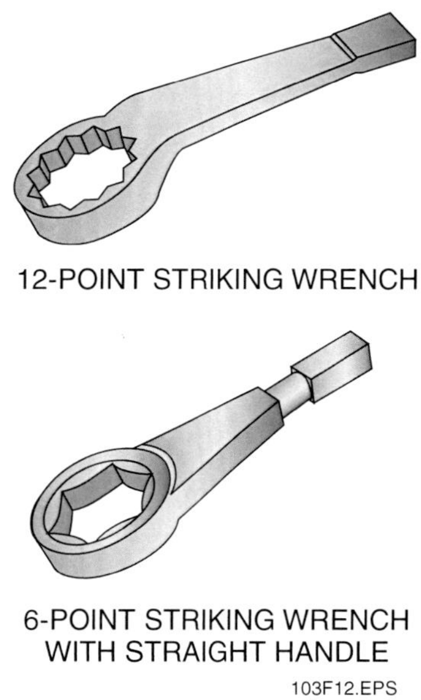

Figure 12 Two commonly used knocker (striking) wrenches.

clean the surface areas of metal prior to welding or installation. The cleaning tools they most often use include wire brushes, scrapers, and files (*Figure 13*).

> **CAUTION**
>
> Because of the flying particles produced when cleaning metals, all persons performing or assisting in cleaning must wear safety glasses, goggles, or face shields. Failure to wear protective equipment can result in personal injury.

3.4.1 Wire Brushes

Wire brushes are primarily used to remove paint and surface corrosion. Wire brushing removes light to medium (loose) corrosion from metals, but it will not remove tight corrosion. Corrosion that is tightly attached to the metal's surface must be removed either by filing, grinding, or sandblasting. Wire brushes also are used to clean weld passes.

Selecting the correct wire brush is critical to the cleaning process. When cleaning stainless steel or aluminum, wire brushes with stainless steel bristles must be used. Standard wire brushes with steel bristles will contaminate the surface of the aluminum with mild steel, causing welding defects.

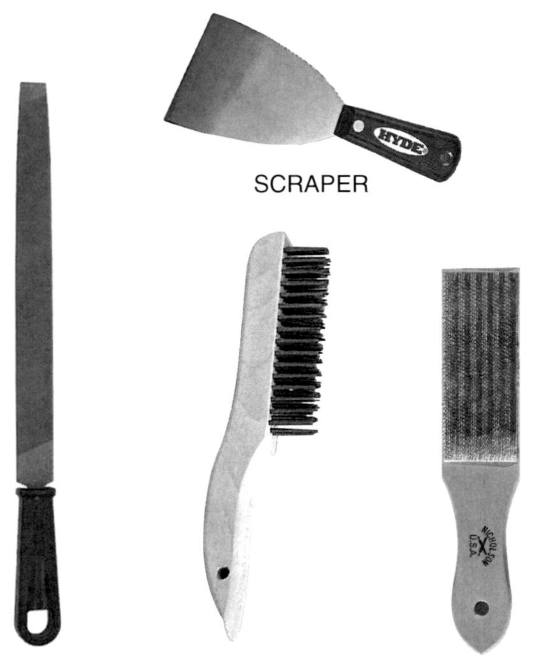

Figure 13 Cleaning tools.

3.4.2 Scrapers

Scrapers are commonly divided into two types: flexible and rigid. Flexible scrapers are used to remove dirt and grease from areas about to be welded. Rigid scrapers are used to remove paint or dirt that has hardened.

3.4.3 Files

Files are used to remove surface corrosion. When using a file, always make sure that a handle is installed on the tang of the file.

> **WARNING**
>
> Using a file without a handle on the tang can result in a serious hand injury.

When filing aluminum or stainless steel materials, avoid using files that have been used on other metals. Filings that stick to the file can contaminate the metal being cleaned, causing weld defects. When filing stainless steel or aluminum, use a fine file, and keep it clean to prevent scratching the base metal's surface. Take care not to damage or scratch the base metal outside the area to be welded.

3.5.0 Cutting Tools

Ironworkers are often required to perform cutting on various forms of metal, such as wire, thin sheet metal or tin, bolts, and small angle iron. Side-cutting pliers and snips and shears are some of the tools used to perform cutting.

3.5.1 Side-Cutting Pliers

Side-cutting pliers (*Figure 14*), also called diagonal cutters, are special pliers used to cut wire, cot-

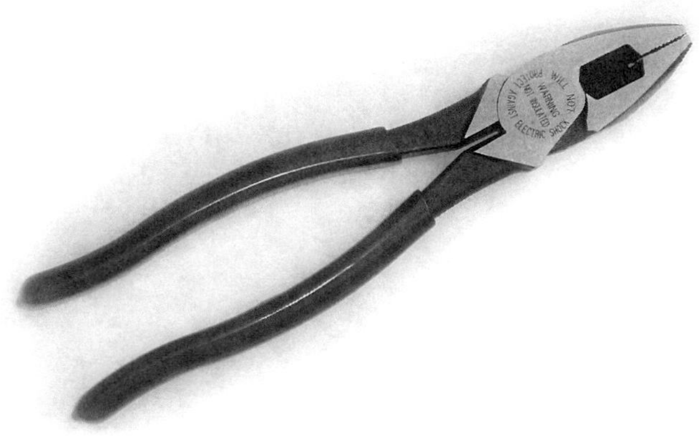

Figure 14 Side-cutting pliers (diagonal cutters).

ter pins, and other items made of light materials, which may not be accessible when using other types of cutters. Since the cutting edges are diagonally offset by approximately 15 degrees from the center of the pliers, side-cutting pliers are also used to cut small objects close to the surface.

3.5.2 Snips and Shears

Snips and shears of various types are useful for cutting thin sheets of brass, aluminum, copper, black and galvanized iron, and stainless steel. Hand snips cannot be used for cutting tempered steel. Snips and shears are separated into two general groups: those designed to make straight cuts and those designed to make circular cuts. Both are available in right- and left-hand styles. Snips and shears come in various types and sizes. With the exception of compound lever shears, these hand-operated snips and shears can only cut metal up to a 20-gauge thickness. Compound lever shears can be used to cut metal up to a 12-gauge thickness. Heavier-gauge metals must be cut with a hacksaw, a chisel, or power tools.

Figure 15 shows some common hand-operated snips.

The following snips are the ones most often used in ironworking activities:

- *Tin snips* – Tin snips are used like scissors to cut thin, soft metals. There are two basic types of tin snip blades: straight blades and combination blades. The straight-blade type has the face of the blade running up from the cutting edges. The combination-blade version is curved back from the cutting edge. The curved surface allows the metal to slip over the top blade when cutting curves. Straight-blade snips have greater strength and can be constructed with longer blades.
- *General purpose snips* – General purpose snips may have either a straight-blade or combination-blade design. They are used for most general purpose cutting of 26-gauge or thinner metals.
- *Aviation snips* – Aviation snips have a compound leverage design that allows them to cut thicker metal than the general purpose snips. The blade

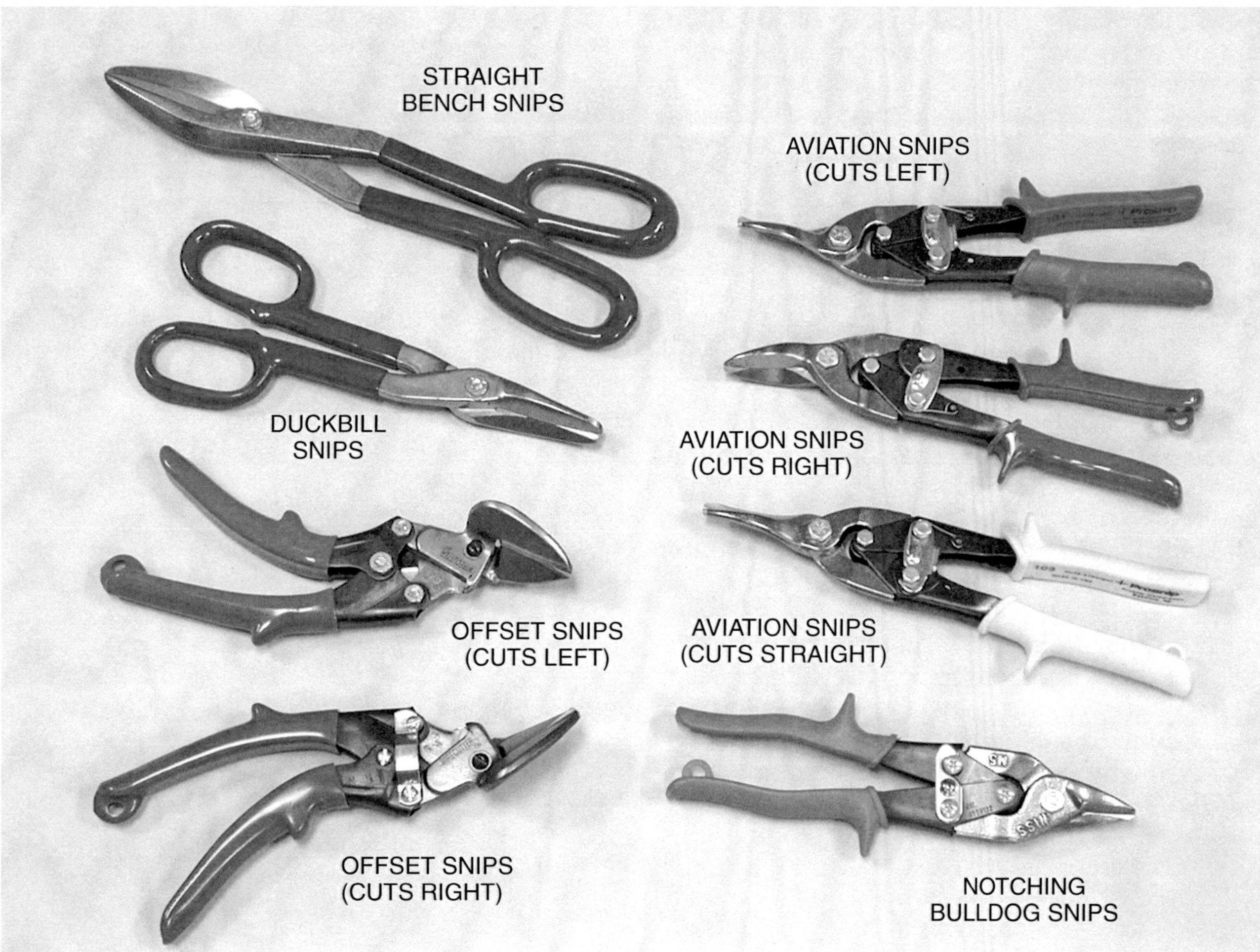

103F15.EPS

Figure 15 Snips.

design allows them to be used for cutting small, irregular curves, and even inside 90-degree corners. Aviation snips are available in straight-line, left-hand, or right-hand cutting models. The blades are serrated, which allows for better gripping of the metal being cut. Aviation snips usually come with insulated (though not dielectric strength) handle grips. The color of the handle grips indicates whether the snips are straight-line (yellow), left-handed (red), or right-handed (green) cutting models. If the handle grip color is not identifiable, the right-hand and left-hand snips can be distinguished by the position of the upper blade. If the upper blade is on the right, the snips are right-handed. If the upper blade is on the left, the snips are left-handed.

- *Circular snips* – Circular snips have blades that curve sideways. They are designed for cutting inside circles and for cutting metal that is close to an obstacle. Most of these snips have been replaced with aviation snips. They are still useful for trimming off sheet metal that is flush with a wall.
- *Double-cutting snips* – Double-cutting snips make two cuts at once. They cut a ⅛-inch wide slit of metal from a sheet. Snips of this type are mainly used for cutting sheet metal pipe because neither side of the metal has to slide over the bottom blade, as is done with conventional snips.

3.6.0 Threading Tools

The threads of a bolt and matching nut must mate perfectly in order for the nut and bolt to securely hold together the objects being fastened. When a nut is tightened onto a bolt, the threads of each are pulled against each other when the proper **torque** (pressure) is applied. This torquing pressure causes the threads of the bolt to match the threads of the nut. The pressure also causes the threads to bend from their original shape. After any bolt and nut combination has been used once under pressure, the threads will never be the same again. This initial mating of the threads can never be duplicated by another nut or bolt. For this reason, nuts and bolts should never be mixed and matched after they have been used the first time. Once mated, they should always be used together until they are no longer needed or are discarded. When installing nuts onto bolts, never overtighten the nuts, because the threads of the bolt or nut will give. When this occurs, the nut and the bolt should be destroyed.

The threads of the bolts and nuts used to fasten metal building materials may occasionally become damaged. If the threads of an installed bolt are damaged, they may need to be repaired before the nut can be removed from the bolt. If the damage is not serious, the ironworker may be able to clean up, straighten, and lubricate the threads so the nut can be more easily removed from the bolt. Special threading tools are used to repair the slightly damaged threads of a bolt. The **thread chaser**, also known as the rethreader die, is used to clean and correct threads on rods and bolts.

After used bolts and nuts have been removed, they may need to be cleaned, inspected, and repaired, if this is allowed. In some industries, slightly damaged nuts and bolts cannot be reused, and new bolts and nuts must be inspected before they are installed.

One way to determine if the threads of a bolt or nut are damaged is to use a thread gauge. If the threads are damaged, they may be repaired using a set of tools called taps and dies.

3.6.1 Thread Gauges

Thread gauges, or screw pitch gauges, are used to measure the pitch of threaded fasteners. Thread gauges can be used to measure both external and internal threads. A thread gauge is a thin, flat metal leaf that has teeth corresponding to standard thread sections. Thread gauges usually come in sets of several sizes, held together by a screw. *Figure 16* shows a thread gauge that measures internal and external threads.

3.6.2 Taps and Dies

Taps and dies are used to cut threads in materials such as metal, plastic, and hard rubber. Taps are used to cut internal threads in materials. Dies are used to cut external threads on bolts and rods. The die that ironworkers use most often is the rethreader die (*Figure 17*), used to clean up damaged threads on a rod or bolt.

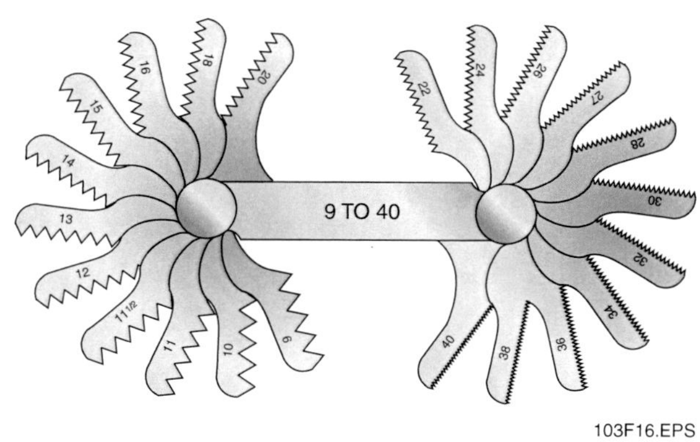

Figure 16 Thread gauge.

3.7.0 Alignment Tools

Many different types of alignment tools are used by ironworkers. The following are the most common types of tools used to align structural members:

- Bull pins
- **Drift pins**
- Erection wrenches
- Sleever bars
- Pinch bars

3.7.1 Bull Pins

Standard or tank bull pins are made of drop-forged alloy steel. The head of the tank bull pin does not allow for added protection against **mushrooming**. Standard bull pins are available in lengths ranging from 8 inches to 15 inches and with a point diameter starting at $\frac{7}{16}$ inch. The top diameter ranges from $1\frac{1}{16}$ inches to $1\frac{5}{16}$ inches.

CAUTION

OSHA requires bull pins to be replaced when the head begins to mushroom. Mushroomed heads on bull pins can cause personal injury.

Broad-head bull pins are considered improved bull pins that increase efficiency by providing a larger striking surface. The design of the head also reduces the chance of the pin slipping from the hand when aligning holes. The angled head and larger radius increase the strength of these bull pins and reduce the danger of chipping. Broad-head bull pins are shorter than standard bull pins, but they can generally be used much longer than the conventional type (*Figure 18*).

103F17.EPS

Figure 17 Rethreader die.

103F18.EPS

Figure 18 Bull pin.

3.7.2 Drift Pins

Drift pins, also known as barrel pins (*Figure 19*), are used for aligning holes in structural steel members and for holding them in place prior to fastening. Avoid using drift pins if they cannot be driven all the way through the connection. Drift pins are approximately 8 inches long. They have a maximum diameter range of $1\frac{1}{16}$ inches to $1\frac{5}{16}$ inches and a point diameter range of $\frac{5}{16}$ inch to $\frac{9}{16}$ inch. These tools are made from drop-forged alloy steel.

3.7.3 Erection Wrenches

Nonadjustable erection wrenches, called spud wrenches, are usually made from drop-forged alloy steel. They are designed to withstand high-leverage and heavy loads. The wrench handle features an extra-deep offset that provides additional hand clearance. The handle has a continuous taper that assists in lining up bolt holes. Erection wrenches are available to fit bolt sizes from $\frac{1}{2}$ inch to 1 inch in $\frac{1}{8}$-inch graduations.

The adjustable-head erection wrench has an adjustable jaw that eliminates the need to carry several fixed-size wrenches. It is designed to fit all nuts and bolts up to $1\frac{1}{2}$ inches (38.1 mm). *Figure 20* shows erection wrenches.

3.7.4 Sleever Bars

Sleever bars (*Figure 21*) are used to provide extra leverage when positioning steel members and aligning bolt holes. These drop-forged fit-up bars are pointed on one end and have a chisel shape

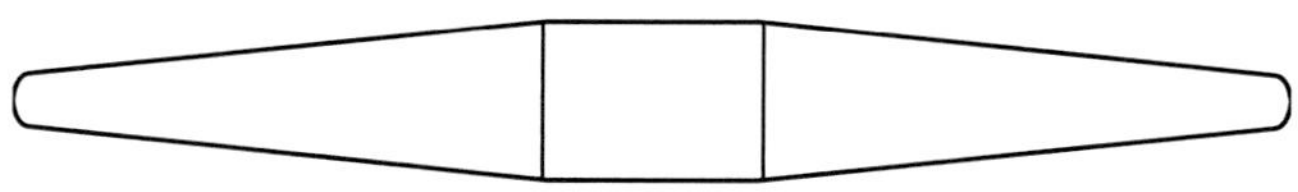

103F19.EPS

Figure 19 Drift pin.

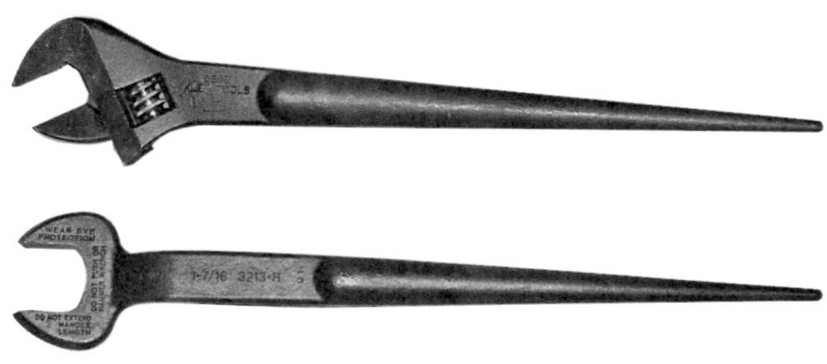

103F20.EPS

Figure 20 Erection wrenches.

103F21.EPS

Figure 21 Sleever bar.

on the other that may either be set at a 15-degree angle or set straight.

3.7.5 Pinch Bar or Pry Bar

The pinch bar or pry bar (*Figure 22*) is about 4 or 5 feet in overall length. It has a round or hexagonal handle and a square, tapered end. The pinch bar is used in situations where extra leverage is required, such as when unloading reinforcing steel from a truck.

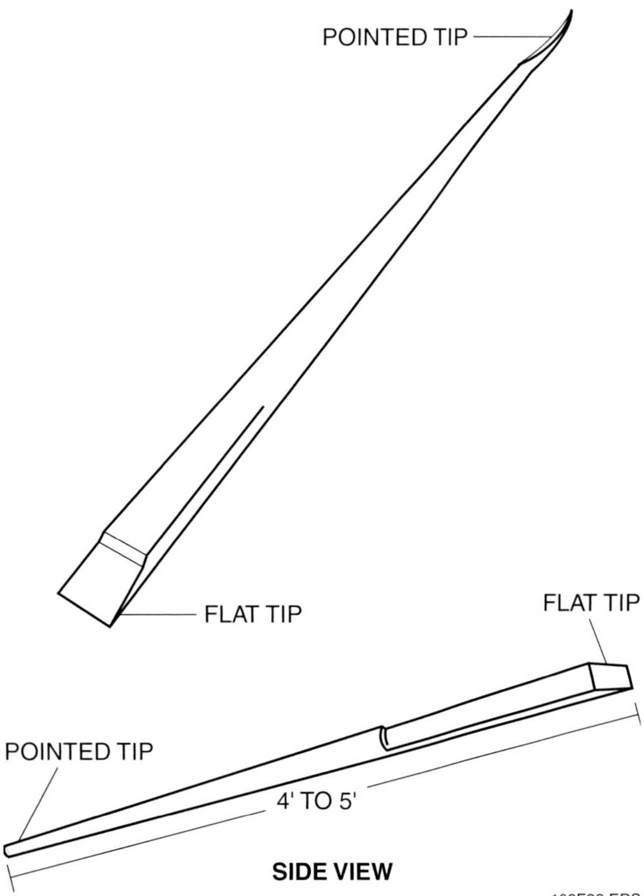

Figure 22 Pinch bar.

3.8.0 Belts and Bags

There are many different types of belts and bags available to hold tools and equipment. Ironworkers select the proper equipment based on the task to be performed. Many times, small tools may be hung on a frog, a small hook or snaplink attached to the belt. The following sections cover belts and bags.

3.8.1 Bolt Bag

A bolt bag (*Figure 23*) is made of canvas or leather. It can attach either to the ironworker's belt or can have a long shoulder strap. Some bolt bags have tool loops on the sides for carrying bull pins.

3.8.2 Structural Tool Belt

To perform their jobs effectively, ironworkers need a convenient way to carry their tools. The following tools comprise the basic tool belt worn by ironworkers doing structural steel work:

- Bolt bags
- Adjustable spud wrenches
- ¾-inch bull pins
- Bull pin holder
- ½-inch drive ratchet
- ½-inch drive extension
- 6- or 8-pound hammer
- 25-foot tape measure
- Striker
- Sleever bar
- Soapstone and holder
- Electrical tape

3.9.0 Welding Tools

Ironworking requires the use of several hand tools related to welding. The following sections describe common welding hand tools used by ironworkers.

3.9.1 Friction Lighter

Always light a torch with a friction lighter (*Figure 24*), also called a strike. Never use matches or a gas-filled lighter to light a torch. For safety reasons, matches and lighters must not be carried by a torch operator. Sparks can cause the matches to ignite or the lighter to explode, causing serious injury.

> **WARNING**
>
> To prevent personal injury, always point the torch in a safe direction when lighting it. To avoid personal injury, only use friction lighters to light a torch.

3.9.2 Tip Cleaners

Tip cleaners (*Figure 25*) are used to clean the slag deposits that form on the openings of cutting tips. The cleaner has a series of wires that correspond to the diameters of the orifices of cutting tips. The wires are packaged in one container for ease of carrying. A small mill file is usually included in a package of tip cleaners. It is used to recondition the face of the cutting tip to ensure efficient operation.

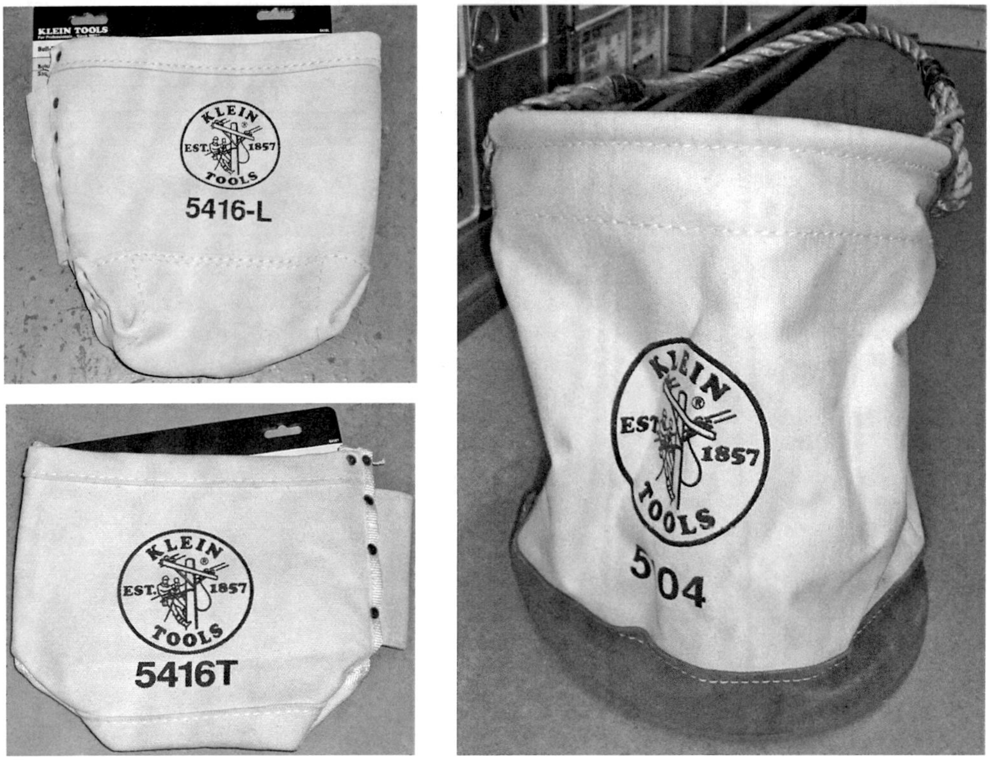

Figure 23 Bolt bags.

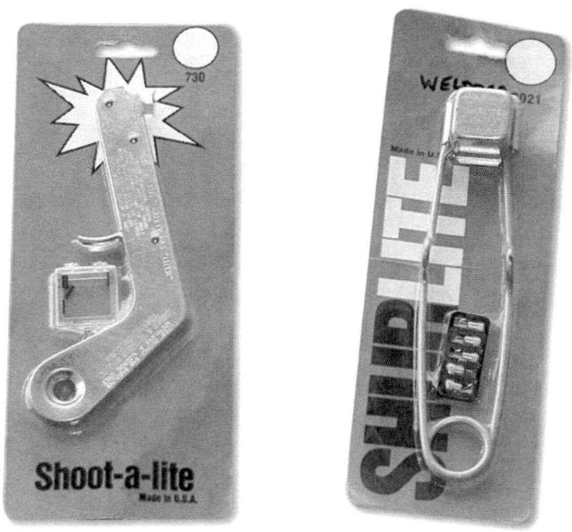

Figure 24 Friction lighters.

3.9.3 Chipping Hammer

Chipping hammers (*Figure 26*) are used to remove cutting and welding slag. The head of a chipping hammer has a point at one end and a chisel at the other. When chipping hammers become dull, they can be sharpened on a grinder. When sharpening a chipping hammer, be careful not to overheat the head. The head is hardened by tempering

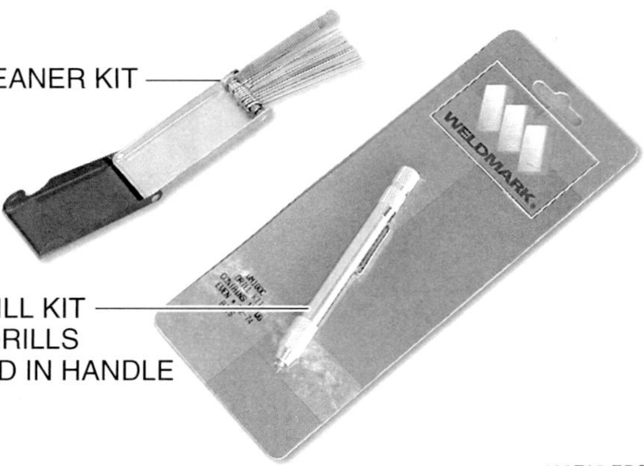

Figure 25 Tip cleaners.

and overheating removes the temper, causing the head to become soft. If the temper is removed from a chipping hammer head, it mushrooms and wears out much faster than a tempered head. To prevent overheating while grinding, plunge the head of the chipping hammer into a pail of water every few seconds.

4.0.0 Power Sources for Powered Hand Tools and Equipment

Portable power tools are divided into groups according to their power sources. A power source can be electric, pneumatic, gasoline, hydraulic, or powder actuated. Many tools, such as saws, drills,

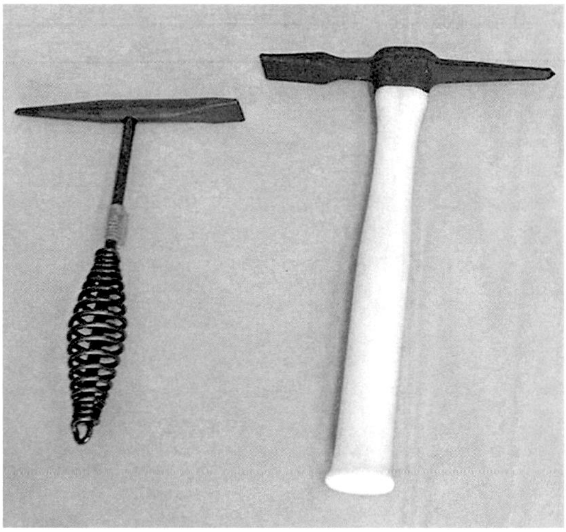

Figure 26 Chipping hammer.

grinders, reamers, and impact wrenches, may be electric, pneumatic, gasoline driven, or hydraulic.

5.0.0 Power Tools

Power tools and equipment designed to assist workers are becoming increasingly efficient and productive. Power tools should only be used for the purposes for which they were designed. When a piece of equipment becomes damaged, time and production are lost. It is essential to keep construction equipment and power tools in good working condition and to observe safe operating procedures.

5.1.0 Rivet Buster and Air Hammer

A rivet buster (*Figure 27*) is a handheld pneumatic tool used for cutting the ends off of rivets. It is also used for other heavy-duty chipping tasks. A rivet buster can be used with a variety of accessories for punching a rivet out of a beam or for cutting.

Air hammers (*Figure 28*) are used for many of the cutting and chipping tasks performed by the heavier rivet busters.

5.2.0 Side Grinder

The side grinder (*Figure 29*) is frequently used when metal needs to be shaped or cut. Side grinders are available in both pneumatic and electrical types. Always handle the grinder with care as it is a very powerful tool. If possible, grind with the sparks coming in your general direction so that if the wheel catches, the grinder will not be pulled into or toward you. At the same time, make sure the sparks are not striking people or flammable materials. Keep a firm grasp on the grinder at all times, and

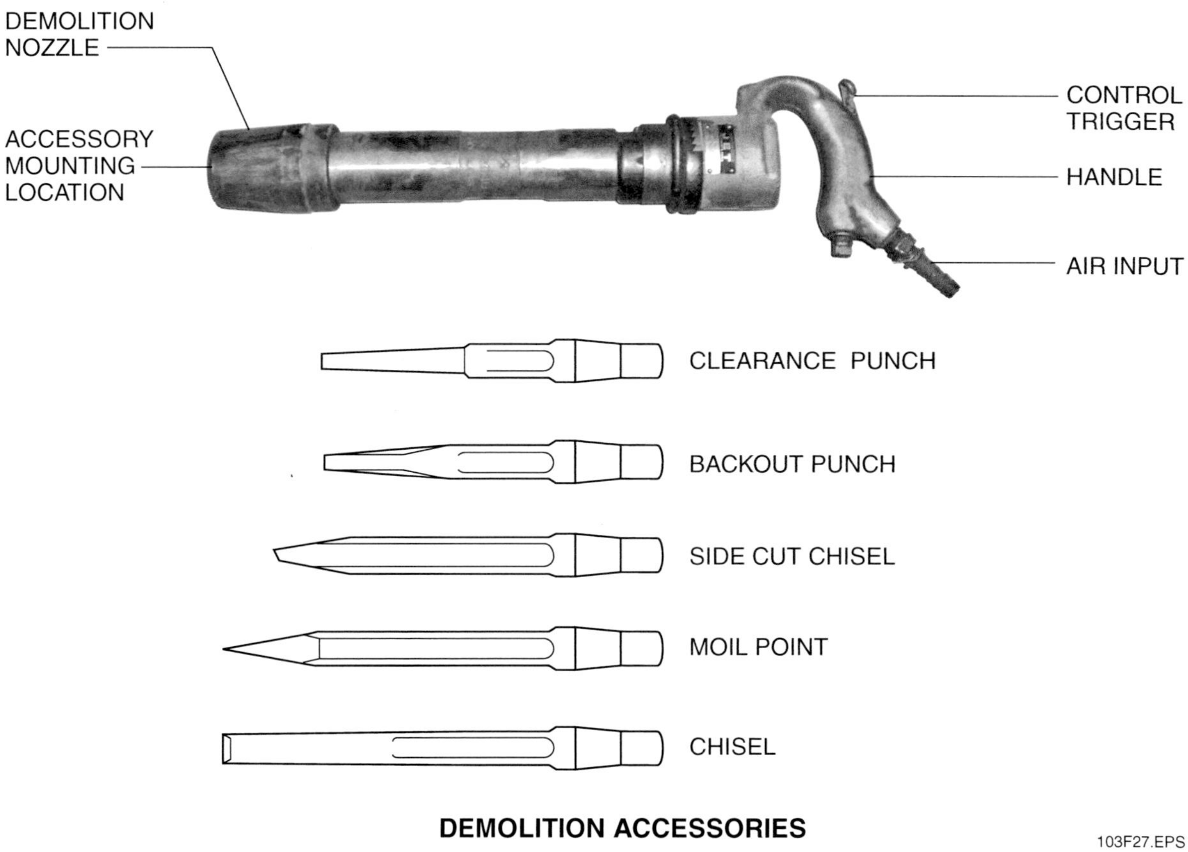

Figure 27 Rivet buster and accessories.

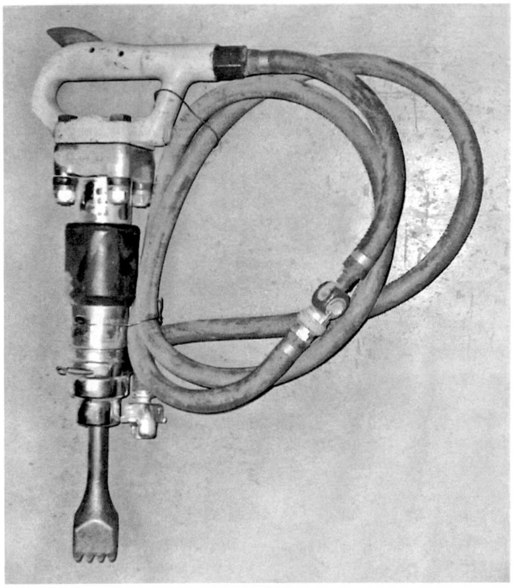

103F28.EPS

Figure 28 Air hammer.

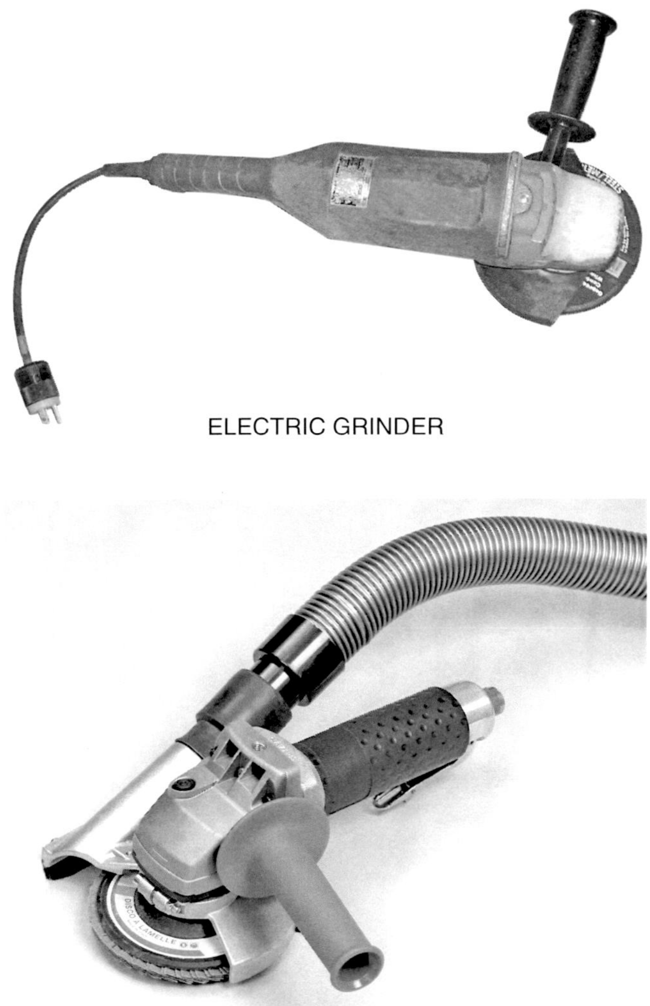

103F29.EPS

Figure 29 Side grinders.

always wait until the wheel stops spinning before setting the tool down. Never allow the spinning wheel to touch clothing or skin. Always use appropriate PPE when operating a grinder, including eye and hearing protection and gloves.

5.3.0 Impact Wrenches

Impact wrenches (*Figure 30*) are designed for continuous, long, hard-fastening service. Types of impact wrenches include electric, air-powered, and hydraulic. Air- and hydraulic-powered wrenches allow the torque to be continuously adjusted from zero to peak power to meet all job requirements. A two-step regulator trigger offers variable low-speed control with feathering ability and progressive second-stage control up to maximum output.

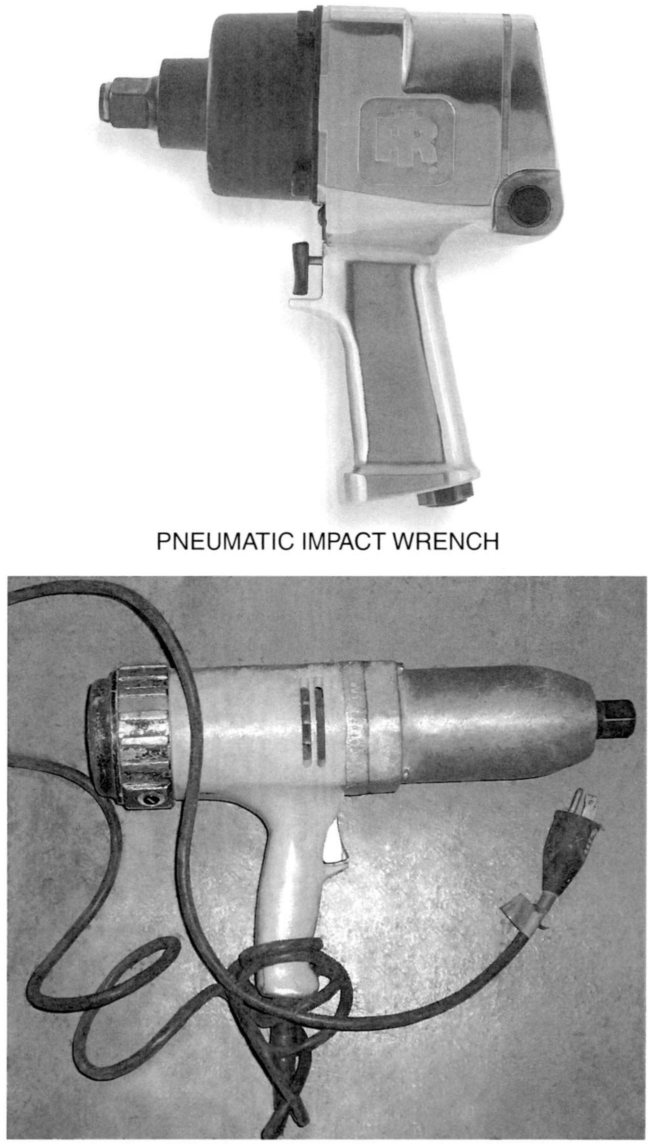

103F30.EPS

Figure 30 Impact wrenches.

These wrenches are reversible and can develop up to 2,200 impacts per minute in either direction. Impact wrenches allow fast, accurate fastening of bolted connections. They can ensure that the turn-of-nut method or the torque control process of fastening high-strength bolts can be attained.

WARNING

To prevent personal injury when using an impact wrench, always wear tight-fitting clothing. Loose or torn apparel may catch in the mechanism.

5.4.0 Powder-Actuated Tools

Powder-actuated tools (*Figure 31*) have been used in the construction industry for over 40 years. All powder-actuated tools use the energy from a gunpowder charge to drive a fastener into steel or concrete. The fastener may be a nail-like fastener or a fastener of special design. The charge may be a metal shell cartridge similar to a gun cartridge, or it may be a powder pellet.

In some powder-actuated tools, the energy from the powder charge is applied directly to the fastener to drive it. In other types, the energy from the powder charge drives a piston, which in turn drives the fastener with a hammer-like blow.

WARNING

An operator's card is required for all persons using powder-actuated tools. Operating powder-actuated tools without an operator's card is dangerous and can result in personal injury.

5.5.0 Reciprocating Saw

The **reciprocating saw** (*Figure 32*), also known as a Sawzall®, is one of the most versatile handheld power tools used in the construction industry today. For the structural ironworker, this saw is used for cutting steel during the fabrication, demolition, and modification (reworking) of existing steel work. The reciprocating saw is available in cordless and electric models. In areas where the risk of fire or explosion exists, the air-powered saw is always preferred.

The following are a few of the many applications of the reciprocating saw:

- Cutting penetrations in steel, aluminum, or fiberglass grating where hot work is not an option.
- Cutting penetrations in steel and aluminum plate and decking where hot work is not an option.
- Trimming and coping beams, channels, tees, angles, etc. where a bandsaw does not fit and hot work is not an option.
- Cutting, coping, or trimming steel, aluminum, and fiberglass handrails where a bandsaw does not fit and hot work is not an option.
- Cutting existing bolts, anchor bolts, and reinforcing bars where a bandsaw is not practical and hot work is not an option.
- Cutting fiberglass and sheet metal wall and roof panels.
- Cutting pipe in which the inside diameter is greater than the extended stroke of the saw.

The reciprocating saw can cold-cut items, such as bolts, pipe, and angle iron, flush with concrete floor surfaces where bandsaws and other mechanical cutting tools cannot. It can also reach between door/window jambs and rough framing to cut through troublesome bolts, nails, or other fasteners.

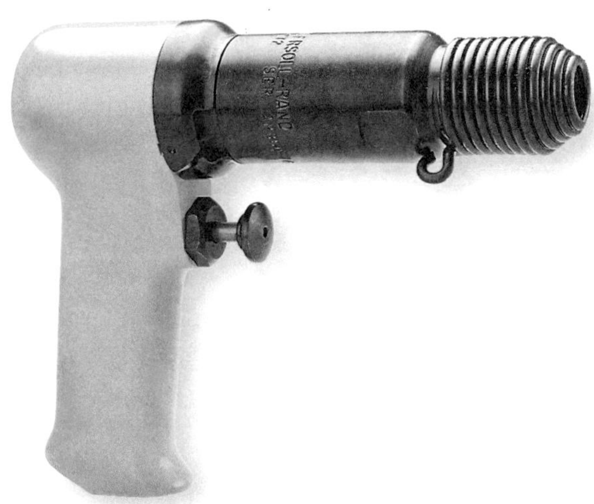

103F31.EPS

Figure 31 Powder fastener.

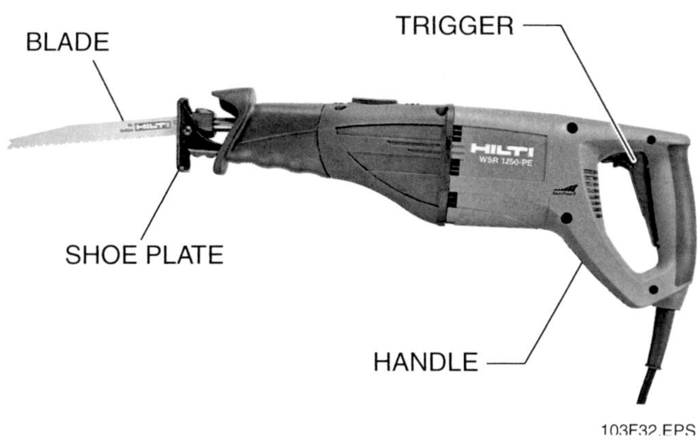

103F32.EPS

Figure 32 Reciprocating saw.

5.6.0 Power Tool Safety

Power tools can save time and energy, but they can also cause severe injuries if not properly used. The following are issues that must be addressed when working with or around power tools.

5.6.1 Housekeeping

Keep the work area clean. Many accidents are caused by material or tools that have been carelessly strewn about. It is very important to keep the work area as free from tripping hazards as possible.

5.6.2 Personal Clothing

Always wear the proper clothing. Do not wear loose-fitting clothing when working with or around power tools. There is always the danger that loose-fitting clothing can become caught in the moving parts of the tool. The job site is no place for jewelry. Rings are especially dangerous.

5.6.3 Appropriate PPE

Wear safety glasses when working with or around power tools. Most power tools eject materials that can cause serious eye injuries. In addition, wear earplugs or earmuffs to protect hearing when using power tools.

5.6.4 Power Tool Hazards and Precautions

All power tools must be used with caution. Refer to the following warnings when using electrical, pneumatic, and hydraulic power sources:

- *Electrical* – Power tools using electricity always present the possibility of electrical shock to the operator if the tool is damaged or improperly used. Electrical power tools must be closely monitored for damage to power cords and housings. The power cords must not have cuts or worn spots, which allow wires to be exposed. Always use the shortest extension cord possible. Power cords must be equipped with ground fault devices that stop the power to the tool if it is damaged and shorted. Always use a power cord that is properly sized for the load. The housing of an electrically-powered tool is sealed and insulated so that the operator will not receive an electrical shock while it is being operated.

  Treat all power cords with care. Avoid placing power cords in situations where they could become frayed or cut. Never yank a power cord from a socket. Always connect to a GFCI-protected power source.
- *Pneumatic* – The word *pneumatic* relates to high-pressure compressed air. Pneumatically powered tools operate by having high-pressure compressed air piped into them through flexible hoses. These high-pressure air hoses must be in good condition and must not be cut or damaged. Connectors and coupling devices must also be in good condition and working as designed. A leak in a hose containing high-pressure compressed air can cut or puncture human skin. If compressed air gets into the bloodstream, it can create health problems. High-pressure air exhausting from a tool may also cause dust and debris to be blown as the tool operates. The flying dust and debris can cause eye irritations and injuries. Personnel using pneumatically-powered tools must wear appropriate PPE, such as safety glasses or goggles and gloves.
- *Hydraulic* – The word *hydraulic* relates to a high-pressure liquid contained within a closed loop system that is used to drive the components of a tool doing work. A hydraulic jack is a good example. Hydraulic power, like pneumatic power, may be piped into the tool through special flexible hoses designed to contain the high-pressure fluid. The connectors and coupling devices on such lines must be securely installed, must not leak, and must be in good condition. The hoses must not be cut, cracked, or contain dry-rot. The end connectors must be securely fastened. The hydraulic liquid or fluid being used may be hot during use. The fluid is under some pressure when at rest and under increased pressure when driving a tool. If the hydraulic fluid escapes the closed loop system and comes in contact with human skin or eyes, the exposure usually results in irritation to the exposed areas.

> **WARNING**
>
> Always check the condition of hoses and connectors. A high-pressure fluid leak can cause personal injury or death.

5.6.5 General Guidelines for Power Tool Usage

Any power tool can be dangerous if it is not used properly. Refer to the following guidelines before using any power tool:

- *Tool selection* – Always use the right tool for the job. Using a light-duty drill to drill through heavy plate will ruin the drill. Learn the capabilities of each tool, and work within those limits. If a tool is not suited for a particular job, select a tool that does the job.

- *Tool overloading* – Never force a power tool. Each power tool is designed to do a specific job at a certain pace. Forcing a tool to work beyond the limits of its design not only wears out the tool prematurely, but it also places the operator in an unsafe situation.
- *Tool storage* – Store tools properly when they are not being used. Leaving power tools out when not in use can cause them to become damaged and can render them unsafe for use.
- *Personal positioning* – Never overreach when using power tools. Always maintain proper footing and balance. Any fall can be hazardous, but one involving a power tool can be even more dangerous.

5.6.6 Tool Changes

When a tool is being changed in a piece of equipment, carelessness can lead to serious injuries. Keep the following points in mind when changing tools:

- *Power disconnect* – Disconnect the power before changing accessories. Though this sounds simple, it cannot be stressed enough. Accidents involving power tools can cause permanent injuries.
- *Change tools* – Remove adjusting keys and wrenches before using power tools. Get in the habit of checking a tool before using it. Make sure that the chuck key is not in the drill prior to use.
- *Cutoff valves* – Some pneumatic tools have a cutoff valve at the tool to shut off air during a tool change.
- *Tool modifications* – Do not modify any tool. Use each tool as it is designed to be used. Do not attempt to improve on the design. Avoid the use of cheater bars to increase leverage.
- *Tool repairs* – Always have tools repaired by a competent technician.

SUMMARY

Ironworkers use a variety of hand and power tools when performing their jobs. They must be familiar with the various tools so they can select the right tool for the job. The proper selection and use of tools greatly increases the likelihood that the work will be completed quickly and easily. Poor workmanship or injury can often be traced back to the improper use of tools. Power tools can make the job even easier, but they present additional safety concerns.

Review Questions

1. To protect against flying debris, wear _____.
   a. a welding helmet
   b. a face shield
   c. a respirator
   d. steel-toed boots

2. The two basic groups of face shields are _____.
   a. welding hoods and cutting goggles
   b. metal chain masks and plastic shields
   c. clear face shields and shields for welding masks
   d. mechanical and automatic face shields

3. To pull two small beams together, use a(n) _____.
   a. basic bar clamp
   b. I-beam bar clamp
   c. locking C-clamp
   d. spring clamp

4. Safe use of a knocker wrench requires that it _____.
   a. should always be pushed and not pulled
   b. should be lightly oiled
   c. should never be lengthened with a pipe
   d. should never be hit with a hammer

5. A wire brush that is used for cleaning aluminum should have bristles made of _____.
   a. mild steel
   b. copper
   c. stainless steel
   d. brass

6. Which tool can cut heavier metal than general purpose snips and can also cut inside 90-degree corners?
   a. Double-cutting snips
   b. Aviation snips
   c. Circular snips
   d. Tin snips

7. Double-cutting snips are mainly used for cutting _____.
   a. sheet metal pipe
   b. air conditioner duct
   c. roof decking
   d. combination decking

8. The die used to clean up damaged threads on a rod or bolt is called a _____.
   a. bottom die
   b. rust remover
   c. rethreader die
   d. thread cleaner

9. Standard bull pins are made of _____.
   a. aluminum
   b. copper
   c. brass
   d. steel

10. If possible, use a side grinder so that the sparks fly _____.
    a. in your general direction
    b. directly away from you
    c. upward
    d. downward

Trade Terms Introduced in This Module

Drift pin: Pin used for aligning holes in structural steel members and for holding them in place prior to fastening.

Frog: A device for carrying tools on a belt.

Mushrooming: The curling or flattening of the end of a tool from continual use.

Reciprocating saw: Saw used for cutting steel during the fabrication, demolition, and modification (reworking) of existing steel work.

Sleever bar: Bar used to provide extra leverage when positioning steel members and aligning bolt holes.

Spud wrench: A type of erection wrench designed to withstand high-leverage and heavy loads. It has a tapered handle that can be used to line up bolt holes.

Thread chaser: A die used for cleaning or repairing the threads on a rod or bolt. Also called a rethreader die.

Torque: A measure of turning force expressed in foot-pounds (ft lb) or Newton-meters (Nm).

Additional Resources

This module presents thorough resources for task training. The following resource material is suggested for further study.

Bolt Depot, Inc. provides charts and tables of information on fasteners and bolts. http://www.boltdepot.com/fastener-information.

Concrete Fastening Systems provides information on concrete anchors and fasteners. http://www.confast.com.

Figure Credits

Topaz Publications, Inc., Module opener, Figures 1-4, 5 (top photo), 6 (photo), 9, 10, 13–15, 17, 18, 20, 21, 23–28 (photos), 29 (top photo), and 30 (bottom photo)

Photo Courtesy of Capital Safety (DBI-SALA and PROTECTA), Figure 5 (bottom photo)

The Stanley Works, Figure 8 (top photo)

Courtesy of Cooper Hand Tools, Figure 8 (middle photo)

IRWIN Industrial Hand Tools, Figure 8 (bottom photo)

W.W. Grainger, Inc., Figures 29 (bottom photo), 30 (top photo), and 31

Courtesy of Hilti North America, Figure 32

NCCER CURRICULA — USER UPDATE

NCCER makes every effort to keep its textbooks up-to-date and free of technical errors. We appreciate your help in this process. If you find an error, a typographical mistake, or an inaccuracy in NCCER's curricula, please fill out this form (or a photocopy), or complete the on-line form at **www.nccer.org/olf**. Be sure to include the exact module ID number, page number, a detailed description, and your recommended correction. Your input will be brought to the attention of the Authoring Team. Thank you for your assistance.

Instructors – If you have an idea for improving this textbook, or have found that additional materials were necessary to teach this module effectively, please let us know so that we may present your suggestions to the Authoring Team.

NCCER Product Development and Revision

13614 Progress Blvd., Alachua, FL 32615

Email: curriculum@nccer.org
Online: www.nccer.org/olf

❑ Trainee Guide ❑ Lesson Plans ❑ Exam ❑ PowerPoints Other ______________________

Craft / Level: Copyright Date:

Module ID Number / Title:

Section Number(s):

Description:

Recommended Correction:

Your Name:

Address:

Email: Phone:

Fastening

30104-11

Trainees with successful module completions may be eligible for credentialing through NCCER's National Registry. To learn more, go to **www.nccer.org** or contact us at **1.888.622.3720**. Our website has information on the latest product releases and training, as well as online versions of our *Cornerstone* newsletter and Pearson's Contren® product catalog.

Your feedback is welcome. You may email your comments to **curriculum@nccer.org,** send general comments and inquiries to **info@nccer.org**, or use the User Update form at the back of this module.

V.1 1/11

30104-11 Fastening

Objectives

When you have completed this module, you will be able to do the following:

1. Recognize and identify A-325 and A-490 bolts, washers, and nuts by their identifying marks.
2. Identify the four common methods of correctly tensioning bolts.
3. Describe how to use the tension control, calibrated wrench, turn-of-nut, and load-indicating washer methods of tightening high-strength bolts.

Performance Tasks

Under the supervision of the instructor, the trainee should be able to do the following:

1. Identify selected high-strength bolts.
2. Demonstrate the turn-of-nut method.
3. Demonstrate calibrated wrench tightening.
4. Demonstrate the proper use of a tension control gun.

Trade Terms

A-325 bolt
A-490 bolt
Bevel washer
Calibrated wrench method
Grip
Skidmore
Snug tight
Stickout
Tension control (TC) bolt
Turn-of-nut method

Contents

Topics to be presented in this module include:

Figures and Tables

1.0.0 Introduction

The strength of certain types of structural connections is determined by the proper torquing of the bolts. Every ironworker must be able to properly and efficiently bolt up structural connections. This module explains how to tighten bolts in a structural connection using **A-325 bolts** or **A-490 bolts**.

2.0.0 Methods for Tensioning Bolts

Structural ironworking components are normally fastened together either by welding or by threaded fasteners. It is important for the ironworker to understand the basic standards that apply to threads and threaded fasteners. Most ironworking assembly uses either A-325 or A-490 bolts, which are heavy hex (six-sided) structural bolts used with hardened washers and heavy hex nuts.

Many types of bolting-up methods are used in structural steel erection. Most of these techniques require the use of special bolts, nuts, and fastening tools. Ironworkers will encounter several of these methods during their work experience. When they do, they are advised to consult the manufacturer's specifications and instructions concerning the particular procedure.

Bolting up is done in a particular pattern. First, the bolts are brought to **snug tight**. That means that a worker tightens the bolt as hard as possible without a tool. The first bolt is often the one either at the top or bottom, so that the weight of the connection is held in place.

Frequently, the first bolt is placed at the top. A drift pin or spud is then used to align the opposite hole before snugging fully, and then the next bolt is installed. Be sure that all of the holes align as well as possible as each bolt is installed, so that there are not any bolts that cannot be inserted.

Since there are commonly at least two rows of bolts, the second bolt to be snugged is the one diagonal to the first one. If the first bolt is in the bottom hole of one vertical row, the second bolt is placed and hand-tightened in the top hole of the other vertical row. The next bolt would be the bottom bolt of the second vertical row, followed by the top bolt of the first vertical row. From there, the alternation continues.

Once the bolts are all in place and snug tight, a socket wrench or impact wrench is used to tighten the bolts, using the same pattern as the initial tightening. Keep tightening until all of the bolts are fairly tight. Then finish by bringing each bolt to the correct specified tension, still using the criss-cross tightening order with which the process began.

There are four methods discussed in this module by which bolts are brought to the correct tension: the **tension control (TC) bolt** method (*Figure 1*), the **calibrated wrench method**, the **turn-of-nut method**, and the load-indicating washer method (*Figure 2*).

3.0.0 Threaded Fasteners

Threaded fasteners (*Figure 3*) are the most commonly used fasteners in ironworking. Many threaded fasteners are assembled with nuts and washers. The following sections describe standard threads used on threaded fasteners, as well as different types of bolts, screws, nuts, and washers.

104F01.EPS

Figure 1 Tension control bolts.

104F02.EPS

Figure 2 Load-indicating washers.

3.1.0 Thread Standards

The many types of threads used for making fasteners are designed for use on different jobs. Threads on fasteners are manufactured to industry-established standards of uniformity. The most common thread standard is the Unified standard, sometimes called the American standard. Unified standards are used to establish thread series and thread classes.

3.1.1 Thread Series

Unified standards are established for three series of threads, depending on the number of threads per inch for a certain diameter of fastener. These three series are as follows:

- *Unified National Coarse (UNC) thread* – Used for bolts, screws, nuts, and other general purposes. Fasteners with UNC threads are commonly used for rapid assembly or disassembly of parts and where corrosion or slight damage may occur.
- *Unified National Fine (UNF) thread* – Used for bolts, screws, nuts, and other applications where a finer thread is desired for a tighter fit.
- *Unified National Extra Fine (UNEF) thread* – Used on thin-walled tubes, nuts, ferrules, and couplings.

3.1.2 Thread Classes

The Unified standards also establish thread classes. Classes 1A, 2A, and 3A apply to external threads only. Classes 1B, 2B, and 3B apply only to internal threads. Thread classes are distinguished from each other by the amounts of tolerance provided. Classes 3A and 3B provide a minimum clearance and Classes 1A and 1B provide a maximum clearance.

The classes most commonly used are 2A and 2B. Classes 3A and 3B are used when close tolerances are required. Classes 1A and 1B are used when quick and easy assembly is needed and a large tolerance is acceptable.

3.1.3 Thread Identification

Thread identification is done using a standard method. *Figure 4* shows how screw threads are designated for a common fastener. The standard method describes a thread using the following characteristics:

- *Nominal size* – The approximate diameter of the fastener.
- *Number of threads per inch (TPI)* – The TPI is standard for all diameters.
- *Thread series symbol* – Specifies the Unified standard thread type (UNC, UNF, or UNEF).
- *Thread class symbol* – Indicates the closeness of fit between the bolt threads and nut threads.
- *Left-hand thread symbol* – Indicated by the letters LH. Unless threads are specified with the LH symbol, the threads are right-hand threads.

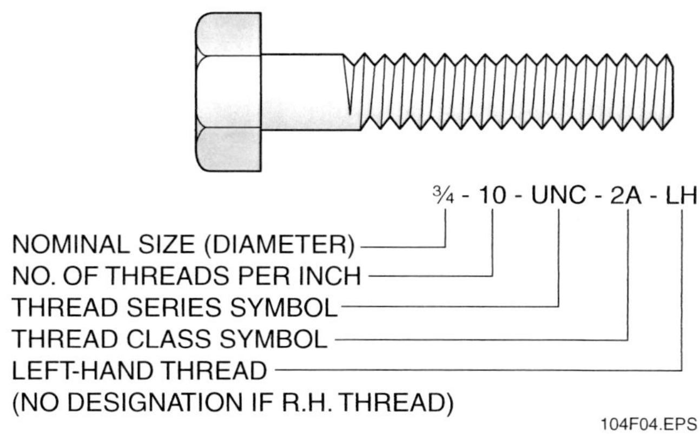

Figure 4 Screw thread designations.

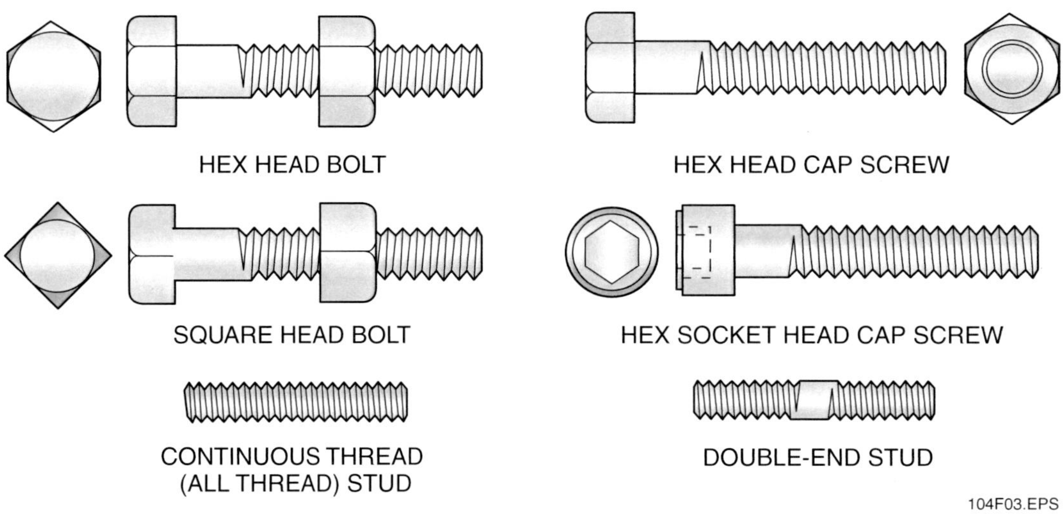

Figure 3 Threaded fasteners.

3.1.4 Thread Design

Several types of thread design are available. The type of thread design used depends on the purpose of the fastener. Power transmission threads are special threads that are used to move machine parts in order to adjust, set, or transmit power from part to part or from machine to part. For example, a power transmission thread may move a tool on a lathe, may set the placement of a tool, or may drive another piece of machinery. Three common transmission threads are the square, acme, and buttress thread designs. *Figure 5* shows an enlargement of the three types.

The buttress thread has one side of the thread cut square and the other side cut at a slant. It has great strength along the thread axis in one direction only. *Figure 5A* shows an enlargement of a buttress thread.

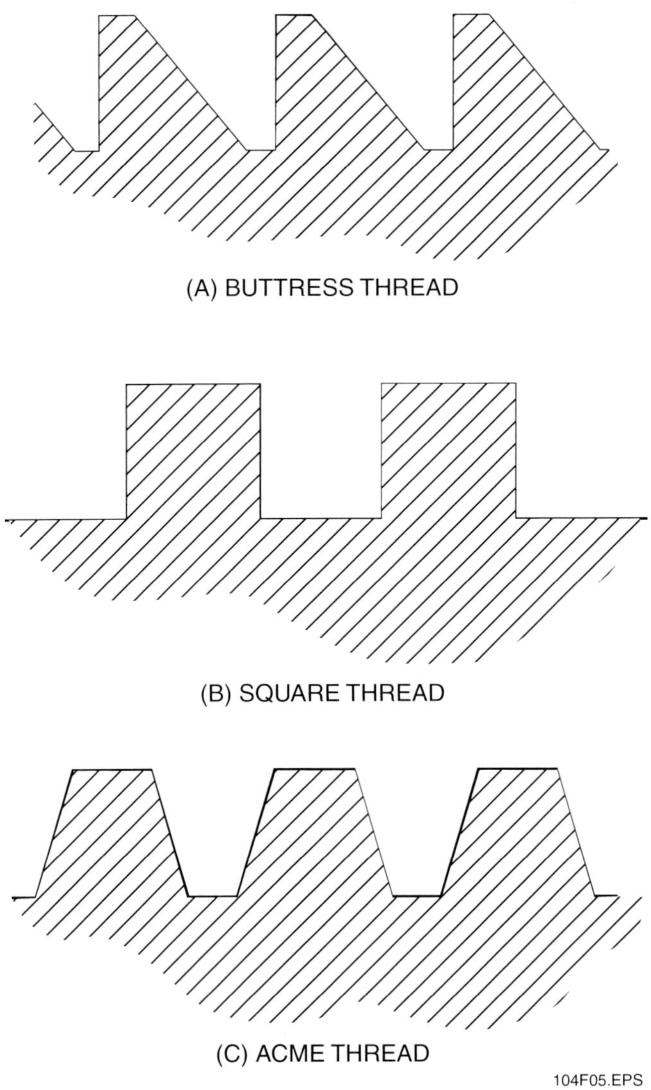

Figure 5 Thread types.

The square thread is the strongest and most useful of all thread forms. It is also the most difficult to make because its parallel sides must be accurately machined. *Figure 5B* shows an enlargement of a square thread.

The acme thread has replaced the square thread in most cases. This is because it is easier to machine. *Figure 5C* shows an enlargement of an acme thread.

3.1.5 Grade Markings

Special markings on the head of a bolt or screw can be used to determine the quality of the fastener. The Society of Automotive Engineers (SAE) and the American Society for Testing and Materials (ASTM) International have developed the standards for these markings. The grade or line markings for steel bolts and screws are shown in *Figure 6*.

Generally, the higher-quality steel fasteners have a greater number of marks on the head. If the head is unmarked, the fastener is usually considered to be made of mild steel, which is steel with a low carbon content.

3.2.0 A-325 Bolts

There are three types of A-325 bolts. Many jobs only require one type; each differs according to the type of steel from which it is made. A separate marking system is used to differentiate the three types of A-325 bolts. *Figure 7* shows the bolt markings for all three types of A-325 bolts.

The three types of A-325 bolts are as follows:

- *Type 1 bolts* – Type 1 bolts are made from medium carbon steel. They are the most widely used of all the A-325 bolts. Type 1 bolts are available either black or galvanized. The designation A-325 is stamped on the bolt along with the manufacturer's symbol. Some manufacturers mark the bolt with three radial lines spaced 120 degrees apart. Type 1 bolts are available in sizes from ½" to 1½".
- *Type 2 bolts* – Type 2 bolts are made from a special type of low-carbon steel. They cannot be galvanized. They are marked with the manufacturer's symbol, the designation A-325, and three radial lines spaced 60 degrees apart.
- *Type 3 bolts* – Type 3 bolts are made to withstand weathering. They have characteristics similar to the various types of weathering steels that are available. The manufacturer's symbol is present on the bolt, and the designation A-325 is underlined. Individual manufacturers of Type 3 bolts may also put a distinguishing mark on the head of the bolt.

ASTM AND SAE GRADE MARKINGS FOR STEEL BOLTS & SCREWS

GRADE MARKING	SPECIFICATION	MATERIAL
	SAE-GRADE 0	STEEL
	SAE-GRADE 1 ASTM-A 307	LOW CARBON STEEL
	SAE-GRADE 2	LOW CARBON STEEL
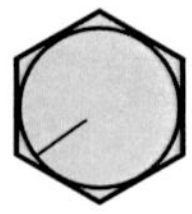	SAE-GRADE 3	MEDIUM CARBON STEEL, COLD WORKED
	SAE-GRADE 5	MEDIUM CARBON STEEL, QUENCHED AND TEMPERED
	ASTM-A 449	
	ASTM-A 325	MEDIUM CARBON STEEL, QUENCHED AND TEMPERED
	ASTM-A 354 GRADE BB	LOW ALLOY STEEL, QUENCHED AND TEMPERED
	ASTM-A 354 GRADE BC	LOW ALLOY STEEL, QUENCHED AND TEMPERED
	SAE-GRADE 7	MEDIUM CARBON ALLOY STEEL, QUENCHED AND TEMPERED ROLL THREADED AFTER HEAT TREATMENT
	SAE-GRADE 8	MEDIUM CARBON ALLOY STEEL, QUENCHED AND TEMPERED
	ASTM-A 354 GRADE BD	ALLOY STEEL, QUENCHED AND TEMPERED
	ASTM-A 490	ALLOY STEEL, QUENCHED AND TEMPERED

ASTM SPECIFICATIONS

A 307 – LOW CARBON STEEL EXTERNALLY AND INTERNALLY THREADED STANDARD FASTENERS.

A 325 – HIGH STRENGTH STEEL BOLTS FOR STRUCTURAL STEEL JOINTS, INCLUDING SUITABLE NUTS AND PLAIN HARDENED WASHERS.

A 449 – QUENCHED AND TEMPERED STEEL BOLTS AND STUDS.

A 354 – QUENCHED AND TEMPERED ALLOY STEEL BOLTS AND STUDS WITH SUITABLE NUTS.

A 490 – HIGH STRENGTH ALLOY STEEL BOLTS FOR STRUCTURAL STEEL JOINTS, INCLUDING SUITABLE NUTS AND PLAIN HARDENED WASHERS.

SAE SPECIFICATION

J 429 – MECHANICAL AND QUALITY REQUIREMENTS FOR THREADED FASTENERS.

104F06.EPS

Figure 6 Grade markings for steel bolts and screws.

When a specific type of A-325 bolt is not specified, Type 1 bolts are normally used. An A-325 bolt may be reused after tightening provided the bolt is not galvanized. According to specifications found in the American Institute of Steel Construction's (AISC) manual, retightening bolts that may have become loosened due to the tightening of adjacent bolts within the connection is not considered a reuse of the bolt.

The hex nuts used with A-325 bolts have the manufacturer's symbol stamped on one face. They may have three equally-spaced lines around the circumference of the nut. They may also be marked with the designation 2 or 2H or by the letters D or DH. *Figure 8* shows the markings on A-325 hex nuts.

Nuts for Type 3 bolts have the number 3 marked on one face. Individual manufacturers may add other distinguishing marks to identify the nut.

Markings may be either raised or depressed. Only depressed markings may be applied to the bearing surface.

Washers designed for use with Type 3 bolts are marked with the number 3 on a face near the outer edge. Individual manufacturers may also add other distinguishing marks to identify the washer. Markings on washers are usually depressed. *Figure 9* shows the markings for Type 3 washers.

3.3.0 A-490 Bolts

A-490 bolts are made of high-strength alloy steel. They are used in connections in which structural members are made of the higher-strength steels. A-490 bolts are marked with the manufacturer's symbol and the designation A-490. These markings may either be raised or depressed, depending on the manufacturer. *Figure 10* shows the markings that are used for A-490 bolts and hex nuts.

> **NOTE**
> Once an A-490 bolt has been tightened to its proper tension, it cannot be reused.

When A-490 bolts are used with certain steels that have a low yield point, hardened washers are required under the bolt and nut. On higher-strength steels, washers are required under the turned element only.

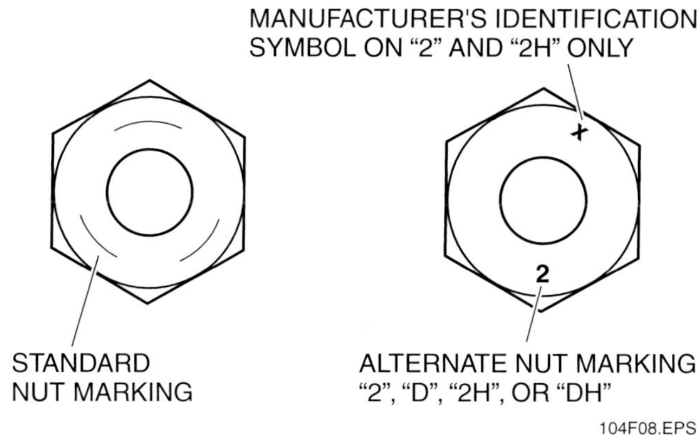

Figure 8 A-325 hex nuts.

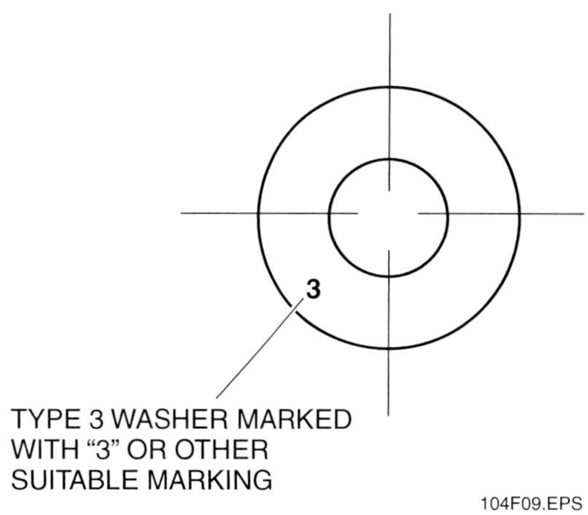

Figure 9 Type 3 washers.

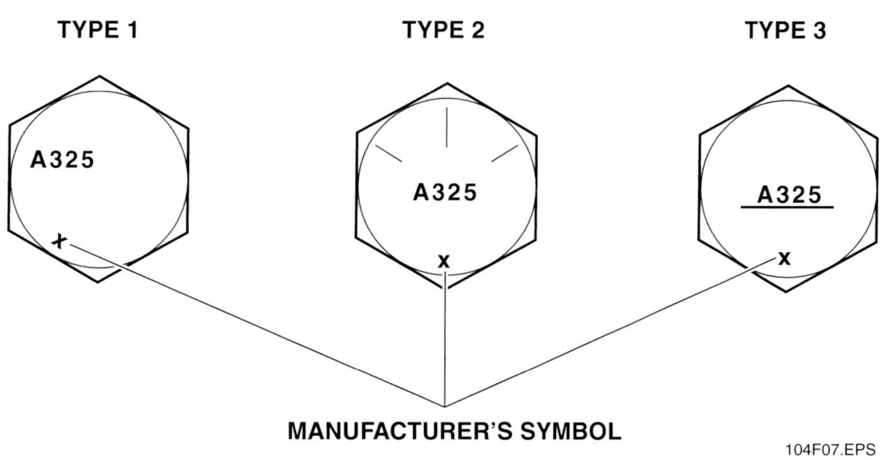

Figure 7 Bolt markings for A-325 bolts.

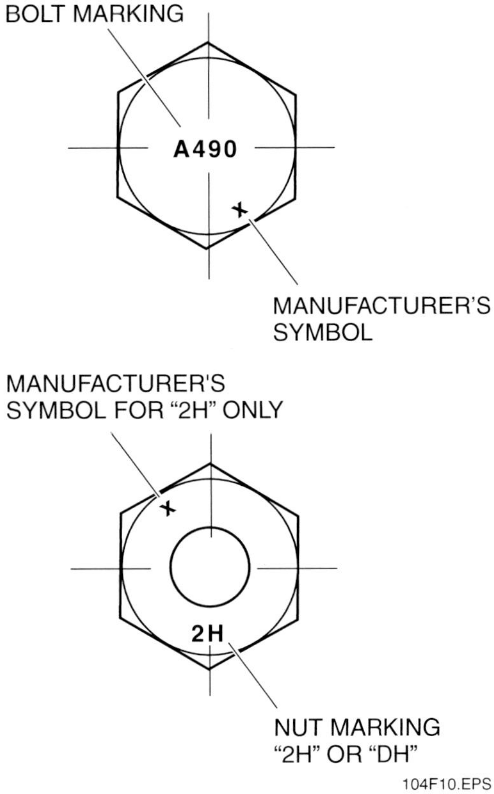

Figure 10 A-490 bolts and hex nuts.

3.4.0 Determining Bolt Length

If a bolt list is not available, the length of bolt required for the connection must be estimated. To determine the bolt length, measure the **grip** distance from steel-to-steel, and add the amount indicated in *Table 1*. Round the required bolt length up to the next standard bolt length. Bolts are available in ¼" increments up to 5", and then in ½" increments through 8". Bolts over 8" in length are typically special ordered.

3.5.0 Bevel Washers

Hardened **bevel washers** provide a square seat for a bolt or nut when it passes through a structural member having a sloping inner face. Bevel washers are required on standard beams and channels having a 1 in 6 flange slope. The length of the bolt must be increased by 5⁄16" when using bevel washers. *Figure 11* shows how a bevel washer must be placed.

3.6.0 Care of Bolts

A bolt list is usually supplied with the job specifications. This list provides information about the sizes, lengths, and types of all the bolts that will be used on the job.

Table 1 Bolt Estimating Table

Diámetro de perno (pulgadas)	Agregue esta cantidad al agarre y aproxime a la longitud del perno siguiente.		
	Ninguna arandela	Una arandela	Dos arandelas
½	11⁄16	27⁄32	1
⅝	⅞	1 1⁄32	1 3⁄16
¾	1	1 5⁄32	1 5⁄16
⅞	1 ⅛	1 9⁄32	1 7⁄16
1	1 ¼	1 13⁄32	1 9⁄16
1 ⅛	1 ½	1 21⁄32	1 13⁄16
1 ¼	1 ⅝	1 25⁄32	1 15⁄16
1 ⅜	1 ¾	1 29⁄32	2 1⁄16
1 ½	1 ⅞	2 1⁄32	2 3⁄16

104T01.EPS

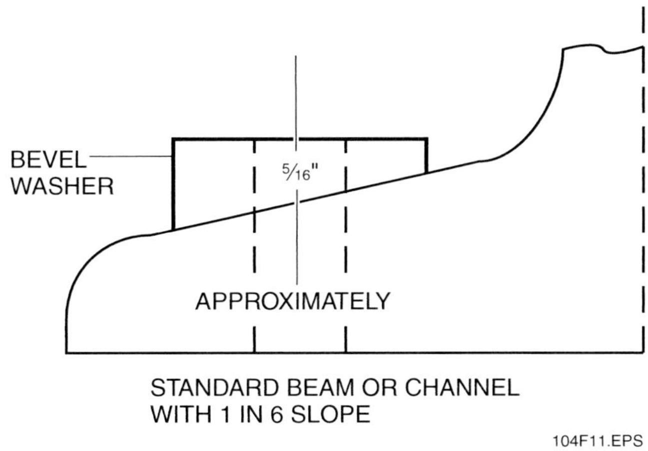

Figure 11 Placement of a bevel washer.

On a large job, there will generally be one person in charge of bolts. It is this person's job to store the bolts and to protect them from damage. The bolts are sorted by type and size. Bolts are often delivered in barrels or other closed containers. All bolts must be protected from the weather. Galvanized bolts must be waxed before installation. Some galvanized bolts are waxed by the manufacturer. Rusty bolts must not be used.

Care of bolts is every ironworker's responsibility. Never take more bolts than are needed, and always return any unused bolts. Never leave bolts out in the weather or lying in an area where the threads could become damaged.

4.0.0 Bolting Up Structural Steel

After the steel has been correctly aligned and plumbed, it is ready to be bolted up. Requirements are set by the design engineer for a job.

Using one of the following methods, high-strength bolts are installed with a developed tension in the bolt shank of approximately 70 percent of the ASTM International tensile strength:

- *Tension control method* – This method uses a wrench specifically designed so that the spline on the special bolt breaks off when the correct tension is applied.
- *Calibrated wrench method* – Either a calibrated torque or impact wrench is used. This method, which is sometimes called the torque control method, requires the use of hardened washers under the turned element to prevent galling and to provide uniform friction.
- *Turn-of-nut method* – The nut is initially tightened to a snug fit. A snug fit is defined as when an impact wrench begins to impact or when a person using a spud wrench has tightened the bolt by hand as much as possible without using a cheater bar. At this point, A-325 and A-490 bolts should be turned an additional $\frac{1}{3}$ turn, provided their length does not exceed four bolt diameters. If their length exceeds four bolt diameters but does not exceed eight bolt diameters, they must be tightened an additional $\frac{1}{2}$ turn. Should their length exceed eight bolt diameters but not 12 bolt diameters, they must be tightened an additional $\frac{2}{3}$ turn. Sockets may be marked to make the additional turns easier to see when using the turn-of-nut method.

Tests on large numbers of bolted joints indicate that either the turn-of-nut or torque control method puts sufficient strain into the bolt shank to produce the desired tension. This procedure also proof-tests the bolt. What cannot be detected is whether adequate strain has been developed. Careful job inspection is therefore necessary, along with frequent checking of the installation equipment. If an A-325 bolt has not been overstressed (if the nut can be turned down the entire length of the threads by hand), it may be reused one or more times. A-490 bolts should not be reused in any situation. Galvanized bolts must be waxed before torquing to correct for the drag of the galvanizing.

4.1.0 Tension Control Bolts

Tension control (TC) or alternative design bolts are designed so that when the tension on the bolt reaches the set level, the end of the bolt snaps off (*Figure 12*). TC bolts may have a round or hex

INSTALLATION PROCEDURES

1. Nut and washer identification markings should face away from the connection. All fasteners in the connection should be snug tight.
2. Fit inner socket over the grooved spline and push the wrench slightly, then engage the outer socket over the nut.
3. Start the wrench. The outer socket rotates the nut relative to the bolt during tightening, and the bolt will be tightened until the required bolt tension is reached. At this point the splined tip shears off.
4. When the installation is complete, remove the socket from the nut and depress the ejection lever to discharge the sheared spline from the inner socket of the wrench.

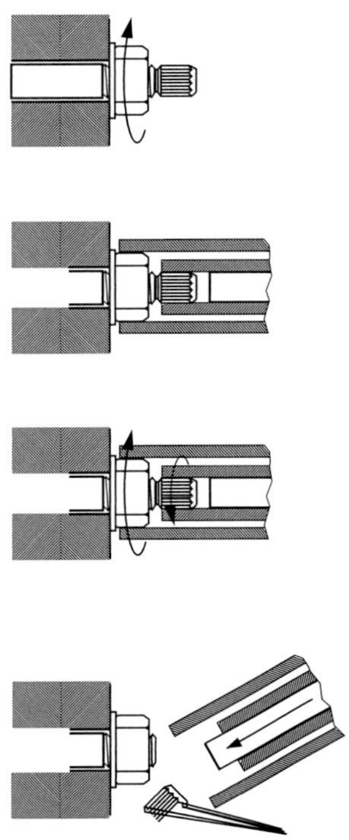

104F12.EPS

Figure 12 Installation of tension control bolts.

head. The other end has a spline at the end of the thread, on which the tool attaches while gripping and turning the nut. The head does not turn. The spline is the part that snaps off when the tension reaches the correct level. The bolt cannot be tightened too much, and a simple visual inspection shows immediately if the bolt has not been tightened enough.

4.2.0 Calibrated Wrench Tightening

The calibrated wrench or torque control method is another method used to tighten structural steel. The wrench may be either manual or power-driven. The following requirements have been established concerning the use of the calibrated wrench method:

- In order to calibrate the wrench, three typical bolts of each diameter used on the job must be tightened in a device capable of indicating actual bolt tension, such as the Skidmore-Wilhelm bolt-tension calibrator.
- The wrench must be set for at least 5 percent over the minimum required bolt tension.
- The wrench must be calibrated at least once each working day for each bolt diameter being installed. Recalibration is necessary if significant changes are made in the equipment or if important differences are noted in the surface condition of the bolts, nuts, or washers.
- If a manual torque wrench is used, nuts shall be in the tightening motion when the torque is measured.

When the calibrated wrench method is used to install several bolts in one connection, the bolts tightened first should be retightened after all the other bolts have been turned to the required torque. This allows for the possible loss of tension in the bolts that were tightened first.

4.3.0 Turn-of-Nut Method

The turn-of-nut method is a relatively simple method that does not require special tools. The procedure for the turn-of-nut method is as follows:

Step 1 Line up the holes.

Step 2 Insert bolts of the proper grade and size in a uniform fashion (with the bolts and nuts facing the same way).

> **NOTE**
> Insert washers, if required. Elongated holes require washers under both turned elements. Never drive the bolts into the holes. Always be sure the bolt is long enough. Grip is the word used to describe the thickness of the steel sections being joined. In certain types of connections, there can be no threads within this area.

Step 3 Hand tighten the bolts and nuts.

Step 4 Snug the bolts and nuts. The joint should be snugged first, starting at the most rigid part. In a joint with a single or double row of bolts, this would be where the steel is already in contact, working toward the end where the steel may not be in contact. If there is solid contact between the steel at all locations, the direction of tightening does not matter. In a bolt pattern with several rows, such as a large web splice plate in a girder, the bolts in the center of the joint should be snugged first. Then proceed to work toward the free edges of the plate.

> **NOTE**
> After the joint has been completely snugged, pretensioning (tightening) of the bolts should follow the same systematic pattern used for snugging.

Step 5 Drive out the drift pins and fill the holes with bolts.

Step 6 Tighten the nuts until the required rotation has been obtained. Make sure the element not being turned does not rotate. To more easily see the amount of rotation, mark the socket of the impact wrench clearly, dividing it into thirds.

Step 7 Check the thread **stickout**. The requirement for stickout is that the end of the bolt must be at least flush with the face of the nut. The bolt is not permitted to extend below the face of the nut after tightening. Excessive stickout indicates that the nut has reached the end of the threads. If this occurs, the nut cannot be turned the required number of turns. Usually six threads are permitted for ½", ⅝", ¾" and 1⅛" bolts. For ⅞", 1", 1¼", and 1⅜" bolts, five threads are permitted. For 1½" bolts, four threads are allowed.

The suggested method for tightening bolts in a structural connection is to tighten them progressively away from the fixed point of the connection to the free edges. Begin at the top of the connection, at the bottom, or even in the center. It is important to establish a sequence for tightening the connection and then to follow that sequence.

If the turn-of-nut method has been correctly performed, the edges of the nut will show the marks of the impact wrench. This is one method of inspection. Another inspection technique is to use a calibrated torque wrench to test bolt tension. A torque wrench is good only if hardened washers have been used.

5.0.0 Load-Indicating Washer

The installation of high-strength bolts using the load-indicating washer (LIW), also known as direct tension indicator, follows the basic procedures for the turn-of-nut or calibrated wrench tightening. The washer has compressible hollow bumps on one side. When the correct tension has been achieved, the hollow bumps collapse, and the gap between the flat face of the washer and the nut or bolt head disappears.

The LIW is placed on the bolt with the protrusions bearing against the underside of the bolt head. If the bolt head is not visible for inspection after installation or if the bolt head must be turned to install properly, then the LIW should be placed at the nut end of the bolt assembly, and an additional hardened round washer must be used. For nut end installation, the LIW is placed on the bolt with the protrusions facing toward the nut. The hardened washer is placed on the bolt against the protrusions, and then the nut is installed. A beveled washer may be used with the load indicator if required by the details.

Figure 13 shows the details of each of the installation methods.

If the LIW must be placed at the head end of the bolt, and if the bolt head must be turned, an additional hardened washer must be placed between the protrusions and the bolt head.

Table 2 shows the gap measurement between the flat surface of the LIW and the hardened surface needed to ensure the required minimum bolt tensions. *Figure 14* shows the amount of gap prior to turning and after proper turning.

Always follow these basic rules of installation for the LIW:

- The protrusions must always bear against a hardened surface.
- The surface contacting the protrusions should not turn during installation. For installations where the LIW is used with a hardened washer, some slight movement of the washer may be noticed and is acceptable.
- An extra ⅛" should be added when calculating bolt grip to allow for the thickness of the installed LIW. An extra 5⁄16" should be allowed if the load indicator is used with a hardened washer.
- Special washer requirements of the *Research Council on Riveted and Bolted Structural Joints* specifications, such as those related to slotted and oversize holes, must be followed. The LIW may not be substituted for required washers, but may be used with them.
- When the LIW is used with A-490 bolts installed in materials having a yield point of less than 40 kilopounds per square inch (ksi), the LIW may be used in place of the hardened washer required under the bolt head, provided the head is not turned during installation.

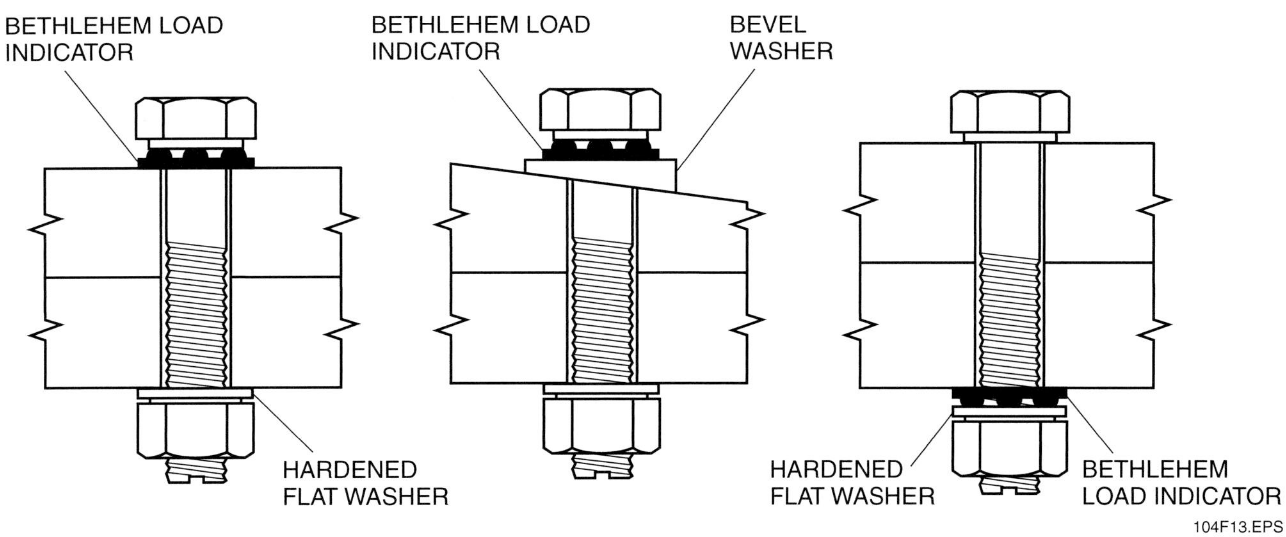

Figure 13 Bolt load indicator installation methods.

Table 2 Gap Measurement

Espacios indicadores de carga para dar la tensión de perno mínima requerida (pernos con acabado negro solamente)	
Ubicación del indicador de carga	**A325 y A490**
Lejos de la pieza torneada	0.015 PUL
Junto a la pieza torneada (con arandela plana endurecida)	0.005 PUL

104T02.EPS

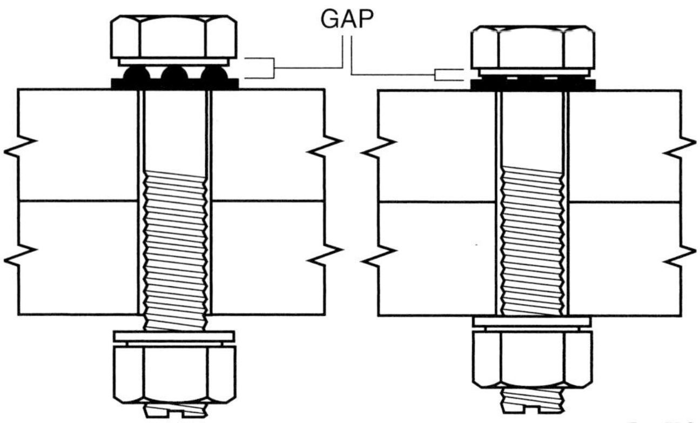

104F14.EPS

Figure 14 Amount of gap before and after turning.

- To avoid excess galling of the moving contact surfaces, it is a good practice to use hardened washers under the turned element, regardless of the location of the LIW.

6.0.0 Skidmore Calibrator

A device that ironworkers use to check the calibration accuracy of bolt-tightening tools is the Skidmore–Wilhelm calibrator, generally known as the **Skidmore** (*Figure 15*). The Skidmore uses an adaptor to fit the bolt being tested. The adaptor and the adaptor plate are mounted on the Skidmore, and the Skidmore is clamped to the leg of the beam or to a plate. The adaptor is fitted over the bolt, and the Skidmore tests the tensile force developed in the bolt by hydraulic force.

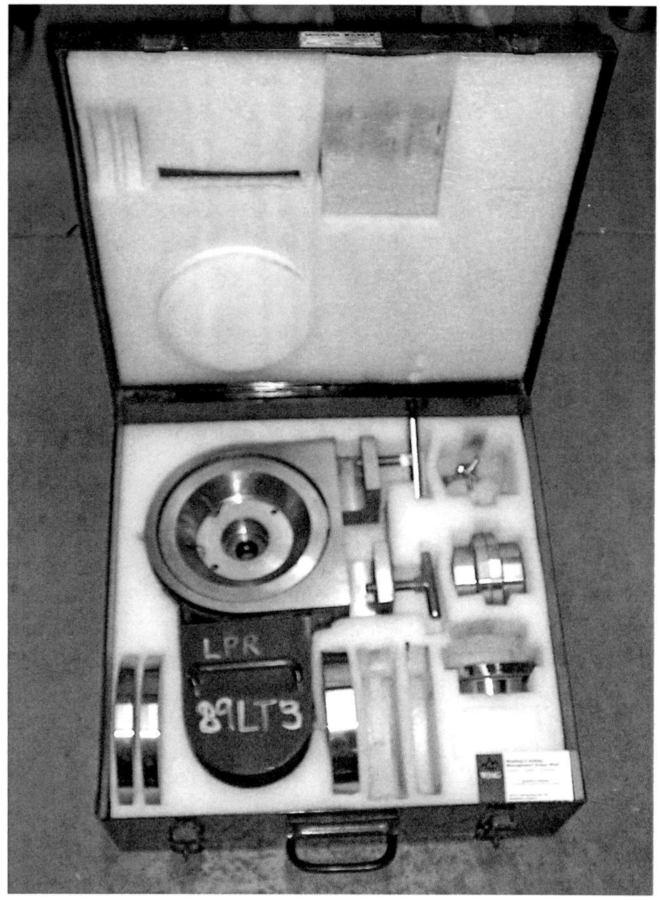

104F15.EPS

Figure 15 Skidmore calibrator.

The force is read out on a gauge in units of pounds force (lbf). If specifications call for kips, one kip is 1,000 pounds force. Follow the manufacturer's instructions to read the lines on the gauge.

For shorter bolts, short bolt plates and short bolt adaptors are used. Companies may also supply special sockets to fit into the recess where the standard socket will not. Spacers are also supplied by the manufacturers for especially long bolts. Tension control (TC) bolts need to be tested with a special adaptor plate and bushing. Direct tension indicator (DTI) washers, both squirter and non-squirter types, can be tested with the calibrator. The washer may be placed under the bolt head or under the nut for testing.

Summary

Two common high-strength bolts are the A-325 and the A-490. There are three types of A-325 bolts. These bolts provide strong connections between structural members when properly tightened. Bolting up is done in a particular pattern. First, the bolts are brought to snug tight. That means a worker tightens the bolt as hard as possible without a tool. Once the bolts are in place and snug tight, a socket wrench or impact wrench is used to tighten the bolts. The ability to tighten high-strength bolts properly, using either the turn-of-nut or the calibrated wrench method, must be part of every ironworker's array of skills.

Review Questions

1. The first step of bolt tightening is to bring all the bolts to _____.
   a. a specified tension
   b. snug tight
   c. 75 percent of tension
   d. torque setting

2. The most common thread standard in the United States is the _____.
   a. Metric
   b. Unified
   c. English
   d. Acme

3. A-325 bolts can be reused after tightening provided they are _____.
   a. properly lubricated
   b. stretched enough
   c. not galvanized
   d. rusted

4. When A-490 bolts are used on steels having a low yield point, hardened washers are required under the _____.
   a. bolt
   b. nut
   c. nut and the bolt
   d. turned element

5. Before torquing, galvanized bolts must undergo which process to correct for the drag of the galvanizing?
   a. Waxing
   b. Oiling
   c. Heating
   d. Greasing

6. What percent must calibrated wrenches be set over the minimum required bolt tension?
   a. At least 5 percent
   b. At least 10 percent
   c. No more than 5 percent
   d. No more than 10 percent

7. When the calibrated wrench method is used to install several bolts in one connection, the bolts tightened first must be _____.
   a. lubricated
   b. heated
   c. retightened after all the bolts have been turned
   d. secured with a lock washer or locking wire

8. When bolting up a flange, the joint should be snugged starting at its _____.
   a. center
   b. top
   c. bottom
   d. most rigid part

9. When installing a load-indicating washer (LIW), the protrusions must bear against _____.
   a. the bolt head
   b. the nut
   c. the bevel washer
   d. a hardened surface

10. During installation, the surface contacting the LIW protrusions must *not* _____.
    a. be smooth
    b. turn
    c. be galvanized
    d. be hardened

Trade Terms Introduced in This Module

A-325 bolt: A high-strength bolt made of medium- or low-carbon steel that is used in bolting up steel.

A-490 bolt: A high-strength bolt made of low-alloy steel that is used in bolting up steel.

Bevel washer: A washer that provides a square seat for a bolt or nut when it passes through a structural member that has a sloping inner surface.

Calibrated wrench method: A technique of tightening a nut and bolt in which a specific amount of torque is applied to the connection.

Grip: The thickness of the steel sections to be joined.

Skidmore: A device that tests the tightness of a bolt.

Snug tight: As tight as a person can tighten a bolt without a tool.

Stickout: The amount of the threaded portion of the bolt that sticks out beyond the face of the nut after tightening.

Tension control (TC) bolt: A special bolt that has a snap-off spline.

Turn-of-nut method: The technique of tightening a nut and bolt connection that requires a specific amount of turning of the nut after it has become snug.

Additional Resources

This module presents thorough resources for task training. The following resource material is suggested for further study.

Concrete Fastening Systems provides information on concrete anchors and fasteners. www.confast.com.

Bolt Depot, Inc. provides charts and tables of information on fasteners and bolts. www.boltdepot.com.

Figure Credits

Topaz Publications, Inc., Module opener and Figure 15

Nucor Corporation, Figures 1 and 12

TurnaSure LLC, Figure 2

NCCER CURRICULA — USER UPDATE

NCCER makes every effort to keep its textbooks up-to-date and free of technical errors. We appreciate your help in this process. If you find an error, a typographical mistake, or an inaccuracy in NCCER's curricula, please fill out this form (or a photocopy), or complete the online form at **www.nccer.org/olf**. Be sure to include the exact module ID number, page number, a detailed description, and your recommended correction. Your input will be brought to the attention of the Authoring Team. Thank you for your assistance.

Instructors – If you have an idea for improving this textbook, or have found that additional materials were necessary to teach this module effectively, please let us know so that we may present your suggestions to the Authoring Team.

NCCER Product Development and Revision

13614 Progress Blvd., Alachua, FL 32615

Email: curriculum@nccer.org

Online: www.nccer.org/olf

❑ Trainee Guide ❑ Lesson Plans ❑ Exam ❑ PowerPoints Other ________________________

Craft / Level: Copyright Date:

Module ID Number / Title:

Section Number(s):

Description:

Recommended Correction:

Your Name:

Address:

Email: Phone:

Mobile Construction Cranes

30105-11

Trainees with successful module completions may be eligible for credentialing through NCCER's National Registry. To learn more, go to **www.nccer.org** or contact us at **1.888.622.3720**. Our website has information on the latest product releases and training, as well as online versions of our *Cornerstone* newsletter and Pearson's Contren® product catalog.

Your feedback is welcome. You may email your comments to **curriculum@nccer.org,** send general comments and inquiries to **info@nccer.org**, or use the User Update form at the back of this module.

V.1 1/11

30105-11
MOBILE CONSTRUCTION CRANES

Objectives

When you have completed this module, you will be able to do the following:

1. Identify and describe common lifting equipment.
2. Identify and explain commonly used construction cranes.
3. Identify and explain crane manuals, record keeping, and safety.
4. Describe the activities involved in assembling construction cranes.
5. Identify and use the correct hand signals to guide a crane operator.

Performance Tasks

Under the supervision of the instructor, the trainee should be able to do the following:

1. Use common lifting equipment.
2. Use crane manuals, perform recordkeeping, and describe crane safety.
3. Use and interpret hand signals.

Trade Terms

Carbody
Counterweights
Dog
Drum
List angle
Luffing
Moment
Outrigger
Swing circle

Contents

Topics to be presented in this module include:

Contents (continued)

Figures and Tables

1.0.0 INTRODUCTION

Ironworkers routinely work with a variety of lifting devices. These devices include manual and powered lifting equipment, as well as stationary and mobile lifts. It is important to understand the safety requirements when lifting loads or working around cranes. Construction cranes are the most commonly used lifting devices in ironworking.

2.0.0 LIFTING EQUIPMENT

The basic hoisting mechanism consists of a length of rope, a rotating **drum**, a fixed block, and a traveling block. As the drum rotates, the rope is passed over the fixed block and wraps around the drum. This causes the traveling block to rise, carrying the load with it. This module introduces some of the basic lifting equipment ironworkers are likely to encounter on the job site.

2.1.0 Derricks

Although cranes and derricks are used to lift loads, they differ in one very important aspect. The hoisting machinery of a crane is built into the structure of the crane itself. The crane possesses an internal power source, and both cranes and derricks have some form of lifting arm or boom. A derrick relies on an external source of power.

There are many different types of derricks (*Figure 1*). All derricks consist of the following four components:

- *Boom* – The boom is a mast that is usually made of steel, but sometimes made of wood. The height and design of the boom determine the lifting capacity of the derrick. The boom must be anchored to some fixed point.
- *Luffing system* – The **luffing** system is used to control boom movement up or down. Moving the boom allows the load to be moved horizontally and vertically. Most derricks have a luffing system.
- *Hoisting system* – The hoisting system usually consists of a standing block, a traveling block, and a length of fiber or wire rope.
- *Power source* – This is usually a single-drum or double-drum hoist. The hoist must be secured to a fixed position.

Derricks usually have a limited horizontal reach and are best suited for vertical lifts. They are commonly used in situations where the work is above the capacity or reach of mobile cranes. A derrick may be used to hoist machinery and equipment onto high-rise buildings or to place exterior stone or facing panels on a building.

Some derricks are capable of swinging movements, but these movements are generally cumbersome and time-consuming.

2.2.0 Cranes

A number of different kinds of mobile cranes have been developed. The following crane types are in widespread use in general construction and erection work:

- Crawler crane
- Truck or motor crane
- Rough-terrain carrier

A mobile crane may have a lattice boom, a luffing boom, or a telescopic boom.

2.2.1 Crawler Cranes

Crawler cranes (*Figure 2*) are available in a variety of configurations including tower, luffing, and heavy lifting. The lifting and rotating assembly is mounted on top of the crawler assembly. The crawler assembly, on which the rotating upper structure is mounted, is called a base or base mounting. The base should not be called a carrier. Only a wheeled crane mounting is correctly referred to as a carrier.

The spread of the crawler tracks affects the stability of the crawler crane. The wider the spread between the treads, the greater lifting capacity the crane has. However, transportation of cranes to job sites can be difficult when the treads are too widely spread. This has led to the development of crawlers that can be extended after reaching the job site. These cranes can be transported while the treads are retracted. When they arrive at the site, the crawlers can be extended to a greater width to achieve increased stability (*Figure 3*). Most crawlers are equipped to extend or retract without assistance. Others may require jacks, blocking, and other equipment.

The base mounting of the crawler crane transmits the crane loads to the ground. To accomplish this without excessive increase in radius under load and without excessive change in radius during swing motions, the base needs to be extremely stiff. The base is stiffened by using complex castings or heavy weldments.

When comparing crawler cranes to other types of cranes of equal lifting capacity, crawler cranes offer lower cost and rental rates, but their transit and erection costs are higher.

2.2.2 Truck Cranes

Truck cranes (*Figure 4*) are available in configurations similar to crawler cranes. They are available with lattice and telescoping booms. Truck-type crane carriers should not be confused with an ordinary commercial truck chassis. A truck-type crane carrier is designed and made for no other purpose than crane service. Some of the larger units may have as many as nine axles.

Truck carriers must be designed for moving around the job site and also for driving on the

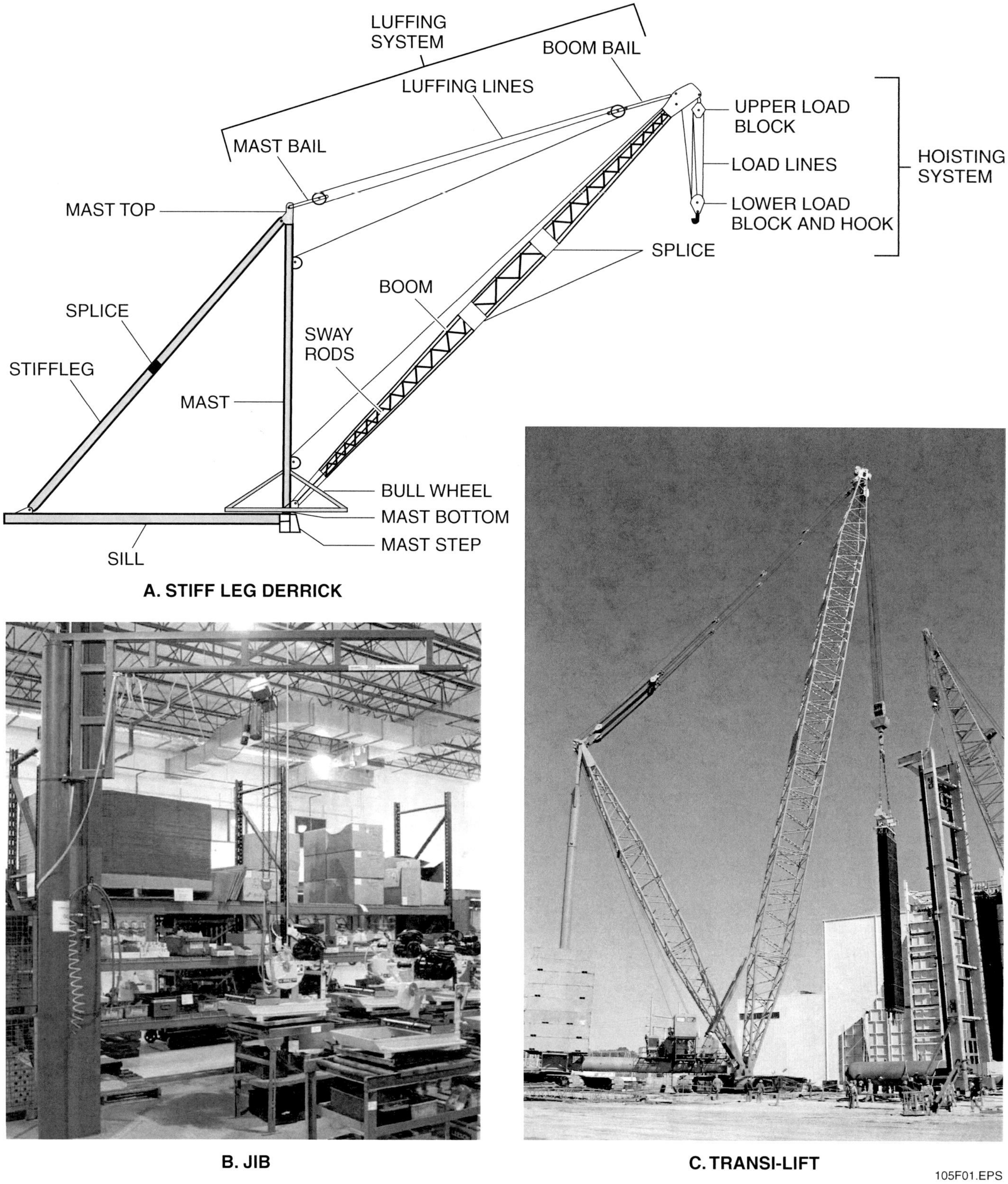

Figure 1 Types of derricks.

highway. To accommodate these extremes, transmissions with as many as 33 forward gears, as well as special creeping gears, are available. The high-creeping gear ratios, as high as 200:1, are used for creeping purposes on reasonably good ground at the job site. Road speeds can reach speeds sufficient for highway travel. Job site ramps of 20 to 40 percent can be climbed by the unit with no load, and the brakes are designed to hold on the maximum grade.

105F02.EPS

Figure 2 Crawler crane.

105F04.EPS

Figure 4 Truck crane.

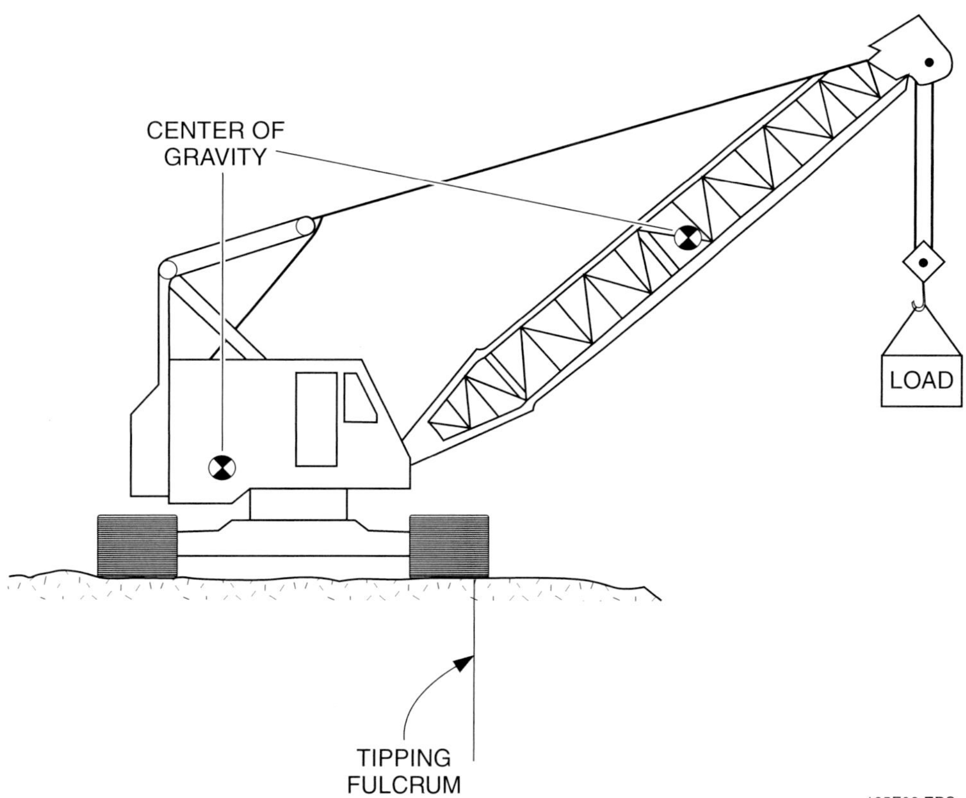

105F03.EPS

Figure 3 Crawler crane tipping axis.

2.2.3 Rough-Terrain Carriers

Rough-terrain mobile cranes are usually two-axled carriers with telescoping cantilevered booms. They are commonly referred to as cherry pickers. **Outriggers** are provided to increase capacity and stability. The heavier lifts are made over the front of the crane. Rough-terrain carriers (*Figure 5*) travel at less than highway speeds (about 30 mph) and are usually transported on low-bed trailers.

2.2.4 General Crane Safety

Cranes have daily, monthly, and annual inspections. Every foot of hoist cable is checked periodically for broken wires and proper lubrication. Brakes and limit switches are also inspected. In addition, audible and visible warning and fire control devices are inspected at regular intervals.

Each crane must have a capacity plate or chart in a conspicuous place in the operator's cab. The capacity chart has a listing of the safe lifting capacities at various radii. On some smaller cranes, limit devices are installed on the load hoist and boom hoist. These devices should be inspected periodically.

Beginning ironworkers may not be directly involved with crane operations, but they should be familiar with the safety factors required for operation on the job site. For example, be aware of the fact that a boom crane must be operated only with the chassis level. In other words, treads or wheels should not be higher on one side than on the other. In some instances, timber mats or a built-up base may be needed to provide a proper operating area. If the crane is out of level, a side loading (bending) of the boom could occur. *Figure 6* shows a crane that is level and one that is not.

For steel erection, it is common to unbolt a two-piece boom and insert an extension. Light-duty extensions called jibs are frequently added to the main boom to increase the lifting height. When operating this type of crane, extreme caution must be exercised to prevent overturning the crane or tipping it over backwards. Boom extensions and jibs must be carefully inspected for damage or missing lattice bracings.

One of the dangers of crane operation is the possibility of the boom or cable coming into contact with power lines. Electronic warning devices are available that can be mounted on the boom. These devices sound an alarm when the boom is brought within a predetermined distance of a high-voltage line.

2.3.0 Helicopters

The increasing use of helicopters in many phases of the construction industry requires workers to become informed about their safe use and about load applications and movement. Helicopters are used for lifting heavy equipment for rooftop placement and loading.

105F05.EPS

Figure 5 Fixed-cab rough-terrain crane.

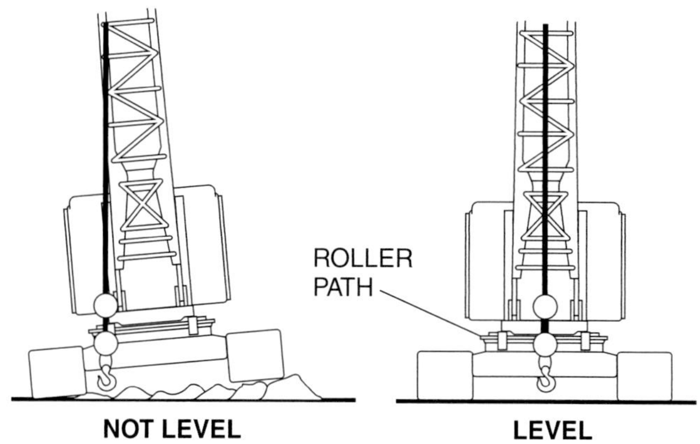

105F06.EPS

Figure 6 Examples of level and unlevel cranes.

2.3.1 General Helicopter Safety

The National Safety Council has developed guidelines for material handling by helicopters. All nonmilitary commercial helicopters used in the United States for external load operations must be registered and licensed by the Federal Aviation Agency (FAA). Pilots of these helicopters, in addition to holding a current rating for the helicopter type, are rated for external load operations. Before using a helicopter for any job, an on-site survey and inspection of the operational area is made by the personnel who will conduct the operation. The plan of operations is submitted for approval to the appropriate FAA district office and any other local, state, or federal agencies having jurisdiction. Other specific guidelines include the following:

> **NOTE**
> Only personnel who have been trained and certified to work with helicopters are permitted to do so.

- The sling attaching points must be carefully selected, considering the center of gravity of the load and in-flight load stability.
- External load slings must be kept to a length that does not allow the load to oscillate or swing; 15 to 20 feet below the helicopter has been found to be the most practical length.
- Load slings must not be allowed to hang free on the hook when flying without a load.
- Only steel cables must be used for slings and load lines.
- All slings must have nonspin hooks.
- The hook assembly on the helicopter must be fixed and should have safety latches.
- Cargo hooks must be equipped with an emergency mechanical control for releasing the load. The hook should be tested before each day's operation.
- Tag lines must only be used with the knowledge and consent of the pilot.
- Ground crews must not wear loose clothing that is likely to be caught on hoist lines or slings. Personnel must not ride on the load, the sling, or on any other attachment at any time during operation. Only personnel essential to the job may ride in the aircraft.
- The loading and landing sites must be kept clear of construction debris.

The following appropriate PPE is recommended for all employees assigned to work near helicopter operations:

- Safety hats with chin straps
- Safety goggles (not glasses) of a closed type with baffled vents
- Self-contained breathing apparatus or respirators if active physical, chemical, or microbial dust is present
- Hearing protection for all ground personnel
- Gloves

2.3.2 Other Precautions

Those engaged in loading or unloading operations must be carefully briefed about their duties and informed of the location of a predetermined safety zone. They should be thoroughly briefed on the overhead danger of the rotor blades. All personnel must be warned of the danger of walking into the tail rotor, which is nearly invisible while operating.

Whenever personnel are approaching or leaving a helicopter with its blades rotating, they should remain in full view of the pilot and should stay in a crouched position. Communications and signal systems should be predetermined between the pilot and signaler and clearly understood before the beginning of the operation. The signaler should be easily distinguishable from other workers by special high-visibility clothing. Only one person should serve as signaler at any one time.

3.0.0 Mobile Construction Cranes

Ironworkers need to understand each mobile construction crane's principles and limitations, as well as how job site environmental conditions affect the load so that it can be safely lifted. Cranes differ greatly from manufacturer to manufacturer. Because of this, ironworkers must familiarize themselves with the information supplied by the manufacturer for the particular crane being used.

3.1.0 Functions and Uses

Crane selection for a particular task depends on the technical requirements of load capacity, reach, clearances, and sitting. Economic considerations very often dictate the type of crane to be used.

3.1.1 Telescopic Cranes

Generally speaking, a telescopic crane is faster and more efficient than an equivalent lattice crane. Little time is lost in its setup, and a minimum number of crew members are required because almost no assembly work is needed. For short transit distances, the lower cost rough-terrain models are most economical.

3.1.2 Rough-Terrain Cranes

Rough-terrain cranes are often used as service cranes on large work sites. This is due to their quick setup, maneuverability, broad range of capabilities, quick movement from assignment to assignment at the site, and the fact that there is a one-person operator. Rough-terrain cranes are suitable for pick-and-carry work and for precise placement of fabricated elements or machinery.

Rough-terrain cranes are usually dual-axle carriers with telescoping cantilevered booms. These are commonly called cherry pickers. A single operator-driver cab can swing with the boom (swing cab) or remain fixed (fixed cab). These types of carriers usually have oversized tires that make travel easier under adverse site conditions.

3.1.3 Boom Truck Cranes

Conventional boom truck cranes usually prove economical for longer-term operations that range from several days to several months. Their purchase price and rental rate are lower than for telescopic truck cranes of similar capacity. The added labor costs for assembly and disassembly make them unsuitable for short-duration work. Boom truck cranes may be used for some short-term tasks in areas where booms of 100 feet or so may be folded for transit.

The lattice boom truck crane is a crane using an open web boom. The lattice boom is based on the strength of the triangle shape. The crane is mounted on an industrial truck, to make transportation easy, and normally has outriggers for stability. The boom is frequently made of at least two pieces and may also have a separate jib.

3.1.4 Crawler Cranes

A typical crawler crane with a heavy-duty boom has tread assemblies called crawler frames or tread members, that can be extended to widen the footprint of the crane. This provides additional stability for the crane when lifting loads at the sides of the crawler frames. The frames can be retracted to transport the crane on a lowboy trailer or railcar. The hoist drum and ring gear drive (for rotation) in this unit are powered by fine-control hydraulic systems that allow precision inching control. The unit has a diesel engine driving individual hydraulic pumps for the hoist drum and swing motors. The engine also powers two hydraulic crawler motors. A clutched transmission connects the engine to the multiple hydraulic pumps. The operator cab provides a six-way adjustable seat with convenient controls and instrumentation.

The lifting and rotating assembly is mounted on top of the crawler assembly. The crawler assembly, on which the rotating upper structure is mounted, is called a base or base mounting. The base mounting of the crawler crane transmits the crane loads to the ground. To accomplish this without excessive increase in operating radius under load, and without excessive change in radius during swing motions, the base needs to be extremely stiff. The base is stiffened by using complex castings or heavy weldments.

The base mounting consists of the **carbody**, **swing circle**, crawler frames, tracks, and the propelling mechanism. The base should not be called a carrier. Only a wheeled crane mounting is correctly referred to as a carrier.

A carbody is the central or main portion of the base, with the swing circle fixed on top. The swing circle is the point where the upper mechanisms are joined to the base. The swing circle must transmit the loads generated by the upper structure to the base and must also serve as the bearing or roller circle for the swing motion. Crawler frames are mounted at the sides of the carbody. The crawler frames hold the crawler track power source, idlers, and crawler tracks.

Most crawler cranes are powered by an engine in the upper structure. Older cranes use a mechanical system to couple power from the engine to the crawler tracks. The newer machines have hydraulic motors mounted on the crawler frames to propel the unit. The hydraulic motors are powered by the engine in the upper structure. However, a few very large machines have separate engines mounted on the base mounting or crawler frame to provide power to the crawler tracks.

When comparing crawler cranes to other types of cranes of equal lifting capacity, crawler cranes offer lower cost and lower rental rates than other cranes, but their transit and erection costs are higher.

If the nature of the work and ground conditions make it necessary to operate a machine from a mat and from a number of different work positions, siting costs must also be considered. The cost of placing a truck crane is usually less than that of placing and leveling a mat.

3.2.0 Terminology

Since ironworkers must routinely work with and around cranes, they need to understand the terminology used to describe crane parts. *Figures 7* and *8* show selected crane parts.

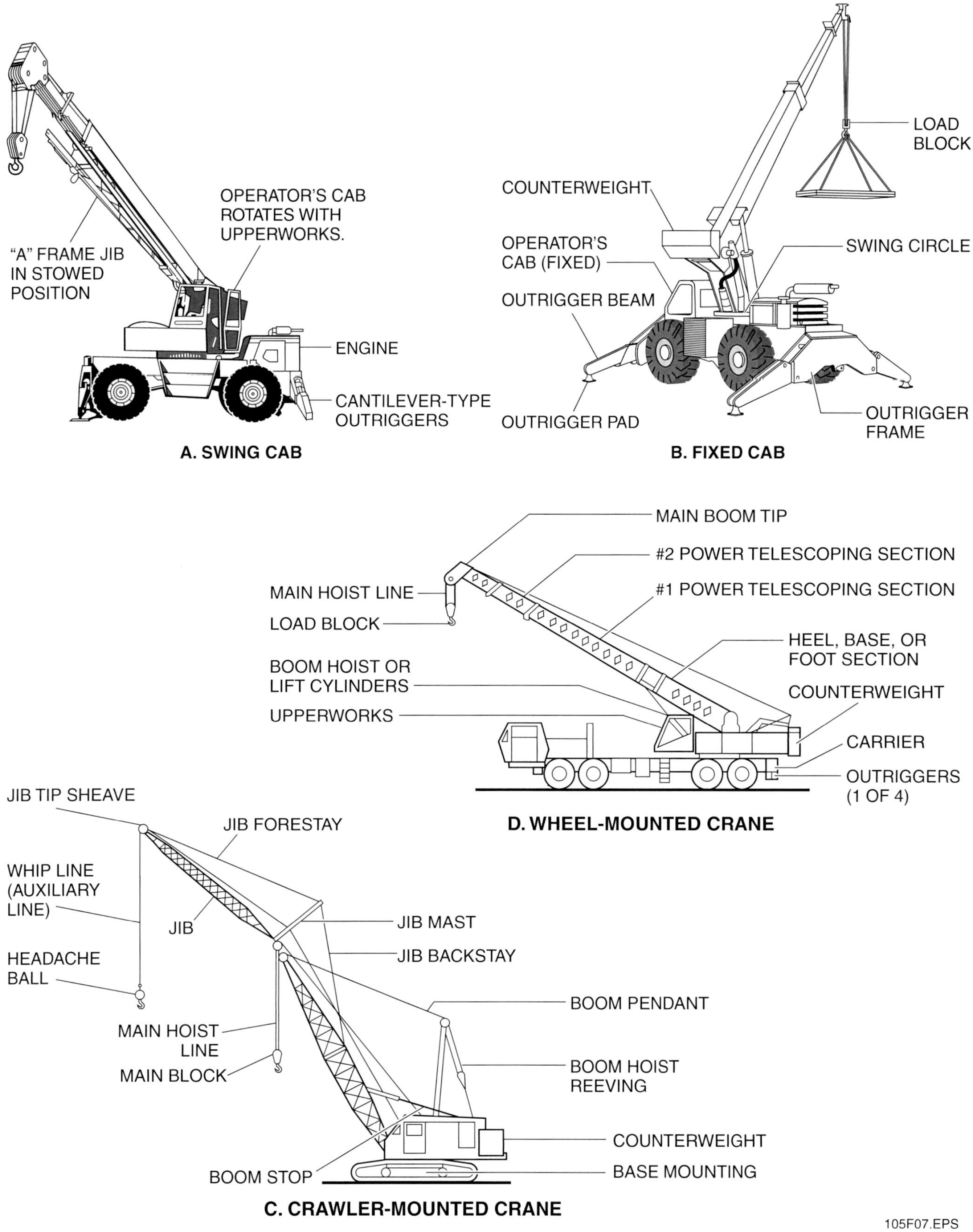

Figure 7 Crawler- and wheel-mounted crane components.

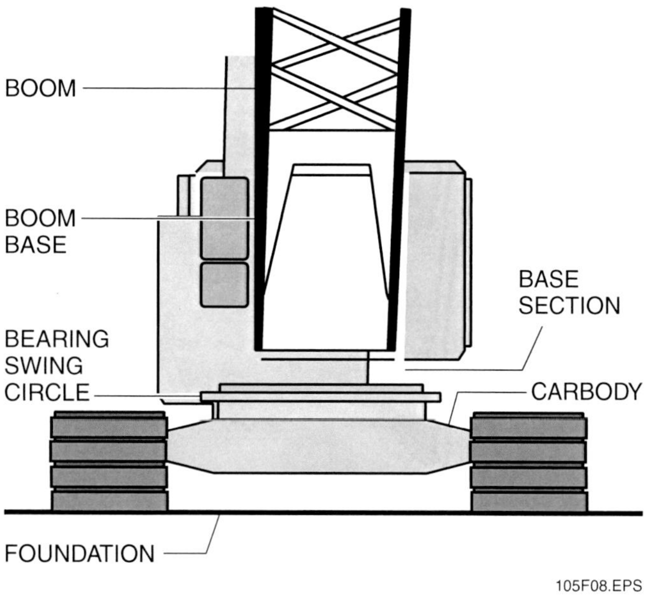

Figure 8 Base mounting carbody.

Some of the terms commonly associated with mobile cranes are as follows:

- *Base mounting* – The structure forming the lowest element of a crane. The base mounting transmits loads to the ground, and it is the carrier or crawler mounting for mobile cranes.
- *Base section* – The lowest section of a telescopic boom, which is attached to the base mounting of a crane. The base section does not telescope, but it contains the boom foot, pin mountings, and the boom-hoist cylinder upper-end mountings.
- *Boom* – A crane member used to project the upper end of the hoisting tackle in reach or in combination of height and reach.
- *Boom base* – The lower-most section of a sectional-latticed boom that has the attachment (boom-foot pins) mounted at its lower end. The boom base is also called the boom butt or butt section.
- *Carbody* – The part of a crawler crane base mounting that carries the upper structure. The crawler side frames are attached to the carbody (*Figure 8*).
- *Fall or parts of line* – The number of running ropes supporting a load; also referred to as parts or falls.
- *Free fall* – The lowering of the hook, load, or boom by gravity. The lowering speed is controlled by a brake.
- *Jib* – An extension added to the boom, which is mounted at the boom tip. The jib (*Figure 9*) may be in line with the boom's long axis, or it may be offset from it. The jib is equipped with its own wire rope suspensions, which are fastened to a mast at the boom tip. The boom tip is supported by jib pendants or backstays.
- *Jib mast or jib strut* – A short strut or frame mounted on the boom head to provide an attachment for the jib support ropes. The jib mast may also be called a gantry, horse, or A-frame.
- *Latticed boom or conventional boom* – A boom that is constructed of four longitudinal corner members, called chords, assembled with transverse and/or diagonal members called lacing, to form a trusswork in two directions. The chords carry the axial (perpendicular) boom forces and bending **moments**, while the lacings resist shear forces.
- *Line pull* – The pulling force in a rope leading off of a rope drum.
- *Loading* – An external physical property that induces force in members of a structure. It may be in the form of a direct physical entity, such as weight on the structure, or as pressure applied by wind, or as force in the form of momentum.
- *Luffing or derricking* – Changing the boom angle by varying the length of the boom suspension ropes. Luffing is also called booming in or out, or topping.
- *Operating radius* – The horizontal distance from the axis of rotation to the center of gravity of a lifted load. This is sometimes called load operating radius (*Figure 10*).
- *Outriggers* – Extendible arms attached to a crane base mounting that relieve the wheels of crane weight and increase stability.
- *Overhauling weight* – Weight added to a load fall to overcome resistance and permit unspooling at the rope drum when no live load is being supported. The overhauling weight is more commonly referred to as a headache ball.
- *Pendant* – A fixed-length rope forming part of the boom suspension system. It may be called

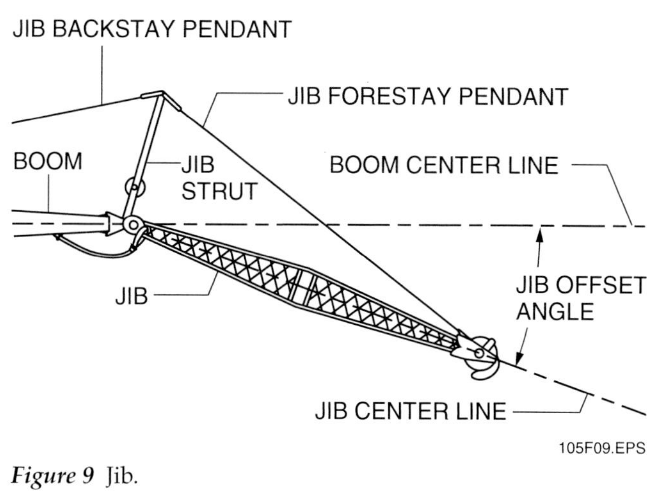

Figure 9 Jib.

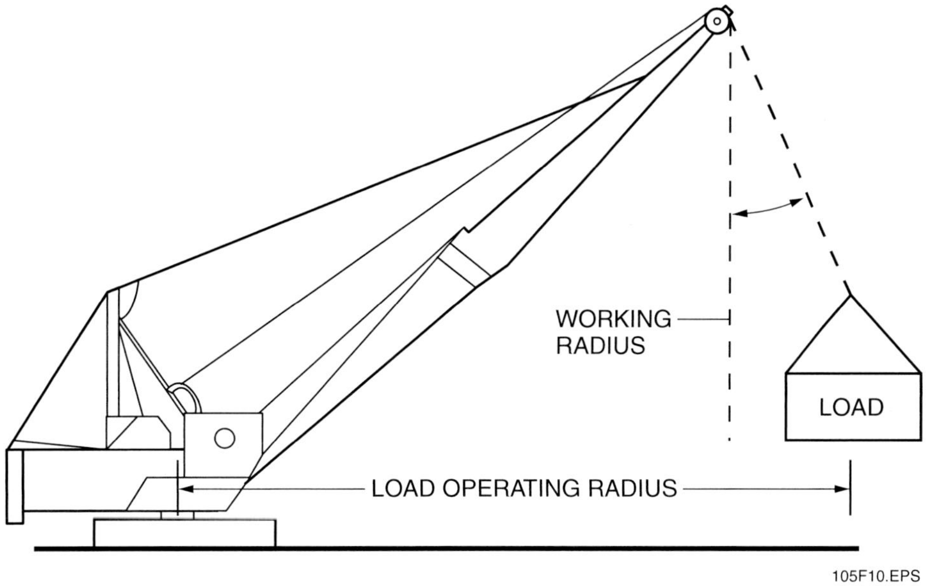

Figure 10 Operating radius of a crane.

the boom guy line, hog line, boom stay, standing line, or stay rope.

- *Power down* – All lowering is done with power. Loads are not allowed to free fall.
- *Reeving diagram* – A diagram indicating the path of the rope through a system of sheaves or blocks. *Figure 11* shows a reeving diagram for hoisting or lowering the boom.
- *Running line* – The rope that moves over sheaves or drums.

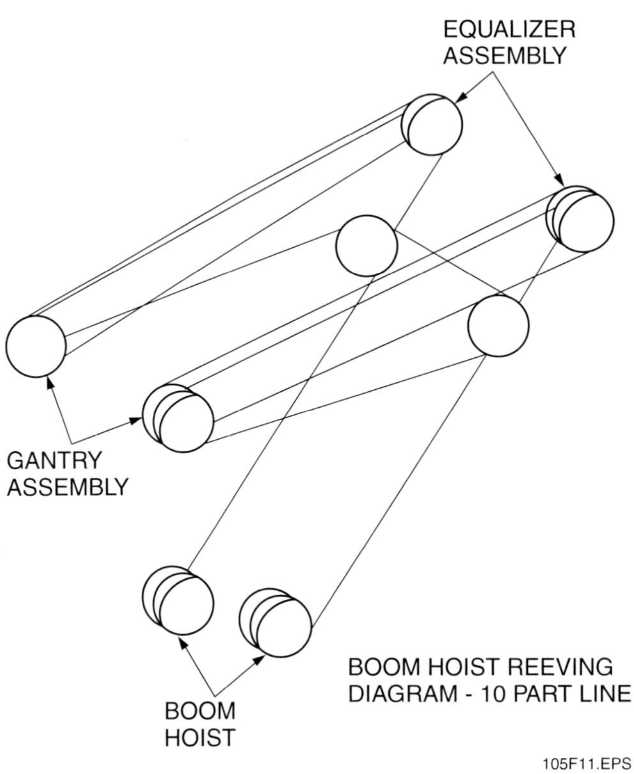

Figure 11 Reeving diagram.

- *Standing line* – A fixed line that supports loads without being spooled on or off of a drum. It is a line in which both ends are dead. It is also referred to as a guy line, stay rope, or pendant.
- *Tag line* – Usually a fiber rope that is attached to the load and is used to control load spin or alignment from the ground. When a tag line is used for clamshell operations, it consists of a wire rope that is needed to retard rotations and the pendulum action of the bucket.
- *Tipping fulcrum* – A horizontal line about which a crane rotates should it overturn. The tipping fulcrum is the point or points on which the entire weight of a crane is imposed during tipping.
- *Upper structure* – The entire rotating structure of the crane less the front end attachment. It may also be referred to as the upper, superstructure, or revolving superstructure.

3.3.0 Load Rating Capacities

The crane load rating is the maximum allowable load for a defined radius with the crane in a specific configuration (length of boom, shape of boom, etc.) while operating under specific conditions. Conditions of operation include the load and wind velocity or forces, speed of operation, levelness or plumbness of the crane mounting, reeving, and any other factors that can create loadings or variations in loadings not anticipated by the design engineer.

The load-lifting capacity of a crane is limited by the following:

- Topping-lift or load-hoist rope strength
- Available line pull at the winch
- Structural strength of the boom
- Structural strength of the mountings

The most important factor controlling load ratings for cranes is stability against overturning and structural failure. Most ratings of mobile cranes are governed by stability. Any crane can overturn if a large enough load is boomed out to a great enough distance. Tipping takes place around a definite line called the tipping fulcrum.

In the United States, load ratings are established by taking a percentage of the tipping load. This percentage is usually the result of cumulative experience over the years rather than mathematical determination. Machine moment remains constant if the machine is not rotated. However, as the radius increases, the boom moment absorbs an increasing proportion of available moment, and the tipping load decreases very rapidly. As maximum radius is approached, the rated load can become very small, and the boom moment can become 95 percent or more of the machine moment. At such high percentages, cranes become quite sensitive. They reach a point where even a small increase in wind velocity can cause overturning. *Table 1* shows the capacity of cranes based on a percentage of tipping weight.

3.3.1 List Angles

List angles are also limiting factors for load ratings on floating cranes. Stability-governed load ratings are usually established for a level machine on firm supports in calm air in the absence of dynamic effects. When one or more of these detrimental effects are present, the rated or lifted load must be reduced in order to maintain the margin of stability. Job site tipping accidents usually result from a failure to make needed reductions.

Table 1 Capacity of Cranes Based on Percentage of Tipping Weight

CAPACITY OF CRANES BASED ON PERCENTAGE OF TIPPING WEIGHT	
Type of Crane	**Percentage**
Locomotive	85%
Crawlers	75%
Mobiles On Rough Terrain	
On Outriggers	85%
On Tires	75%
Commercial Boom Trucks	
On Stabilizers	85%

*CHECK YOUR CRANE. ABOVE PERCENTAGES ARE NOT USED BY ALL MANUFACTURERS.

105T01.EPS

3.3.2 Center of Gravity

The weight and center-of-gravity location of various mobile crane members are determined by calculation or by weighing. This data then becomes the basis upon which stability ratings are calculated. The ratings and the accuracy of the weight and center-of-gravity data are then confirmed by testing. The maximum true lifting capacity (greatest stability margin) is typically obtained when a rough-terrain crane's boom is operated directly over the front.

Prior to making a lift, the operator and rigger should be thoroughly prepared. All the factors that affect lifting the load should be known, including the following:

- The weight of the load, including the weights of the load block, slings, and handling devices
- The shape of the load
- The height to which the load will be raised
- The position of the load when raised

Figure 12 shows the tipping axis for different lifting locations.

3.3.3 Capacity Chart

Other determinations that must be made include the boom capacity limitations related to length and different radii. The method of the lift, as well as the radius range through which the load can be carried, may be decided by reviewing the capacity chart. Range diagrams are useful for planning lifts. The proper range diagram and capacity chart should remain with the crane and should be consulted for every lift (*Figure 13*).

3.3.4 Reeving

The reeving of the boom hoist and equalizer must be done in compliance with the manufacturer's recommendations. Twisted lines must be avoided. Whenever possible, shackles or slings should be used as a link between the load and the hook to prevent undue rope wear.

3.3.5 Balancing

The footing of the lift area must be checked to make sure it can support the bearing weight of both the machine and the load being lifted. If there is doubt, mats should be used or a competent engineer consulted. A load should always be balanced and should not be lifted until it is. The load may be raised only a few inches off the ground to determine if it is balanced. If it is not balanced, the lifting slings should be adjusted.

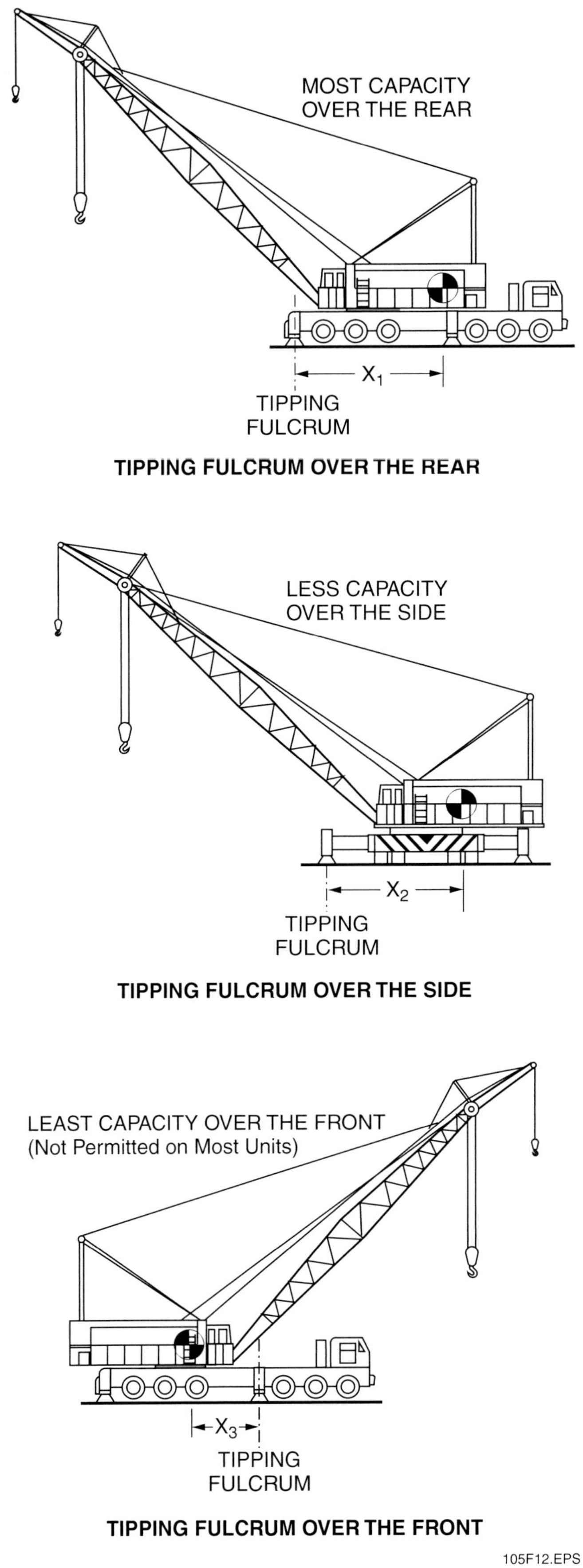

Figure 12 Crane leverage distances for each position.

3.3.6 Signals and Lift Areas

To avoid swinging into people or obstructions, signal persons and standard crane signals should be used. Accessible areas within the rear swing radius of the crane must be barricaded to prevent anyone from being struck by the crane when it is swinging. Lift areas must be designated by signs and barricades to minimize traffic. The operator must make sure that there are no obstructions in the way before swinging the load.

3.3.7 Power Lines

Contact with power lines is the leading cause of crane-related fatalities. The boom must always be kept clear of electrical power lines. If the boom accidentally touches overhead wires, the operator must stay in the cab until the boom can be cleared or the electrical power cut off. Riggers should use insulated handling sticks when working near electrical power lines.

Follow these recommendations when working near power lines:

- Maintain safe distances from power lines when moving a crane. *Table 2* shows the minimum clearances required.
- Maintain a safe operating distance.
- Use a signaler (watch person).
- Slow down operation near power lines.
- De-energize lines when possible.
- Use insulators/guards.
- Use synthetic slings when possible.
- Use tag lines only to control the load (dry polypropylene is best).

3.3.8 Wind

The wind's effect on a lifted load (*Figure 14*) can produce a load sufficient to overstress the boom or jib structure and cause these members to collapse. This type of wind condition can be observed when the suspended load will not remain in line with the boom chords. The longer the boom, the more serious this condition becomes.

3.3.9 Other Considerations

When the load is in the air, operators must have their hands and feet on the appropriate controls. Workers guiding the load must use tag lines and must never get between the load and any other large object. No one should ever be permitted to ride hoisting lines. Equipment should never be parked directly behind a crane. Because slings and chokers can be hand traps, a rigger should

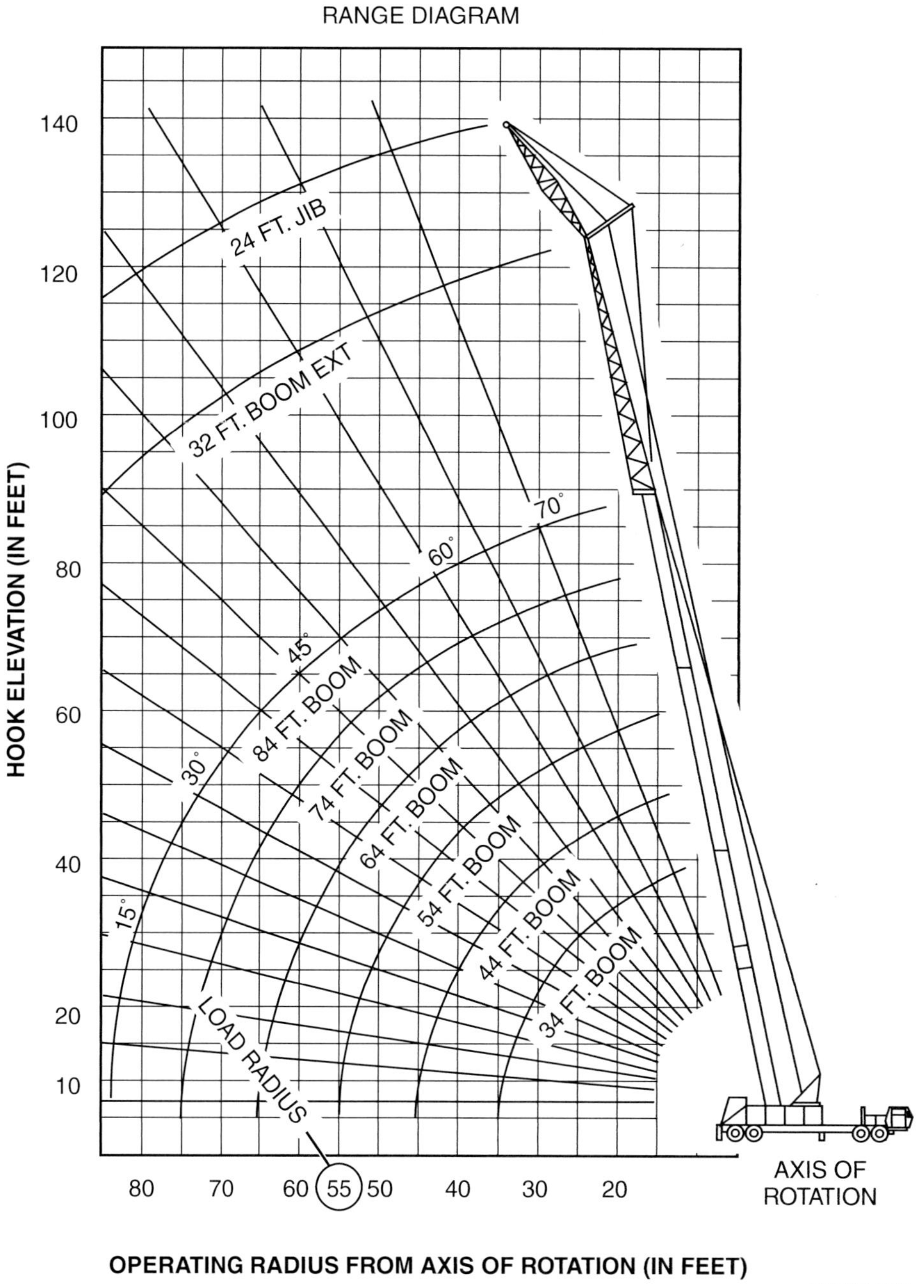

Figure 13 Sample range diagram and capacity chart (1 of 2).

never signal the operator to lift the load until both hands and the rest of the body are clear of the load.

3.3.10 Crane Safety Features

The built-in safety features of mobile cranes usually include level indicators, boom stops, and outriggers. The use of standardized control systems that include braking systems, locks, clutch control systems, and locking pawls provides for safer crane operation.

The most important aspect of crane safety are the skills of the operator and the riggers performing a lift. Preventive safety checks are performed each morning before a machine is put into operation. The most important factors in safe rigging and hoisting operations are well-planned lift procedures and common sense.

3.4.0 Inspecting the Crane

Each crane should have a thorough inspection before it is used for the first time and periodic inspections after that. A visual inspection can

MAXIMUM ALLOWABLE LOADS IN POUNDS						
BOOM LENGTH IN FEET	RADIUS IN FEET	BOOM ANGLE IN DEG.	BOOM POINT PIN HEIGHT (FT-IN)	OUTRIGGERS SET OVER SIDE OR REAR	WITHOUT OUTRIGGERS	
					OVER SIDE	OVER REAR
40	12	78	46' 6"	* 220,000	100,400	139,700
	15	74	45' 9"	* 185,000	71,800	97,400
	20	66	44' 0"	* 149,000	48,100	54,600
	25	58	41' 3"	118,000	35,800	47,900
	30	49	37' 6"	86,000	28,300	37,800
	35	39	32' 6"	67,000	23,200	31,100
60	15	79	66' 3"	* 180,000	71,000	97,000
	20	74	65' 0"	* 147,000	47,300	63,900
	25	69	63' 6"	* 118,500	35,000	47,200
	30	64	61' 3"	86,200	27,500	37,100
	35	59	58' 6"	67,400	22,400	30,300
	40	53	55' 3"	54,700	18,800	25,500
	50	40	45' 9"	39,400	13,900	19,100
80	20	78	95' 9"	* 145,000	46,700	63,300
	25	75	84' 6"	118,200	34,400	46,600
	30	71	83' 0"	85,900	26,900	36,500
	35	67	81' 0"	67,000	21,800	29,700
	40	63	78' 9"	54,300	18,200	24,900
	50	55	72' 9"	38,900	13,300	18,500
	60	45	64' 3"	29,900	10,100	14,300
	70	34	52' 3"	24,000	8,400	11,500
90	20	80	95' 9"	* 143,000	46,400	63,100
	25	76	94' 9"	* 117,200	34,100	46,300
	30	73	93' 6"	85,900	26,600	36,200
	40	66	89' 9"	54,100	17,900	24,600
	50	59	84' 6"	38,600	13,000	18,200
	60	51	77' 9"	29,600	9,900	14,100
	70	43	68' 3"	23,800	7,700	11,200
	80	32	55' 3"	19,600	6,100	9.150

105F13B.EPS

Figure 13 Sample range diagram and capacity chart (2 of 2).

be very revealing when the crane is checked as follows:

- Inspect the boom and jib for straightness, cracks, and signs of abrasion.

WARNING

Cracking or flaking paint may indicate bending of metal, which may lead to eventual failure. A bent boom or jib could collapse and cause personal injury or death.

- Check boom pins for missing cotter pins.
- The lattices and chords of the boom must be straight. Never attempt to hammer a bent lattice back into shape, and never apply heat to a lattice to straighten it. Lattices must be replaced according to the manufacturer's exact specifications, particularly if any welding is to be done.
- Check all welds for cracks and indications of corrosion. If any welding must be done on the boom, follow the manufacturer's specified procedure.
- The boom length should be marked on all boom inserts. Make sure that the overall length of the boom is known.
- All exposed moving parts must be guarded.
- **Counterweights** must be secure.
- Weight and capacity should be clearly stamped on jibs, blocks, and equalizer beams.
- Inspect all wire ropes for signs of wear.
- Inspect sheaves, drums, hooks, shackles, and fittings for signs of wear.

Table 2 High-Voltage Power Line Clearances

CRANE IN OPERATION (1)	
POWER LINE (kV)	BOOM OR MAST MINIMUM CLEARANCES (feet)
0 to 50	10
50 to 200	15
200 to 350	20
350 to 500	25
500 to 750	35
750 to 1000	45
CRANE IN TRANSIT (with no load and the boom or mast lowered) (2)	
POWER LINE (kV)	BOOM OR MAST MINIMUM CLEARANCES (feet)
0 to 0.75	4
0.75 to 50	6
50 to 345	10
345 to 7500	16
750 to 1000	20

Note 1: For voltages over 50kV, clearance increases 5 feet for every 150 kV.
Note 2: Environmental conditions such as fog, smoke, or precipitation may require increased clearances.

105T02.EPS

> **WARNING**
>
> Any defective sheaves, drums, hooks, shackles or fittings must be destroyed.

- Check connecting rods and actuator arms back to the brakes. Examine the turnbuckles to make sure they are not loose.
- When using headache balls with swivels, periodically check the pin inside the swivel. This pin has a tendency to shear.
- Check the becket on the dead end of the boom reeving when using a single-drum boom hoist. The wire rope inside the becket is subject to internal breaking.

3.5.0 Positioning the Crane

Before operating a crane, every aspect of its positioning must be carefully planned and considered. More than half of the accidents involving mobile cranes are caused by one or more of the following factors:

- Failure to level the crane
- Failure to use outriggers
- Failure to extend outrigger beams fully and get the wheels completely off the ground
- Poor ground conditions
- Improper blocking beneath the outrigger floats

3.5.1 Leveling a Crane

The most important prerequisite to safe crane operation is that the crane be exactly level before making any lift. All capacity charts are based on the machine being within one percent of level. *Table 3* shows the difference in capacity that results from the crane being out of level. Note that with a long boom at its minimum radius, a three-degree error causes a fifty percent reduction in the crane's listed capacity.

Initially level the crane with a carpenter's level. Then raise the boom and lower the load line. The load line should lie directly in the center of the boom. If it does not, recheck the level of the crane before attempting any lifts (*Figure 15*).

3.5.2 Cribbing

When using mobile cranes equipped with outriggers, make sure that the outriggers are properly blocked. Cribbing is required to distribute the load more evenly, even when the crane is sitting on concrete. If wood cribbing is used, always use hardwood. The cribbing area needs to be at least three times the float area and should completely support the float. The float ram should be pinned to the float. A crawler crane needs to be positioned on a crane mat. Mats should be 4 feet longer than the crawler track and 4 feet wider than the overall crawler width.

Before determining how to block the crane, check the condition of the soil. Soil strength increases with an increase in the size of particles. Fine-grained soils can present problems. An excess of organic material also causes problems because it makes the soil resilient. Good soil is compact, well-graded, and composed of rough, uneven particles of sand and gravel. *Table 4* shows the bearing capacities of different soils.

After the crane has been blocked, there will be some initial settling of the soil, which must be allowed. Check the level of the crane shortly after putting it into operation. Also remember to allow for seasonal variations in ground conditions. If the crane is set up for a long-term project, periodic checks must be made and the level of the crane adjusted accordingly.

WIND FORCE	
MILES PER HOUR	FORCE PER SQUARE FOOT (LBS.)
1	0.005
5	0.123
10	0.492
15	1.107
20	1.969
25	3.075
30	4.429

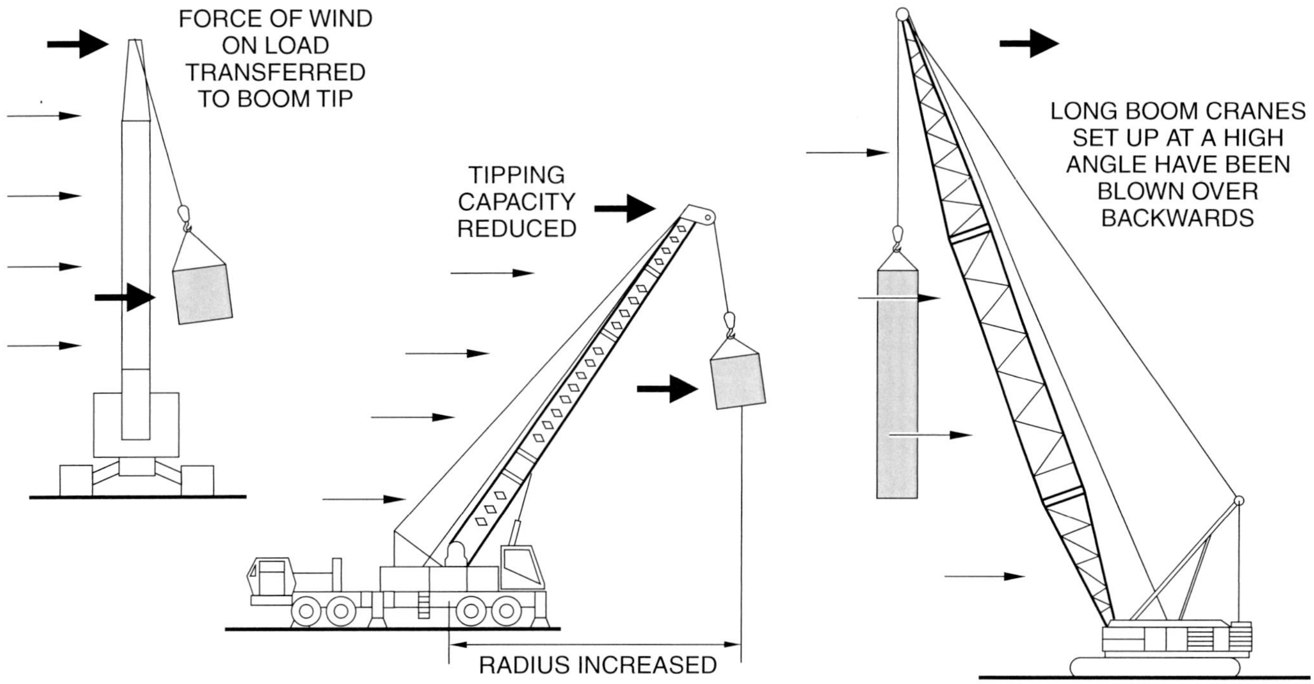

Figure 14 Wind effects on suspended loads.

Table 3 Lifting Capacity

BOOM LIFT AND LIFT RADIUS	Capacity Lost When Crane Is Out Of Level By:		
	1 Degree	2 Degrees	3 Degrees
SHORT BOOM			
Minimum Radius	10%	20%	30%
Maximum Radius	8%	15%	20%
LONG BOOM			
Minimum Radius	30%	41%	50%
Maximum Radius	5%	10%	15%

105T03.EPS

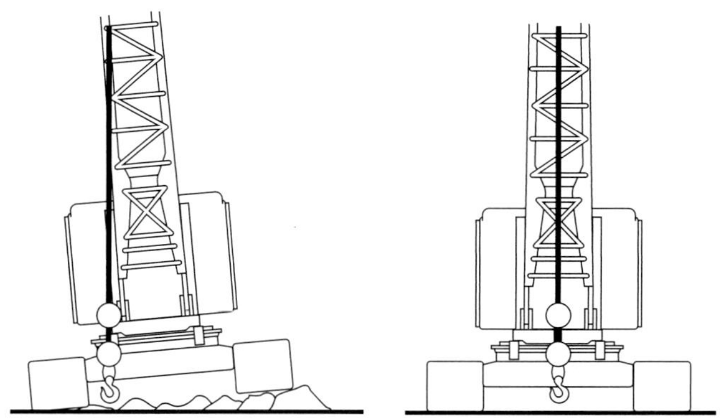

Figure 15 Examples of load line position.

Table 4 Soil Bearing Capacities

SOIL BEARING CAPACITIES	
Soil	**Capacity**
CEMENTED SAND AND GRAVEL	135psi
SAND AND GRAVEL (Compact)	110psi
SAND, COARSE TO MEDIUM (Compact)	60psi
SAND, FINE TO SILTY (Compact)	54psi
SILT (Compact)	40psi
CLAY (Compact)	54psi

105T04.EPS

3.5.3 Positioning Safety

Each crane application requires different precautions. However, the following are some general safety guidelines to observe when positioning cranes:

- Allow a minimum of 2 feet of clearance between the counterweight and the nearest obstacle. Make sure that proper barricades are erected around the crane to keep unauthorized personnel a safe distance away.
- Special precautions must be taken when using cranes near shoring locations, excavations, slopes, trenches, or embankments.
- Special precautions must be taken when using cranes over cellars, buried pipes, or water or gas mains if it is unsure whether there is danger of collapse. Locate and mark buried utilities clearly.
- Operators must be able to see other cranes working near them and communicate with the other operators.
- Special precautions must be taken when using a crane to work over areas of public access. When such work is unavoidable, warning signs must be posted, and a signal person must be assigned to give ample warning.
- Special precautions must be taken when using a crane within a boom's length of an overhead power line.
- If a crane is to be working in a fixed location near electrical hazards, ground the crane with a ground rod or ground mat. Place the rod as close to the machine as possible. First connect the ground lead to the rod, and then connect it to the machine. Instruct all personnel to keep clear of both the lead and the rod. To remove the rod, first disconnect the lead from the machine, and then disconnect it from the rod.
- Also ground the crane if it is to be positioned near a radio transmission tower. The boom can act like an antenna and become electrically charged.
- Always use the outriggers according to manufacturer's specifications.
- Use the crane only if its boom angle, boom length, load radius, and rated capacity are known.
- Position the crane as close to the load as possible.
- Make sure that the weight of the load is known.
- Ensure that the hook is directly above the load's center of gravity.
- Verify that the crane is set up level.
- Make sure that the rigging is correct.
- Extend hydraulic boom sections evenly.

3.5.4 Traveling With Suspended Loads

Cranes are often used to move objects from place to place, but this can present problems. There are no capacity charts for traveling with a load. This is because too many variables can affect the operation. Be sure to check the manufacturer's specifications about traveling with loads. The following safety precautions can be used as guidelines:

- Never travel over unstable ground if the stability of the machine is likely to be endangered.
- Never travel with capacity (or near capacity) loads without proper supervision.
- Carry the boom in line with the direction of motion.
- Do not use a high boom angle without proper supervision.
- Avoid sudden starts and stops.
- Use tag lines to control the swinging of the load.
- Keep the load as close to the ground as possible. This reduces the impact loading on the boom due to the elasticity of the wire rope.
- When using a crawler crane on soft ground, it may be necessary to travel backward in order to raise the leading end of the crawlers. When traveling backward, snub the load to the machine to prevent swinging, and travel with extreme caution.
- Never use a rubber-tired crane to carry a load suspended over the side.
- Use the crane only if its boom angle, boom length, load radius, and rated capacity are known.
- Position the crane as close to the load as possible.
- Make sure that the load's weight is known.
- Ensure that the hook is directly above the load's center of gravity.
- Verify that the rigging is correct.
- Extend hydraulic boom sections evenly.

- Make the boom as short as possible.
- Keep the boom as low as possible.
- Tire size (if applicable), type, and inflation pressure are per manufacturer's specifications.
- Maximum speed while traveling is per manufacturer's specifications.

3.5.5 Good Lifting Practices

After the crane has been properly assembled, positioned, inspected, and leveled, determine how to rig the load. There are safe ways to rig loads that ensure the safety of each lift. All sharp corners should be padded (*Figure 16*). Steel can easily cut through wire rope.

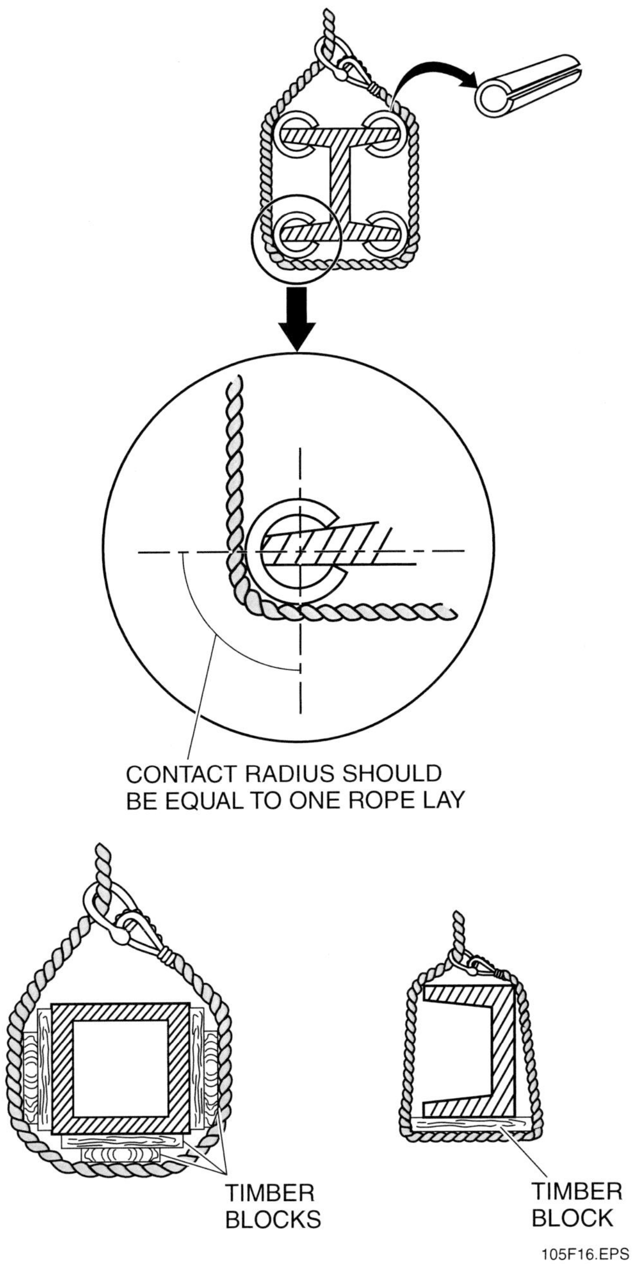

Figure 16 Padding of sharp corners.

When using hooks, always turn the openings of the hooks out, facing away from each other, so that the load pulls into the hook. To keep the load balanced when lifting long loads, use a spreader beam, or at least use double straps.

3.5.6 Hand Signals

Standard hand signals, which are detailed later in this module, have been developed that are used for all types of hoisting equipment. From the viewpoint of personal safety and common sense, it is important that everyone working with hoisting equipment knows and understands these signals. The following rules apply concerning signals:

- Only one person should be designated to give signals. If it is necessary to change a signal person, this must be done with the knowledge and consent of the boom equipment operator. It is very important for the operator and signal person to have a close working relationship.
- Any variation on the standard hand signals must be worked out with the operator of the boom equipment. Any such variations should not conflict with the standard signals.
- No response should be made to improper signals.
- The Stop signal can be given by anyone.
- The person designated to give signals should be placed in a position where the operator of the boom equipment can clearly see him/her.
- Signals should be given away from the body so they can be easily understood (not be misinterpreted).
- Always check the position of the sun before beginning a lift to ensure that the operator can see the signaler.

3.5.7 Audible Signals

Signals may be given by whistle, bell, horn, or other audible means. The following audible signals may be used when moving the crane:

- *Stop* – One short blast
- *Go ahead* – Two short blasts
- *Back up* – Three short blasts

WARNING

Audible signals should not be used when there are surrounding noises of the same, or nearly the same, frequency and octave level.

4.0.0 Handbooks and Records

Manuals published by the manufacturer contain all of the important information related to operation and maintenance of a specific model of crane. These manuals are provided with each machine.

4.1.0 Crane Manuals and Handbooks

Each crane must be supplied with the correct manufacturer's handbook and/or company manual that explains the proper use of that particular machine. Crane manuals and handbooks should include the following information:

- The equipment designation or type
- Name of the manufacturer
- Equipment model and serial number
- Year of original sale
- Total weight of the unit and ground bearing pressure on tracks, tires, or outriggers
- Weight of the boom sections
- A copy of the load chart along with any and all rated combinations and variations in capacity and geometry
- Inspection and maintenance procedures that include the material specifications and the welding specifications on booms and outriggers, as well as the bolting and torquing specifications
- Rigging specifications
- Erection procedures
- Operating precautions

4.2.0 Crane Records

If a log is not supplied with the piece of equipment, then one should be started. The log must be kept on the work site for regular recording of all inspections, tests, repairs, maintenance, and hours of service.

All log entries should be dated and signed by the operator, the repair person, and the supervisor. The log book must remain with the machine and be kept up-to-date throughout its working life. No machine should be leased, rented, or purchased if it has been modified or altered from the original manufacturer's specifications. If the unit requires repairs, they must be done by a capable repair person and be certified by a qualified authority.

4.3.0 General Crane Safety

The identification of safety hazards usually begins with machine inspection and proper maintenance procedures. Prior to operation, the operator should take a few minutes to go through a preventive safety-check procedure. A daily preventive maintenance checklist is presently required on all new cranes by *ASME Standard B30.5*. The list should be reviewed and all items checked before operating the crane. Dated records of the reports must be kept on file.

4.3.1 Inspection

Before daily operations begin, the operator should inspect around and under the crane for visible evidence of potential safety hazards or operational problems. Check for loose, worn, or damaged reeving. These and any other unsafe conditions should either be corrected immediately or reported. Crane operations should not proceed until the conditions have been corrected. Annual inspections must be done in accordance with *ASME Standard B30.5*. The following is a basic list of things to check in any inspection:

- Engine performance and gauges
- Oil pressure and levels
- Housekeeping
- Fire extinguisher
- Audible warning device
- Hydraulic lines and fluid
- Windshield and wipers, mirrors, and lights
- Batteries and charging system
- Hand signal and load chart posted
- Guards and clutches
- Outriggers, tracks, wheels, and tires
- Brakes, **dogs**, locks, and safety devices
- Drive chains, steering, and all rollers
- Sheaves, drums, and all cables
- Block and hook
- Boom, gantry, jib, and extension
- Carrier assembly
- All controls, including hoisting, swing, travel, and the boom

4.3.2 Lubrication

Lubrication and greasing must always be done in compliance with the manufacturer's recommendations. All control linkage should be oiled and should work easily. All mud, ice, grease, and oil spots should be wiped up to prevent falls. This is the operator's responsibility.

4.3.3 Fire Extinguishers

A fire extinguisher is required in the cab of each crane to comply with OSHA regulations. The fire extinguisher must be in good operating condition, and should be tested at least once a month. It needs to be readily accessible, and its location well marked.

4.3.4 Sheaves

Inspect sheaves periodically for spots that could cause excess wear on wire rope. Frayed rope ends should be properly seized and the frayed sections cut off. Wire rope must be inspected daily for broken wire and periodically for distortion, evidence of heat damage, or for a substantial decrease in diameter.

4.3.5 Booms

Booms should be given a periodic inspection for bent or broken lacings. If any lacings are bent or broken, that insert should not be used until the lacings have been replaced. Dents, gouges, or rust could indicate a fatigue crack. Any cracked or fatigued members must be repaired following the manufacturer's procedures. Pendants should also be regularly inspected for broken wires and defective end connections.

4.3.6 Ropes

The entire rope should be lightly lubricated with oil, especially at the socket. Wire rope must be properly spooled and lubricated, even during storage. Lifting equipment should also be stored properly.

4.3.7 Hooks

All hooks must have a safety latch or be moused to prevent the bridle slings from coming unhooked. Even the weight ball hook on the whip line should have a safety latch. Keep in mind that a safety latch is not foolproof, and its existence does not guarantee a safe hook. All hooks must be inspected (*Figure 17*). This must be done at frequent intervals—daily or monthly—in order to check for deformation or cracks. Hooks must be discarded when the throat has increased 15 percent from the original size or when a 10-degree twist exists from the plane of the unbent hook. The capacity of hooks must be known and not exceeded.

4.3.8 Boom Stops and Angle Indicators

Boom stops and angle indicators are safety devices installed by the manufacturer that must be kept operable at all times. The boom angle indicator gives the operator a quick way to determine the approximate working radius, but the actual radius should be measured with a tape. On critical lifts, it is a good practice to have someone other than the operator check the radius. The operator may not see the actual boom radius from his or her vantage point. Some automatic boom stops cut off the air supply to the boom hoist and stop the boom at a preselected boom angle. The service manual must be studied and understood by each person responsible for the operation and maintenance of the machine. All warning plates both inside and outside the cab must be readable and in place.

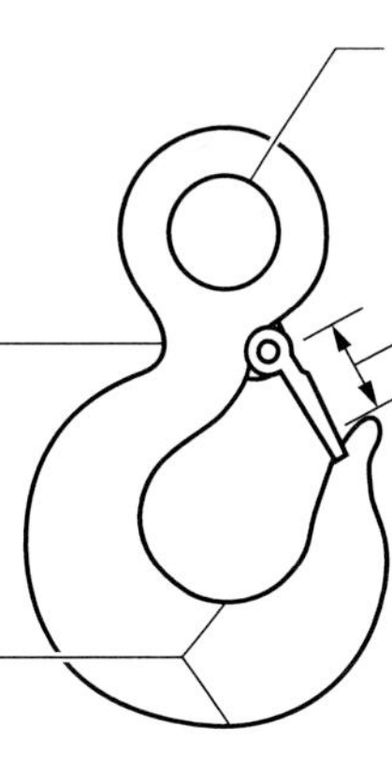

Figure 17 Inspection locations for hooks.

5.0.0 CRANE ASSEMBLY

Cranes often have unique details of assembly that are provided in the documentation. These instructions must always be carefully observed and carried out. The following rules apply to the assembly of cranes in general:

- The crane operators must be qualified to operate the types of equipment being assembled. This requirement also applies to any helper cranes being used.
- The crane must be assembled according to the manufacturer's instructions.
- Make sure that the handbook for the crane is available at the site.
- Before assembling the boom, determine the weight of the largest load and necessary rigging, as well as the height of the maximum lift and the radius of operation. Never use more boom than is necessary.
- Make sure that the counterweight is sufficient to get the boom off the ground. This is especially important when assembling a long boom. It may be necessary to install extra counterweights, which may have to be removed before operation.
- Before adding or removing a counterweight from a carrier-mounted crane, make sure that the crane is on firm, level ground and that the swing lock is engaged. Make sure that the boom is not too high.

- Use only the correct sections for the crane being used. All components must be clearly marked and installed in the proper sequence. The boom sections must assemble easily, without being forced.
- Cotter pins must be installed in every connecting pin.
- Use only the correct size and quality of nuts, bolts, and pins. Install them in the correct manner and sequence provided by the manufacturer so that the lacing pattern is maintained throughout the entire length of the boom.
- When possible, install short sections close to the boom to improve machine stability. This may vary by manufacturer.
- Block each end of each section of the boom when assembling it.
- Know the maximum length of boom that can be lifted during boom assembly. Do not exceed this limit.
- Make sure that the pendants are of the proper length for the boom. Pendants that are slightly too long or slightly too short can cause serious loading problems. Both pendants must be of the same size and length.
- On bolted booms, a washer goes under the bolt head. Place the nut toward the point of the boom so that it is visible in case it should fall out.
- Think about having a manufacturer's representative or consultant oversee assembly.
- Consider the area of assembly in relation to the location where the crane will actually be used.
- Identify any limiting factors, overhead obstructions, power lines, etc., that are close to the assembly area.

5.1.0 Assembling a Short Boom

Construction cranes can be configured with either short or long booms. The difference between short and long booms is the number of components involved. *Figure 18* shows short boom assembly.

The illustration of the assembly process is meant only as an aid to the general understanding of the procedure and should not be used to assemble a specific boom. Always refer to the manufacturer's instructions to assemble a boom.

The following procedure applies to the assembly of a short boom:

Step 1 Line up two upper pin connection holes.

Step 2 Insert the top boom pins, and install the cotter pins.

Step 3 Lay out the remaining sections, draw them together, and insert all the top pins.

Step 4 Boom up slightly, and insert all bottom pins.

Step 5 Lower the boom, and move the pendants out to the boom tip.

5.2.0 Assembling a Long Boom

The procedure for assembling a long boom (*Figure 19*) is similar to the procedure for assembling a short boom. The main difference is that the boom pendants must be moved out to the last pinned section during assembly. Always block beneath the end of each section of the boom during assembly and disassembly. Never stand on or beneath the boom during this operation.

The procedure for assembling a long boom is as follows:

Step 1 Line up two upper pin connection holes.

Step 2 Insert the top boom pins, and install cotter pins.

Step 3 Draw the next section in, and install the top boom pins and cotter pins.

Step 4 Boom up slightly, and install the bottom boom pins and cotter pins.

Step 5 Move the pendants to the last pinned section.

Step 6 Repeat this procedure for each remaining boom section.

Step 7 Install pendant spreader bars on long booms. This prevents the main pendants from rubbing against the sides of the boom or the boom connecting pins.

Step 8 Install midpoint suspension on long booms, especially those longer than 150 feet.

5.3.0 Disassembly of the Boom

The disassembly of a boom is not actually a reversal of the assembly process. The disassembly is much more dangerous than the assembly, and the steps must be followed carefully and in order.

WARNING

Do not allow any part of your body to be under the boom at any time. Remember that steel-toed boots, for example, will not protect your foot from injury if a boom section lands on the toe box.

To disassemble a boom, follow these steps:

Step 1 Rest the boom tip on the dunnage while removing the top and bottom pins or bolts. All of the section must be supported so that there is no movement when the pins are removed.

Step 2 Move the pendant to the next section joint.

Step 3 Rest the next section so that the stress is taken off the bottom pins at the section joint. Remove the bottom pins or bolts.

Step 4 Boom down so that the last section is resting completely flat on dunnage with no stress on the top pins at the section joint. Remove the top pins or bolts.

Step 5 Repeat this process until all of the sections are resting safely on dunnage.

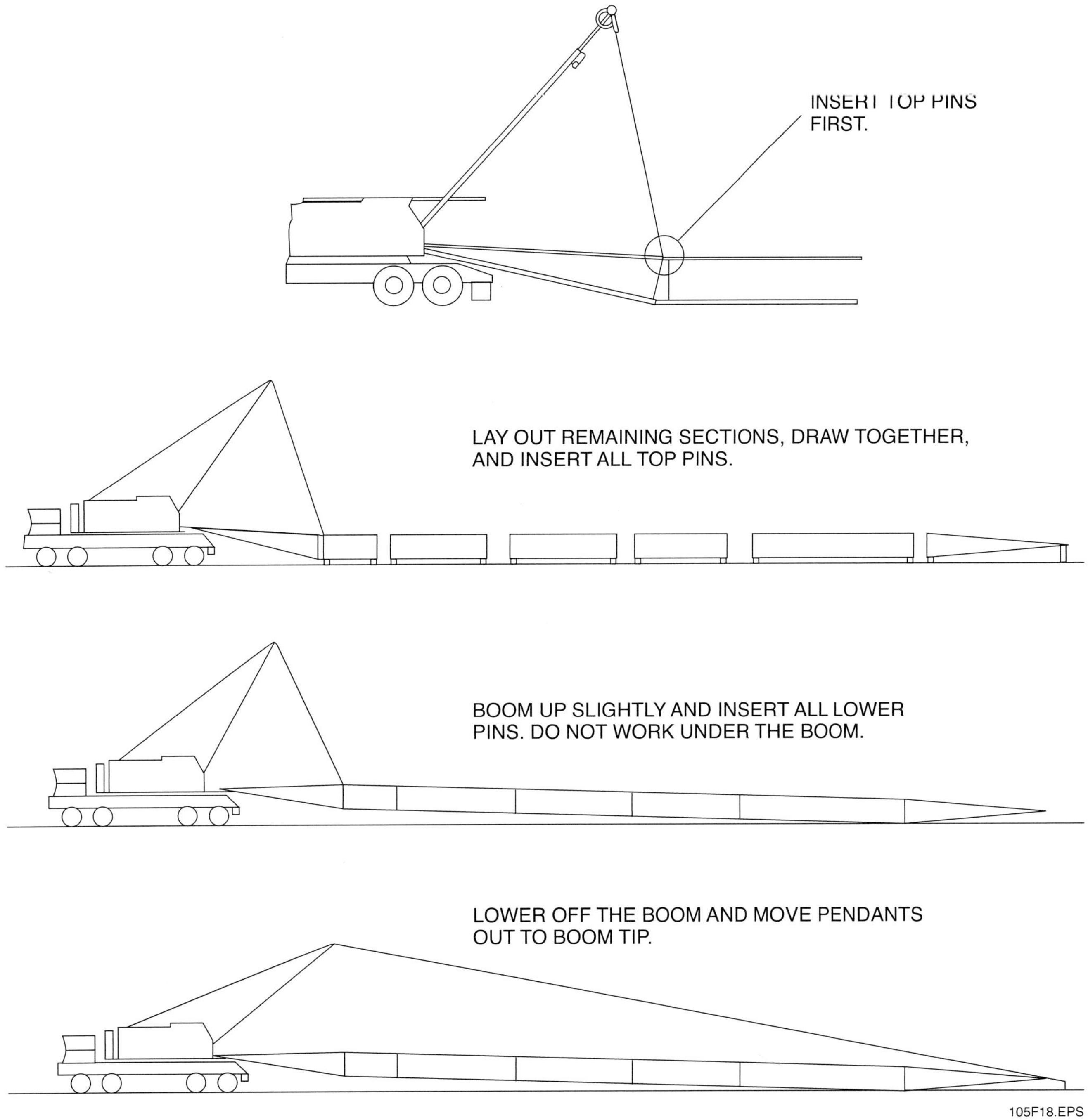

Figure 18 Assembly of a short boom.

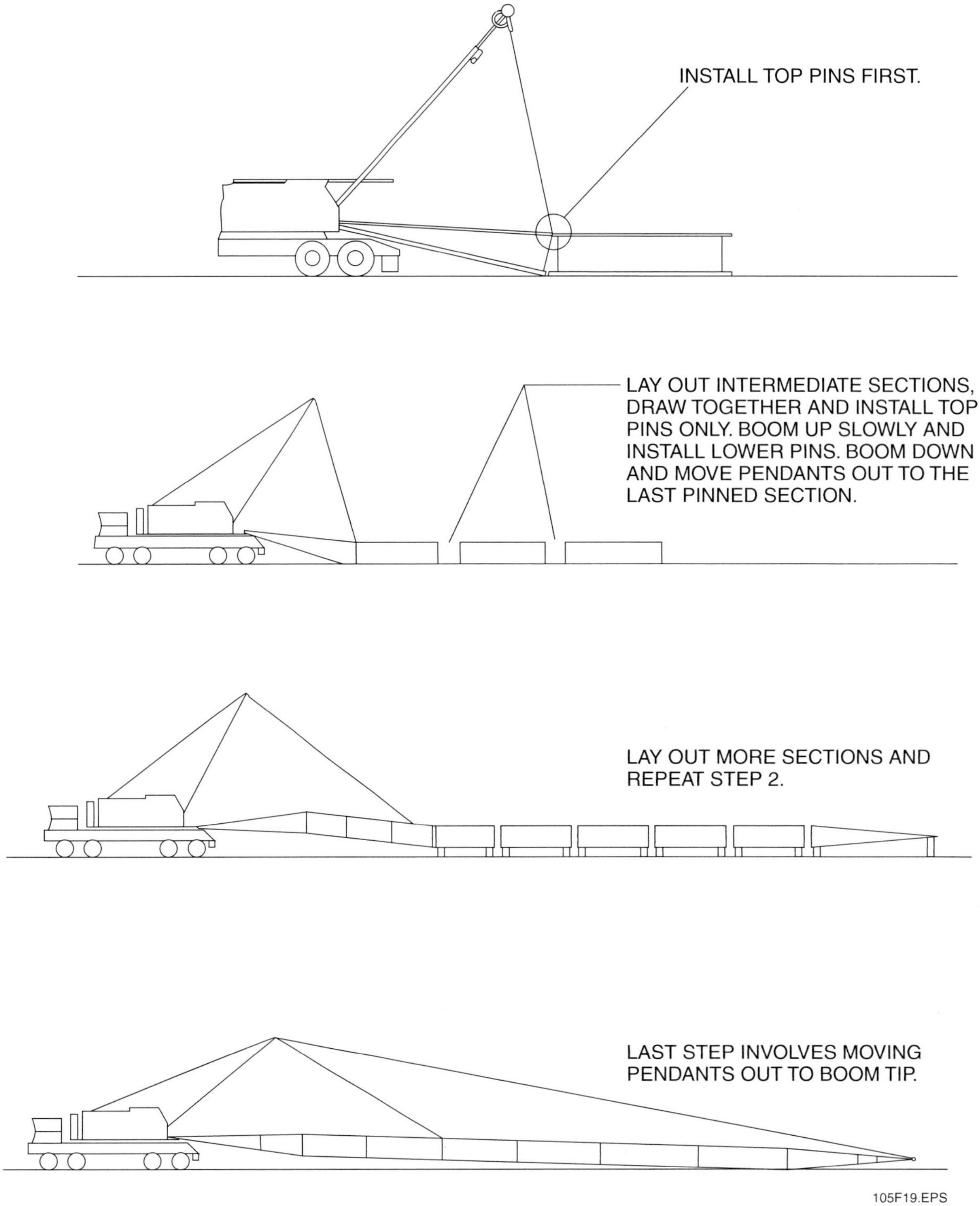

Figure 19 Assembly of a long boom.

6.0.0 Methods and Modes of Communication

The methods and modes of communication vary widely in lifting operations. The method refers to whether the communication is verbal (spoken) or nonverbal. The mode is what is used to facilitate the communication. Modes can include, for example, a bullhorn, a radio, hand signals, or flags.

6.1.0 Verbal Modes of Communication

Verbal modes of communication vary depending on the requirements of the situation. One of the most common modes of verbal communication is a portable radio (walkie-talkie). Compact, low-power, inexpensive units enable the crane operator and signal person to communicate verbally. These units are rugged, dependable and widely used on construction sites and in industrial plants.

There are some disadvantages, however, to using low-power and inexpensive equipment in an industrial setting. One disadvantage is interference. With low-power, inexpensive units, the frequency used to carry the signal may have many other users. This crowding of the frequency could disrupt the signal from a more powerful unit. Another disadvantage is when high background noise is present. In attempting to send a signal in a high-noise area, the person sending the signal may transmit unintended noise, resulting in a garbled, unintelligible signal for the receiver. On the receiving end, the individual may not be able to hear the transmission due to a high level of background noise in the cab of the crane.

There are several solutions to the problems associated with radio use. To overcome the shortcomings associated with low-power units, more expensive units with the ability to program specific frequencies and transmit at a higher power level may be needed. Some of these more expensive units may require licensing. To overcome the background noise problem, the use of an ear-mounted, noise-canceling microphone/headphone combination may be required (*Figure 20*).

Another solution is the use of an optional throat microphone. This device feeds the transmitted sound directly to the ear and picks up the voice communication from the jawbone at the ear junction. This prevents noise from entering the microphone and blocks out any background noise when listening. To avoid missed communication, the signal person's radio is usually locked in transmit so that the crane operator can tell if the unit is not transmitting. In any event, a feedback method should be established between the signal person and the crane operator so that the signal person knows the crane operator has received the signal.

Another mode of verbal communication is a hardwired system (*Figure 20*). These units overcome some of the disadvantages of radio use. When using this type of system, interference from another unit is unlikely because this system does not use a radio frequency to transmit information. Like a telephone system, occasional interference may be encountered if the wiring is not properly shielded from very strong radio transmissions. A hardwired unit is not very portable or practical when the crane is moved often. These units can also use an ear-mounted noise-canceling microphone/headphone combination to minimize the effects of background noise.

6.2.0 Nonverbal Modes of Communication

Nonverbal communication is the most common type of communication used when performing crane operations. There are several nonverbal modes of communication. One mode is the use of signal flags, which may mean different colored

VOICE-OPERATED (VOX) OR PUSH-TO-TALK (PTT) RADIO SYSTEM WITH THROAT MICROPHONE

DUPLEX HARDWIRED SYSTEM WITH IN-LINE AMPLIFIER AND EXTENSION CORDS FOR SIMULTANEOUS TWO-WAY COMMUNICATION

105F20.EPS

Figure 20 Electronic communications systems.

flags or a specific positioning of the flags to communicate the desired message. Another mode is the use of sirens, buzzers, and whistles in which the number of repetitions and duration of the sound convey the message. The disadvantage of these two modes is that there is no established meaning to any of the distinct signals unless they are prearranged between the sender and receiver. Also note that when sirens, buzzers, and whistles are used, background noise levels can be a problem.

The most common mode of communication reference used during crane operations is the *ASME B30.5 Mobile and Locomotive Cranes* hand signals (see *Figures 21* through *37*). In accordance with *ASME B30.5*, the operator must use standard hand signals unless voice communication equipment is utilized. It also requires that the hand signal chart be posted conspicuously at the job site. The advantage to using these hand signals is that they are well established and published in an industry-wide standard. This means that these hand signals are recognized by the industry as the standard hand signals to be used on all job sites. This helps ensure that there is a common core of knowledge and a universal meaning to the signals when lifting operations are conducted. Agreed-upon signals eliminate significant barriers to effective communication.

WARNING

If it is apparent that the operator is not following the signal, immediately signal a stop.

Additions or modifications may be made for operations not covered by the illustrated hand signals, such as deployment of outriggers. The operator and signal person must agree upon these special signals before the crane is operated, and these signals should not be in conflict with any standard signal.

If you need to give instructions verbally to the operator instead of by hand signals, all crane motions must be stopped before doing so.

When a mobile crane is being moved without direction from the signal person, audible travel signals must be given using the crane's horn:

- *Stop* – One audible signal
- *Forward* – Two audible signals
- *Reverse* – Three audible signals

RAISE LOAD OR HOIST UP
Fist up with pointer finger pointing straight up.
Move hand in small horizontal circles.

EXPECTED MACHINE MOVEMENT
The load attached to the block or ball rises vertically, accelerating and decelerating smoothly.

105F21.EPS

Figure 21 Hoist.

LOWER LOAD OR HOIST DOWN
Fist down with pointer finger pointing straight down. Move hand in small horizontal circles.

EXPECTED MACHINE MOVEMENT
The load block or ball smoothly lowers vertically.

105F22.EPS

Figure 22 Lower.

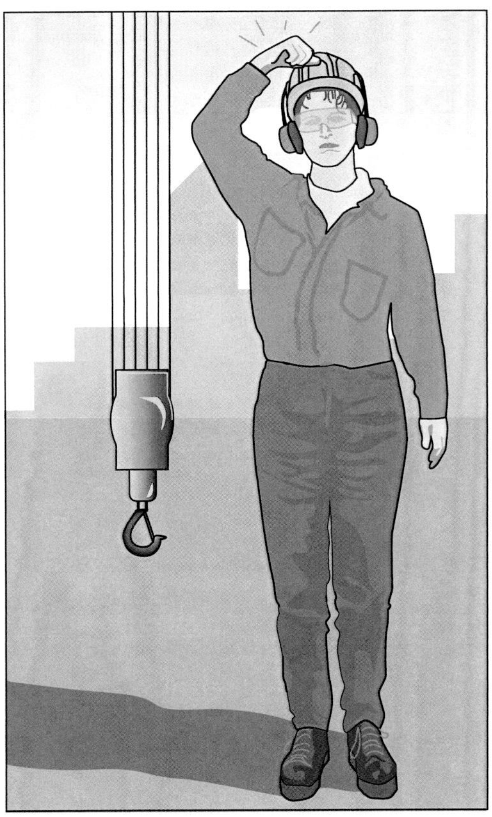

USE MAIN HOIST
Rap on hard hat with closed fist.

EXPECTED MACHINE MOVEMENT
None. This signal is used only to inform the operator that the signal person has chosen the main hoist for the action to be performed as opposed to the auxiliary hoist.

105F23.EPS

Figure 23 Use main hoist.

USE WHIP LINE (AUXILIARY HOIST)
Tap elbow with open palm of one hand, then use regular hand signal.

EXPECTED MACHINE MOVEMENT
None. This signal is used only to inform the operator that the signal person has chosen the auxiliary hoist for the action to be performed as opposed to the main hoist.

105F24.EPS

Figure 24 Use whip line.

RAISE BOOM
Arm extended, fingers closed, thumb pointing upward.

EXPECTED MACHINE MOVEMENT
The boom rises, increasing the hook height and reducing the overall machine height clearance. The operating radius is slowly decreased, thus possibly increasing machine capacity and stability.

105F25.EPS

Figure 25 Raise the boom.

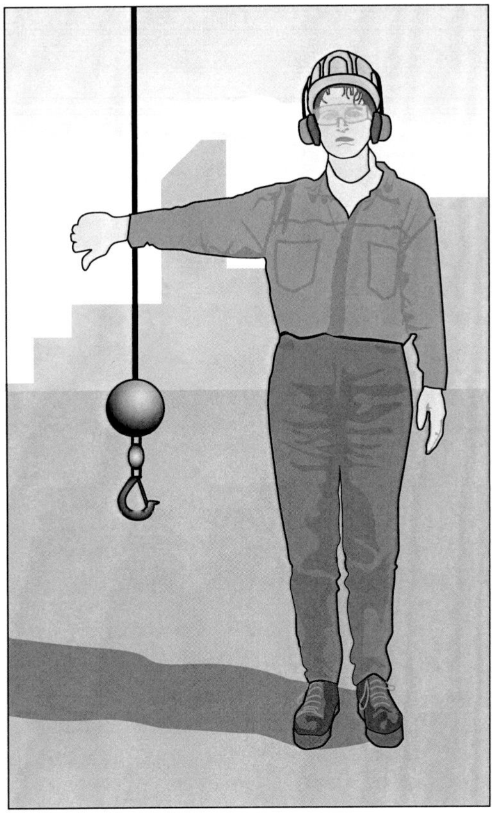

LOWER THE BOOM
Arm extended, fingers closed, thumb pointing downward.

EXPECTED MACHINE MOVEMENT
The boom will lower, decreasing the hook height and reducing the overall machine horizontal clearance. The operating radius is slowly increased, thus possibly decreasing machine capacity and stability.

105F26.EPS

Figure 26 Lower the boom.

MOVE SLOWLY
Use one hand to give any motion signal and place the other hand motionless over the hand giving the motion signal.

EXPECTED MACHINE MOVEMENT
Machine movement will vary depending on the signal being given.

105F27.EPS

Figure 27 Move slowly.

RAISE BOOM AND LOWER LOAD
Arm extended, fingers closed, thumb pointing up, flex fingers in and out as long as load movement is desired.

EXPECTED MACHINE MOVEMENT
The boom rises, reducing overall machine height clearance, as the load moves horizontally toward the crane. The operating radius is slowly decreased, thus possibly increasing machine capacity and stability.

105F28.EPS

Figure 28 Raise the boom and lower the load.

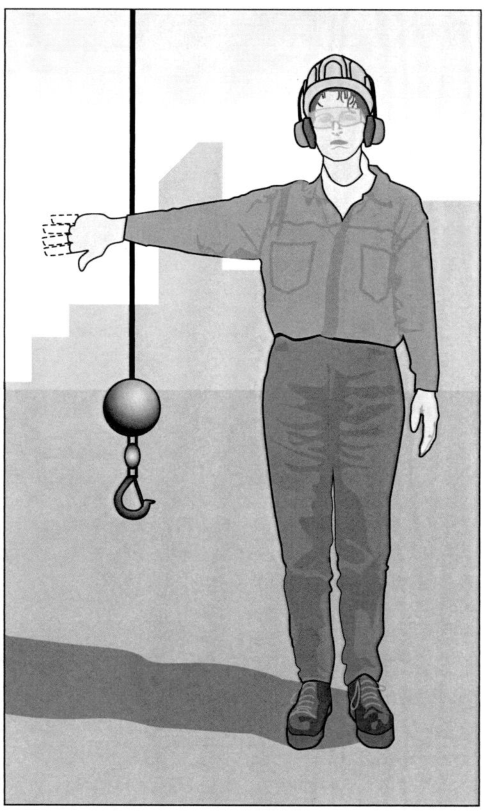

LOWER BOOM AND RAISE LOAD
Arm extended, fingers closed, thumb pointing down, flex fingers in and out as long as load movement is desired.

EXPECTED MACHINE MOVEMENT
The boom lowers, increasing overall machine height clearance, as the load moves horizontally away from the crane. The operating radius is slowly increased, thus possibly reducing both machine capacity and stability.

105F29.EPS

Figure 29 Lower the boom and raise the load.

SWING
Arm extended, point with fingers in direction of boom swing. (Swing left is shown as viewed by the operator.) Use appropriate arm for desired direction.

EXPECTED MACHINE MOVEMENT
The boom moves about the center of rotation with the load (block or ball) swinging in an arc, either toward the right or left, while remaining approximately equidistant to a level plane.

105F30.EPS

Figure 30 Swing.

STOP
Arm extended, palm down, move arm back and forth horizontally.

EXPECTED MACHINE MOVEMENT
None. All movement of the machine ceases.

105F31.EPS

Figure 31 Stop.

EMERGENCY STOP
Both arms extended, palms down, move arms back and forth horizontally.

EXPECTED MACHINE MOVEMENT
None. All movement of the machine ceases.

105F32.EPS

Figure 32 Emergency stop.

DOG EVERYTHING
Clasp hands in front of body.

EXPECTED MACHINE MOVEMENT
None.

105F33.EPS

Figure 33 Dog everything.

TRAVEL (BOTH TRACKS)
Position both fists in front of body and move them in a circular motion, indicating the direction of travel (forward or backward).

EXPECTED MACHINE MOVEMENT
Machine travels in the direction chosen.

105F34.EPS

Figure 34 Travel (both tracks).

TRAVEL (ONE TRACK)
Lock track on side of raised fist. Travel opposite track in direction indicated by circular motion of other fist, rotated vertically in front of body.

EXPECTED MACHINE MOVEMENT
Machine turns in the direction chosen.

105F35.EPS

Figure 35 Travel (one track).

A. TWO-HANDED

EXTENDING BOOM (TELESCOPING BOOM)
Both fists in front of body with thumbs pointing outward.

EXPECTED MACHINE MOVEMENT
Boom sections telescope out. The load radius is increased, possibly decreasing machine capacity and stability. The load (block or ball) rises.

B. ONE-HANDED

EXTENDING BOOM (TELESCOPING BOOM)
One fist in front of body with thumb pointing toward body.

EXPECTED MACHINE MOVEMENT
Boom sections telescope out. The load radius is increased, possibly decreasing machine capacity and stability. The load (block or ball) rises.

105F36.EPS

Figure 36 Extend boom.

RETRACT BOOM (TELESCOPING BOOM)
Both fists in front of body with thumbs pointing toward each other.

EXPECTED MACHINE MOVEMENT
Boom sections retract. The load radius is decreased, possibly increasing machine capacity and stability. The load (block or ball) lowers.

A. TWO-HANDED

RETRACT BOOM (TELESCOPING BOOM)
One fist in front of body with thumb pointing outward.

EXPECTED MACHINE MOVEMENT
Boom sections retract. The load radius is decreased, possibly increasing machine capacity and stability. The load (block or ball) lowers vertically.

B. ONE-HANDED

105F37.EPS

Figure 37 Retract boom.

There are certain requirements that mandate the presence of a signal person. When the operator of the crane can't see the load, the landing area, or the path of motion, or can't judge distance, a signal person is required. A signal person is also required if the crane is operating near power lines or another crane is working in close proximity.

Not just anyone can be a signal person. Signal persons must be qualified by experience, be knowledgeable in all established communication methods, be stationed in full view of the operator, have a full view of the load path, and understand the load's intended path of travel in order to position themselves accordingly. In addition, they must wear high-visibility gloves and/or clothing, be responsible for keeping everyone out of the operating radius of the crane, and never direct the load over anyone.

WARNING

Do not give signals to the crane operator unless you have been designated as the signal person.

Although personnel involved in lifting operations are expected to understand these signals when they are given, it is acceptable for a signal person to give a verbal or nonverbal signal to an operator that is not part of the *ASME B30.5* standard. In cases where such non-standardized signals are given, it is important that both the operator and the signal person have a complete understanding of the message that is being sent.

SUMMARY

A crane is a piece of hoisting or lifting equipment designed to pick up a load, transport it, and land it again. Mobile cranes include the crawler crane, the truck crane, and the rough-terrain crane. Each type has advantages that are taken into account when planning lifting operations at a construction site. Economics is an important factor governing crane selection for specific job applications. Manufacturers design cranes according to strict engineering and safety principles. Because of this, operators and riggers should refer to their specifications when setting up, lifting, or traveling to the various units. Safety is the single most important issue during crane lifting operations. All appropriate safety precautions and procedures must be strictly observed at all times.

One of the important tasks a rigger performs is guiding the crane operator during a lift. In some instances, radio communications will be used. In others, it will be necessary to use the *ASME B30.5* consensus standard hand signals. Both the rigger and operator must know these signals by heart.

Review Questions

1. The up or down movement of a derrick boom is controlled by the _____.
   a. luffing system
   b. hoisting system
   c. power source
   d. boom

2. Truck-type crane carriers are cranes that are mounted on _____.
   a. ordinary commercial truck frames
   b. frames designed and manufactured for them
   c. special military surplus truck frames
   d. the right type of tracked vehicles

3. The crawler assembly of a crawler crane is called the _____.
   a. base or base mounting
   b. swing circle
   c. carbody
   d. deck

4. The lowest section of a telescopic boom is called the _____.
   a. boom base
   b. carbody
   c. base section
   d. base mounting

5. The pulling force that can be achieved in the rope leading from a drum is called _____.
   a. line pull
   b. loading
   c. luffing
   d. parts of line

6. What should be the minimum clearance between the counterweight of a crane and the nearest obstacle?
   a. 2 feet
   b. 52 feet
   c. 122 feet
   d. 202 feet

7. When using a crane to move material from place to place, the boom should be _____.
   a. kept at a high angle
   b. transported as quickly as possible
   c. at right angles to the direction of motion
   d. carried in line with the direction of motion

8. The weight of a boom section used on a crane can be found in the crane's _____.
   a. bill of sale
   b. operations log
   c. manuals or handbooks
   d. preventive maintenance log

9. The crane log book should remain with the machine for _____.
   a. 1 year
   b. 5 years
   c. 20 years
   d. the working life of the machine

10. Wire rope used on cranes must be inspected for broken wires _____.
    a. on each shift
    b. daily
    c. weekly
    d. monthly

11. The main advantage of a hardwired communication system is _____.
    a. low cost
    b. portability
    c. less radio interference
    d. ease of use

12. What action is signaled by tapping your fist on your head, then using regular hand signals?
    a. Use main hoist
    b. Lower
    c. Use whip line
    d. Raise boom and lower load

13. Which of the following is a mode of nonverbal communication?
    a. Buzzer
    b. Portable radio
    c. Bullhorn
    d. Hardwired radio

14. When the signal person gives the Dog Everything signal, the operator is expected to _____.
    a. turn the machine
    b. travel the machine
    c. retract the boom sections
    d. make no machine movement

15. If it is apparent that the crane operator has misunderstood a hand signal, the signal person should _____.
    a. climb up on the cab and knock on the door
    b. find a supervisor
    c. signal the operator to stop
    d. get away from the crane as quickly as possible

Trade Terms Introduced in This Module

Carbody: The part of a crawler crane base mounting that carries the rotating upper structure.

Counterweights: Weights added to the opposite side of the crane boom to offset the weight of the load.

Dog: A safety mechanism to prevent operation.

Drum: The part of a crane that holds the unused lifting line and that transmits the line pull to the lifting line.

List angle: The tipping angle measured from the level of a floating crane.

Luffing: Changing a boom angle by varying the length of the boom suspension ropes.

Moment: The rotation produced in a body when a force is applied; used to determine tipping load.

Outrigger: An extension that projects from the main body of a crane to add stability and support.

Swing circle: The place where the upper mechanisms of a crane are joined to the base.

Additional Resources

This module presents thorough resources for task training. The following resource material is suggested for further study.

Crane Safety: A Guide to OSHA Compliance and Injury Prevention. Carl O. Morgan, Ph.D. Lanham, MD: Rowan & Littlefield Publishing Group, 1999.

OSHA Standards for the Construction Industry (29 CFR Part 1926). Latest edition. Washington, DC: Occupational Safety & Health Administration.

http://www.osha.gov/pls/publications/publication.html.

Figure Credits

Morrow Equipment Co., L.L.C., Module opener

Gorbel, Inc., Figure 1 (jib)

LPR Construction, Figure 1 (guy derrick)

Link-Belt Construction Equipment Co., Figures 2, 4, and 5

AEARO Company, Figure 20

NCCER CURRICULA — USER UPDATE

NCCER makes every effort to keep its textbooks up-to-date and free of technical errors. We appreciate your help in this process. If you find an error, a typographical mistake, or an inaccuracy in NCCER's curricula, please fill out this form (or a photocopy), or complete the online form at **www.nccer.org/olf**. Be sure to include the exact module ID number, page number, a detailed description, and your recommended correction. Your input will be brought to the attention of the Authoring Team. Thank you for your assistance.

Instructors – If you have an idea for improving this textbook, or have found that additional materials were necessary to teach this module effectively, please let us know so that we may present your suggestions to the Authoring Team.

NCCER Product Development and Revision

13614 Progress Blvd., Alachua, FL 32615

Email: curriculum@nccer.org
Online: www.nccer.org/olf

❑ Trainee Guide ❑ Lesson Plans ❑ Exam ❑ PowerPoints Other ______________________

Craft / Level: Copyright Date:

Module ID Number / Title:

Section Number(s):

Description:

Recommended Correction:

Your Name:

Address:

Email: Phone:

Rigging Equipment

30106-11

Trainees with successful module completions may be eligible for credentialing through NCCER's National Registry. To learn more, go to **www.nccer.org** or contact us at **1.888.622.3720**. Our website has information on the latest product releases and training, as well as online versions of our *Cornerstone* newsletter and Pearson's Contren® product catalog.

Your feedback is welcome. You may email your comments to **curriculum@nccer.org,** send general comments and inquiries to **info@nccer.org**, or use the User Update form at the back of this module.

V.1 1/11

30106-11
Rigging Equipment

Objectives

When you have completed this module, you will be able to do the following:

1. Identify and describe the uses of common rigging hardware and equipment.
2. Perform a safety inspection on hooks, slings, and other rigging equipment.
3. Describe common slings, and determine sling capacities and angles.
4. Select, inspect, use, and maintain special rigging equipment, including:
   - Block and tackle
   - Chain hoists
   - Come-alongs
   - Jacks
   - Tuggers
   - Wire rope
   - Chain
5. Inspect heavy rigging hardware.
6. Tie knots used in rigging.

Performance Tasks

Under the supervision of the instructor, the trainee should be able to do the following:

1. Perform a safety inspection on hooks, slings, and other rigging equipment.
2. Select, inspect, and use special rigging equipment, including:
   - Block and tackle
   - Chain hoists
   - Come-alongs
   - Jacks
   - Tuggers
   - Wire rope
   - Chain
3. Tie knots used in rigging.

Trade Terms

Anneal
Bird caging
Bull's tail (Christmas tree)
Equalizer beam
Fixed block
Hauling line
Hydraulic
Kinking
Parts of line
Seize
Shackles
Sling angle
Spreader beams
Wire rope

Contents

Topics to be presented in this module include:

Figures and Tables

1.0.0 INTRODUCTION

Rigging is the planned movement of an object from one place to another (*Figure 1*). However, rigging does not necessarily apply to the actual operation of a crane. Rigging may also include moving an object from one position to another or tying down an object. For example, tools and equipment must often be lifted several floors to get them where they are needed. This moving operation must be done safely and effectively.

The rigger and the crane operator work together in preparing a lift. The rigger must make sure that the crane operator understands what is to be lifted, what it weighs, and any complexities involved in the lift. At the same time, the rigger must be aware of the crane's capabilities and limitations in order to ensure that it is able to safely make the lift. As a team, the rigger and crane operator should examine the area of the lift for any potential safety hazards, such as power lines and structures, as well as any possible risks to nearby workers.

2.0.0 RIGGING HARDWARE

Hardware used in rigging includes hooks, **shackles**, eyebolts, **spreader beams** and **equalizer beams**, and blocks. These hardware items must be carefully matched to the lifting capabilities of the crane or hoist and the slings used in each application. Mixing items that have different safe working loads makes it impossible to determine if the lift will be safe. Careful inspection and maintenance of all lifting hardware are essential for continuous operation. Hardware should always be inspected before each use.

106F01.EPS

Figure 1 A rigging application.

> **NOTE**
>
> All rigging equipment must be properly marked or tagged with the manufacturer's ratings and capacity information. If a piece of equipment is not properly marked, do not use it. Also, rigging hardware must never be modified. Modifications could affect the hardware's capacity.

2.1.0 Hooks

A rigging hook can be attached directly to a sling or to a load. There are six basic types of rigging hooks. The eye-type hook is the most commonly used. Rigging hooks are equipped with safety latches to prevent a sling attachment from coming off the hook when a sling is slackened. The capacity of a rigging hook, in tons, is determined by its size and physical dimensions. Information about a specific hook's capacity is readily available from the hook manufacturer.

The safe working load of the hook is accurate only when the load is suspended from the saddle of the hook. If the working load is applied anywhere between the saddle and the hook tip, the safe working load will be considerably reduced, as shown in *Figure 2*. The hash marks show the places where different proportions of the load may safely be applied. Some manufacturers do not permit their hooks to be used in any position other than with the load suspended from the saddle.

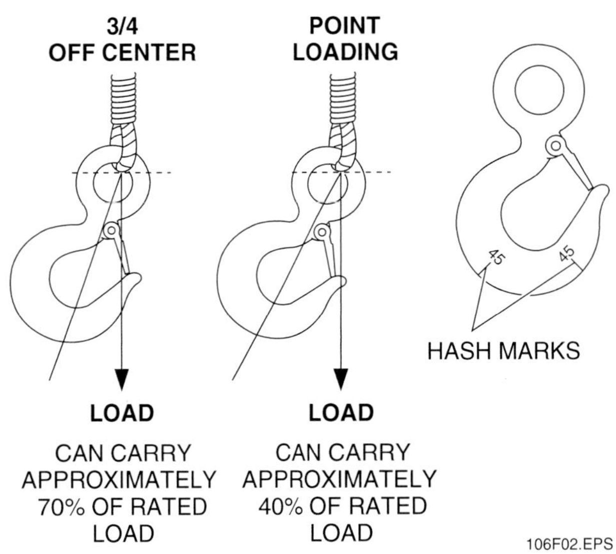

106F02.EPS

Figure 2 Rigging hook rated load versus load location.

Always inspect hooks before each use. Look for wear in the saddle of the hook. Also look for cracks, severe corrosion, and twisting of the hook body. The safety latch should be in good working order. Measure the hook throat opening. If a hook has been overloaded or if it is beginning to weaken, the throat will open. OSHA regulations require the hook to be replaced if the throat has opened 15 percent from its original size or if the body has been twisted 10 degrees from its original plane. *Figure 3* shows the location of the inspection points for a hook.

Never use a sling eye over a hook that has a body diameter larger than the natural width of the eye. Never force the eye onto the hook. Always use an eye with at least the nominal diameter of the hook.

2.2.0 Shackles

A shackle is used to attach an item to a load or to attach slings together. It can also be used to attach the end of a **wire rope** to an eye fitting, hook, or other type of connector. Shackles are made in several configurations (*Figure 4*).

Shackles used for overhead lifting are made of forged alloy steel. They are sized by the diameter of the steel in the bow section rather than the pin size. When using shackles, make sure that all pins are straight, all screw pins are completely seated, and cotter pins are used with all round pin shackles. A shackle must not be used in excess of its rated load. Flat pin shackles are most commonly used in marine applications.

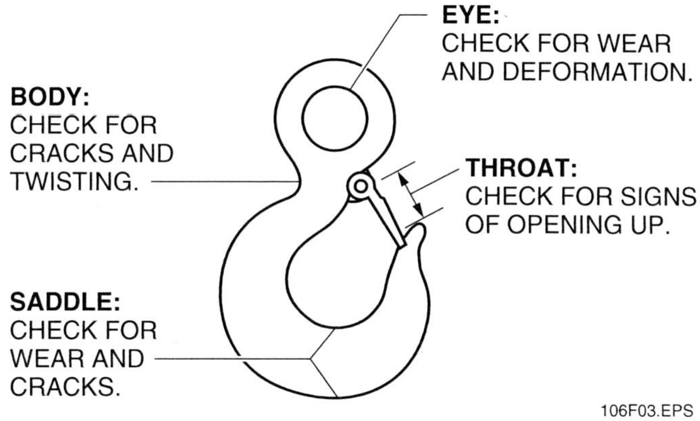

Figure 3 Rigging hook inspection points.

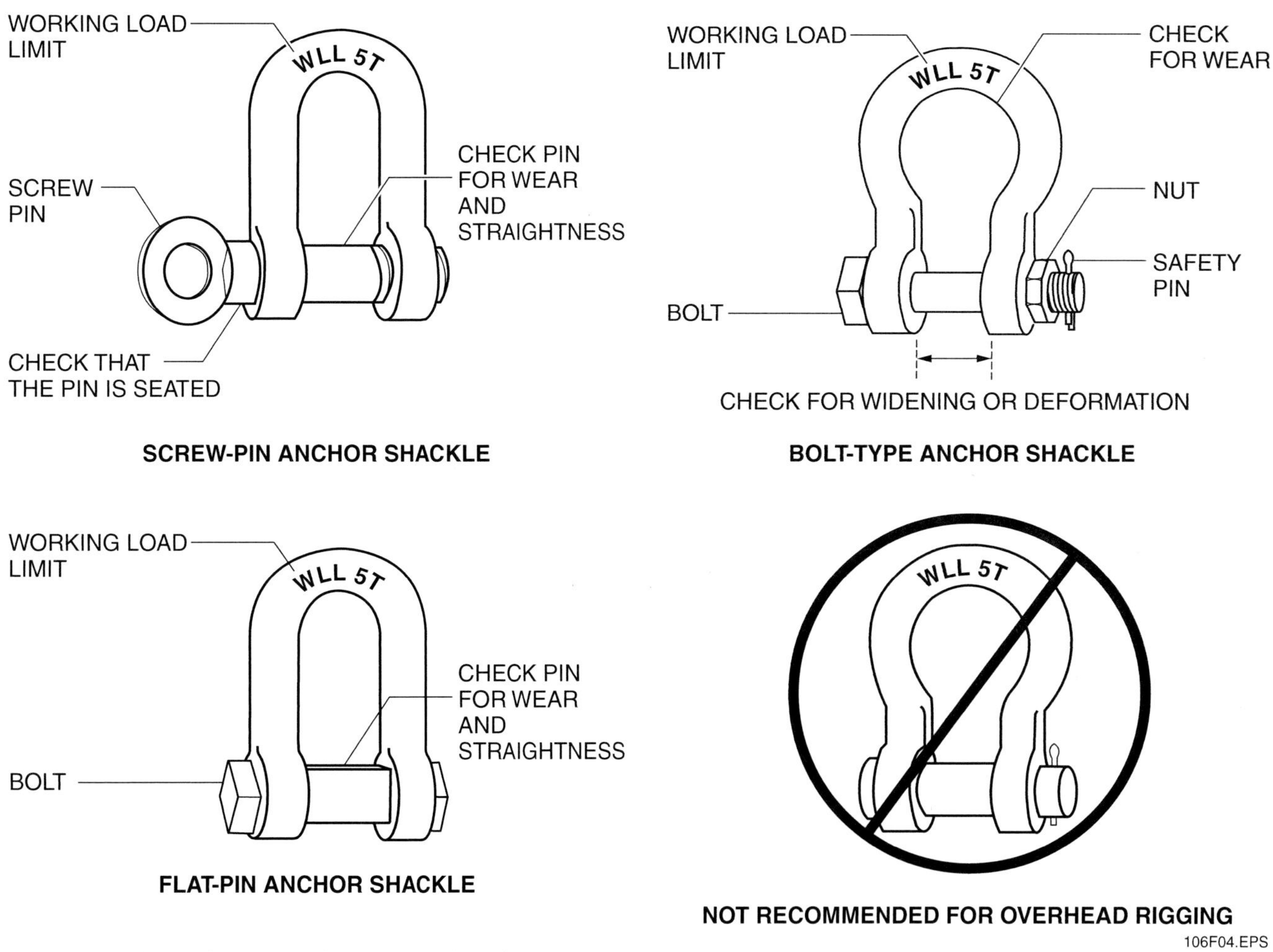

Figure 4 Round pin, flat pin, and screw pin shackle.

Shackle pins should never be replaced with a bolt. Bolts cannot take the bending force normally applied to a pin. Shackles that are stretched or that have crowns or pins worn more than 10 percent should be destroyed and replaced. If a shackle is pulled out at an angle, the capacity is reduced. Use spacers to center the load being hoisted on the pin. Only shackles of suitable load ratings can be used for lifting. When using a shackle on a hook, the pin of the shackle should be hung on the hook. Spacers can be used on the pin to keep the shackle hanging evenly on the hook (*Figure 5*). Never use a screw pin shackle in a situation where the pin can roll under load (*Figure 6*).

WARNING

Shackles may not be loaded at an angle unless approved by the manufacturer. Contact the manufacturer for any angular loading.

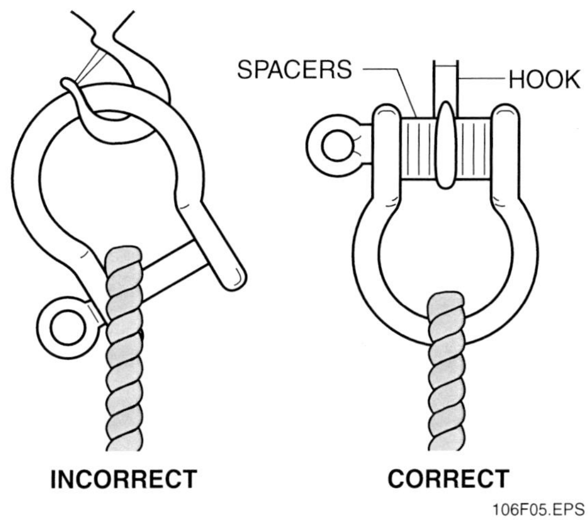

Figure 5 Spacers used with a shackle to keep it hanging evenly.

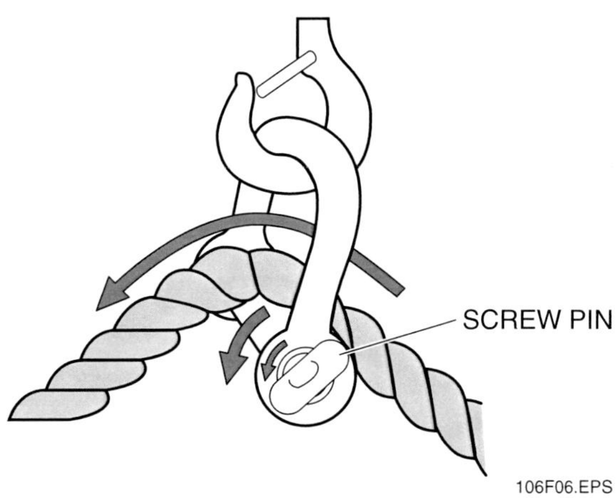

Figure 6 Example of how a rope can cause a shackle screw pin to roll under load.

2.3.0 Eyebolts

Eyebolts (*Figure 7*) are often attached to heavy loads by a manufacturer to aid in hoisting the load. One type of eyebolt, called a ringbolt, is equipped with an additional movable lifting ring. Eyebolts and ringbolts can be either the shoulder or shoulderless type. The shoulder type is recommended for use in hoisting applications because it can be used with angular lifting pulls. The shoulderless type is only designed for lifting loads vertically. The safe working load of shoulder eyebolts and ringbolts is reduced with angular loading. Loads should always be applied to the plane of the eye to reduce bending. This procedure is especially important when bridle slings are used.

When installed, the shoulder of the eyebolt must be at right angles to the axis of the hole and must be in full contact with the working surface when the nuts are properly fastened. Washers or other suitable spacers may be used to ensure that the shoulders are in firm contact with the working surface. Tapped holes used with screwed-in eyebolts should have a minimum depth of 1½ times the bolt diameter. Swivel eyebolts may be installed instead of fixed eyebolts. These devices swivel to the desired lift position and do not require any load rating reduction.

2.4.0 Lifting Lugs

Lifting lugs (*Figure 8*) are typically welded, bolted, or pinned by a manufacturer to the object to be lifted. They are designed by engineers and located to balance the load and support it safely. Lifting lugs should be used for straight, vertical lifting only, unless specifically designed for angular loads. This is because they can bend if loaded at an angle.

WARNING

All lifting lugs must be designed and approved by an engineer. They must not be modified.

Lifting lugs should be inspected before each use for deformation, cracks, corrosion, and defective welds. A lifting lug must be removed from service if any of these conditions is found.

2.5.0 Turnbuckles

Turnbuckles are available in a variety of sizes. They are used to adjust the length of rigging connections. Three common turnbuckles are the eye, jaw, and hook ends (*Figure 9*). Turnbuckles

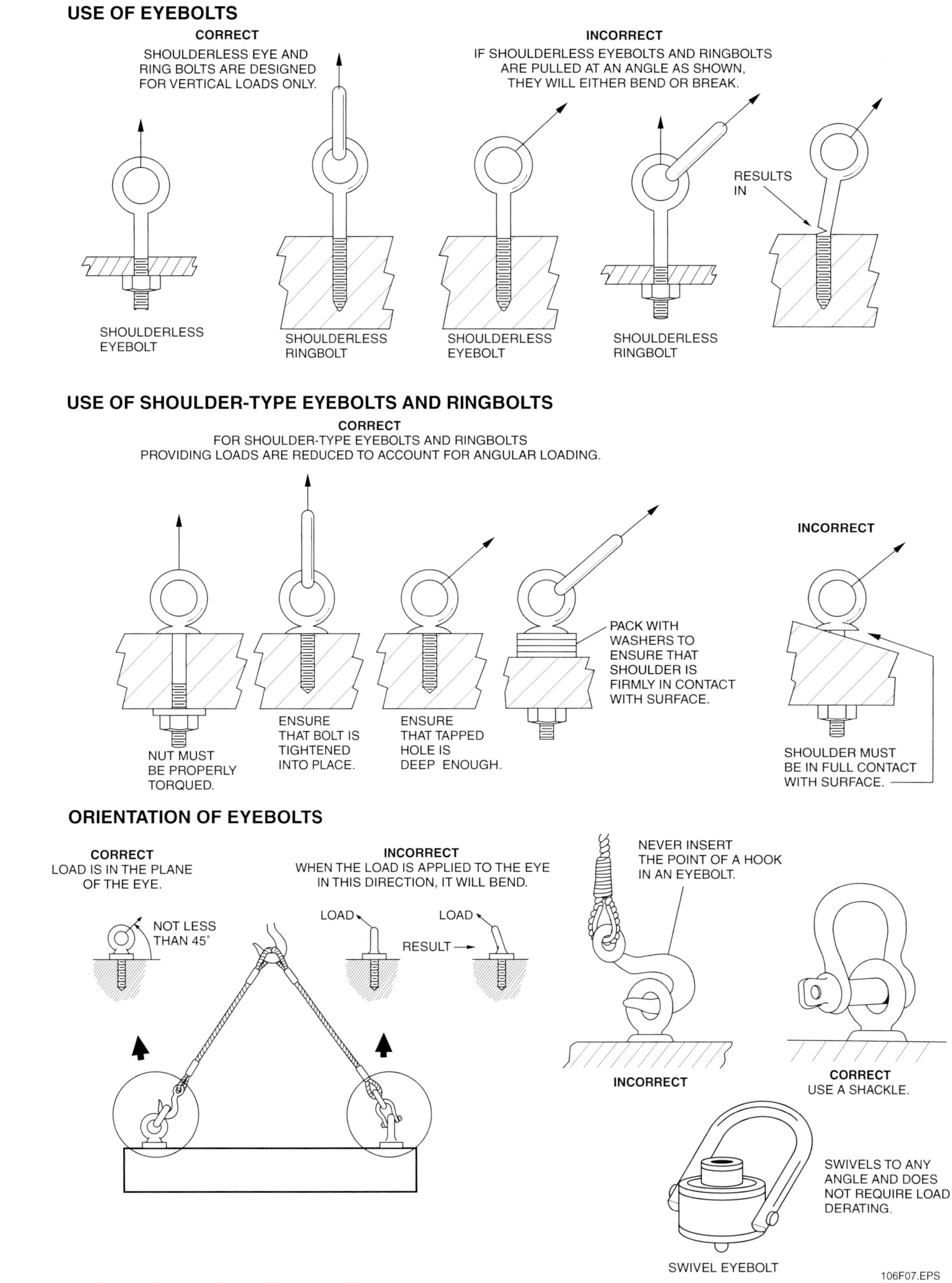

Figure 7 Eyebolt and ringbolt installation and lifting criteria.

can be used in any combination. The safe working load for turnbuckles is based on the diameter of the threaded rods. The safe working load can be found in the manufacturer's catalog. The safe working load of turnbuckles with hook ends is less than that of the same size turnbuckles with other types of ends. The ratings of turnbuckles must be known before they are used.

Consider the following guidelines when selecting turnbuckles for rigging:

- Turnbuckles should be made of alloy steel and should not be welded.
- When using turnbuckles with multi-legged slings, do not use more than one turnbuckle per leg.
- Do not use jamb nuts on turnbuckles that do not come equipped with them.
- Turnbuckles should not be overtightened. Do not use a cheater to tighten them. Perform tightening with a wrench of the proper size, using as much force as can be achieved by hand.
- Turnbuckles must not be used for angular loading unless permitted by the manufacturer, or if the angular loading has been calculated by an engineer.

Turnbuckles should be inspected as shown in *Figure 10*. If a turnbuckle is damaged, remove it from service.

2.6.0 Beam Clamps

Beam clamps (*Figure 11*) are used to connect hoisting devices to beams so that the beams can be lifted and positioned in place. The following are guidelines for using beam clamps:

- Do not use homemade clamps unless they are designed, load tested, and stamped by an engineer.
- Ensure that the clamp fits the beam and is of the correct capacity.
- Make sure that beam clamps are securely fastened to the beam.

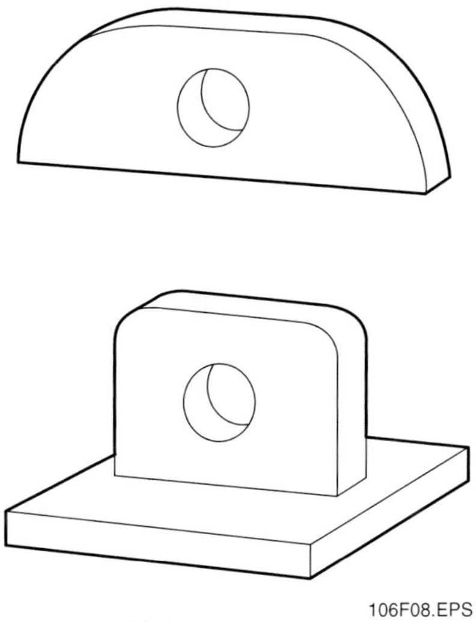

Figure 8 Two types of lifting lugs.

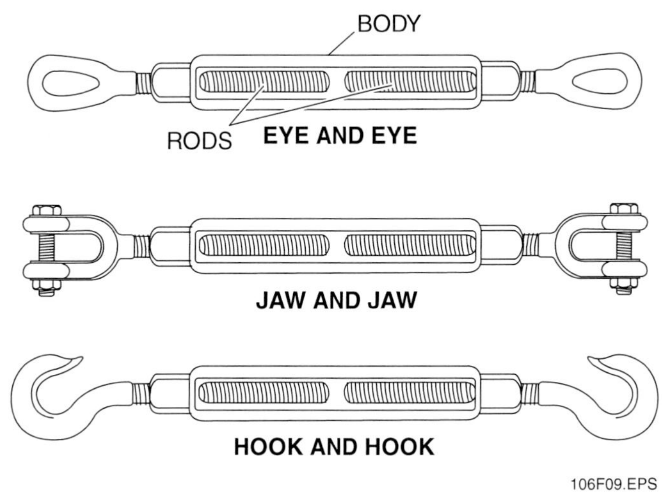

Figure 9 Turnbuckles.

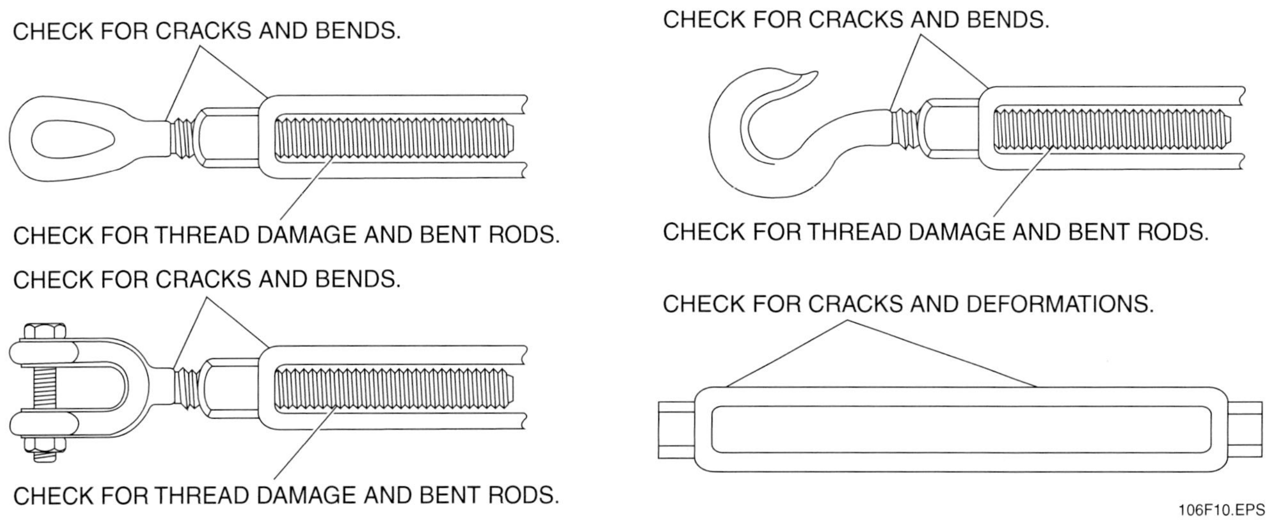

Figure 10 Turnbuckle inspection point.

106F11.EPS

Figure 11 Typical beam clamp application.

- Be careful when using beam clamps in situations where angle lifts will be made. Most beam clamps are designed for essentially straight vertical lifts only. However, some manufacturers allow two clamps to be used with a long bridle sling if the angle of lift does not exceed 25 degrees from the vertical.
- Verify that the capacity appears on the beam clamp.
- Do not place a hoist hook directly in the beam clamp's lifting eye.
- Never overload a beam clamp beyond its rating capacity.

Beam clamps must be inspected before each use and removed from service if any of the following conditions is observed:

- The jaws of the beam clamp have been opened more than 15 percent of their normal opening.
- The lifting eye is bent or elongated.
- The lifting eye is worn.
- The capacity and beam size are unreadable.

2.7.0 Plate Clamps

Plate clamps (*Figure 12*) attach to structural steel plates to allow for easier rigging attachment and handling of the plate. The two basic kinds of plate clamps are the serrated jaw type and the screw type.

Serrated clamps are designed to grip a single plate for hoisting. They are available with a locking device. Serrated clamps are used for vertical lifting. Screw clamps are considered the safer of the two types. They rely on the clamping action of a screw against the plate to secure it. Screw clamps can be used from a horizontal position through 180 degrees.

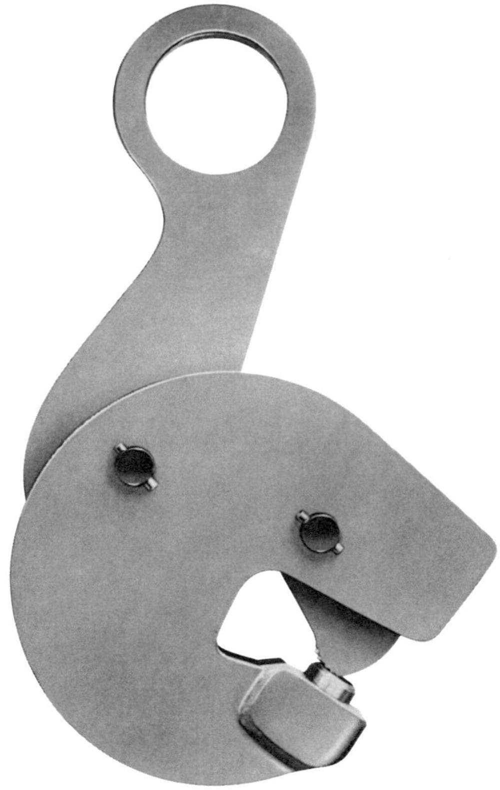

106F12.EPS

Figure 12 Standard lifting clamp.

Plate clamps are designed to lift only one plate at a time. Always follow the manufacturer's recommendations for use and safe working load.

Remove plate clamps from service if any of the following conditions is present during inspection:

- Changes in the opening at the jaw plate or wear of cam teeth
- Cracks in the body
- Loose or damaged rivets
- Worn, bent, or elongated lifting eye
- Excessive rust or corrosion
- Unreadable capacity size

2.8.0 Rigging Plates and Links

Rigging plates and links (*Figure 13*) are made for specific uses. The holes in the plates or links may be of various sizes, and they may be placed in different locations in the plates for different weights and types of lifts. Plates with two holes are called rigging links. Plates with three or more holes are called equalizer plates. Equalizer plates can be used to level loads when the legs of a sling are unequal. Plates are attached to the rigging with high-strength pins or bolts.

Inspect rigging plates and links, and remove them from service if any of the following conditions is present:

- Cracks in body
- Worn or elongated lifting eye
- Excessive rust or corrosion

2.9.0 Spreader and Equalizer Beams

Spreader beams (*Figure 14*) are used to support long loads during lifting operations. Spreader beams must be tagged for capacity. When used properly, spreader beams help to eliminate the hazards of load tipping, sliding, or bending. They reduce low **sling angles** and the tendency of the slings to crush loads. Equalizer beams are used to balance the load on sling legs and to maintain equal loads on dual hoist lines when making tandem lifts. *ANSI/ASME B30.20* applies to spreader and equalizer beams. Documentation of proof loading must be on site.

Lifting beams, spreader beams, and spider frames (frames that allow the crane to lift circular or rectangular structures) all require engineering calculations. They must be used as they have been designed to be used. A **bull's tail (Christmas tree)** is a special rig consisting of a pair of lines with hooks attached, allowing a series of beams or joists to be lifted in the sequence they are to be hung. The hooks are set at the same intervals along the lines, keeping the structural members level as they are lifted and placed.

Spreader and equalizer beams are usually fabricated to suit a specific application. They are often made of heavy pipe, I-beams, or other suitable material. Custom-fabricated spreader and equalizer beams must be designed by an engineer and must have their capacity clearly stamped on the side. They should be tested at 125 percent of rated capacity. Information related to the beams should be kept on file. The capacity of beams designed for use with multiple attachment points depends on the distance between attachment points.

Before use, a spreader or equalizer beam should be inspected for the following:

- Solid welds
- No cracks, nicks, gouges, or corrosion
- Condition of the attachment points
- Capacity rating
- Sling angle tension at the points of attachment
- Documentation of load rating

3.0.0 Slings

Common sling types include wire rope slings, synthetic slings, chain slings, and metal mesh slings. Metal mesh slings are typically made of wire or chain mesh. They are similar in appearance to web slings. Metal mesh slings are suited for situations in which the loads are abrasive, hot, or tend to cut other types of slings. They resist abrasion and cutting, grip the load firmly without stretching, and can withstand temperatures up to 550°F. They are also smooth, conform to irregular shapes, do not kink or tangle, and resist corrosion.

106F14.EPS

Figure 14 Typical use of a spreader beam.

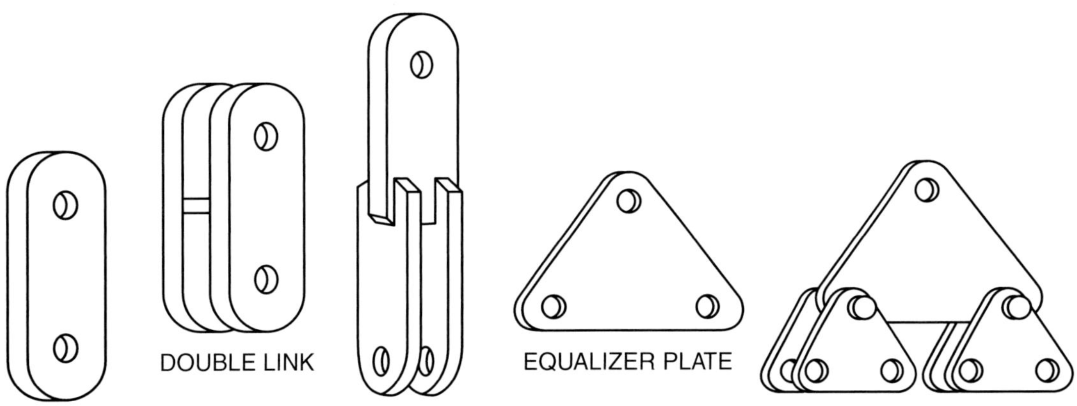

106F13.EPS

Figure 13 Rigging plates and links.

Metal mesh slings are available in several mesh sizes. They can be coated with a variety of substances, such as rubber or plastic, to help protect the load they are handling.

3.1.0 Sling Capacity

Sling capacity depends on the sling material, construction, size of the hitch configuration, number of slings, and the angle for the specific type of sling being used. The amount of tension on the sling is directly affected by the angle of the sling (*Figure 15*). For this reason, proper sling angles are crucial to safe rigging. This information, along with other important rigging information, is available from rigging equipment manufacturers and rigging trade organizations in the form of easy-to-use pocket guides, such as those shown in *Figure 16*. Sling capacity information is also given in *OSHA Standard 29 CFR 1926.251, Rigging Equipment for Material Handling*. *Table 1* shows a typical capacity table used with wire rope slings.

WARNING

Never use a sling that is untagged.

Going from left to right, the vertical columns in *Table 1* provide the following information: the International Wire Rope Class; the size of the wire rope; the allowable load for a vertical lift over the eyebolt; the allowable load for a choker on the eye; the allowable loads for the given angle of the basket hitch; and the eye dimensions to be used.

The D/d ratio applies to wire rope slings. It refers to the relationship between the diameter of the surface (D) that the sling is around and the diameter of the sling (d). This relationship, specifically the severity of the bends in the sling, has an impact on the capacity of the sling. Efficiency charts have been developed to help determine the actual sling capacity. *Figure 17* shows load and hook diameter examples and a D/d ratio chart.

For example, consider a 1" diameter rope sling in a basket hitch around a 2" shackle pin. The capacity of the sling in the basket hitch is listed as 17 tons. Look across the bottom of the chart for the 2/1 ratio, and find the line that meets the ratio curve. It is opposite 65 percent efficiency. Multiply the rated load of 17 tons by 65 percent and you will get an actual capacity of 11 tons. Sling eyes with pin or body sizes that are at least equal to the rope diameter should be used with shackles and hooks. Always refer to the tag when selecting a wire rope sling.

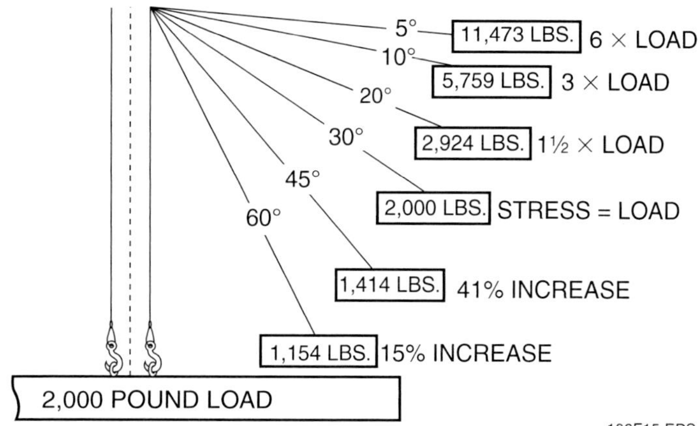

Figure 15 Sling angles.

Figure 16 Rigging pocket guides.

3.2.0 Sling Care and Storage

The inspection criteria for slings are covered in *Basic Rigging*. The following are some important reminders for sling inspection, care, and storage:

- Store slings in a rack to keep them off the ground. The rack should be in an area free of moisture and away from acid or acid fumes and extreme heat.
- Never let slings lie on the ground in areas where heavy machinery could run over them.
- Slings should be inspected at each use for broken wires, kinks, rust, or damaged fittings. Any slings found to be defective should be disposed of.

Table 1 Example of a Wire Rope Sling Capacity Table

Wire Rope EIPS/IWRC	Mechanical Splice in Pounds					
Size in Inches			90°	60°	45°	30°
¼	1,300	960	2,600	2,200	1,820	1,300
5/16	2,000	1,480	4,000	3,400	2,800	2,000
⅜	2,800	2,200	5,600	5,000	4,000	2,800
7/16	3,800	2,800	7,600	6,800	5,400	3,800
½	5,000	3,800	10,000	8,800	7,200	5,000
9/16	6,400	4,800	12,800	11,000	9,000	6,400
⅝	7,800	5,800	15,600	13,600	11,000	7,800
¾	11,200	8,200	22,400	19,400	15,800	11,200
⅞	15,200	11,200	30,400	26,000	22,000	15,200
1	19,600	14,400	39,200	34,000	28,000	19,600
1⅛	24,000	18,000	48,000	42,000	34,000	24,000
1¼	30,000	22,500	60,000	52,000	42,000	30,000
			MULTIPLIER ⇨	1.00	0.75	0.60

Sling Length = Total distance between pick points × multiplier.

106T01.EPS

3.2.1 Wire Rope Slings

If in doubt, wire rope slings (*Figure 18*) should be visually examined before every use to make sure they are free from damage. Slings must be discarded if any of the following conditions is discovered:

- *Broken wires* – Report to a supervisor if you find any broken wires in one rope lay, or any broken wire at any fitting of a wire rope sling.
- *Diameter reduction* – Some wear is normal, but if the wear exceeds one-third of the original rope diameter, the sling must be destroyed. This type of wear is often caused by excessive abrasion of the outside wires, loss of core support, internal or external corrosion, inner wire failure, loosening of the rope lay, or stretching due to overloading.
- *Heat damage* – This is usually indicated by the discoloration and pitting associated with high temperature sources.
- *Corrosion* – Any noticeable rusting, pitting, or discoloration indicates corrosion in the rope or at the end fittings. Corrosion around end fittings is typically associated with broken wires near those end fittings. Corrosion can be difficult to detect when it develops internally in areas hidden from outside inspection.
- *Damaged end fittings* – This type of damage is easily detected. It includes cracks, gouges, nicks, severe corrosion, and evidence that the end fitting is creasing into the rope.
- *Rope distortion* – This includes **kinking**, crushing, and **bird caging**.

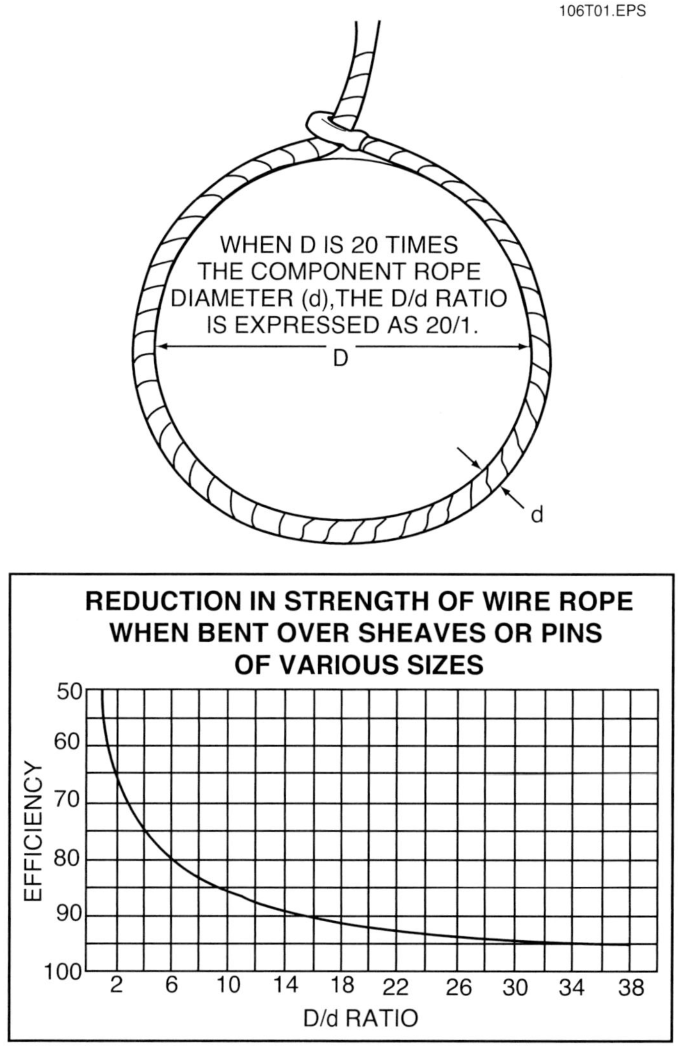

106F17.EPS

Figure 17 Load and hook diameter and D/d ratio chart for wire rope.

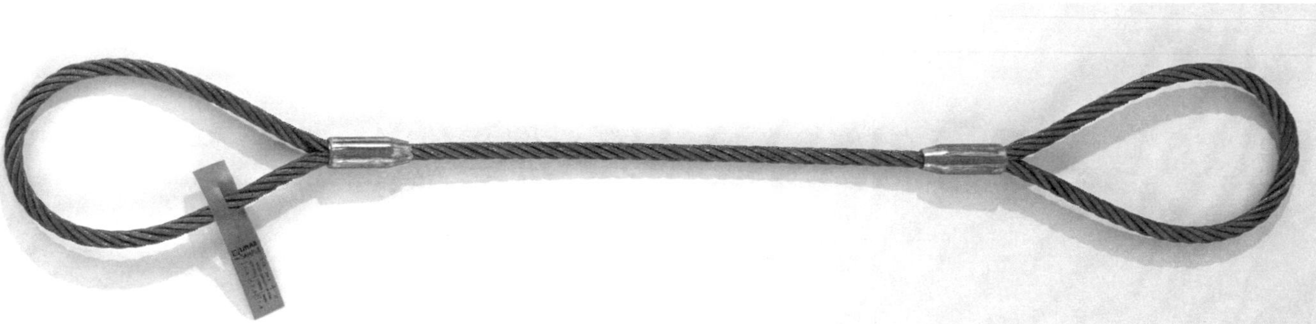

106F18.EPS

Figure 18 Wire rope sling.

- *Core protrusion* – This damage consists of the core sticking out from between the strands or the strands separating and exposing the core.
- *Unlaying of a splice* – The unraveling of a splice.

3.2.2 Synthetic Web and Round Slings

When inspecting synthetic and round slings, check for the following types of damage or missing items. If any of these conditions is found, discard the slings immediately:

- Acid or caustic burns
- Melting or charring of any part of the sling
- Holes, cuts, tears, or snags
- Broken or worn stitching in loadbearing splices
- Excessive abrasive wear (to the point where the colored yarns are showing)
- Knots in any part of the sling
- Excessive pitting or corrosion, or cracked, distorted, or broken fittings
- Other visible damage that causes doubt as to the strength of the sling
- Missing or illegible tag

3.2.3 Metal Mesh Slings

Although metal mesh slings withstand abrasive or hot loads, they are still vulnerable to damage. Metal mesh slings must be discarded if any of the following conditions is detected during inspection:

- A broken welded or brazed joint along the sling edge
- A broken wire in any part of the mesh
- A reduction in wire diameter of 25 percent due to abrasion or 15 percent from corrosion
- Lack of flexibility caused by distortion of the mesh
- Distortion of the choker fitting so that the depth of the slot is increased by 10 percent
- Distortion of either end fitting so that the width of the eye opening is decreased by more than 10 percent
- A 15 percent reduction in the original cross-sectional area of metal at any point around the sling

3.3.0 Chain Slings

Some jobs are better suited for chain slings than for wire rope or web slings. The use of chain slings is recommended when lifting rough castings that would quickly destroy wire slings by bending the wires over rough edges. Chain slings are used in high-heat applications or where wire chokers are not suitable. They are also used for dredging and other marine work because they withstand abrasion and corrosion better than cable. Information about sling angles and sling capacities, described earlier for rope and web slings, also applies to chain slings. *Figure 19* shows some common configurations of chain slings and hooks.

A chain link consists of two sides. The failure of either side causes the link to open and drop the load. Wire rope is frequently composed of as many as 114 individual wires, all of which must fail before it breaks. Chains have less reserve strength and should be more carefully inspected. Chains must never be used for routine lifting operations.

Chains stretch under excessive loading. This causes elongating and narrowing of the links until they bind on each other, giving visible warning signs. When overloading is severe, the chain fails with less warning than a wire rope. When a weld breaks, there is little, if any, warning.

WARNING

Chains must be carefully inspected for weak or damaged links. The failure of one link can cause the entire chain to fail.

Chains used as slings should be **annealed** every year. After being annealed six times, the chain should be destroyed. Steel chains should not be heat treated after leaving the factory.

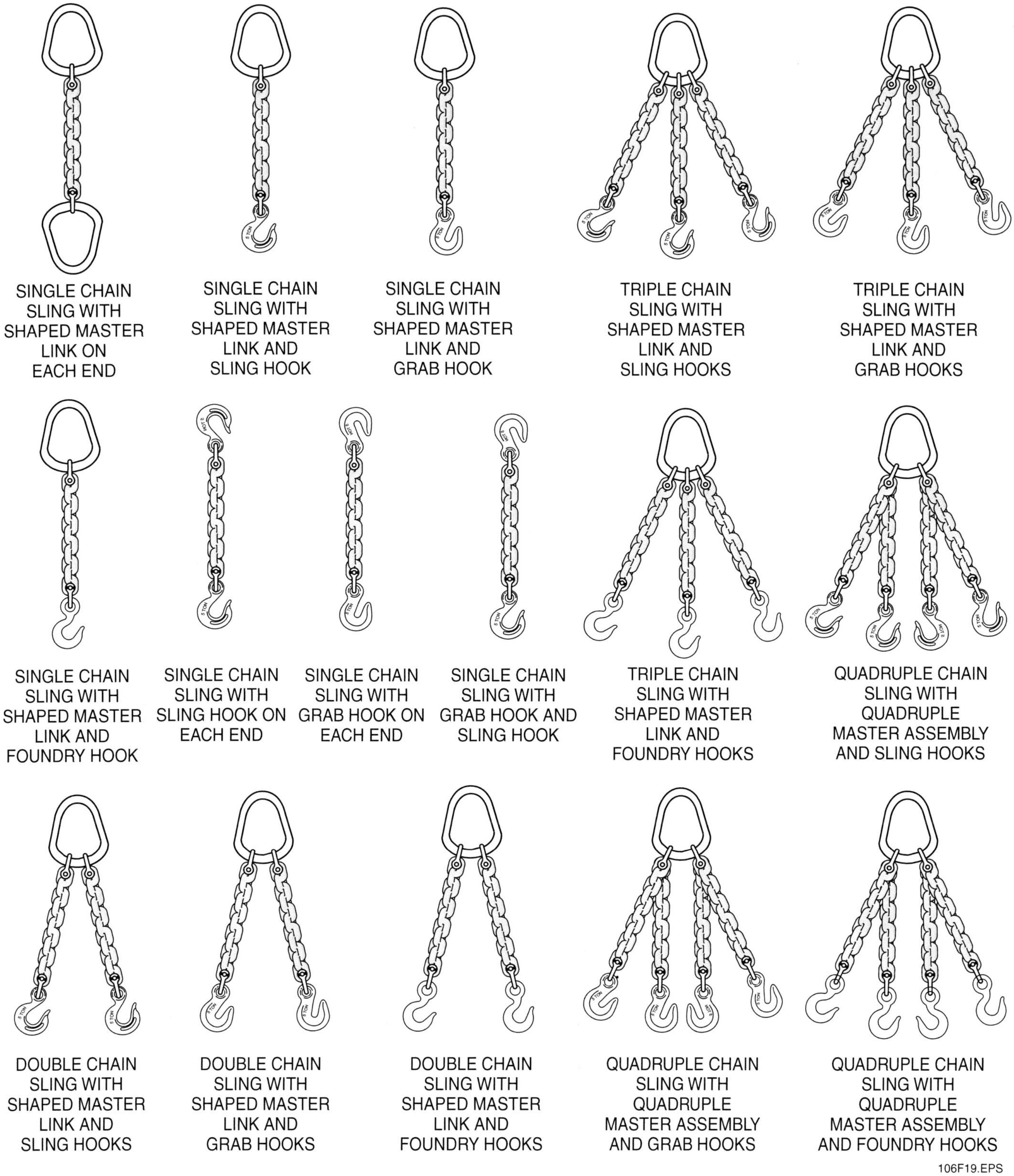

Figure 19 Common chain slings and hooks.

Note that only Grade 8 or higher chain is permitted for overhead lifting.

3.3.1 Chain Sling Storage

To store chains properly, hang them on racks inside a building or vehicle to reduce deterioration from rust and corrosion from the weather. Never allow chain slings to lie on the ground in areas where heavy machinery could run over them.

Be aware that some manufacturers suggest lubrication of alloy chains while in use. However, slippery chains increase handling hazards. Chains coated with oil or grease accumulate dirt and grit, which may cause abrasive wear. Chains that will be stored in exposed areas should be coated with a film of oil or grease for rust and corrosion protection.

3.3.2 Chain Sling Care and Inspection

Chain slings should be visually inspected before every lift. Annual inspections are also required, and records of inspections must be maintained. Discard chain slings if any of the following conditions is found during inspection (*Figure 20*):

- *Wear* – A chain sling that has any portion of the chain worn by 15 percent or more should be removed from service immediately. Wear usually occurs at the points at which chain links bear on each other or on the outside of the link barrels.
- *Stretch* – Compare the chain with its rated length or with a new length of the same type of chain. Any length increase means wear or stretch has occurred. If the length has increased 3 percent, further careful inspection is needed. If the length has increased more than 5 percent,

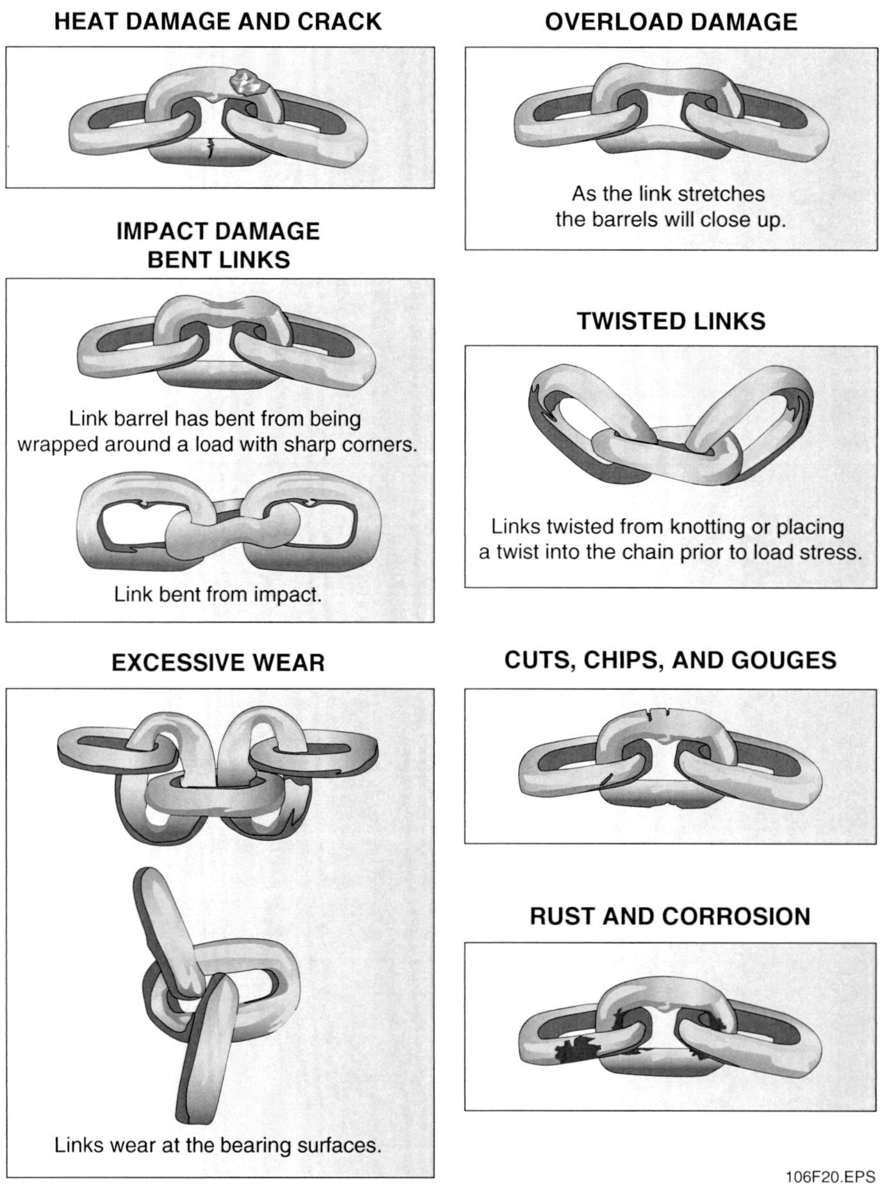

Figure 20 Types of chain link damage.

the chain should be removed from service immediately. Significantly stretched links have an hourglass shape, and links tend to bind on each other. Be sure to check for localized stretching because a chain link can be easily overlooked.

- *Link condition* – Look for links that are twisted, bent, cut, gouged, or nicked.
- *Cracks* – Discard the chain if any cracks are found in any part.
- *Link welds* – Lifted fins at the weld edges indicate overloading.
- *End fittings* – Check for signs of stretching, wear, twisting, bending, opening up, and corrosion.
- *Capacity tag* – Check for missing or illegible tags.

4.0.0 Tag Lines

Tag lines are usually natural fiber or synthetic rope lines used to control the swinging of the load in hoisting activities (*Figure 21*). Improper use of tag lines can turn the simplest hoisting operation into a dangerous situation. Tag lines should be used to control swinging of the load when a crane is traveling. They should also be used when the crane is rotated, if rotation of the load is hazardous. A nonconductive tag line is required during operation of a mobile crane in the vicinity of power lines.

When using a tag line to control swinging in a large lift, do not try to stop the swing in one movement. Pull hard when the object is swinging away from you, wait while it swings back toward you, and maintain a rhythmic pull, and relax until you have damped the swing. If the load starts to spin, keep a steady strain on the line, and be careful not to let your rope get fouled or to let yourself be pulled off balance.

When selecting a tag line, take certain factors into consideration. Natural fiber rope is inexpensive when compared to synthetic rope, and it is the most common type of rope used for tag lines. However, natural fiber rope is notably weaker than synthetic fiber ropes (nylon, polyester, polypropylene, or polyethylene) of the same size. In addition, natural fiber rope is subject to ultraviolet deterioration and damage from heat and chemicals. Manila rope can also become a conductor of electricity when wet. Synthetic ropes are light and strong for their size, and most are resistant to chemicals. When dry, synthetic ropes are poor conductors of electricity, although some readily absorb water and conduct electricity. Dry polypropylene line is preferred for use as a tag line.

106F21.EPS

Figure 21 Use of a tag line to control swinging of a suspended load.

The diameter of a tag line should be large enough so that it can be gripped well even when wearing gloves. Rope having a diameter of ½" is common, but ¾"- and 1"-diameter rope is sometimes used on heavy loads or when the tag line must be extremely long. Tag lines should be long enough to allow control of the load from its original lift location until it is safely placed, or until control is taken over by co-workers. Special consideration should be given to situations in which a long tag line would interfere with the safe handling of loads, such as during steel erection or catalyst pours.

Tag lines should be attached to loads at a location that gives the best mechanical advantage for controlling the load. For long loads, attach tag lines as close as possible to the ends. The tag line should be placed in a location that allows personnel to remove it easily after the load has been placed. Knots used to attach tag lines should be tied properly to prevent slipping or accidental loosening, but keep in mind that they may need to be easily untied after the load has been placed. Some recommended knots are the clove hitch with overhand safety and the bowline. Riggers use different knots for different purposes. *Figure 22* shows several knots commonly used in rigging. Whenever possible, tag lines should be of one continuous length, free of knots and splices, and **seized** on both ends. When joining two tag lines together is necessary, this should be done by splicing the lines. Knots tied in the middle or with free ends of tag lines can create problems.

To properly handle a tag line, determine the mechanical advantage intended by the tag line, and stand away from the load in order to have a clear area. When possible, stand where you can see the crane being used for lifting. Keep yourself and the tag line in view of the crane operator. Stay alert. Do not become complacent or distracted during the lift. Be aware of the location of any excess rope, and do not allow it to become fouled or entangled on anything.

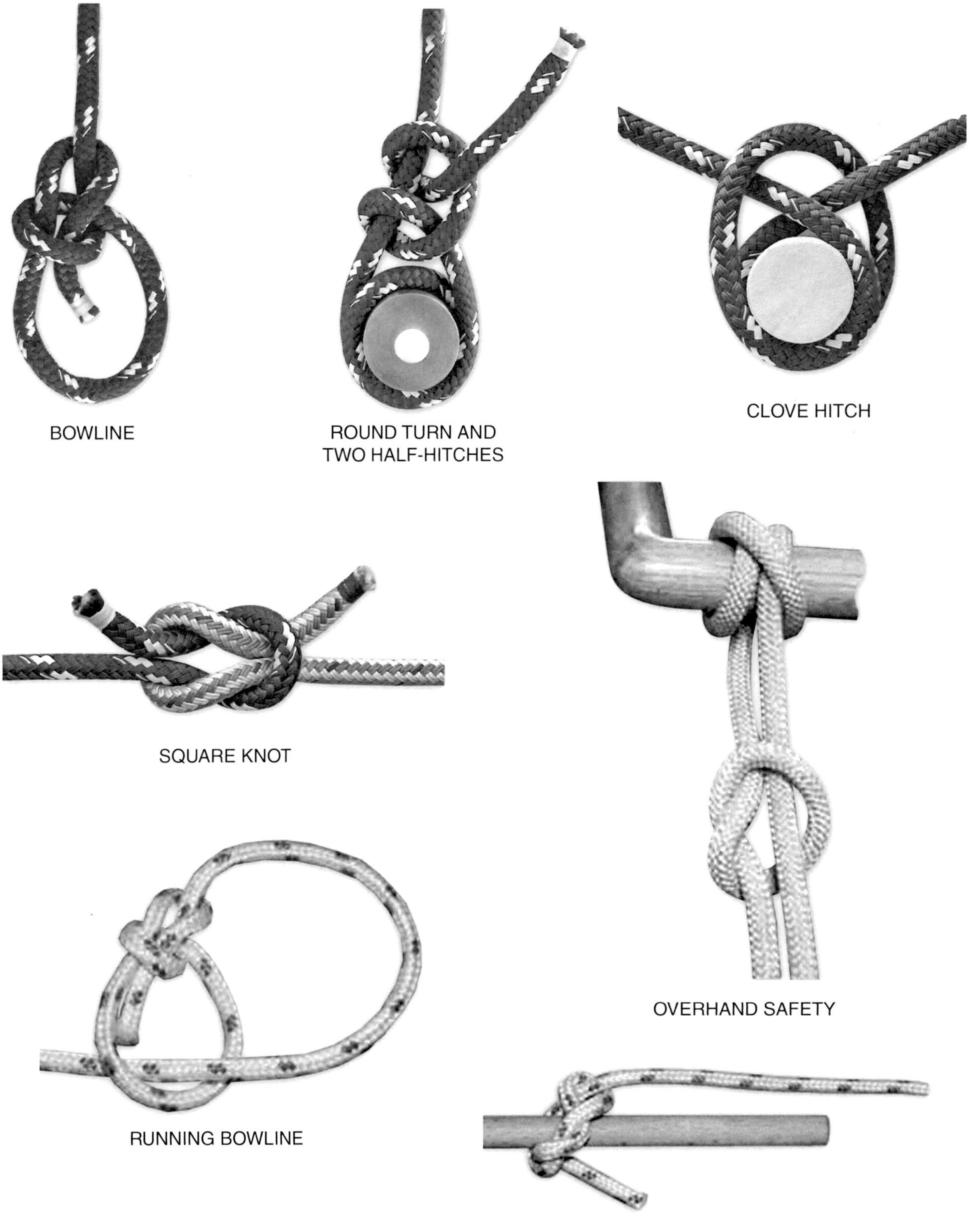

Figure 22 Common rigging knots.

Large loads often require the use of more than one tag line. In such cases, tag line personnel must work as a team and coordinate their actions.

5.0.0 BLOCK AND TACKLE

The block and tackle is the most basic lifting device. It is very old technology and is not often used. However, it is necessary to know what a block and tackle is and to be able to use one if required to do so. A block and tackle is used to lift or pull light loads. A block consists of one or more sheaves or pulleys fitted into a wood or metal frame with a hook attached to the top. The tackle is the line that runs through the block, which is used for lifting and pulling. Some block and tackle rigs have a brake that holds the load once it has been lifted; others do not. The types that lack a brake require continuous pull on the **hauling line**; or the hauling line must be tied off to hold the load. There are two types of block and tackle rigs: simple and compound.

A simple block and tackle (*Figure 23*) consists of one sheave and a single line. It is used to lift or pull very light loads. The line hook is attached to the load, and the line is pulled by hand to lift the load. The load capacity of a simple block and tackle is equal to the capacity of the load line. The block must be attached to a building's structure or to some other support by a method that provides enough load capacity to support the load and the tackle.

A compound block and tackle (*Figure 24*) uses more than one block. It has an upper, **fixed block** that is attached to the building's structure or other support and a lower, movable block that is attached to the load. Each block may have one or more sheaves. The more sheaves the blocks have, the more **parts of line** the block and tackle has, and the higher the lifting capacity. The compound block and tackle can handle heavier loads than the simple block and tackle, and it requires less effort to raise the load.

6.0.0 CHAIN HOISTS

A chain hoist, also called a chain fall, is another very useful and commonly used lifting device. It is used for many of the same purposes as the block and tackle. Chain hoists should be used for straight, vertical lifts only. They are designed for straight lifts and may become damaged if used for angled lifts, unless the load has been calculated beforehand. Always use chokers with chain hoists. The load chain of a chain hoist should never be wrapped around an object and used as a choker.

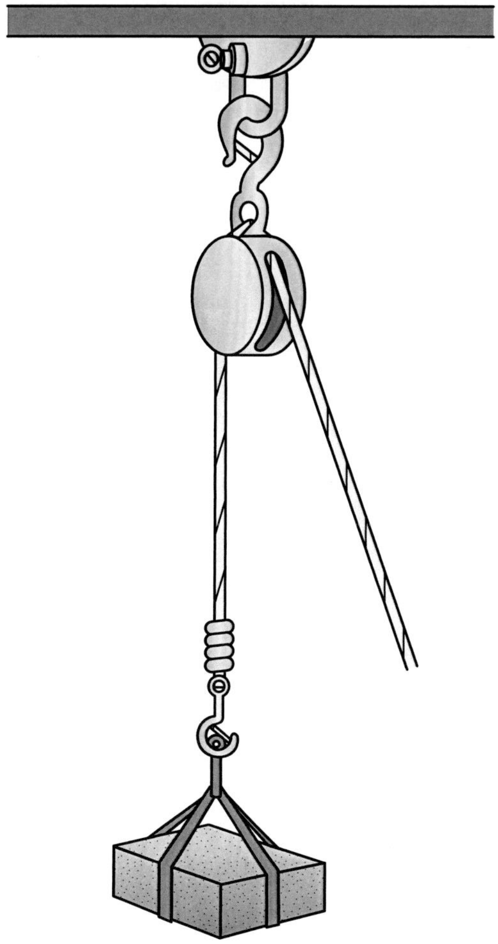

106F23.EPS

Figure 23 Simple block and tackle.

106F24.EPS

Figure 24 Compound block and tackle.

The load capacity of any chain hoist should be marked on the device. The capacity of a chain hoist must never be exceeded. Chain hoists are standard equipment in most shops and rigging departments. They are dependable, portable, and easy to use. Some common chain hoists include the spur-geared and electric types. The hoist should be load tested once a year.

6.1.0 Spur-Geared Chain Hoists

The spur-geared chain hoist (*Figure 25*) has two chains. An endless chain, the hand chain, drives a single pocketed sheave, which in turn drives a gear-reduction unit. The load chain is fitted to the gear-reduction unit. It has a hook that attaches to the load, and it must have a mechanical brake. Roller chain is sometimes used as the load chain instead of link-type chain. The spur-geared chain hoist is the most efficient type of manual chain hoist. It is most commonly used for heavier loads.

6.2.0 Electric Chain Hoists

Electric chain hoists (*Figure 26*) work much like manual chain hoists. The main difference is that they have an electric motor instead of a pull chain to raise and lower the load. Electric chain hoists are the fastest and most efficient type of chain hoist. Common electric chain hoists are available in capacities of ¼ ton to 5 tons. Special-application electric chain hoists are available in much higher capacities. Electric chain hoists are controlled from a handheld pushbutton control that is suspended on a wire from the hoist.

Special care should be taken when raising a load with an electric chain hoist because it is hard to tell how much force is being exerted by the electric motor. If the load were to get caught on something, the chain hoist could become overloaded and be damaged. The electrically driven pneumatic hoist is operated by a rotary air pump; in this type of hoist, the motor stalls before the chain or hoist mechanism can break.

6.3.0 Care of Chain Hoists

Like other lifting devices, chain hoists must be carefully inspected before each use. The chain, in particular, must be checked to ensure that it has no defects. Follow these steps to select, inspect, use, and maintain a chain hoist:

Step 1 Select a chain hoist of adequate capacity to handle the load.

Step 2 Inspect the load chain and hook to ensure that they are not excessively worn, bent, or deformed in any way.

106F25.EPS

Figure 25 Spur-geared chain hoist.

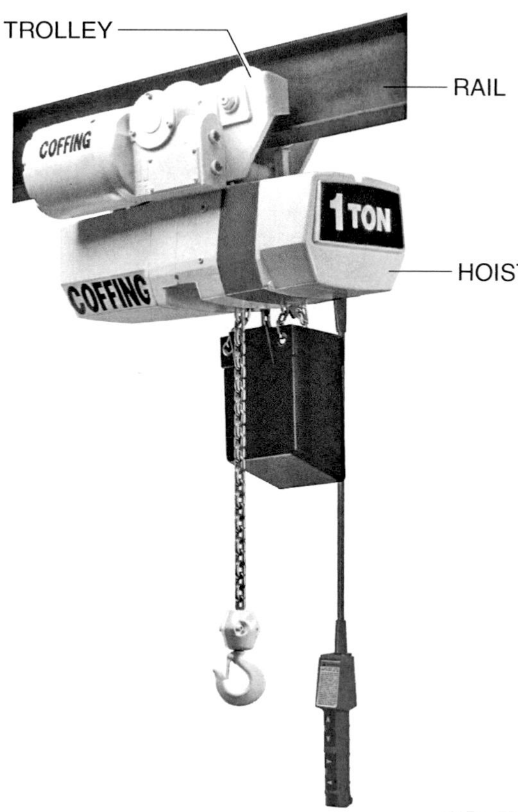

106F26.EPS

Figure 26 Electric chain hoist.

Step 3 Check the sheaves, and make sure that they are not bent or excessively worn.

Step 4 Make sure that the chain hoist has proper lubrication.

Step 5 Hang the chain hoist on a suitable support using the proper rigging.

Step 6 Lower the load hook, and connect it to the load using the proper rigging.

> **WARNING**
> Make sure the voltage in the plug matches the motor voltage.

Step 7 Raise and place the load.

> **NOTE**
> To raise the load using a manual chain hoist, pull the hand chain. To raise the load using an electric chain hoist, press and hold the Up pushbutton on the handheld control.

Step 8 Disconnect the load hook from the load.

Step 9 Remove the chain hoist from its support.

Step 10 Coil the chain so that it will not get tangled.

Step 11 Store the chain hoist in its proper place.

7.0.0 Ratchet-Lever Hoists and Come-Alongs

Ratchet-lever hoists and come-alongs (*Figure 27*) are used for short pulls on heavy loads. The term *come-along* is widely used to identify both types of devices, but they are not the same. A come-along uses a cable, while a ratchet-lever hoist uses a chain. Ratchet-lever hoists can be used for vertical lifts, but cable-type come-alongs can only be used for horizontal pulls. Come-alongs and ratchet-lever hoists are portable and easy to store and use. They are available in capacities ranging from ½ ton to 6 tons.

> **WARNING**
> Never use a come-along for vertical lifts. Come-alongs do not have the same safety braking mechanisms as the ratchet-lever hoist to prevent the load from slipping.

A ratchet-lever hoist consists of the following parts:

- *Suspension hook* – A steel hook used to hang the hoist.
- *Load hook* – A steel hook that attaches to the load.
- *Load chain* – The chain that attaches to the load hook and lifts the load.
- *Ratchet handle* – A handle that operates the ratchet that takes up the chain and lifts or lowers the load.
- *Fast-wind handle* – A handle that takes up or lets out the chain without using the ratchet handle. It cannot be used to raise or lower the load.
- *Ratchet release* – A device that releases the ratchet so the chain can be pulled out. It also switches the hoist to the up or down position.

Follow these steps to select, inspect, use, and maintain a ratchet-lever hoist:

Step 1 Select a device of adequate capacity to handle the load.

Step 2 Inspect the chain and hooks to ensure that they are not excessively worn, bent, or deformed in any way.

Step 3 Inspect the device to make sure that it is not damaged in any way.

Step 4 Hang the device on a suitable support.

Step 5 Turn the ratchet release to the middle position.

Step 6 Pull the chain out enough to be able to attach it to the load.

Step 7 Attach the load hook to the load using the proper rigging.

Figure 27 Come-along and ratchet-lever hoist.

Step 8 Turn the fast-wind handle to take the slack out of the chain.

Step 9 Turn the ratchet release to the Up position.

Step 10 Pump the ratchet handle to raise and position the load. Stop and check the brake as soon as the load clears the ground, then continue the lift.

Step 11 Turn the ratchet release to the Down position.

Step 12 Pump the ratchet handle to lower the load into position and until there is slack in the chain.

Step 13 Disconnect the load hook from the load.

Step 14 Store the device in its proper place.

8.0.0 JACKS

A jack is a device used to raise or lower equipment. Jacks are also used to move heavy loads a short distance, and they offer good control over the movement. The three basic types of jacks are ratchet, screw, and **hydraulic**.

106F28.EPS

Figure 28 Ratchet jack.

8.1.0 Ratchet Jacks

The ratchet jack (*Figure 28*), also called a railroad jack, is only used to raise loads less than 25 tons. The ratchet jack uses the lever-and-fulcrum principle. The downward stroke of the lever raises the rack bar one notch at a time. A latching mechanism, called a pawl, automatically springs into position, holding the load and releasing the lever for the next lifting stroke.

Ratchet jacks allow safe lifting, lowering, and leveling. They can lift full jack capacity on the toe or on the cap.

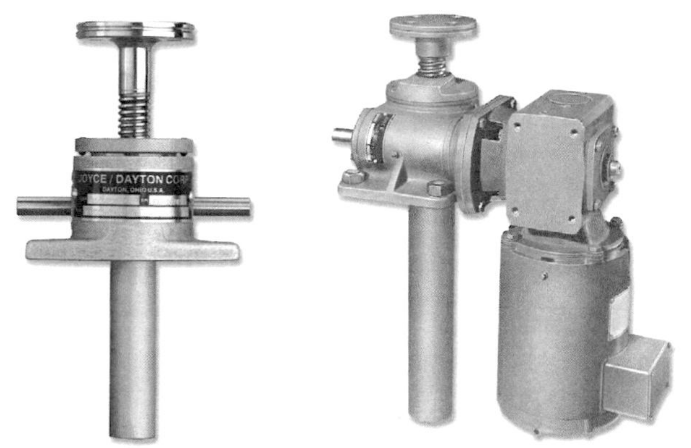

UPRIGHT SCREW JACK MOTORIZED SCREW JACK

106F29.EPS

Figure 29 Upright and motorized screw jacks.

8.2.0 Screw Jacks

Screw jacks (*Figure 29*) are used to lift heavier loads. There are two general types of screw jacks: upright and inverted. The screw jack uses the screw-and-nut principle. For lighter loads, a simple lever applies enough power to turn the screw. For heavier loads, gear-reduction units and ratchet devices are used to increase the operator's strength. In the heaviest jacks, the screw jack is operated by an air motor for faster lifting and lowering.

8.3.0 Hydraulic Jacks

Hydraulic jacks (*Figure 30*) are the most useful general-purpose jacks. They are operated by the pressure of an enclosed liquid. There are two kinds of hydraulic jacks. One type has the pump built into the jack. The other uses a separate pump. Both jack types can lift heavy loads with little effort. The hydraulic jack that has a separate pump provides the following advantages:

- The person operating the pump can do so at a safe distance from the load.
- The ram provides the greatest amount of travel for a given closed height.
- The pumps can be powered by hand, electric motor, air motor, or gasoline engine.

100 TON CAPACITY

2 TO 10 TON CAPACITY

106F30.EPS

Figure 30 Hydraulic jacks.

The hydraulic jack is not as fast as the ratchet jack, but it is excellent for lifting heavy loads.

Portable hydraulic jacks, called porta-powers after a common brand name, consist of a hand hydraulic cylinder and any one of a number of tools, including jacks, clamps, and other equipment.

8.4.0 Inspecting and Using Jacks

Over time, a jack can become damaged or weakened and can fail under a load. To avoid such failures, all jacks should be carefully inspected before each use. Always follow these guidelines when using jacks:

- Inspect jacks before using them to ensure that they are not damaged in any way.
- Use wood softeners when jacking against metal.
- Use the proper jack handle, and remove it from the jack when it is not in use.
- Never step on a jack handle to create additional force.
- Use blocking or cribbing under the load when jacking.
- Never leave a jack under a load without having the load blocked up.
- Position jacks properly, and raise the load evenly to prevent it from shifting or falling.
- Never place jacks directly on the ground when lifting; always use a solid footing.
- Position jacks so that the direction of force is perpendicular to the base and the surface of the load.
- Never exceed the load capacity of the jack.
- Do not use extensions to the jack handles.
- Brace loads to prevent the jacks from tipping.
- Lash or block jacks when using them in a horizontal position to move an object.
- Never jack against rollers.
- Match ratchet jacks for uniform lifting.
- Keep fingers away from ratchet parts when raising or lowering an object.
- Thoroughly clean all hydraulic hose connectors before connecting them.

9.0.0 TUGGERS

Tuggers are hydraulic, electric, or pneumatically powered winches. They can be used for lifting and lowering and for pulling. The tugger must be securely anchored to the floor or to a steel beam using a method that will support any load placed on the tugger. When securing a tugger to a steel beam, use a softener on the choker. Tuggers can be located in almost any convenient location in a building. The load line can be run through a series of blocks to the area where the lift will be made. The blocks and all hardware must be of adequate load capacity to handle the load being lifted. The controls may be located on the tugger, or the tugger may have remote controls. The tugger is often in a location from which the operator cannot see the load being lifted. In this case, a signaler is needed to direct the operator, using the same hand signals that are used when signaling a crane. The tugger has a band-type brake to hold the load. *Figure 31* shows a common tugger.

Follow these steps to select and operate a tugger:

Step 1 Verify that the tugger is of adequate capacity to handle the load.

Figure 31 Tugger.

Step 2 Make sure that an adequate air supply is connected to the tugger and that it is in service.

Step 3 Push the control handle toward the Down position, and reel off enough cable to attach to the load.

Step 4 Attach the cable to the load using the proper rigging.

> **NOTE**
> Often the load is far removed from the tugger. When this is the case, the tugger operator will require the help of a rigger to assist in attaching the cable to the load and in directing the movement of the object.

Step 5 Ensure that the cable is running properly through all blocks.

Step 6 Push the control handle toward the Up position to raise the load.

> **WARNING**
> When operating the tugger, always stand to the side of it, behind the safety shield.

> **NOTE**
> A tugger is a variable-speed device. Pushing the control handle slightly toward the Up position raises the load slowly. Start slowly, and then push the handle farther toward the Up position to speed up. Slow down as the load approaches its destination.

Step 7 Push the control handle toward the Down position to position the load at its destination.

Step 8 Push the control handle toward the Down position to slacken the cable.

Step 9 Disconnect the load hook from the load.

Step 10 Operate the control handle to position the cable in its normal stored position.

Summary

The process called rigging includes selecting and setting up hoisting equipment, hooking cables to the load to be lifted or moved, and directing the load into position. Performing the rigging process safely and efficiently requires the selection of the proper equipment for each hoisting job. It also calls for an understanding of safety hazards and a knowledge of how to prevent accidents. Riggers must have the greatest respect for the hardware and tools of their trade because their lives may depend on having them work correctly.

Review Questions

1. Mixing rigging hardware of different lifting capacities is permitted as long as the capacity of one of the hardware items exceeds the weight of the load.
   a. True
   b. False

2. The safe working load of a rigging hook is accurate only when the load is suspended from the part of the hook called the _____.
   a. body
   b. eye
   c. throat
   d. saddle

3. OSHA requires a hook to be replaced if the throat has opened from its original size by _____.
   a. 5 percent
   b. 10 percent
   c. 15 percent
   d. 20 percent

4. A hook that is point-loaded can carry approximately what percent of the rated load?
   a. 20
   b. 40
   c. 60
   d. 75

5. Slings are attached together using _____.
   a. spacers
   b. sling eyes
   c. shackles
   d. lifting lugs

6. Tapped holes used with screwed-in eyebolts should have a minimum depth equal to the bolt diameter multiplied by _____.
   a. 1¼
   b. 1½
   c. 1¾
   d. 2

7. When the legs of a sling are unequal, the device that can be used to level loads is a(n) _____.
   a. rigging link
   b. plate clamp
   c. equalizer plate
   d. beam clamp

8. Custom-fabricated spreader and equalizer beams should be tested at what percent of their rated capacity?
   a. 100 percent
   b. 125 percent
   c. 150 percent
   d. 200 percent

9. During lifting operations, long loads are supported using _____.
   a. spreader beams
   b. equalizer beams
   c. serrated clamps
   d. compound equalizer plates

10. One of the slings best suited for situations in which the load is abrasive, hot, or tends to cut is a(n) _____.
    a. wire rope sling
    b. synthetic web sling
    c. endless grommet sling
    d. metal mesh sling

11. Because of its resistance to corrosion and abrasion, the type of sling that is required for dredging and other marine work is the _____.
    a. chain sling
    b. synthetic web sling
    c. wire rope sling
    d. mesh sling

12. A chain sling should be discarded if any portion of the chain is worn by _____.
    a. 5 percent or more
    b. 10 percent or more
    c. 15 percent or more
    d. 20 percent or more

13. Which of the following statements about a compound block and tackle is correct?
    a. The fixed block is attached to the load.
    b. It has greater capacity than the simple type but requires more effort.
    c. The more sheaves the blocks have, the greater is the lifting capacity.
    d. It is always necessary to tie off the hauling line to keep the load from falling.

14. The capacity range of come-alongs is _____.
    a. ½ ton to 6 tons
    b. 2 to 5 tons
    c. 3 to 7 tons
    d. 8 to 10 tons

15. The most useful general-purpose jack is the _____.
    a. inverted screw jack
    b. hydraulic jack
    c. upright screw jack
    d. ratchet jack

Cornerstone of Craftsmanship

Bryan McClure

LPR Construction
Training Manager

How did you get started in the construction industry?
My father had been an ironworker his whole life. I used to work weekends at his job sites during my summers off from high school. Of course, this was before OSHA really cracked down on underage workers.

Who inspired you to enter the industry? Why?
My father inspired me to enter the industry. Originally, ironworking was not my career choice or my father's first choice for me. After my first year of college, though, I had stacked up plenty of bills and ironworking paid really well.

What do you enjoy most about your job?
Looking back, I am very happy with my career choice. Being an ironworker has afforded me opportunities to travel and make good money. It has also given me a unique career path. I have gone from ironworker to foreman, crane operator to superintendent, and now I am a training manager.

Do you think training and education are important in construction? If so, why?
Every day the demand from owners for quality production and safety goes up. People want things bigger, faster, and better. The only way to provide this is to constantly train our workforce so they can adapt to the ever-changing needs of the industry.

How important are NCCER credentials to your career?
NCCER credentials provide another level of accomplishment for all the hard work our apprentices put in. It also means they don't have to constantly prove themselves at new companies, because the credentials are portable.

How has training/construction impacted your life and your career?
I am very proud of the fact that I am an ironworker. The only thing that I have ever found that rivals that passion is teaching the next generation of ironworkers.

Would you suggest construction as a career to others? If so, why?
Absolutely! I feel that those of us in the industry need to do a better job of making the upcoming generations aware of all the career choices in construction. There is more to construction than just swinging a hammer.

How do you define craftsmanship?
Craftsmanship is going about your business every day as a professional. Complete each task like you are signing your name to it, because the world will be judging you for the work you do.

Trade Terms Introduced in This Module

Anneal: To soften a metal by heat treatment.

Bird caging: A deformation of wire rope that causes the strands or lays to separate and balloon outward, similar to the vertical bars of a bird cage.

Bull's tail (Christmas tree): A double strand of wire rope with a series of hooks, used to carry a series of beams or joists and lift them into place sequentially.

Equalizer beam: A beam used to distribute weight on multi-crane lifts.

Fixed block: The upper block of a block and tackle; the block that is attached to the support.

Hauling line: The line of a lifting device that is pulled by hand to raise the load.

Hydraulic: Operated by fluid pressure.

Kinking: Bending a rope so severely that the bend is permanent and individual wires or fibers are damaged.

Parts of line: The number of ropes between the upper and lower blocks of a block and tackle. Parts of line, also called falls, are the lines that carry the load.

Seize: To bind a rope in order to prevent it from untwisting after it has been cut.

Shackle: A piece of rigging equipment consisting of a horseshoe-shaped piece with a pin across the opening, used to connect slings, chains, and other lifting lines to loads, or to join slings together.

Sling angle: The angle formed by the legs of a sling with respect to the horizontal when tension is put upon the load.

Spreaders: Beams used to distribute weight bearing along longer loads.

Wire rope: A rope made of steel wires, used for cranes and rigging.

Additional Resources

This module presents thorough resources for task training. The following resource material is suggested for further study.

OSHA Standards for the Construction Industry (29 CFR Part 1926). Latest edition. Washington, DC: Occupational Safety & Health Administration. http://www.osha.gov/pls/publications/publication.html.

Figure Credits

Topaz Publications, Inc., Module opener, Figures 1, 11, 14, and 16

J.C. Renfroe & Sons, Inc., Figure 12

Lift-All Company, Figure 18

Cianbro Corporation, Figure 21

Alan W. Grogono, M.D., Figure 22 (bowline, round turn and two half-hitches, clove hitch, and square knot)

Dave Root, Figure 22 (timber hitch and running bowline)

Lincoln Fire & Rescue, Lincoln, NE, Figure 22 (overhand safety)

Garrett Wade Company, Figure 24

Chester Hoist Company, Figure 25

Coffing Hoists, Figure 26

Lug-All Corporation, Figure 27 (come-along)

JET Tools & Equipment, Figure 27 (ratchet-lever hoist)

Duff-Norton, Figures 28 and 30

Joyce/Dayton Corp., Figure 29

Vernon Smith, Figure 31

NCCER CURRICULA — USER UPDATE

NCCER makes every effort to keep its textbooks up-to-date and free of technical errors. We appreciate your help in this process. If you find an error, a typographical mistake, or an inaccuracy in NCCER's curricula, please fill out this form (or a photocopy), or complete the online form at **www.nccer.org/olf**. Be sure to include the exact module ID number, page number, a detailed description, and your recommended correction. Your input will be brought to the attention of the Authoring Team. Thank you for your assistance.

Instructors – If you have an idea for improving this textbook, or have found that additional materials were necessary to teach this module effectively, please let us know so that we may present your suggestions to the Authoring Team.

NCCER Product Development and Revision

13614 Progress Blvd., Alachua, FL 32615

Email: curriculum@nccer.org

Online: www.nccer.org/olf

❑ Trainee Guide ❑ Lesson Plans ❑ Exam ❑ PowerPoints Other ______________________

Craft / Level: Copyright Date:

Module ID Number / Title:

Section Number(s):

Description:

Recommended Correction:

Your Name:

Address:

Email: Phone:

Rigging Practices

30107-11

Trainees with successful module completions may be eligible for credentialing through NCCER's National Registry. To learn more, go to **www.nccer.org** or contact us at **1.888.622.3720**. Our website has information on the latest product releases and training, as well as online versions of our *Cornerstone* newsletter and Pearson's Contren® product catalog.

Your feedback is welcome. You may email your comments to **curriculum@nccer.org,** send general comments and inquiries to **info@nccer.org**, or use the User Update form at the back of this module.

V.1 1/1

30107-11
Rigging Practices

Objectives

When you have completed this module, you will be able to do the following:

1. Determine the center of gravity of a load.
2. Identify the pinch points of a crane.
3. Identify site and environmental hazards associated with rigging.
4. Properly attach rigging hardware for routine lifts.
5. Identify the components of a lift plan.
6. Perform sling tension calculations.
7. Determine the weight of beams and basic weight estimation.
8. Explain D/d.

Performance Tasks

Under the supervision of the instructor, the trainee should be able to do the following:

1. Determine the center of gravity of a load.
2. Properly attach rigging hardware for routine lifts.
3. Perform sling tension calculations.
4. Perform a weight/volume calculation.

Trade Terms

Anti-two-blocking devices
Blocking
Center of gravity
Cribbing
Lifting lugs
Softener
Spur track
Yarding

Contents

Topics to be presented in this module include:

Figures and Tables

1.0.0 INTRODUCTION

Rigging involves the lifting and moving of heavy, and often bulky, objects. For that reason, safety must be the foremost consideration in any rigging job.

One important job that a rigger performs is guiding the crane operator. This is often done using a standard set of hand signals, which everyone involved in rigging must memorize.

Anyone doing rigging work must know how to select the right equipment, with the right capacity, for every rigging job. Failure to use the right equipment and correct rigging method can lead to serious consequences, including death or injury to workers and damage to equipment or materials.

Every crane and lifting device has capacity limits and conditions under which they are designed to operate safely. If you are not aware of these limits, you could make a fatal error. If you make the effort to learn the right and safe way to do the job, you will be successful. Always err on the side of caution.

This module is intended to instruct you on the basic safety precautions and rigging practices that will enable you to assist in rigging equipment and materials.

2.0.0 GENERAL RIGGING SAFETY

Riggers must be aware of the unavoidable hazards associated with the work. Riggers will be directing the movement of loads above and around other workers, where falling equipment and material can present a grave safety hazard (*Figure 1*). The rigger may also work during extreme weather conditions where winds, slippery surfaces, and unguarded work areas exist. When working near cranes, always look up and be mindful of the hazards above and around you.

As a result of tighter regulations established by OSHA, and more lost work days experienced by construction employees, individual construction trade contractors have put stringent safety policies into place.

Safety consciousness is extremely important on the job site. The earning ability of injured employees may be reduced or eliminated for the rest of their lives when safety is not exercised. The number of employees injured can be reduced if each employee is committed to safety awareness. Full participation in the employer's safety program is your personal responsibility.

Safety consciousness is the key to reducing accidents, injuries, and deaths on the job site. Accidents can be avoided because most result from human error. Mistakes can be reduced by developing safe work habits derived from the principles of on-the-job safety.

2.1.0 Personal Protection

Workers on the job have responsibilities for their own safety and the safety of their fellow workers. Management has a responsibility to each worker to ensure that those who prepare and use the equipment, and those who work with or around it, are well trained in operating procedures and safety practices.

Always be aware of your environment when working with cranes. Stay alert and know the location of equipment at all times when moving about and working within the job area.

Standard personal protective equipment, including hard hats, safety shoes, and barricaded work areas (*Figure 2*) are among the important safety requirements at any job site.

2.2.0 Equipment and Supervision

Your employer is responsible for ensuring that all hoisting equipment is operated by experienced, trained operators. Rigging workers must also be

107F01.EPS

Figure 1 Overhead hazards.

107F02.EPS

Figure 2 Personnel must be protected at all times.

capable of selecting suitable rigging and lifting equipment, and directing the movement of the crane and the load to ensure the safety of all personnel. Rigging operations must be planned and supervised by competent personnel who ensure the following:

- Proper rigging equipment is available.
- Correct load ratings are available for the material and rigging equipment.
- Rigging material and equipment are well maintained and in good working condition.

A rigging supervisor is responsible for the following functions:

- Proper load rigging
- Crew supervision
- Ensuring that the rigged material and equipment meet the required capacity and are in safe condition
- Ensuring that the lifting equipment and rigging materials are installed correctly
- Guaranteeing the safety of the rigging crew and other personnel

2.3.0 Basic Rigging Precautions

The most important rigging precaution is determining the weight of all loads before attempting to lift them. When the assessment of the weight load is difficult, safe-load indicators or weighing devices should be attached to the rigging equipment. It is equally important to rig the load so that it is stable and the **center of gravity** is below the hook.

The personal safety of riggers and hoisting operators depends on common sense. Always observe the following safety practices:

- Examine all equipment and rigging before use.
- Immediately report defective equipment or hazardous conditions to the supervisor. Someone in authority must issue orders to proceed after safe conditions have been ensured.
- Know the safe working load of the equipment and rigging, and never exceed the limit.
- Determine the weight of loads, including rigging and hardware, before rigging. Site management must provide this information if it is not known.

NOTE

Weight of beams is easily found. The beam description, such as W24X76, gives the nominal size of the beam and the weight per linear foot. Multiply the length in feet by the weight per linear foot. A ten-foot beam would then be 760 pounds.

- Stop hoisting or rigging operations when weather conditions present hazards to property, workers, or bystanders, such as when winds exceed manufacturer's specifications; when the visibility of the rigger or hoist crew is impaired by darkness, dust, fog, rain, or snow; or when the temperature is cold enough that hoist or crane steel structures could fracture upon shock or impact. Any time that temperatures are below 10°F, equipment should not be stressed heavily until it has warmed up for at least an hour, as the metal is relatively brittle.
- Recognize factors that can reduce equipment capacity. Safe working loads of all hoisting and rigging equipment are based on ideal conditions. These conditions are seldom achieved during a typical workday.
- Remember that safe working loads of hoisting equipment are applicable only to freely suspended loads and plumb hoist lines. Side loads and hoist lines that are not plumb can stress equipment beyond design limits, and structural failure can occur without warning.
- Ensure that the safe working load of equipment is not exceeded if it is exposed to wind. Avoid sudden snatching, swinging, and stopping of suspended loads. Rapid acceleration and deceleration greatly increase the stress on equipment and rigging.
- Follow all manufacturer's guidelines, and consult applicable standards to stabilize mobile cranes properly (*Figure 3*).

107F03.EPS

Figure 3 Mobile crane with outriggers.

2.4.0 Load Path, Load Control, and Tag Lines

Physical control of the load beyond that of the ability of the crane may be required. Tag lines are used to limit the unwanted or inadvertent movement of the load as the load reacts to the motion of the crane. They are also used to allow the controlled rotation of the load for positioning (*Figure 4*). Tag lines are attached before the load is lifted. The rigger verifies that the load is balanced after it is lifted.

WARNING

Make sure that the tag line is free of all obstructions and not wrapped around anything.

107F04.EPS

Figure 4 Using a tag line.

2.5.0 Barricades

Barricades should always be used to isolate the area of an overhead lift to prevent the possibility of injuring personnel who may walk into the area. Always follow the individual site requirements for proper barricade erection. If in doubt as to the proper procedure, ask your supervisor for guidance before proceeding with any overhead lifting operation. It is important to remember that accessible areas within the swing radius of the rear of the crane's rotating structure must be barricaded in such a manner as to prevent anyone from being struck or crushed by the crane (*Figure 5*).

2.6.0 Load-Handling Safety

The safe and effective control of the load involves the rigger's strict observance of load-handling safety requirements. This includes making sure that the swing path or load path is clear of personnel and obstructions (*Figure 6*). Keep the front and rear swing paths (*Figure 7*) of the crane clear for the duration of the lift. The rear swing path is a pinch point. Most people watch the load when it is in motion, which prevents them from seeing the back end of the crane coming around.

Make sure the landing zone is clear of nondesignated personnel. Also make sure that the necessary **blocking** and **cribbing** for the load are in place before you position the load for landing. The practice of lowering the load just above the landing zone and then placing the cribbing and blocking can be dangerous. No one except designated personnel should work under the load. The layout of the cribbing can be completed in the landing zone before setting the load. Blocking of the load may have to be done after the load is

107F05.EPS

Figure 5 Use of a barrier to isolate the swing circle area of a crane.

107F06.EPS

Figure 6 Front swing path.

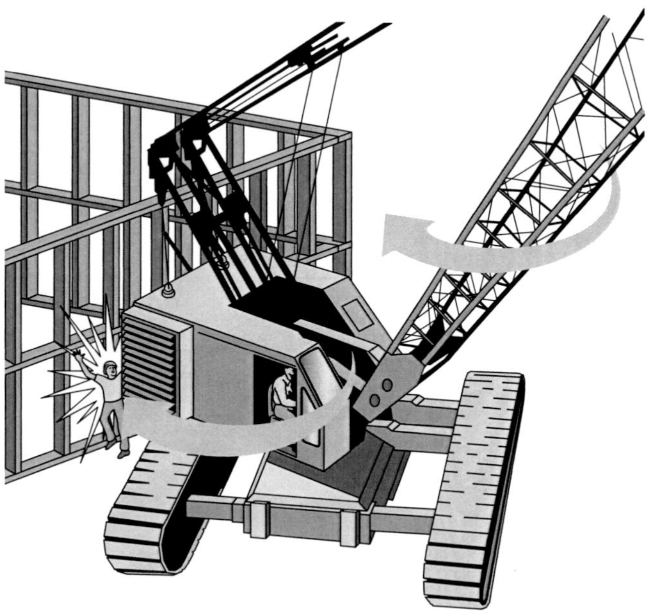

107F07.EPS

Figure 7 Rear swing path.

set. In this case, do not take the load stress off the sling until the blocking is set and secured. Do not attempt to position the load onto the cribbing by manhandling it.

After the rigging has been set and whenever loads are to be handled, follow these procedures:

- Before lifting, make sure that all loads are securely slung and properly balanced to prevent shifting of any part.
- Use one or more tag lines to keep the load under control.
- Safely land and properly block all loads before removing the slings.
- Only use lifting beams for the purpose for which they were designed. Their weight and working load abilities must be visible on the beams.
- Never wrap hoist ropes around the load. Use only slings or other adequate lifting devices.
- Do not twist multiple-part lines around each other.
- Bring the load line over the center of gravity of the load before starting the lift.
- Load and secure any materials and equipment being hoisted to prevent movement which, in turn, could create a hazard.
- Keep hands and feet away from pinch points as the slack is taken up.
- Wear gloves when handling wire rope.
- Never use a crane to pull a choker out from under a load.
- Never ride on a load.
- Always alert personnel of the path of the load.
- Never work under a suspended load unless you are a designated person.
- Never leave a load suspended in the air when the hoisting equipment is unattended.
- Never make temporary repairs to any lifting equipment.
- Always secure unused legs of a multi-leg sling before lifting.
- Ensure that all slings are made from the same material when using two or more slings on a load.
- Remove or secure all loose pieces from a load before it is moved.
- Lower loads onto adequate blocking to prevent damage to the slings.

3.0.0 Working Around Power Lines

A competent signal person must be stationed at all times to warn the operator when any part of the machine or load is approaching the minimum safe distance from a power line. The signal person must be in full view of the operator at all times. *Figure 8* and *Table 1* show the minimum safe distance from power lines.

The preferred working condition is to have the owner of the power lines de-energize and provide grounding of the lines that are visible to the crane operator. When that is not possible, observe the following procedures and precautions if any part of the crane can reach the power line:

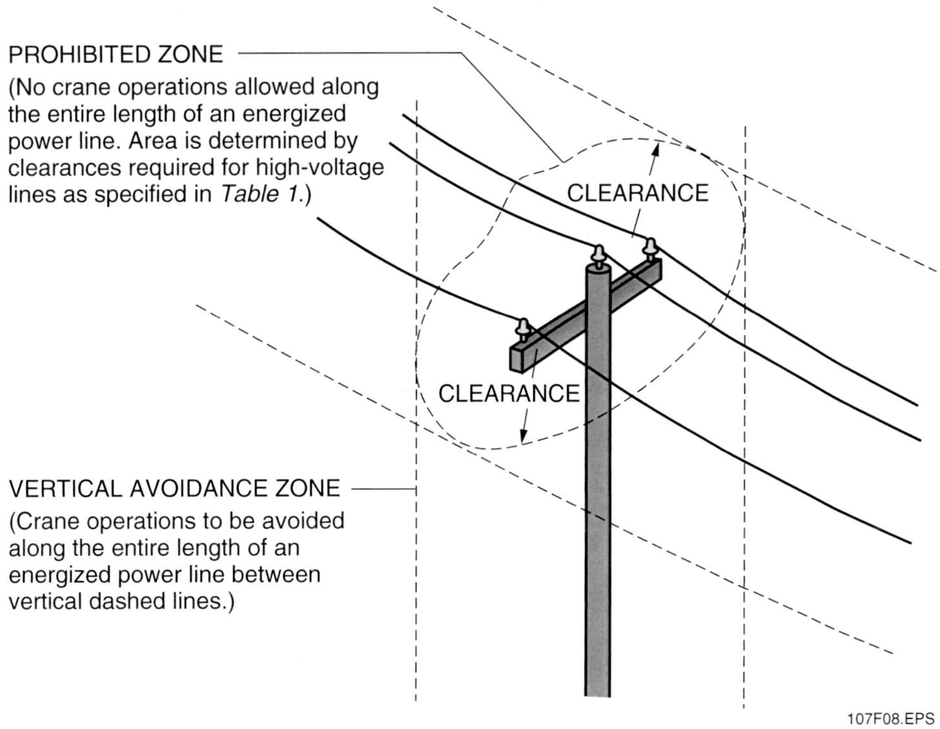

Figure 8 Prohibited zone and avoidance zone.

- Make sure a power line awareness permit (see *Appendix*) or equivalent has been prepared.

Table 1 High-Voltage Power Line Clearances

CRANE IN OPERATION (1)	
POWER LINE (kV)	BOOM OR MAST MINIMUM CLEARANCES (feet)
0 to 50	10
50 to 200	15
200 to 350	20
350 to 500	25
500 to 750	35
750 to 1,000	45
CRANE IN TRANSIT (with no load and the boom or mast lowered) (2)	
POWER LINE (kV)	BOOM OR MAST MINIMUM CLEARANCES (feet)
0 to 0.75	4
0.75 to 50	6
50 to 345	10
345 to 7500	16
750 to 1000	20

Note 1: For voltages over 50kV, clearance increases 5 feet for every 150 kV.
Note 2: Environmental conditions such as fog, smoke, or precipitation may require increased clearances.

107T01.EPS

- Erect non-conductive barricades to restrict access to the prohibited work areas under power lines.
- Use tag lines of a non-conductive type if necessary for load control.
- The qualified signal person(s), whose sole responsibility is to verify that the proper clearances are established and maintained, shall be in constant contact with the crane operator.
- The person(s) responsible for the operation shall alert and warn the crane operator and all persons working around or near the crane about the hazards of electrocution or serious injury, and instruct them on how to avoid these hazards.
- All non-essential personnel shall be prohibited from the crane work area.
- No one shall be permitted to touch the crane or load unless the signal person indicates it is safe to do so.

If a crane or load comes in contact with or becomes entangled in power lines, assume that the power lines are energized unless the lines are visibly grounded. Any other assumption could be fatal.

The following guidelines should be followed if the crane comes in contact with an electrical power source:

- The operator should stay in the cab of the crane unless a fire occurs.

- Do not allow anyone to touch the crane or the load.
- If possible, the operator should reverse the movement of the crane to break contact with the energized power line.
- If the operator cannot stay in the cab due to fire or arcing, the operator should jump clear of the crane, landing with both feet together on the ground. Once out of the crane, the operator must take very short steps or hop with feet together until well clear of the crane.
- Call the local power authority or owner of the power line.
- Have the lines verified as secure and properly grounded within the operator's view before allowing anyone to approach the crane or the load.

WARNING

Transmitters such as radio towers can also represent a hazard. These transmitters generate high-power radio frequency (RF) energy that can induce a hazardous voltage into the boom of a nearby crane. OSHA requires that the transmitter be de-energized in such situations, or that appropriate grounding methods be used to protect workers. Always consult your supervisor or site safety officer in such situations.

While induced RF voltages from transmitters are not an electrocution hazard like power line voltages, they can result in sparks to any object or person. The sparks can cause fires or explosions of combustible or flammable materials. They can also result in painful burns and shocks to personnel that come in close contact with the crane or the load being lifted by the crane.

4.0.0 Site Safety

It takes a combined effort by everyone involved in crane operations to make sure no one is injured or killed. Site hazards and restrictions as well as crane manufacturer's requirements must be observed for safe crane operation.

4.1.0 Site Hazards and Restrictions

There are many site hazards and restrictions related to crane operations. These hazards include the following:

- Underground utilities such as gas, oil, electrical, and telephone lines; sewage and drainage piping; and underground tanks
- Electrical lines or high-frequency transmitters
- Structures such as buildings, excavations, bridges, and abutments

WARNING

Power lines and environmental issues such as weather are common causes of injury and death during crane operations.

The operator and riggers should inspect the work area and identify hazards or restrictions that may affect the safe operation of the crane. This includes the following actions:

- Ensuring the ground can support the crane and the load
- Checking that there is a safe path to move the crane around on site
- Making sure that the crane can rotate in the required quadrants for the planned lift

5.0.0 Emergency Response

Operators and riggers must react quickly and correctly to any crane malfunction or emergency situation that might arise. They must learn the proper responses to emergency situations. The first priority is to prevent injury and loss of life. The second priority is to prevent damage to equipment or surrounding structures.

5.1.0 Fire

Judgment is crucial in determining the correct response to fire. The first response is to cease crane operation and, if time permits, lower the load and secure the crane. In all cases of fire, evacuate the area even if the load cannot be lowered or the crane secured. After emergency services have been notified, a qualified individual may judge if the fire can be combated with a fire extinguisher. A fire extinguisher can be successful at fighting a small fire in its beginning stage, but a fire can get out of control very quickly. Do not be overconfident, and keep in mind that priority number one is preventing loss of life or injury to anyone. Even trained firefighters using the best equipment can be overwhelmed and injured by fires.

5.2.0 Malfunctions During Lifting Operations

Mechanical malfunctions during a lift can be very serious. If a failure causes the radius to increase unexpectedly, the crane can tip or the structure could collapse. Loads can also be dropped during a mechanical malfunction. A sudden loss of load on the crane can cause a whiplash effect that can tip the crane or cause the boom to fail.

The chance of these types of failures occurring in modern cranes is greatly reduced because of system redundancies and safety backups. However, failures do happen, so stay alert at all times.

If any mechanical problem arises, the operator should secure the crane, tag the controls out of service, and report the problem to a supervisor. The crane should not be operated until it is repaired by a qualified technician.

5.3.0 Hazardous Weather

Mobile crane operations generally take place outdoors. Under certain environmental conditions, such as extreme hot or cold weather or in high winds, work can become uncomfortable and possibly dangerous. For example, snow and rain can have a dramatic effect on the weight of the load and on ground compaction. During the winter, the tires, outriggers, and crawlers can freeze to the ground. This may lead the operator or rigger to the false conclusion that the crane is on stable ground. In fact, as weight is added during the lifting operation, it may cause an outrigger float, tire, or crawler to sink into the ground below the frozen surface. Severe rain can cause the ground under the crane to become unstable, resulting in instability of the crane due to erosion or loss of compaction of the soil. There are other specific things to be aware of when working under these adverse conditions.

5.3.1 High Winds and Lightning

High winds and lightning may cause severe problems on the job site (*Figure 9*). They are major weather hazards and must be taken seriously. Crane operators and riggers must be prepared for extreme weather in order to avoid accidents, injuries, and damages. It is very rare for high winds or lightning to arrive without some warning. This gives operators and riggers time to react appropriately.

Figure 9 Wind and lightning hazards.

High winds typically start out as less dramatic gusts. Operators and riggers must be aware of changing weather conditions, such as worsening winds, to determine if the weather is becoming hazardous. If high winds develop, the operator must secure crane operations as soon as it is practical. This involves placing the boom in the lowest possible position and securing the crane. Once this is done, all personnel should seek indoor shelter away from the crane.

Because cranes' booms extend so high and are made of metal, they are likely targets for lightning. Operators and riggers must be constantly aware of this threat. Lightning can usually be detected when it is several miles away. As a general rule of thumb, when you hear thunder, the lightning associated with it is not more than 6 to 8 miles away. Be aware, however, that successive lightning strikes can touch down up to 8 miles apart. That means once you hear thunder or see lightning, it is close enough to present a hazard.

In some high-risk areas, proximity sensors provide warnings when lightning strikes within a 20-mile radius. Once a warning is given or lightning is spotted, crane operations must be secured as soon as practical, following the crane manufacturer's recommendations for doing so.

WARNING

Even with the boom in the lowest position, it may be taller than surrounding structures and could still be a target for lightning strikes.

Once crane operations have been shut down, all personnel should seek indoor shelter away from the crane. Always wait a minimum of 30 minutes from the last observed instance of lightning or thunder before resuming work.

6.0.0 Using Cranes to Lift Personnel

Although using cranes to lift people was common in the past, OSHA regulations, as spelled out in *29 CFR 1926.550*, now discourage the practice. Using a crane to lift personnel is not specifically prohibited by OSHA, but the restrictions are such that it is only permitted in special situations where no other method is suitable. When it is allowed, certain controls must be in place, including the following:

- The rope design factor is doubled.
- No more than 50 percent of the crane's capacity, including rigging, may be used.
- Free-falling is prohibited.
- **Anti-two-blocking devices** are required on the crane boom.
- The platform must be specifically designed for lifting personnel.
- Before the personnel basket is used, it must be tested with appropriate weight, and then inspected.
- Every intended use must undergo a trial run with weights rather than people.

Figures 10 and *11* illustrate the requirements for personnel lifts.

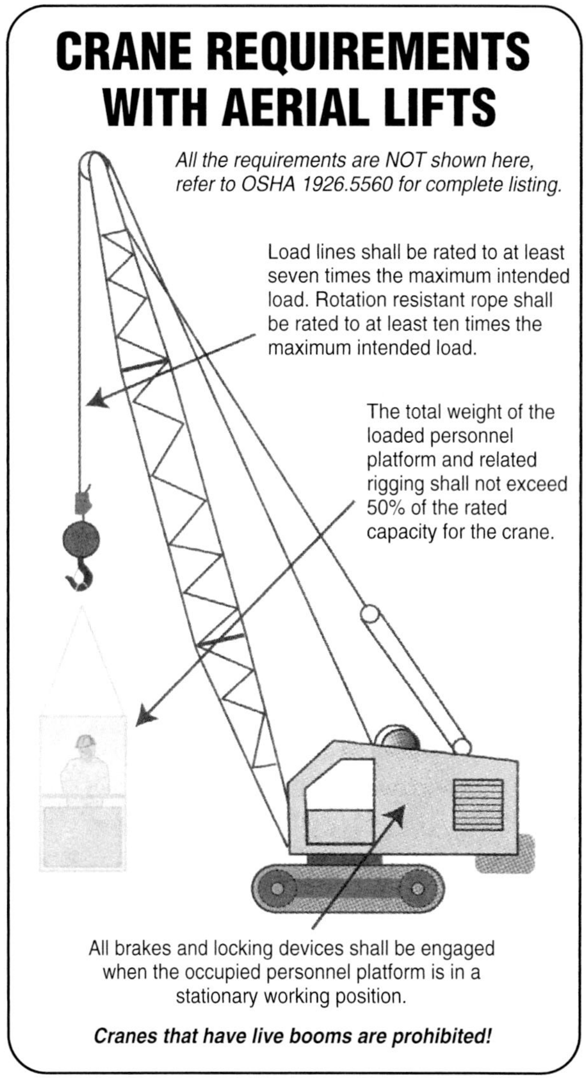

This information provides a generic, non-exhaustive overview of the OSHA standard on suspended aerial lifts. Standards and interpretations change over time, you should always check current OSHA compliance requirements for your specific requirements.

29 CFR 1926.550 addresses the use of personnel hoisting in the construction industry, and 29 CFR 1910.180 addresses the use of personnel hoisting in general industry.

107F10.EPS

Figure 10 Crane requirements with personnel platforms.

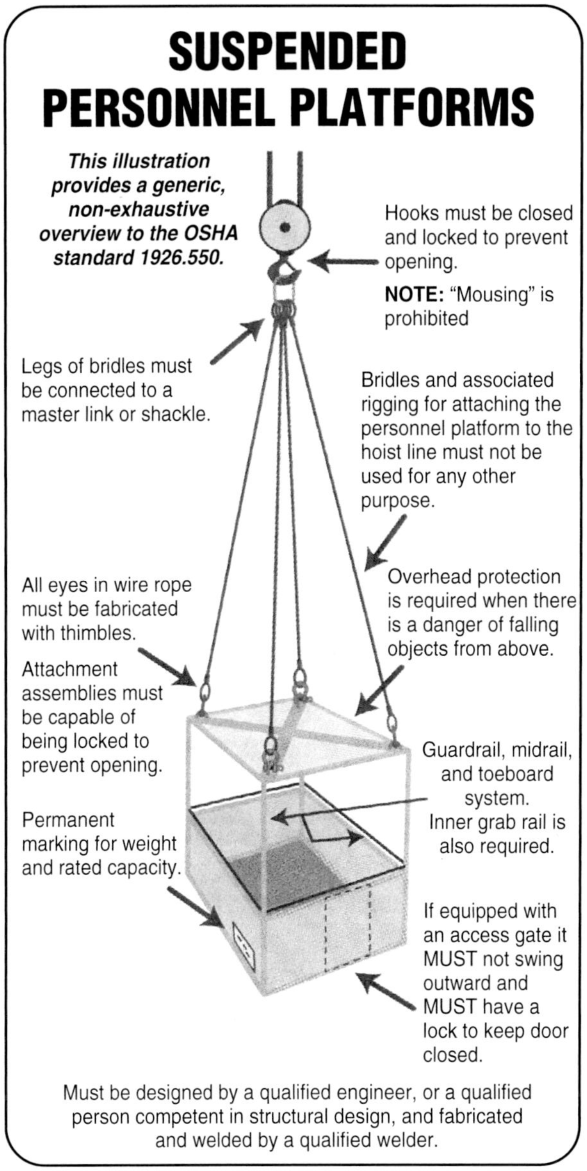

The OSHA rules on crane suspended personnel platforms contain many specifics that are not covered in this book. Refer to 29 CFR 1926.550 for the currrent OSHA compliance requirements.

107F11.EPS

Figure 11 Suspended personnel platforms.

6.1.0 Personnel Platform Loading

The personnel platform must not be loaded in excess of its rated load capacity. The number of employees, along with material, occupying the personnel platform must not exceed the limit established for the platform and the rated load capacity or the maximum intended load.

Personnel platforms must be used only for employees, their tools, and the materials necessary to do their work, and must not be used to hoist only materials or tools when not hoisting personnel.

Materials and tools for use during a personnel lift must be secured to prevent displacement. These items must be evenly distributed within the confines of the platform while the platform is suspended.

6.2.0 Personnel Platform Rigging

When a wire-rope bridle sling is used to connect the personnel platform to the load line, each bridle leg shall be connected to a master link or shackle in such a manner as to ensure that the load is evenly divided among the bridle legs (*Figures 12* and *13*).

Hooks on headache ball assemblies, lower load blocks, or other attachment assemblies shall be of a type that can be closed and locked, eliminating the hook throat opening. Alternatively, an alloy anchor type shackle with a bolt, nut, and retaining pin may be used.

Wire rope, slings, shackles, rings, master links, and other rigging hardware must be capable of supporting, without failure, at least five times the maximum intended load applied or transmitted to that component. Where rotation-resistant rope is used, the slings shall be capable of supporting without failure at least ten times the maximum intended load. All eyes in wire rope slings shall be fabricated with thimbles.

Bridles and associated rigging for attaching the personnel platform to the hoist line shall be used only for the platform and the necessary employees, their tools, and the materials necessary to do their work, and they shall not be used for any other purpose when not hoisting personnel.

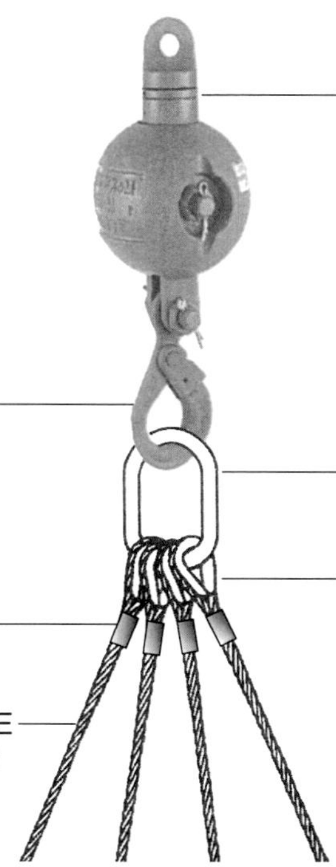

107F12.EPS

Figure 12 Bridle sling using a master link.

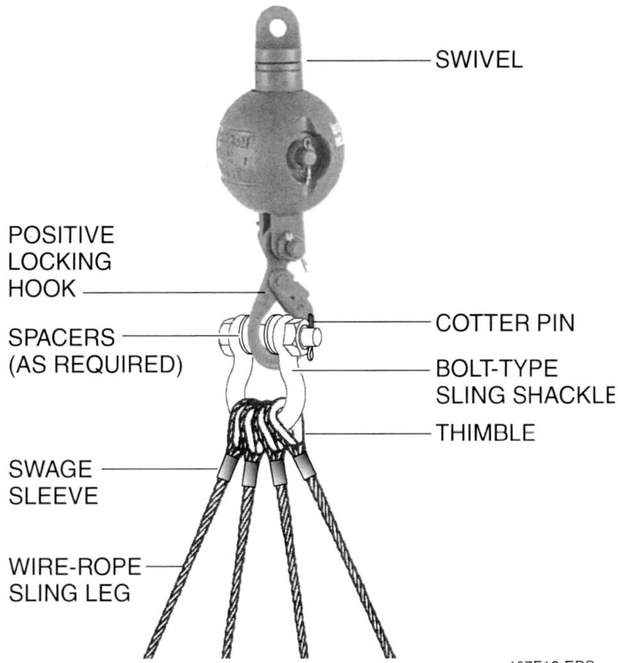

Figure 13 Bridle sling using a shackle.

7.0.0 LIFT PLANNING

Before conducting a lift, most construction sites or companies require that a lift plan be completed and signed by competent personnel. Lift plans are mandatory for steel erection and multiple-crane lifts. A lift plan contains information relative to the crane, load, and rigging, and it lists any special instructions or restrictions for the lift. It is important to remember that any deviation from the original lift plan, no matter how small, requires a new lift plan.

7.1.0 Lift Plan Data

Critical lifts always require a lift plan. As a matter of policy, all personnel involved in a lift should require or prepare a plan for all lifts. The rigger is particularly concerned with crane placement, rope, and sizing of slings. A typical lift plan for either a critical or ordinary (minor) lift breaks down the following elements and information:

- Weight
  - Equipment condition
  - Weight empty
  - Weight of block
  - Weight of lifting bar
  - Weight of slings and shackles
  - Weight of jib (if erected or stowed and not used)
  - Weight of overhaul ball on jib
  - Weight of wire rope
  - Allowance for unaccounted material in equipment
- Jib
  - Jib to be used
  - Length of jib
  - Angle of jib
  - Rated capacity of jib
- Crane placement
  - Smooth solid foundation within one percent of grade
  - Electrical hazards in area
  - Obstacles or obstructions to lift or swing
  - Swing direction and degree
  - Area of operation (over-the-front, over-the-rear, over-the-side)
- Rope
  - Number of parts of rope
  - Size of rope
  - Reeving capacity
- Sizing and types of slings
  - Sling selection
  - Shackle selection
  - Sizing and types
- Crane
  - Type of crane
  - Crane capacity
  - Lifting arrangement
- Other
  - Special instructions or restrictions for crane, rigging, or lift
  - Diagrams for crane, load placement, and rigging configuration
  - Signature of job supervisor, lift plan checker, and operator
  - Derigging plan

8.0.0 LOAD BALANCING

Load balancing is equalizing the weight of the load at various lifting points. If the weight is concentrated at one point, the object being lifted may bend, break, or fall. Distributing the weight over several lifting points ensures that the load is balanced and lifted safely.

8.1.0 Center of Gravity

The load's center of gravity is the point around which its weight is concentrated. The location of a load's center of gravity directly affects its stability.

To understand the relationship between a load's center of gravity and stability, you must understand the principle of leverage. The principle of leverage can be illustrated with a simple teeter-totter (*Figure 14*). When the weight of the heavy load multiplied by the distance (X) from its center of gravity to the tipping fulcrum (the point where the load is carried) is equal to the weight of the lighter load multiplied by the distance (Y) from its center of gravity at its tipping fulcrum, a condition of balance has occurred.

This relationship is illustrated again with a single load in *Figure 15.*

The weight and center of gravity of various loads are determined by calculation or by weighing. This data then becomes the basis upon which stability ratings are calculated. The ratings and accuracy of the weight and center-of-gravity data are then confirmed by testing.

The center of gravity can change depending on the crane's quadrant of operation. When the crane is in over-the-side, over-the-front, or over-the-rear configurations, the center of gravity is shifted either closer or farther away from the tipping fulcrum. This, in turn, may increase or decrease the crane's leverage. For a rough-terrain crane, maximum true lifting capacity (greatest stability margin) is obtained when the crane boom is operated directly over the front. Maximum true lifting capacity of a truck crane is obtained when the boom is over the rear.

8.2.0 Sling Angles

The angle formed by the legs of a sling with respect to the horizontal when tension is put upon the load is called the sling angle. The amount of tension on the sling is directly affected by the angle of the sling. For this reason, proper sling angles are crucial to safe rigging. *Figure 16* shows the effect of sling angles on sling loading. To actually determine the load on a sling, a factor table is used. *Table 2* shows a sling factor table.

In the example shown in *Figure 16,* two slings are being used to lift 2,000 pounds. When the slings are at a 45-degree angle when measured from the horizontal (90 degrees between the legs), there are 1,414 pounds of tension on each sling. This can be determined mathematically by taking the weight of the load (2,000 pounds) and dividing it by the number of slings (2); this equals 1,000 pounds. Then, multiply 1,000 pounds by the corresponding factor in *Table 2* for 45 degrees (1.414). The result is 1,414 pounds of tension on each sling based on the sling angle.

Optimum sling angles fall between 90 degrees and 60 degrees to horizontal (0 degrees and up

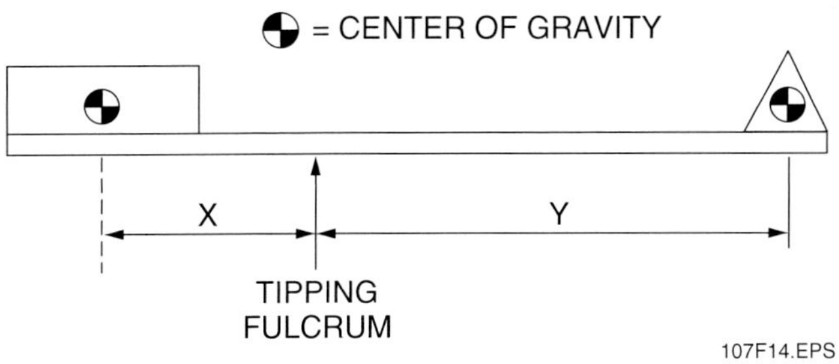

Figure 14 Teeter-totter demonstrating leverage.

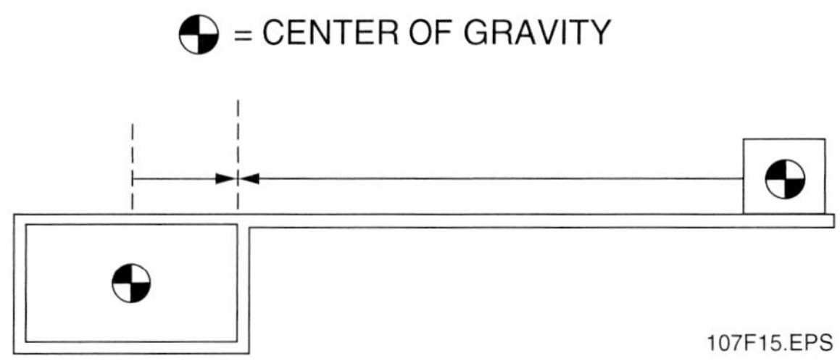

Figure 15 Load hung from center of gravity.

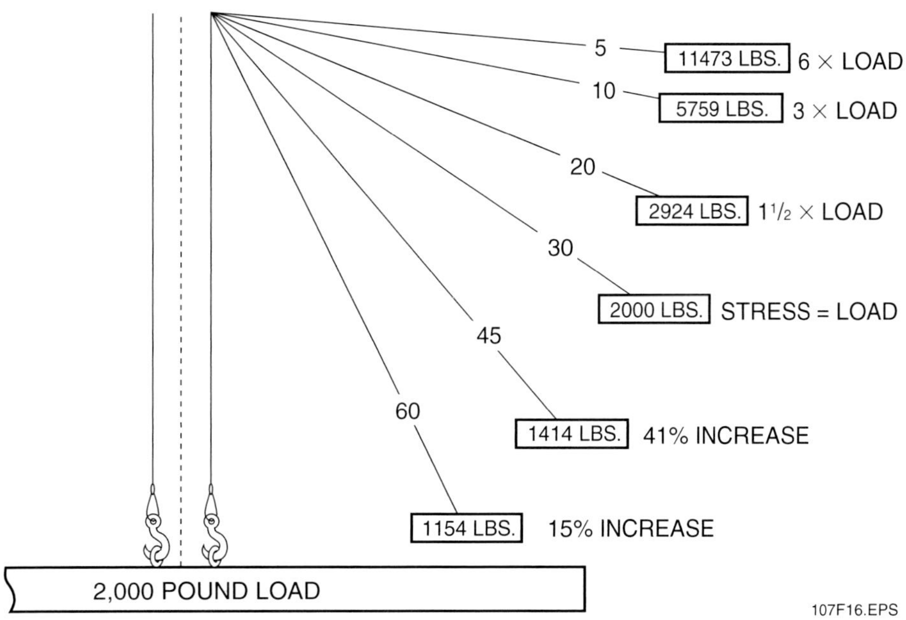

Figure 16 Sling angles.

Table 2 Sling Factor Table

SLING ANGLE	LOAD ANGLE FACTOR
5	11.490
10	5.747
15	3.861
20	2.924
25	2.364
30	2.000
35	1.742
40	1.555
45	1.414
50	1.305
55	1.221
60	1.155
65	1.104
70	1.064
75	1.035
80	1.015
85	1.004
90	1.000

107T02.EPS

to 60 degrees between sling legs). Further examination of *Figure 16* shows that the tension on the sling legs is higher when the legs are positioned at an angle of 30 degrees relative to the horizontal (120-degree angle between the legs) than when the legs are at an angle of 60 degrees relative to the horizontal (60-degree angle between the legs). Angles beyond this range are considered hazardous.

8.3.0 Lifting Connectors

Many pieces of equipment and machinery come with lifting connectors already attached at proper locations for balanced lifting. Lifting connectors may be manufactured or field-fabricated only if the lifting connectors have been engineered. Field-fabricated connectors must be fabricated of approved material and installed using approved methods to ensure that they are of adequate capacity for the load being lifted. Some common lifting connectors are **lifting lugs** and lifting eyebolts.

8.3.1 Lifting Lugs

Lifting lugs are welded, bolted, or pinned to the object to be lifted. They are designed and located to balance the load and support it safely. Unless angular lifts are specifically allowed by the manufacturer, lifting lugs should be used for straight, vertical lifting only, because they will bend if loaded at an angle. *Figure 17* shows a lifting lug.

8.4.0 Choking

When an item more than 12 feet long is being rigged, the general rule is to use two choker hitches spaced far enough apart to provide the stability required to transport the load.

To lift a bundle of loose items, or to maintain the load in a certain position during transport, a double-wrap choker hitch (*Figure 18*) may be useful. The double-wrap choker hitch is made by wrapping the sling completely around the load, and then wrapping the choke end around again and passing it through the eye like a conventional choker hitch. This enables the load weight to produce a constricting action that binds the load into the middle of the hitch, holding it firmly in place throughout the lift.

Forcing the choke down will drastically increase the stress placed on the sling at the choke point. To gain gripping power, use a double-wrap choker hitch. The double-wrap choker uses the load weight to provide the constricting force, so there is no need to force the sling down into a tighter choke.

A double-wrap choker hitch is ideal for lifting bundles of items, such as structural steel. It will also keep the load in a certain position, which makes it ideal for equipment installation lifts. Lifting a load longer than 12 feet requires two of these hitches.

Use the following guidelines when choking a load of beams:

- Beams of the same length should be lifted together. If beams of greatly different lengths

107F17.EPS

Figure 17 Lifting lug.

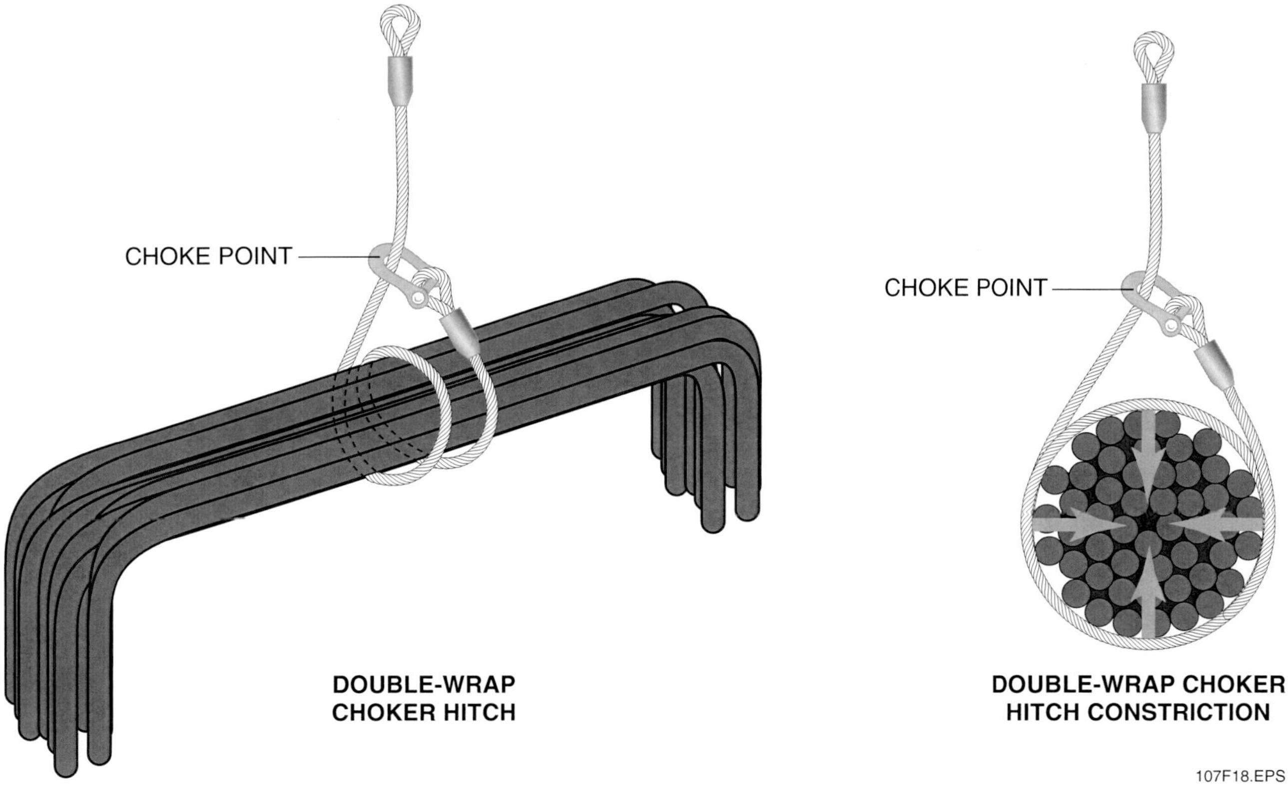

Figure 18 Double-wrap choker hitch and double-wrap choker hitch constriction.

must be lifted, lift each different length separately.

- Lay the beams on blocks so the chokers can be wrapped around the load.
- Use two choker hitches to lift a load of beams. This allows a level lift to be made and ensures that the beams will not slip out of the sling. Make sure both chokers grab all beams.
- Ensure that the open part of the hook is facing away from center when using slings with fixed or sliding hooks.
- Do not use carbon steel chokers on painted or stainless steel beams because of cross-contamination of the metals. Use nylon web slings on stainless steel beams.

When rigging any piece with sharp corners, the lifting ropes or slings should be protected with **softeners**. These may be pieces of wood, or rounded material such as split pipe or folded heavy cardboard, pinched under the rope or sling.

8.4.1 D/d Ratio

The D/d ratio (*Figure 19*) is the relationship between the pitch diameter of the sheave and the rope diameter. This factor is important in establishing the fatigue resistance or relative service life of the wire rope. When the tread D/d ratio is too small, it can increase the rope and sheave wear significantly. This wear can be directly attributed to the increased bending stress applied to the rope and sheave when the wire rope is forced into too tight a bend.

8.5.0 Lifting

Follow these guidelines when lifting a load:

- Use the hook and lifting line of the crane as the center line with which to line up the load.
- Be sure to check the boom angle of the crane to make sure that the load will not be moved forward or backward when tension is put on the line.
- Stand clear of the load and where the operator can clearly see you, and signal any necessary adjustments, using the proper hand signals. Never stand underneath the load unless you are a designated person.
- As soon as the load clears the ground, check its orientation. If it is not level, signal the operator to stop lifting the load, and guide the load back onto the blocks to correct the position of the chokers.
- Handle all loads with a tag line.
- Keep the load as low to the ground as possible.

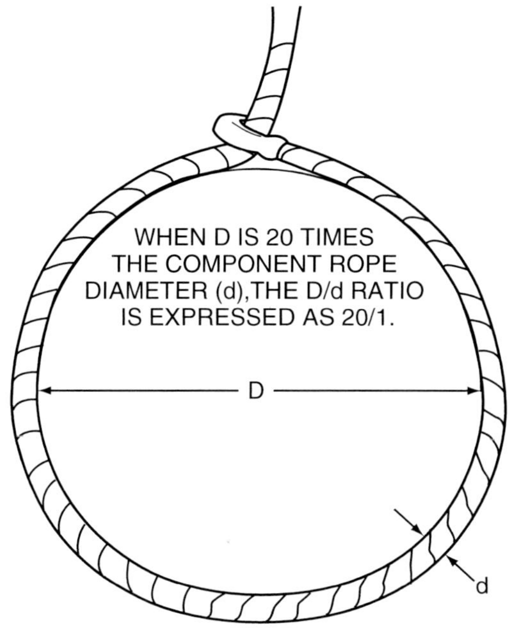

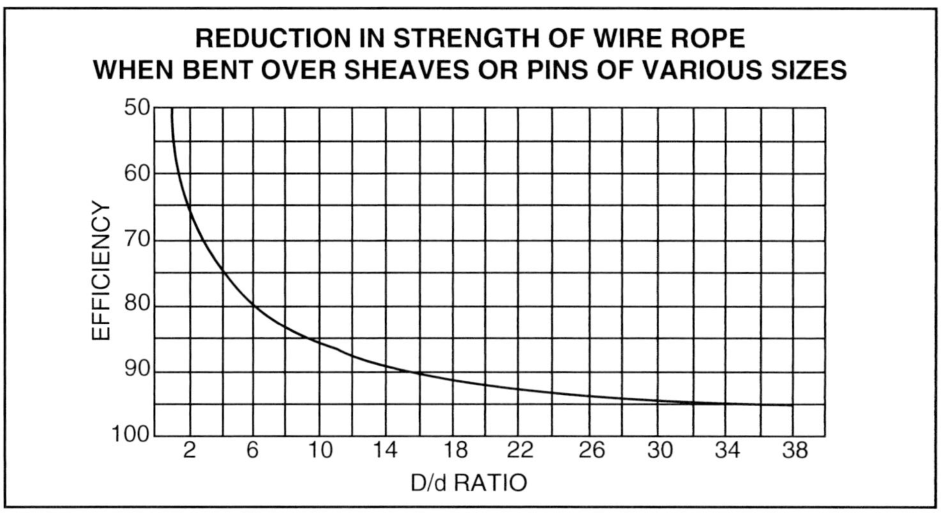

107F19.EPS

Figure 19 Load and hook diameter and D/d ratio chart for wire rope.

8.6.0 Block Twisting

Block twisting is frequently encountered when using a load block. This problem is directly associated with the construction of wire rope. When the load is placed on the wire rope, it has a tendency to unlay or remove the twist inserted during the manufacturing process. As the load increases, the amount of unlaying will increase. This untwisting or unlaying will rotate the load block. The rotation of the load block can cause severe problems during a lift, up to and including binding of the load block and sheaves.

Several actions can be taken to minimize the amount of block twisting that occurs. First, reduce the wire rope length. Long wire rope is more susceptible to block twisting than shorter ropes. Second, reduce the load. Block twisting increases at higher loads. Reducing the load and the amount of shock loading decreases block twisting.

Another way to reduce block twisting is to remove the odd part of line by reeving the block to an even number of parts. The dead end of the rope should be terminated at the boom to increase its separation from the other parts of line. Another action to increase the separation of the parts of line is to use larger sheaves; this reduces the amount of torque applied to the block.

An additional method is to use a tag line attached to the load. The block must be locked. This provides a stabilizing counter-torque to the block.

Rotation-resistant rope is used in some applications with the manufacturer's approval, but it does not guarantee that the rope won't twist.

9.0.0 GUIDELINES FOR UNLOADING AND YARDING MATERIALS

Structural steel components are ordered from the fabricator. Delivery of these materials is usually scheduled to coincide with the construction schedule. Structural steel usually arrives at the job site on flatbed trucks or tractor trailers. If a **spur track** is available at the job site, shipments may be received by flat or gondola railroad cars. Where delivery of steel is scheduled to meet daily placement requirements, truckloads are delivered to the points of placement. Workers should be aware of safety factors that are necessary to achieve safe, efficient handling, storage, and hoisting of structural materials.

9.1.0 Calculating Weight of an Object

It is frequently necessary for workers to calculate the weight of an object, especially when the object must be rigged for transport. To do so, calculate the volume of the object, in cubic feet, and multiply the volume by the weight per cubic foot shown for that material in *Table 3*. For example, the volume and weight of a rolled steel bar 1 foot wide, 2 feet in height, and 15 feet long can be found by calculating the volume of the rectangle using the formula:

$$V = LWH$$

Now substitute values into the equation:

$$V = 15 \times 2 \times 1 = 30 \text{ cubic feet}$$

Now that you have the volume of the bar in cubic feet, determine the weight of the bar in pounds by multiplying the volume by the weight of a cubic foot:

$$W = 30 \times 495 = 14{,}850 \text{ pounds, or } 7.425 \text{ tons}$$

If a solid form has a large hollow area, first calculate the volume of the entire object, then subtract the volume of the open area from the total volume. For example, a housing that is rectangular might have a 24-inch OD shaft passing through the middle. The housing is 4 feet wide, 4 feet deep, and 10 feet long. The calculation of the volume of the rectangular solid is:

$$4 \text{ feet} \times 4 \text{ feet} \times 10 \text{ feet} = 160 \text{ cubic feet}$$

The shaft is 24 inches in diameter; the formula to calculate its volume is:

$$V = \pi r^2 h$$

The radius (r) is half the diameter. Substituting the numbers,

$$\pi \times 1 \times 1 \times 10 = 31.416 \text{ cubic feet}$$

Subtract the shaft volume from the total volume, and you find the housing volume to be 128.584 cubic feet. If the housing is cast iron, it weighs:

$$128.584 \times 450 = 57{,}863 \text{ pounds}$$
(rounded to the nearest pound)

Table 3 Weight of Common Materials

Material	Weight in lbs/cu ft
Aluminum	165
Cast Brass	134
Cast Copper	542
Cast Iron	450
Concrete	150
Dense Earth	125
Lead	708
Rolled Nickel	541
Rolled Steel	495
Tungsten	1224
Water	62.4

9.2.0 Unloading

Trucks should be promptly unloaded to prevent job-site clutter and potential safety hazards. Structural materials that are stored on the ground must be placed on timbers or other suitable blocking to keep them free from mud and to allow safe and easy rehandling.

Tag lines are required when bundles are being hoisted. Tag lines guide the load as it is being hoisted. Fiber rope is normally used for this purpose. Safe hoisting requires a skilled crane operator and an authorized signal person. Operators take their directions from the signal person using hand signals, radios, or telephone headsets.

9.3.0 Using Slings

Slings must be used to lift bundles. Choose a sling appropriate to the material being lifted. Check the weight of each bundle on the bill of lading, and ensure that the sizes of the wire rope slings, hooks, and shackles are safe. Chokers and slings of sufficient strength must be selected to lift the load.

The load on each sling depends on the number of slings or chokers, the angle of the choker, and the total load. Although the total weight lifted can be divided between the supporting slings, the force of the load is down. The greater the angle of the sling, the greater the tension on the sling. If the hoisting line and the slings are the same material and size, the tension on each sling must be less than or equal to the tension in the hoist line. This means that the strength of the hoist line determines the maximum load-lifting capability of the combination.

If two slings are used to lift bundles of structural iron, always thread the loops in the same direction. If possible, double wrap the slings to help prevent slipping of the slings and items in the bundle.

10.0.0 Unloading Joists

Most joists are delivered to the job site by truck. Unloading procedures vary depending on the type of truck used for the final delivery. Joists should be lifted carefully and placed onto suitable cribbing. They should never be dropped. Proper equipment must be used for unloading, and only load-rated rigging should be used. Whenever possible, joists should be unloaded by bundles.

WARNING

To avoid personal injury or death, never lift bundles by their strapping. Always make sure that the load has not shifted before removing or securing hardware.

The strapping is not strong enough to safely support the weight of the bundle of joists. The purpose of the strapping is to contain the joists in a bundle during shipping and storage. Thread cables underneath the bottom chords (which are facing up) of the joists at approximately ⅓ points and attach the cables to the crane (*Figure 20*).

CAUTION

Do not attach cables, chains, or hooks to the web members. Lifting the joists by the web members can bend the web and ruin the joists.

Slowly lift the bundle to determine the load balance. If the load is unbalanced, lower the bundle, adjust the cables, and lift again. Lifting bundles with a single cable is dangerous, because it requires nearly perfect balance. If balance is not obtained, one end of the bundle falls and the other rises. Then, the bundle tends to shift, creating greater imbalance and an unmanageable and dangerous load.

10.1.0 Flatbed Trailers

Chains or webbing are wrapped around the entire load of joists to prevent them from falling off the side of the trailer and from shifting back and forth during transit. The crane is positioned over the trailer and the lifting cables are threaded under the bottom chords of the top bundle of joists. The chains (or webbing) are then repositioned over the next layer of joists before lifting the top bundles from the truck. This process of rigging the top layer and repositioning the straps or chains is repeated as each layer is removed from the trailer.

11.0.0 Miscellaneous Iron

Most miscellaneous or ornamental metal stock is received in banded packages or bundles. If cranes are not available for unloading, lift trucks can be used. When metal stock is delivered in banded bundles, two pieces of 4 × 4 wood blocking are usually banded to the bottom of the package for support in loading and handling.

Always inform the supplier about what sizes and methods of banding or crating to use for miscellaneous metal stock. These specifications depend on the capacity and type of unloading and handling equipment available. Sometimes, 4-inch

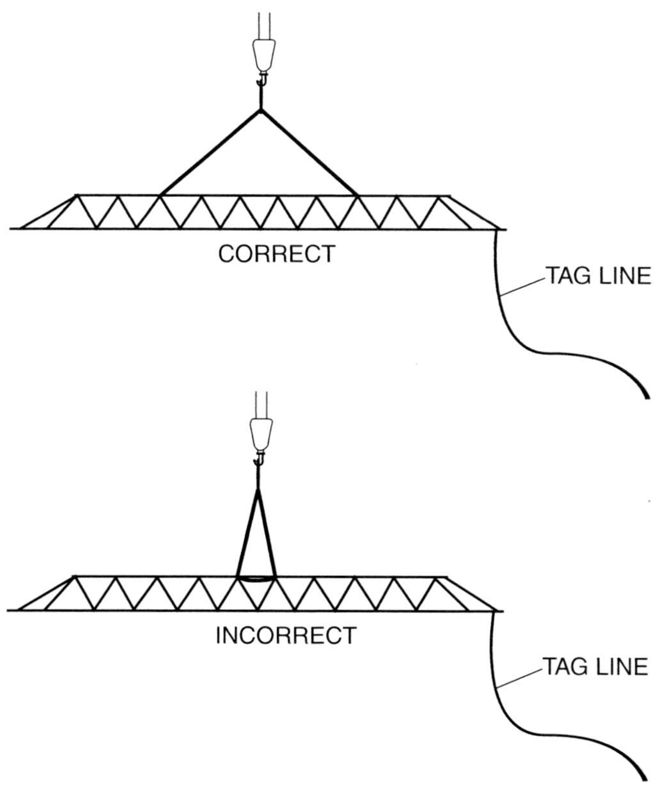

Figure 20 Correct and incorrect methods of rigging a bundle of joists.

blocks are placed under the attached skids so that the lift trucks or plate lifter attachment for cranes or hoists can be easily placed under the stock for unloading.

When miscellaneous metal is handled by crane, always place slings securely and use standard signaling procedures. Loads should not be carried over machinery work areas or workers.

12.0.0 Structural Iron

On large jobs, the construction site is big enough to have a storage yard where steel can be stored until needed. In congested areas where the building fills nearly all of the land area, a nearby storage site must be set up. If a yard is used for unloading and storing shipments before they are needed at the site or for the assembly of trusses, girders, or other components, the yard must be set up in time to be ready for the first steel delivery.

Steel is normally shipped by rail or highway. A yard some distance from the site of the structure is often set up for unloading the steel. A storage yard may be a necessity if the fabricator must ship steel before it is needed at the site. The erector may also need a yard to sort the steel into areas or smaller bundles. Store steel in a level position whenever possible. **Yarding** requires consideration of access to individual materials and safe storage and transportation of material as it is needed.

12.1.0 Unloading

Tools and equipment needed to set up the unloading equipment should be delivered well in advance of the first shipment of steel. The unloading and assembly schedule must accommodate the arrival schedule. Qualified personnel must be available to operate all unloading equipment.

Individual pieces must be placed on the ground with identifying marks or numbered sides visible from the same direction so that they can be found when needed. Beams are often identified by the nature of their end connections, then verified by the markings. It is good to have a complete set of shop drawings at the job storage and handling site.

If the components are small, the steel can be unloaded by hand. The hoist boom swings the piece to rest on blocking that has been prepared. This makes it easier to pass chokers under each piece at a later time. Cribbing is often used for blocking. Blocking should be arranged so that the materials are left in a safe condition. When lifting a load, remember that it will tend to drift away from the crane. Personnel on the ground should stay on the crane side of the load. Cut open banded bundles while the load is suspended, then lower the load onto the cribbing.

When pieces are piled on top of each other, or girders or trusses are stood upright, be sure to block or land them so that they cannot fall and injure workers. They should be placed so that girder dogs, slings, or other lifting devices can be used later without additional handling of the pieces.

12.2.0 Storage

All materials must be stored so that they are accessible for delivery and erection. Columns should be landed so that if piled two or three layers high with 4×4 timber blocking between layers, the top ones are those that are needed first and the bottom ones used last. Beams should be piled with blocking between drafts in the same order. Bolts and shear studs are often delivered in kegs and should be kept sealed, dry, and out of the weather.

If any assembly of trusses or girders is necessary, it should be performed in a level, clear area. Timber blocking is laid so that the truss or girder will be level and the pieces readily assembled. Sorting is accomplished by placing pieces that will be together in the structure into groups. These groups are then moved to a location near the erection location.

After all steel has been delivered to the yard, reloaded, and sent to the job site, the unloading equipment can be released to the construction site or shipped away to the next job. After this is complete, the yard area should be restored to its original condition.

Summary

The selecting and setting up of hoisting equipment, hooking cables to the load to be lifted or moved, and directing the load into position are all part of the process called rigging. Performing this process safely and efficiently requires selecting the proper equipment for each hoisting job, understanding safety hazards, and knowing how to prevent accidents. Riggers must have the greatest respect for the hardware and tools of their trade because their lives may depend on them working correctly.

The size of a crane and the huge loads it handles create a large potential for danger to anyone in the vicinity of the crane. The crane operator and rigging workers must be vigilant to ensure that the crane avoids power lines and that site personnel are not endangered by the movement of the crane or the load. Because of tipping hazards, it is important to make sure a crane is on level ground with outriggers extended, if applicable, before lifting a load. Although using cranes to lift personnel was once a common practice, it is discouraged by OSHA, and can only be done in special situations, and under special conditions.

Review Questions

1. Accidents can be reduced because most result from _____.
   a. human error
   b. faulty equipment
   c. false economy
   d. excess caution
2. If each employee is committed to safety awareness, the number of employees injured can be _____.
   a. cut in half
   b. studied
   c. maintained
   d. reduced
3. The most important rigging precaution is determining the _____.
   a. number of pieces being lifted
   b. shape of the loads
   c. weight of all loads
   d. owner of each piece
4. When the load weight is difficult to assess, attach _____.
   a. hooks to bands
   b. safe-load indicators
   c. shackles to chains
   d. extra slings and shackles
5. It is important that the load to be rigged is stable.
   a. True
   b. False
6. When calculating the weight of the load, the rigging and related hardware should *not* be considered.
   a. True
   b. False
7. Safe working loads of hoisting equipment are applicable only to plumb hoist lines and _____.
   a. side loads
   b. freely suspended loads
   c. extremely heavy loads
   d. loads stabilized by tag lines
8. When safe working loads are determined, they are based on _____ conditions.
   a. ideal
   b. average
   c. rare
   d. worst-case
9. To allow controlled rotation of the load for positioning, a(n) _____ is used.
   a. tag line
   b. equalizer beam
   c. hoist line
   d. safe-load indicator
10. Tag lines are attached after the rigger verifies that the load is balanced.
    a. True
    b. False
11. When using two or more slings on a load, the rigger must ensure that _____.
    a. the sling angle is greater than 30 degrees relative to the horizontal
    b. all slings are made from the same material
    c. only chain slings are used
    d. only wire rope slings are used
12. For a crane in operation, the minimum safe distance from power lines of up to 50,000 volts (50kV) is _____ feet.
    a. 10
    b. 15
    c. 20
    d. 25
13. If a crane boom comes into contact with a power line, the operator should _____.
    a. immediately jump off the crane
    b. stop and turn off the crane
    c. try to back the crane away from the power line
    d. find out if the line is energized
14. High winds and lightning, which pose danger to cranes, often arrive without warning.
    a. True
    b. False

15. When cranes are used to lift people, the lift capacity of the crane is _____ percent as a safety factor.
    a. increased 10
    b. increased 20
    c. decreased 20
    d. decreased 50

16. When two slings positioned at a sling angle of 60 degrees are being used to lift a 1,000-pound load, what is the tension on each sling? Assume a sling angle load factor of 1.155.
    a. 577.5 pounds
    b. 652.5 pounds
    c. 777.5 pounds
    d. 1,000 pounds

17. A lifting device that is welded, bolted, or pinned to an object to be lifted is a(n) _____.
    a. equalizer
    b. seizing
    c. lifting lug
    d. beam clamp

18. A piece of material placed to prevent abrasion of slings by the sharp corners of loads is a _____.
    a. sheave
    b. softener
    c. choker
    d. housing

19. Special lines of railroad tracks that allow cars to be left at a site for unloading are _____.
    a. yard tracks
    b. spur tracks
    c. car tracks
    d. spare tracks

20. The purpose of joist strapping is to _____.
    a. provide lifting points
    b. support the weight of the bundle during lifting
    c. provide attachments for lifting cables
    d. contain the joists in a bundle during shipping and storage

Trade Terms Introduced in This Module

Anti-two-blocking devices: Devices that provide warnings and prevent two-blocking from occuring. Two-blocking occurs when the lower load block or hook comes into contact with the upper load block, boom point, or boom point machinery. The likely result is failure of the rope or release of the load or hook block.

Blocking: Pieces of hardwood used to support and brace equipment.

Center of gravity: The point at which an object is perpendicular to and balanced in relation to the earth's gravitational field.

Cribbing: Timbers stacked in alternate tiers. Used to support heavy loads.

Lifting lugs: Metal attachments welded to a metal item. The lug has a hole to which a shackle can be attached, allowing the piece to be lifted.

Softener: A piece of material that can be placed so as to prevent abrasion or cutting of slings or wire ropes by the sharp corners of loads.

Spur tracks: Special lines of railroad tracks that allow flat cars of material to be left at a site for unloading, so as not to interfere with railroad operations. Spur tracks may be single-ended or double-ended, depending on whether the spur returns to the main line or not.

Yarding: The unloading and placement of material in the site area, including placement, storage and movement of material.

POWERLINE AWARENESS PERMIT

POWER LINE AWARENESS PERMIT

Today's Date ___________ **Job Number** ___________________________________

Contractor Name			
Job Address			
Telephone Number		**Fax Number**	

Emergency Contact Number	

Survey

Before beginning any project, you must first survey your work area to find power lines at the job site. (See job-site sketch on reverse side.)

Identify

After finding all of the power lines at your site, identify the activities you'll be doing that may put you or your workers at risk. Mark one or more of the following:

- ☐ Cranes (mobile or truck mounted)
- ☐ Drilling rigs
- ☐ Backhoes/excavators
- ☐ Long-handled tools
- ☐ Other tools/high-reaching equipment
- ☐ Concrete pumper
- ☐ Aerial lifts
- ☐ Dump trucks
- ☐ Ladders
- ☐ Material handling & storage
- ☐ Scaffolding
- ☐ Other___________________

Eliminate or Control

After identifying the power line and high-risk activities on the job site, determine how to eliminate or control the risk of electrocution (a successful determination is often reached only after consultation with the utility).

Mark one or more of the following:

- ☐ Move the activity
- ☐ Change the activity
- ☐ Have the utility de-energize the power line
- ☐ Have the utility move the power line
- ☐ Use barrier protection (insulated sleeves)
- ☐ Use an observer
- ☐ Use warning lines with flags
- ☐ Use non-conductive tools
- ☐ Use a protective technology:
  - ☐ Insulated link
  - ☐ Boom cage guard
  - ☐ Proximity device

Always maintain your minimum safe clearance distance from the power line, except when the utility has de-energized and visibly grounded the power line.

Voltages	Distance from Power Line
Less than 50 kV	10 Feet
More than 50 kV	10'+(0.4")(# of kV over 50 kV)

WARNING!
It is unlawful to operate any piece of equipment within 10' of energized lines

CONSTRUCTION SAFETY COUNCIL

107A01.EPS

Job-Site Sketch

(Draw in location of power lines and their proximity to construction site, including such things as proposed excavations, location of heavy equipment, scaffolding, material storage areas, etc.)

Completed by______________________________ **Date**____________

Title______________________________

Approved by______________________________ **Date**____________

Title______________________________

CONSTRUCTION SAFETY COUNCIL

107A02.EPS

Additional Resources

This module presents thorough resources for task training. The following resource material is suggested for further study.

Crane Safety on Construction Sites, 1998. Task Committee on Crane Safety on Construction Sites. Reston, VA: ASCE.

Rigging Handbook, 2003. Jerry A. Klinke. Stevensville, MI: ACRA Enterprises, Inc.

Figure Credits

Topaz Publications, Inc., Module opener, Figures 1, 2, 5, and 17

Link Belt Construction Equipment Co., Figure 3

Cianbro Corporation, Figure 4

Rigging Handbook, 2003, by Jerry Klinke, ACRA Enterprises, Stevensville, MI, Figures 10 and 11

The Crosby Group, Inc., Figures 12 and 13 (photos)

Wire Rope Technical Board, Figures 12 and 13 (illustrations)

Construction Safety Council, Appendix

NCCER CURRICULA — USER UPDATE

NCCER makes every effort to keep its textbooks up-to-date and free of technical errors. We appreciate your help in this process. If you find an error, a typographical mistake, or an inaccuracy in NCCER's curricula, please fill out this form (or a photocopy), or complete the online form at **www.nccer.org/olf**. Be sure to include the exact module ID number, page number, a detailed description, and your recommended correction. Your input will be brought to the attention of the Authoring Team. Thank you for your assistance.

Instructors – If you have an idea for improving this textbook, or have found that additional materials were necessary to teach this module effectively, please let us know so that we may present your suggestions to the Authoring Team.

NCCER Product Development and Revision

13614 Progress Blvd., Alachua, FL 32615

Email: curriculum@nccer.org

Online: www.nccer.org/olf

❑ Trainee Guide ❑ Lesson Plans ❑ Exam ❑ PowerPoints Other ______________________

Craft / Level: Copyright Date:

Module ID Number / Title:

Section Number(s):

Description:

Recommended Correction:

Your Name:

Address:

Email: Phone:

Trade Drawings One

30108-11

Trainees with successful module completions may be eligible for credentialing through NCCER's National Registry. To learn more, go to **www.nccer.org** or contact us at **1.888.622.3720**. Our website has information on the latest product releases and training, as well as online versions of our *Cornerstone* newsletter and Pearson's Contren® product catalog.

Your feedback is welcome. You may email your comments to **curriculum@nccer.org,** send general comments and inquiries to **info@nccer.org**, or use the User Update form at the back of this module.

V.1 1/11

30108-11
Trade Drawings One

Objectives

When you have completed this module, you will be able to do the following:

1. Identify the materials used in steel-frame buildings.
2. Name the parts of steel frames.
3. Interpret symbols used on plans and drawings, including symbols for:
   - Structural steel
   - Ornamental ironwork
   - Welding

Performance Tasks

Under the supervision of the instructor, the trainee should be able to do the following:

1. Identify job plans and drawings used for ironworking jobs.
2. Identify the symbols used on selected ironworking plans and drawings.
3. Identify selected structural steel symbols and applications on job plans and drawings.
4. Identify selected ornamental ironwork and welding symbols and applications on job plans and drawings.

Trade Terms

Bent
Connections
Purlin
Shop drawings
Spandrel
Truss

Contents

Topics to be presented in this module include:

Figures and Tables

1.0.0 INTRODUCTION

Large industrial and commercial buildings are constructed around a framework of reinforced concrete or structural steel. Many of today's steel-framed structures are designed with the aid of a computer, which can consider all kinds of loading combinations, wind loads, temperature changes, and the vertical live and dead loads. The wind loads are based on hurricane-force winds up to 100 mph. The forces are carried through the complete structure to the ground. All structures must be designed for any given geographical location and its prevailing weather patterns. The architectural designer determines the function, size, and shape of the form.

All structural materials have special methods or systems for fastening and erection that must be considered in the drafting process. The basic structural materials used in contemporary structures are steel, wood, concrete, structural clay products, and stone masonry.

2.0.0 STEEL STRUCTURES

Steel structures include bridges, buildings, transmission towers, storage tanks, sign supports, and even art. Buildings are commonly classified according to use, such as industrial and multistory buildings. Little use is made of steel in residential construction except in multistory apartment buildings.

Industrial buildings may include a steel frame or have a roof supported by steel members resting on loadbearing walls. The steel skeleton of the building may be rigid or pinned, a two- or three-hinged arch, or a **truss**-on-column system. The truss may also be rigid or pinned. The steel frame of a pre-engineered structure can be erected quickly because it is assembled with bolts.

2.1.0 Steel Shapes

Steel shapes used in construction are produced in the steel-rolling mills and are shipped to fabrication shops where they are cut to specific lengths and the **connections** are prepared. Some of the basic shapes include the wide-flanged W-shape which is used as a beam or column, the S-shape, the C-shape, and the L-shape. These steel shapes have framed the skyscrapers of our cities for almost 75 years. The American Institute of Steel Construction (AISC) maintains guidelines for the design and construction of substructures and superstructures.

Table 1 shows the most common steel shapes, listing their description, symbol, use of the symbol, and a pictorial of the shape.

Table 2 shows more details about the shapes including the group name, an example of the billing information on detail drawings, and the conventional way of showing the shape on detail drawings.

2.1.1 Drawing Symbols

The symbols used to represent structural shapes correspond to sections of the actual shapes. When drawn to small scale, section symbols are shaded in solid black. Large-scale section symbols are lined with pairs of lines drawn at a 45-degree angle. Smaller scale drawings do not show all details.

- *Wide-flanged shapes* – These are called W-shapes, with flanges having parallel surfaces. Bearing piles are usually wide-flanged shapes and are designated HP shapes in any notation. Bearing piles are normally used as piling to support foundations, and unlike H shapes, their flanges are wider than the depth of the member. The notations on drawings are sequenced and shown as: nominal depth × weight per foot × length. *Figure 1* shows a typical wide-flanged shape symbol.
- *American Standard shapes* – These are generally called S-shapes, with their flange inner surfaces tapered. The notation sequence shown on drawings is: nominal depth × weight per foot × length. *Figure 2* shows the American Standard shape symbol.
- *Joists and junior shapes* – These are lightweight members called M-shapes, similar in contour to W-shapes, and noted in the same sequence. The symbols are the same as the American Standard shapes.
- *American Standard channels* – These are called C-shapes. The notation sequence shown on drawings is: depth × weight per foot × length. The inner surfaces of the flanges are tapered like the S-shapes. *Figure 3* shows the symbol for the American Standard channel.
- *H beam* – These beams are used for pilings for bridges and retaining walls.
- *Angles* – Angles consist of two legs set at right angles and are represented by the group symbol L. Details made to a scale of 1" = 1' or smaller do not show rounded-off toes or the interior fillet between legs. The notation sequence for equal-length angles is: the width of each leg × thickness × length in inches. With an unequal leg angle, the long leg should be billed first. *Figure 4* shows the symbols for equal and unequal leg angles.

Table 1 Common Steel Shapes, Symbols, and Pictures

DESCRIPTION	SYMBOL	EXAMPLE USE	PICTORIAL
GAUGE PITCH	G P	G = 3½ P = 2½	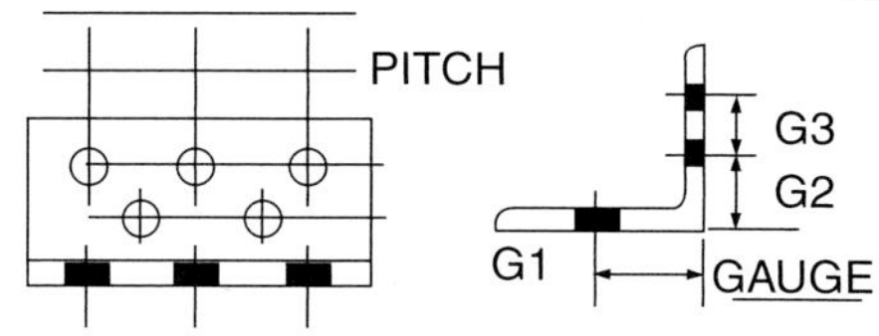
SHAPES WIDE - FLANGE AMERICAN STANDARD JOISTS AND JUNIOR BEARING PILE	W S M HP	W24 × 76 S15 × 42.9 M12 × 36.6 HP10 × 57	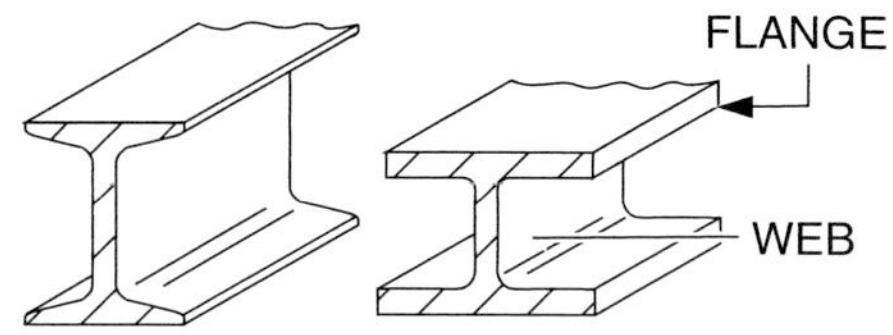
CHANNELS AMERICAN STANDARD	C	C8 × 11.5	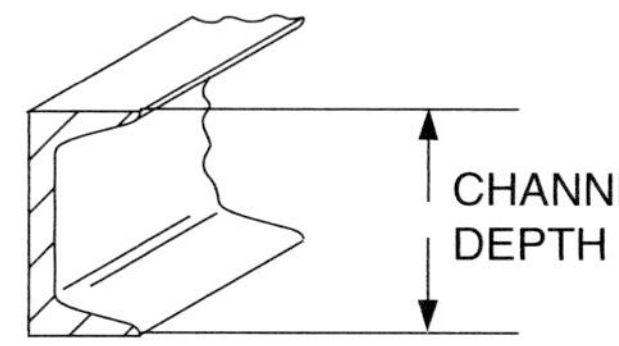
ANGLES EQUAL LEG UNEQUAL LEG	L L	L3 × 3 × ¼ L7 × 4 × ½	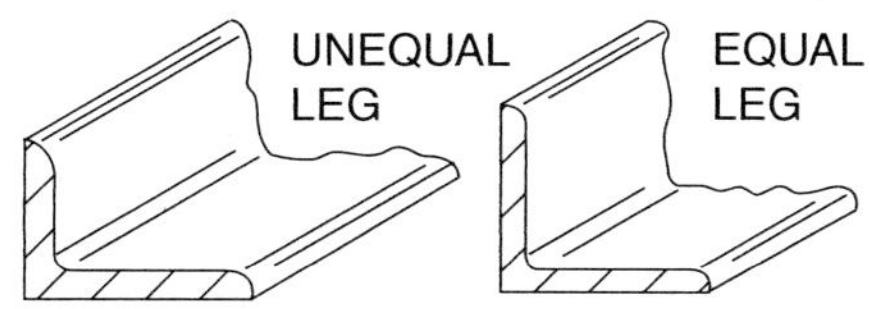
TEES	WT MT ST	WT8 × 15.5 MT5 × 4.5 ST5 × 12.7	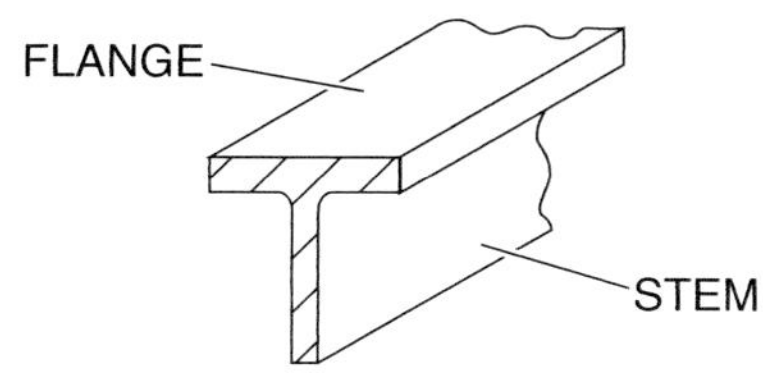
PLATE CHECKERED PLATE	PL CK PL	PL ½ × 18 × 2'-6 CK PL ½	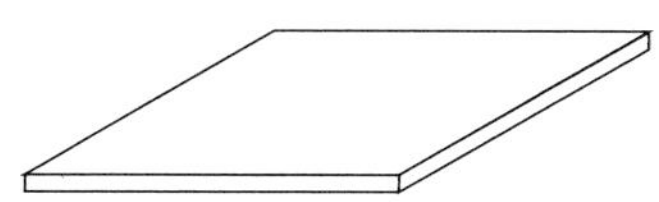
BARS FLAT SQUARE ROUND	BAR BAR ϕ BAR	BAR 2½ × ½ BAR 2½ × 1 BAR 2½ ϕ	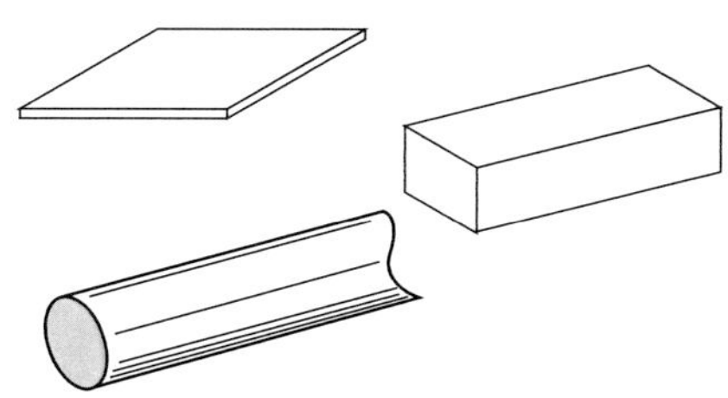
PIPE	PIPE	PIPE 4 STD	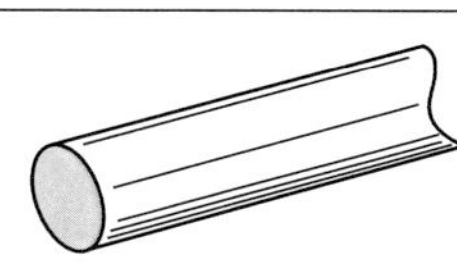
H BEAM	HP	PILINGS 2 × 84	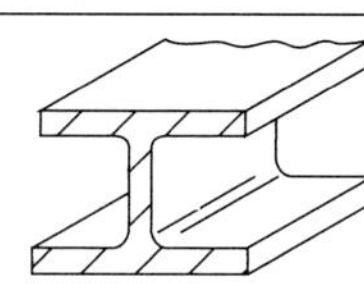

108T01.EPS

Table 2 Shape Billing Information and Detail Drawing (1 of 3)

GROUP	EXAMPLE OF BILLING ON DETAIL DRAWING	CONVENTIONAL WAY OF SHOWING ON DETAIL DRAWINGS AND IDENTIFICATION OF MAJOR COMPONENT PARTS
RAILS	40 Ø ASCE RAIL × 30' -0 104 Ø BETH. RAIL × 16' -0 105 Ø USS RAIL × 21' -0 Wgt. per Yd. in Pounds; Profile Type; Group Symbol; Length in Ft. and In.	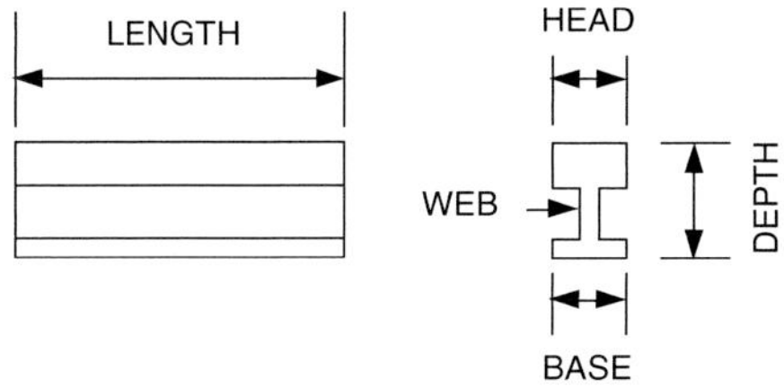
REMARKS (Example)	On Details Made to Scale of 1" = 1' Or Less, Show Profile about as Shown Herein, Disregarding All Rounded Corners, Fillets, Curved and Sloping Surfaces.	
<u>PLATES</u> PLAIN - SHEARED PLAIN - UNIVERSAL MILL RAISED PATTERN	PL ⅜ × 62½ × 9'-3 UM PL ½ × 14 × 18'-11 PL 5/16 × 31 × 10'-0 (*) Producer's Designation; Group Symbol; Thickness in Inches; Width in Inches; Length in Ft. and In.	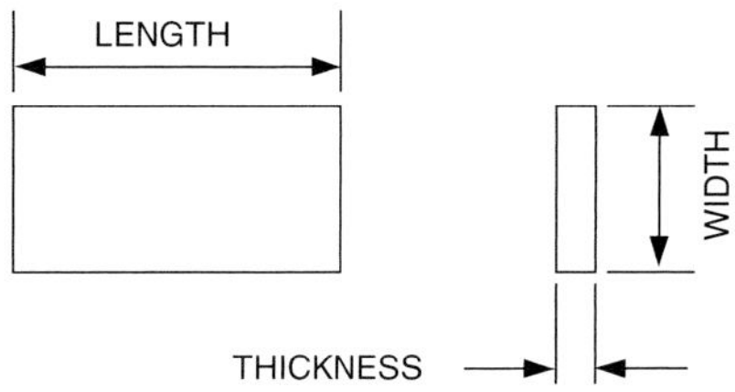
FLAT BARS	BAR 4½ × ½ × 7'-2 Group Symbol; Width in Inches; Thickness in Inches; Length in Ft. and In.	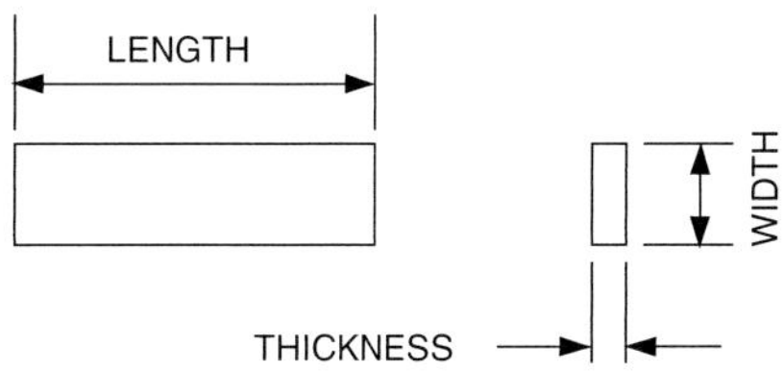
REMARKS (Example)	On Details to Scale Of 1" = 1' or Less, Exaggerate Thickness of Plates And Bars to Suit. See Manual Part 6, Under "Wire and Sheet Metal Gauges" for Size Classification of Bars & Plates; See the Discussion in Manual Part 1 on "Bars And Plates Product Availability" for General Information	
SQUARE BARS	BAR 1½ ⧄ × 13'-4 Group Symbol; Size in Inches; Convention for "Square"; Length in Ft. and In.	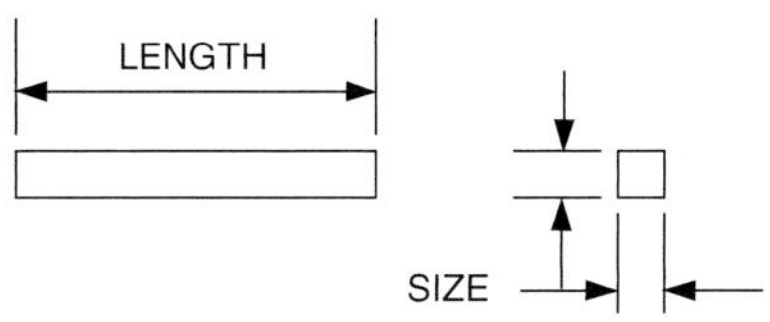
ROUND BARS	BAR 1-1/2 ∅ × 12'-6 Group Symbol; Size in Inches; Convention for "Circle"; Length in Ft. and In.	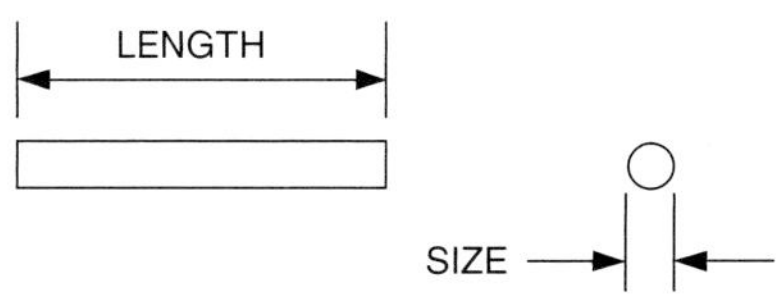

108T02A.EPS

Table 2 Shape Billing Information and Detail Drawing (2 of 3)

GROUP	EXAMPLE OF BILLING ON DETAIL DRAWING	CONVENTIONAL WAY OF SHOWING ON DETAIL DRAWINGS AND IDENTIFICATION OF MAJOR COMPONENT PARTS
PIPES	PIPE 3 STD. × 8'-6 PIPE 3 x-STRONG × 8'-6 PIPE 3 xx-STRONG × 8'-6 Group Symbol (PIPE); Nominal Diameter in Inches (3); Weight Type (STD., x-STRONG, xx-STRONG); Length in Ft. and In. (8'-6)	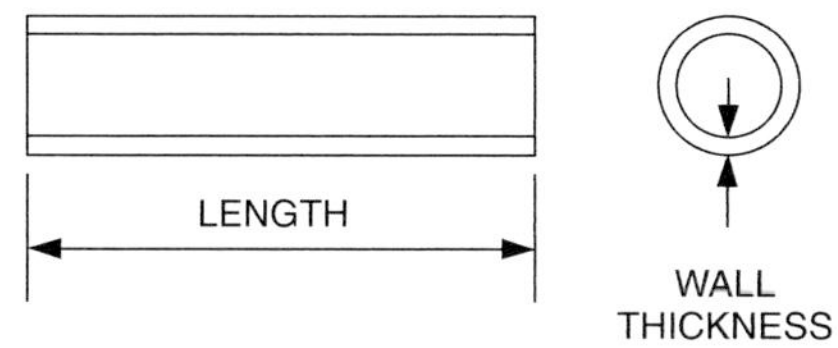
REMARKS (Example)	See Pipe, Manual Part 1, for Actual Diameter and Wall Thickness	
TUBES SQUARE RECTANGULAR	TS 4 × 4 × 0.375 × 9'-1 TS 4 × 4 × 0.375 × 9'-1 Group Symbol (TS); Widths in Inches (4 × 4); Wall Thickness in Inches (0.375); Length in Ft. and In. (9'-1)	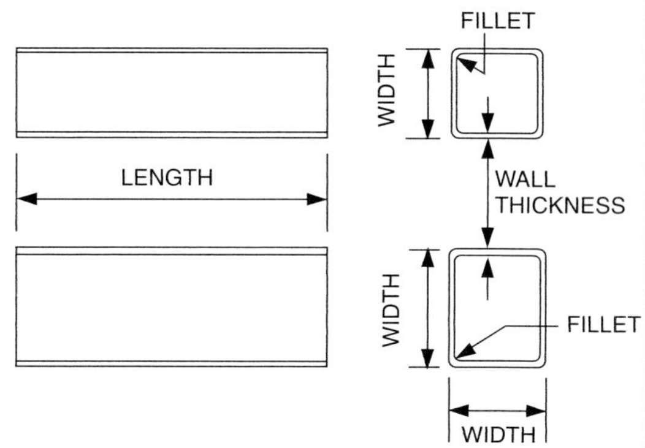
TUBES CIRCULAR	TS 3 OD × 250 × 9'-1 Group Symbol (TS); Outside Diameter in Inches (3 OD); Wall Thickness in Inches (250); Length in Ft. and In. (9'-1)	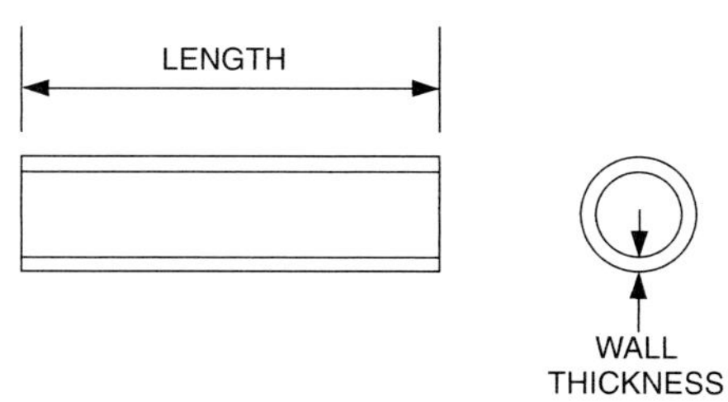
REMARKS (Example)	Wall Thickness May Be Shown as a Fractional Dimension	

108T02B.EPS

Table 2 Shape Billing Information and Detail Drawing (3 of 3)

GROUP	EXAMPLE OF BILLING ON DETAIL DRAWING	CONVENTIONAL WAY OF SHOWING ON DETAIL DRAWINGS AND IDENTIFICATION OF MAJOR COMPONENT PARTS
EQUAL LEG ANGLE	L 3½ × 3½ × ¼ × 5'-6 L: Group Symbol 3½: Leg Width in Inches 3½: Leg Width in Inches ¼: Thickness in Inches 5'-6: Length in Ft. and In.	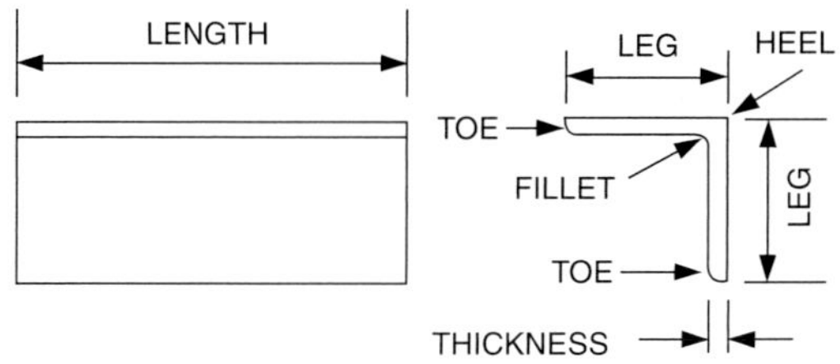
UNEQUAL LEG ANGLE	L 6 × 4 × ⅜ × 5'-6 L: Group Symbol 6: Long Leg in inches 4: Short Leg in Inches ⅜: Thickness in Inches 5'-6: Length in Ft. and In.	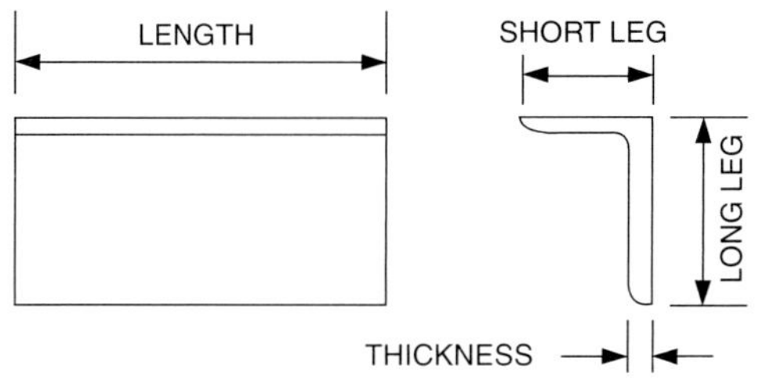
REMARKS (Example)	On Details Made to Scale of 1"=1' or Smaller, Do Not Show Rounded-Off Toes of Angles or Interior Fillet Between Legs. Bill Long Leg of Unequal Angles First. Exaggerate Leg Thickness to Suit.	
WIDE - FLANGE BEAMS & COLS. AM. STD. BEAMS MISC. SHAPES BEARING PILES	W27 × 94 × 26'-10 S15 × 50 × 16'-3½ M8 × 18.5 × 9'-3 HP12 × 74 × 20'-3 Group Symbol and Nominal Depth in Inches Weight per Ft. in Pounds Length in Ft. and In.	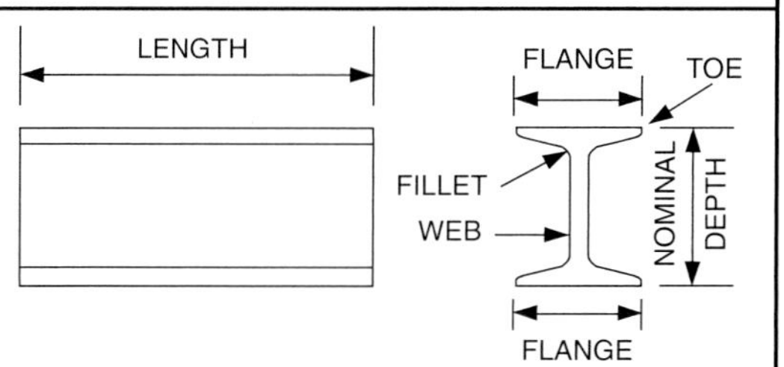
AM. STD. CHANNELS MISC. CHANNELS	C10 × 15.3 × 18'-8 MC13 × 31.8 × 9'-0 Group Symbol and Nominal Depth in Inches Weight per Ft. in Pounds Length in Ft. and In.	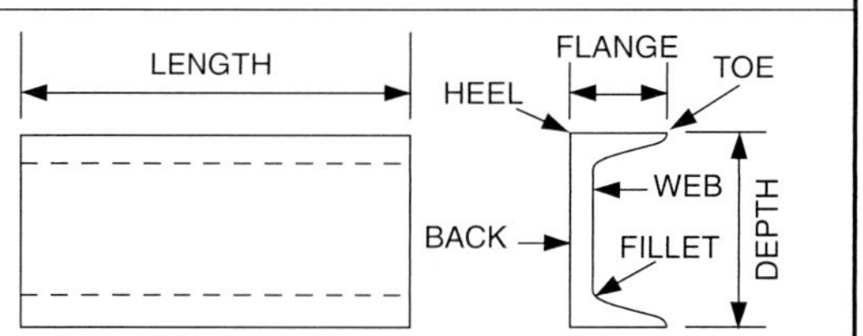
STRUCTURAL TEES W SHAPES S SHAPES M SHAPES	WT18 × 150 × 34'-6 ST6 × 17.5 × 8'-3 MT4 × 3.25 × 9'-0 Group Symbol and Nominal Depth in Inches Weight per Ft. in Pounds Length in Ft. and In.	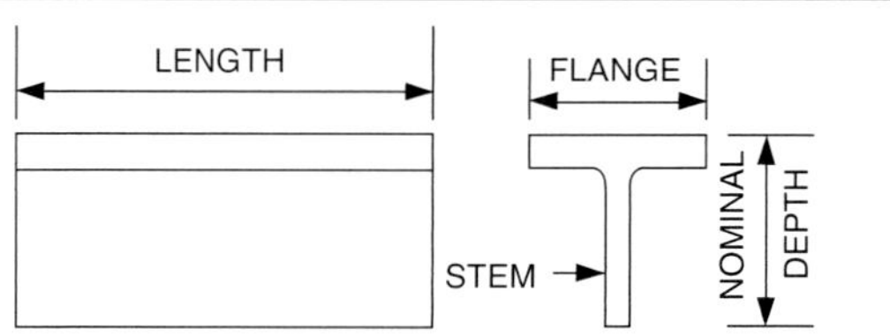
REMARKS (Example)	On Details Made to Scale of 1"=1' or Smaller, Do Not Show Rounded-Off Toes of Flanges or Interior Fillets Between Web and Flanges. Do Not Show Flange Slope for Channels or Beams with Sloping Flanges. Exaggerate Web and Thickness to Suit.	

108T02C.EPS

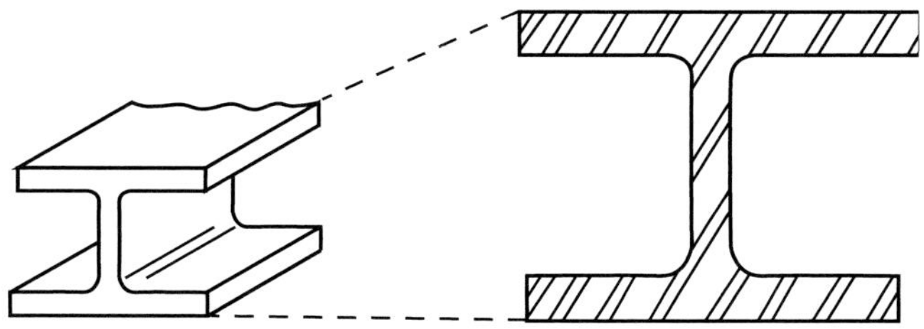

108F01.EPS

Figure 1 Wide-flanged shape symbol.

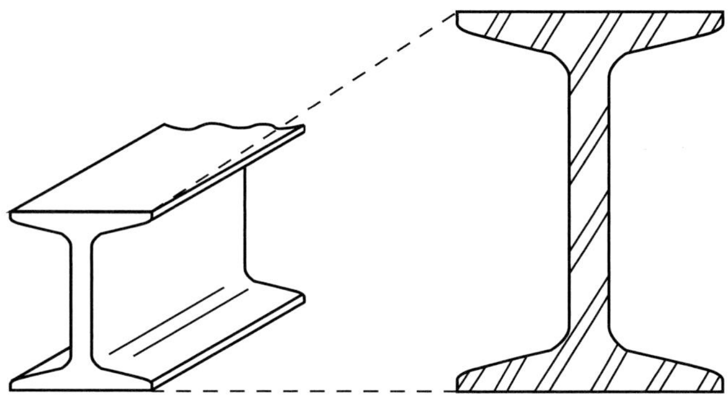

108F02.EPS

Figure 2 American Standard shape symbol.

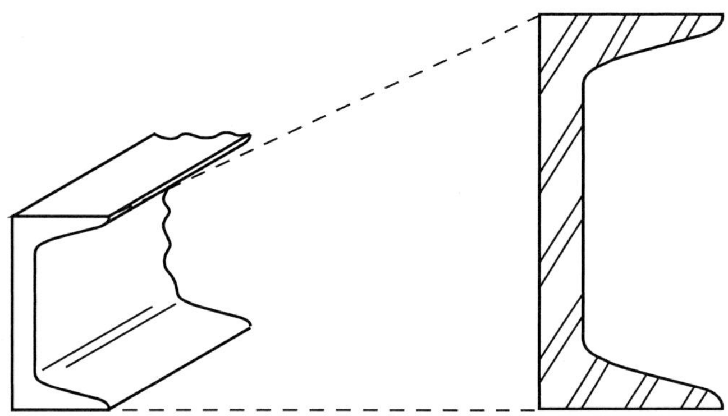

108F03.EPS

Figure 3 American Standard channel symbol.

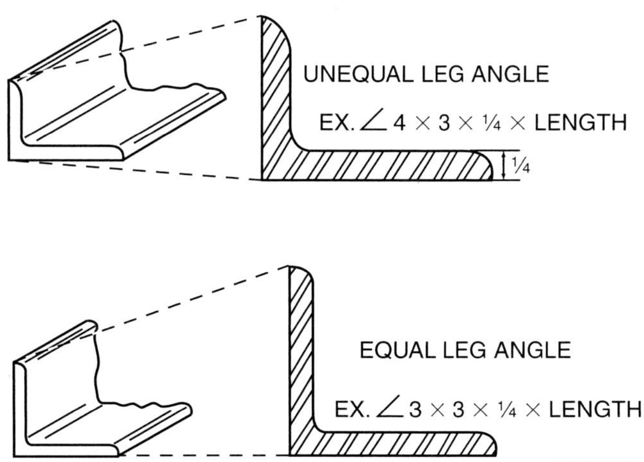

108F04.EPS

Figure 4 Equal and unequal leg angle symbols.

- *Tees* – Structural tees are made by splitting standard W-, S-, and M-shapes through the centers of their webs, thus forming two T-shapes from each one. Group symbols for the structural tees combine the appropriate shape symbol with the letter T such as WT, ST, and MT. The web of the split shape becomes the stem of the tee. *Figure 5* shows the symbol for a structural tee.
- *Plates* – These are represented by the combined letters PL and are noted in the following sequence: thickness × width × length. On details to scale of 1" = 1' or less, the thickness should be exaggerated to suit. *Figure 6* shows the symbol for a plate.
- *Bars* – Three bar classifications are commonly used: flat, square, and round. The word BAR alone represents the flat bar in notation. A box (□) is added for a square bar and a circle (○) for a round bar. The notation sequence for a flat bar is: width × thickness × length. Flat bar is available up to 8" wide. Round and square bars are noted as size × length. *Figure 7* shows the symbol for flat bar, round bar, and square bar.
- *Rails* – Rails are noted with the word RAIL. Details made to scale of 1" = 1' or less show profiles disregarding all rounded corners, fillets, curved, and sloping surfaces. *Figure 8* shows the symbol for a crane rail.

2.2.0 Detail Drawings

While shop detail drawings are absolutely imperative in the fabrication of structural steel, they are also used by inspectors to make sure that members are being made as detailed. The details have lasting value to the owner of the structure because they show the materials the structure contains if any alterations or additions are required at some later date.

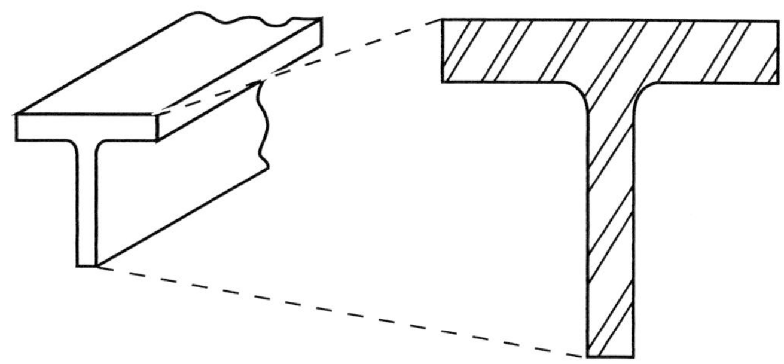

108F05.EPS

Figure 5 Structural tee symbol.

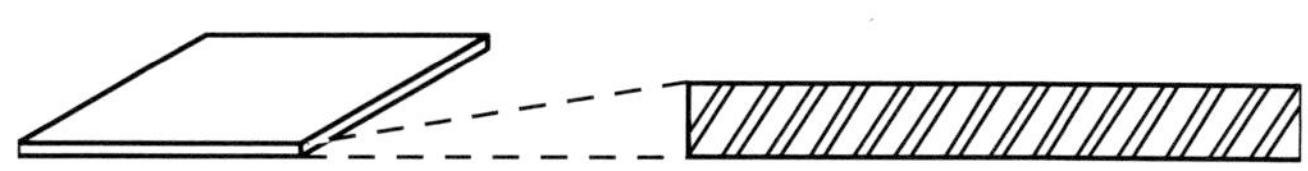

108F06.EPS

Figure 6 Plate symbol.

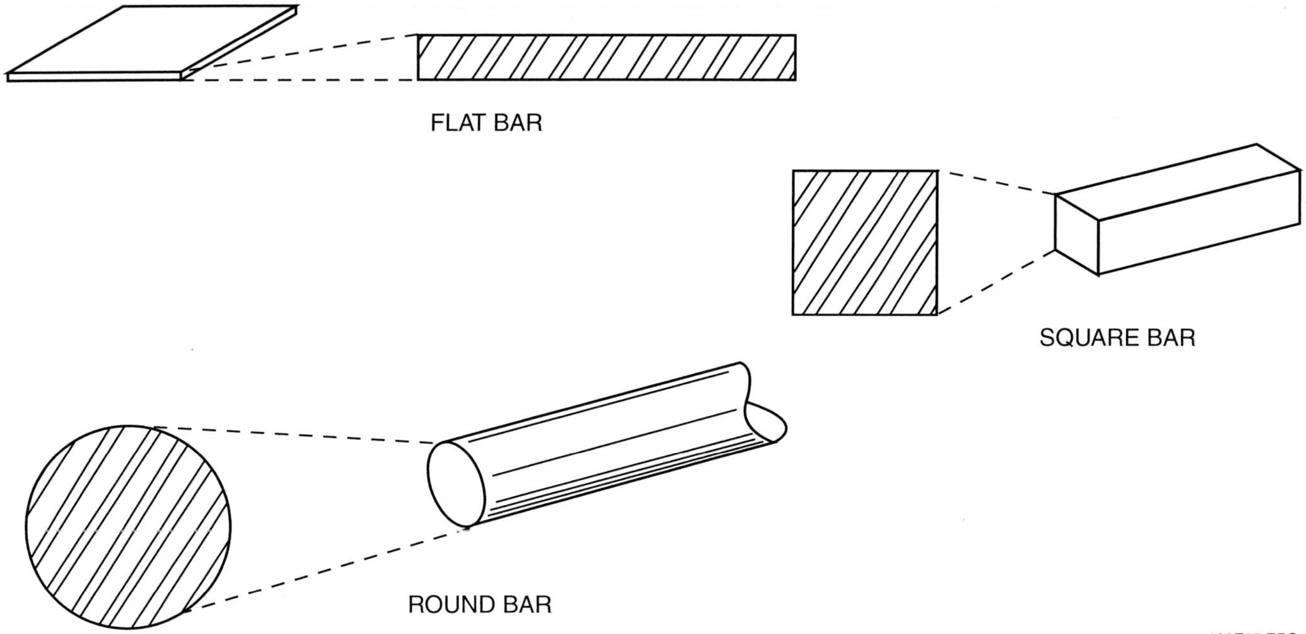

108F07.EPS

Figure 7 Flat bar, round bar, and square bar symbols.

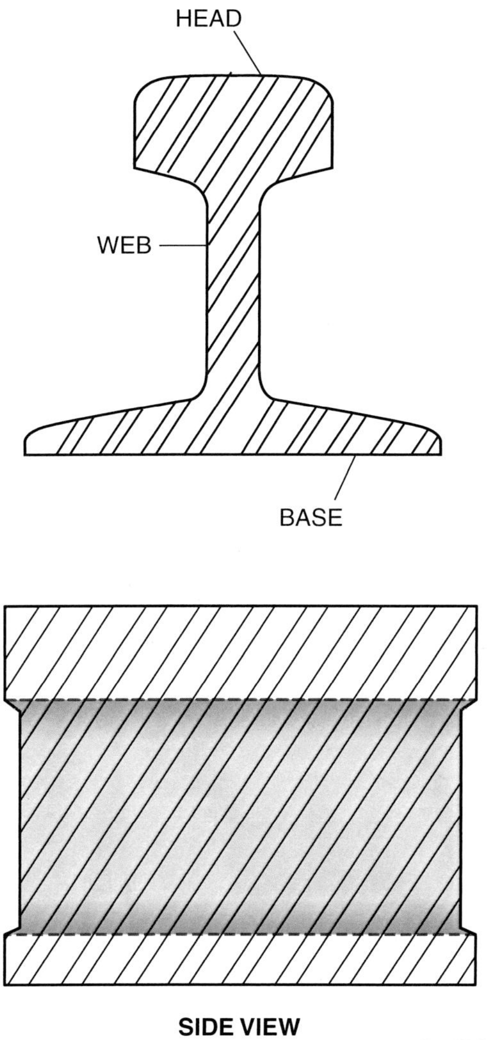

108F08.EPS

Figure 8 Crane rail symbol.

To enable detailers to do their job, the designer should provide the following information:

- *Simple-beam connections* – Reactions of beams should be shown on design drawings, particularly when the fabricator must develop the connections. For unusual or complicated connections, it is good practice for the designer to consult with a fabricator during the design stages of a project to determine what information should be included in the design drawings.
- *Rigid beam-to-column connections* – Some fabricators prefer to be furnished the movements and forces in such connections. With this data, fabricators can develop an efficient connection best suited to their practices.
- *Welding* – Weld sizes and electrode types should be shown on design drawings. Designers unfamiliar with welding can gain a lot by consulting with a fabricator, preferably while the project is being designed.
- *Fasteners* – The type of fastening must be shown in design drawings. When specifying high-strength bolts, designers must indicate whether the bolts are to be used in slip-critical or bearing-type connections.
- *Tolerances* – If unusual tolerances for dimensional accuracy exist, they must be clearly shown on the design drawings. Unusual tolerances are more stringent than tolerances required in the general specifications. The following codes apply for the different types of structures under consideration:
  - *AISC 303-05, Code of Standard Practice for Steel Buildings and Bridges*

- *AISC 360-05, Specification for Structural Steel Buildings Allowable Stress Design and Plastic Design*
- *AISC S342L, Load and Resistance Factor Design Specification for Structural Steel Buildings*
- *AASHTO HB-17, Standard Specifications for Highway Bridges*
- *ASTM A6, General Requirements for Delivery of Rolled Steel Plates, Shapes, Sheet Piling, and Bars for Structural Use*

2.3.0 Frames

Building frames are three-dimensional skeletons. They are usually considered to be rigid in only one plane. The frame element that gives a structure its rigidity is called a **bent** and may be one or more stories in height. The term *bent* is used for all frames whether rigid, truss-on-column, rafters-on-column, or other members are used to span between columns in the principal plane. The spacing between bents is the bay spacing. **Spandrel** and floor beams are used to span the bays in multistory buildings. Girders are usually heavier members than floor beams and are used to span between columns or bents. *Figure 9* shows a metal or pre-engineered building frame with the various components identified.

2.4.0 Roofs

The roof system of all buildings consists of framing, decking, and a waterproof covering. The main roof framing consists of the rafters or trusses in any bent. **Purlins**, spaced 2' to 6' or more on centers, are used to span the bay distance. Sag rods are provided as additional support for the purlins used on sloping roofs. The roof deck rests on the purlins and may be precast concrete slabs, wood planking, or plywood sheets.

3.0.0 ASSEMBLY DRAWINGS

The process of reading construction drawings moves back and forth between the erection and the detail drawings. The erection drawing shows where the columns are placed with respect to each other. The detail shows how the columns are fastened together, or how each column is mounted on a base plate. **Shop drawings** tell the fabricator how to make the beam, and how end angle clips will be attached, so that the pieces will go together.

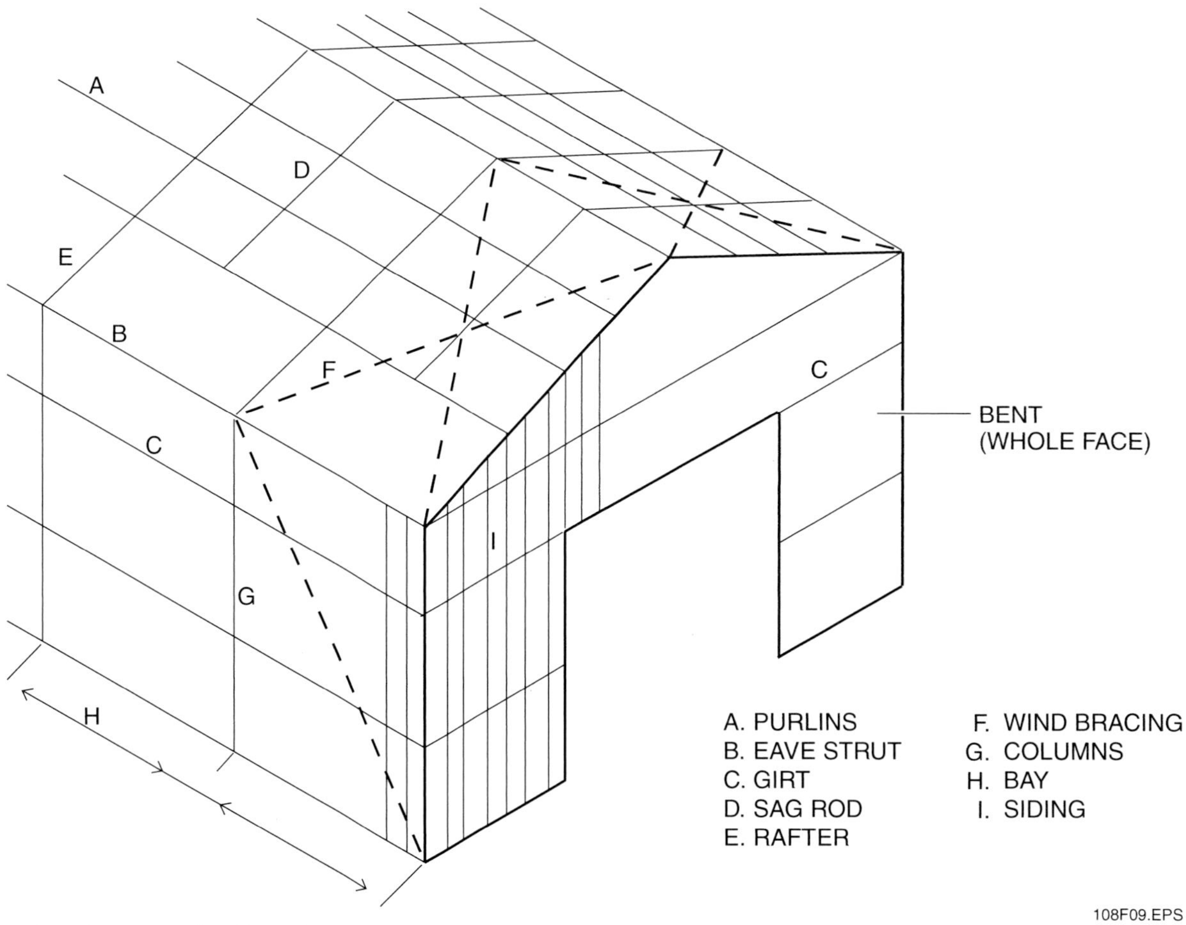

Figure 9 Identification of building frame components.

3.1.0 Base Plate Drawings

The beams are assembled into a building, beginning with the placement of a set of base plates on a slab. The base plates on this set of plans are placed according to drawing AB1, shown in *Figure 10.* The slab is shown, with the distances between the column center lines. The orientation of the columns is also displayed, that is, the way the webs and flanges are turned with respect to each other. The General Notes give the specifications for the types of material required for components, including pipes, beams, and base plates. The list of symbols gives the intended meaning of different abbreviations, such as TYP for typical. The Anchor Bolt Setting Plan gives the elevation of the bottom of the base plates, and explains where to find the anchor bolt details.

Each of the columns has a circle at the end of a line next to it, which identifies the detail from that drawing that is applicable. These details of the anchor bolt settings are shown on *Figure 11.* Only one of the details shown on *Figure 11* is from detail 2. That is the column that appears in the center part of the slab on AB1. The exterior columns, shown as Detail 1/AB1, show a W12 × 45 column set on a base plate with anchor bolts center lined 8" off the beam center line lengthwise, and 6" each side of the beam center line the other way.

The anchor bolts are described on details 1 and 2 as 1" diameter by 1'-5½" long. In the materials supplied, these bolts will be identified by being marked as 1AB. There will be a total of 28 bolts. Be sure to have the right number before beginning the setting, so that the procedure is not held up.

The center of the base plates rests on shim plates that are 3" × 3", producing a 2½" gap between the top of the pad and the bottom of the plate.

3.2.0 Framing Plan

The framing plans for the first and roof levels of these drawings identify the beams for the structure of the building. As shown on the drawings in *Figure 12,* the columns from AB1 are tied into the levels above, and are braced by the crossbeams and spandrels. In the notes under the title of the First Level Framing Plan, the elevation of the top of steel (T.O.S.) for this level is 104', thus placing it 4' above the bottom of the base plates shown on AB1.

The beams are identified along the line of each beam. Notice that the spandrel beams around the outside are larger than the floor beams inside. Notice, also, the beam across the section to the column in the middle is somewhat larger than the other cross beams in that direction. Each beam is also provided with a specific identifier, a combination of letters and numbers that tells the installer which beam that is and where it goes in the assembly. The beam will be marked with that identifier, usually with a marker that shows through the paint used as a primer on the beam by the fabricator.

Starting at the upper left-hand corner of the drawing and reading clockwise, the perimeter beams are: 18B, which is a W24 × 68; 12B1, which is a W27 × 84; 10B2, a W24 × 55; 14B1, another W27 × 84; 20B, a W24 × 62; and 17B, another W27 × 84. The other beams are identified both by the mark and size, while trusses are shown for the right-hand half of the diagram.

On the roof plan (*Figure 12B*), the beams are all identified. One piece that should be noticed is the presence of two diagonal pieces made up of doubled 4 × 4 × ⅜ angle iron at the middle wall. The western piece is identified as 57HB1, running from an exterior column to the intersection of 45B and 56B2. The eastern piece is identified as 57HB2, running from the southwestern exterior corner column to the intersection of 43B and 56B1. The remainder of the beams are W-shapes.

The two sections indicated are vertical frames A/E1 and B/E1, shown on *Figure 13.* Notice that the T.O.S. at the top of this elevation is 119-0, so that the elevation is showing the frame from the first level up to the roof level. The distance from the bottom of the base plate to the top of the steel at the first level is shown as 3'-9½" on elevation A and on the two horizontal beams at north and south sides of the first level framing plan, and as 4' above the bottom of the base plate on the east and west beams, so the W24 × 55 beam on the east end will fit into the W27 × 84 so as to be 2½" above the T.O.S. of the latter, allowing room for the decking and slab to rest on the frame on the north and south sides.

The cross-bracing on elevation B is described as HSS6 × 6 × ⅜. That is a hollow structural section unit, essentially a rectangular tube 6" × 6" with a ⅜" wall. The shorter length involved makes the HSS a useable choice. Detail 5 on *Figure 14* shows the connection.

The floor level and roofing level framing plans have section arrows, numbered 1 through 5, with arms indicating the area covered by the section. Those section drawings appear on *Figure 14.* They are identified by the symbols used to identify the arrows. To look at an example of the sections, Detail 1 is a beam to column connection. The beam is connected to the column by a ½" thick shear plate welded to the flange of the column with a ¼" fillet on both sides. The shear plates are attached to the beam with bolts in slotted holes.

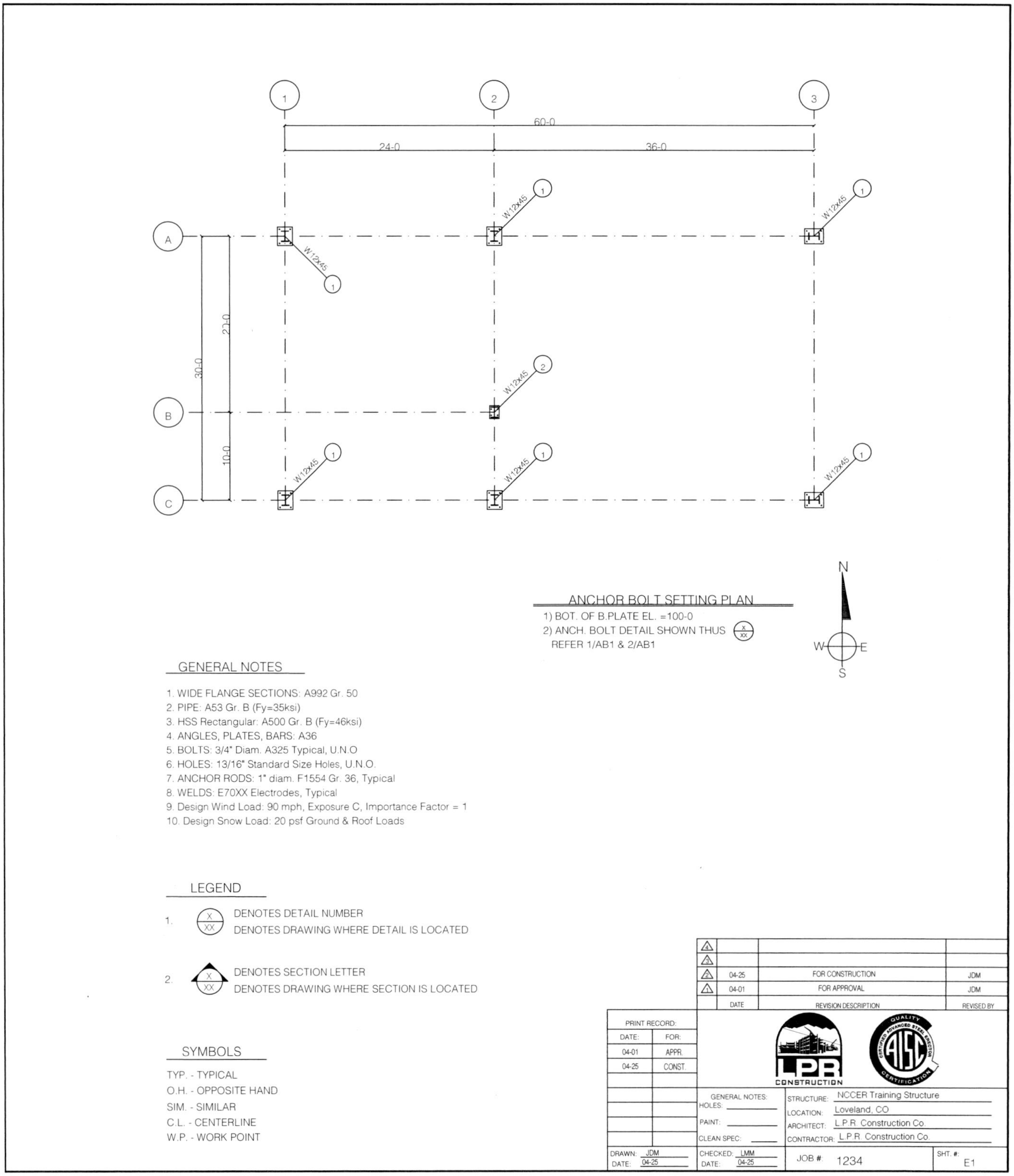

108F10.EPS

Figure 10 Base plate drawing AB1.

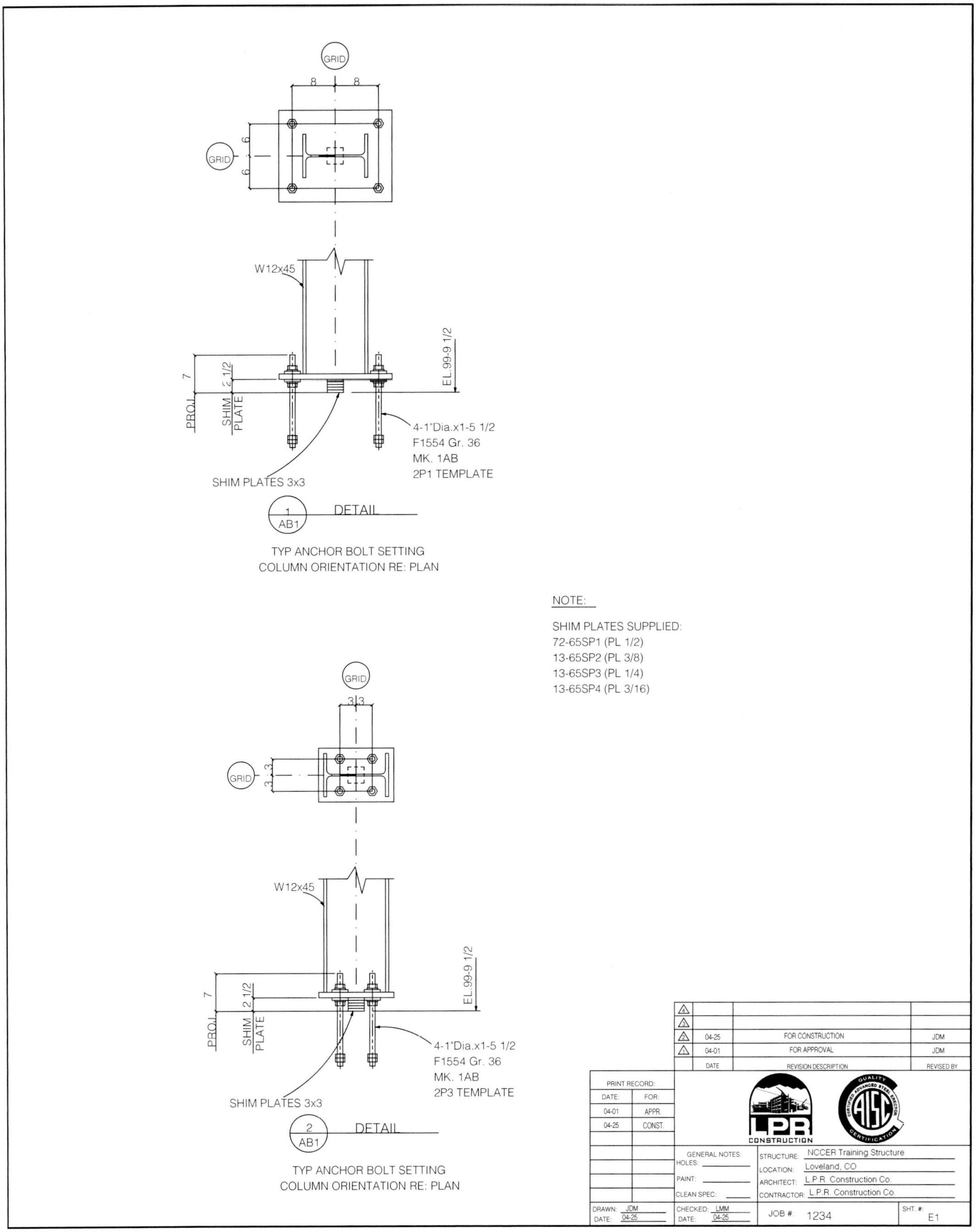

108F11.EPS

Figure 11 Pre-engineered building drawing.

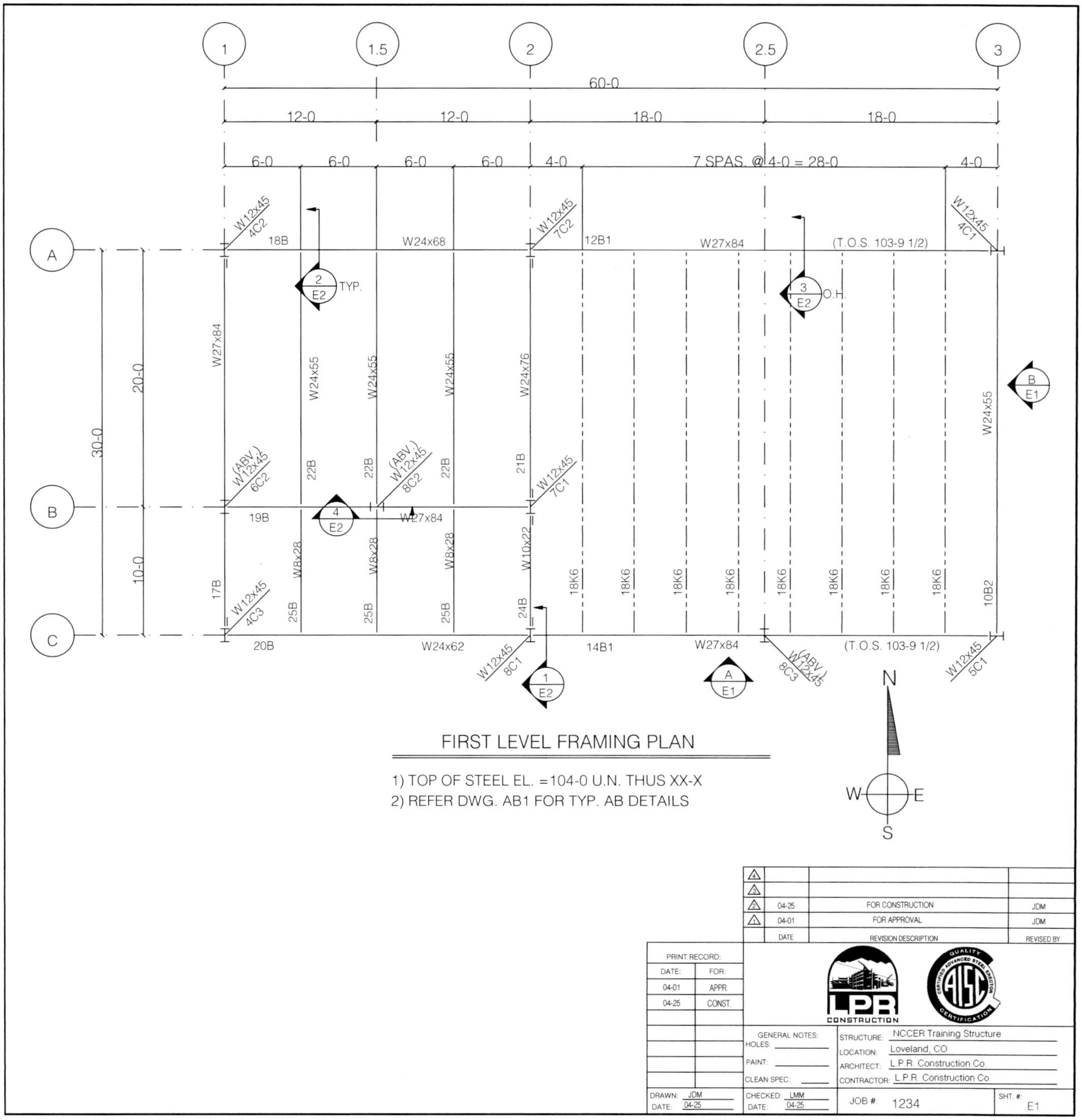

108F12A.EPS

Figure 12 Framing plans (1 of 2).

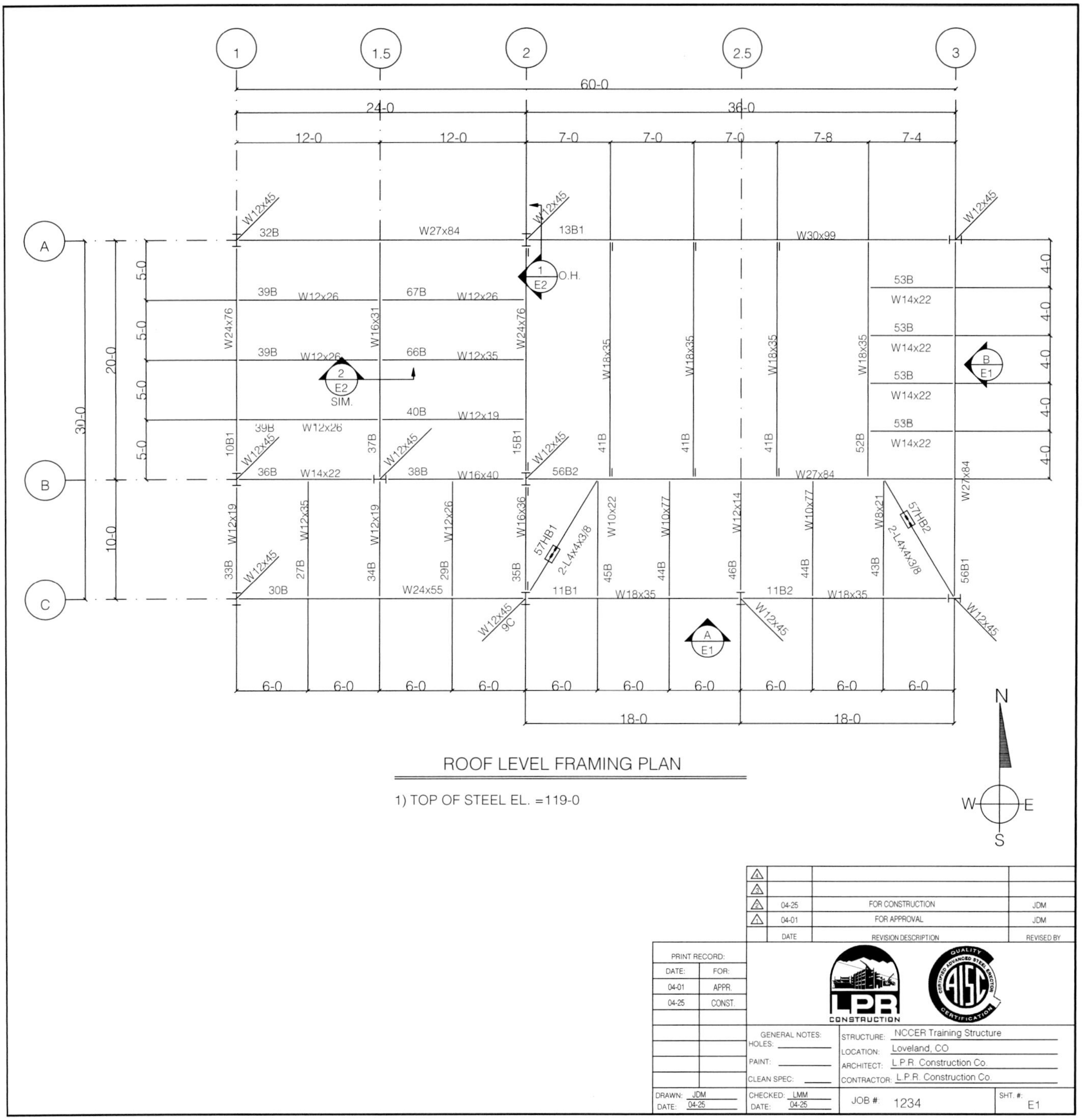

108F12B.EPS

Figure 12 Framing plans (2 of 2).

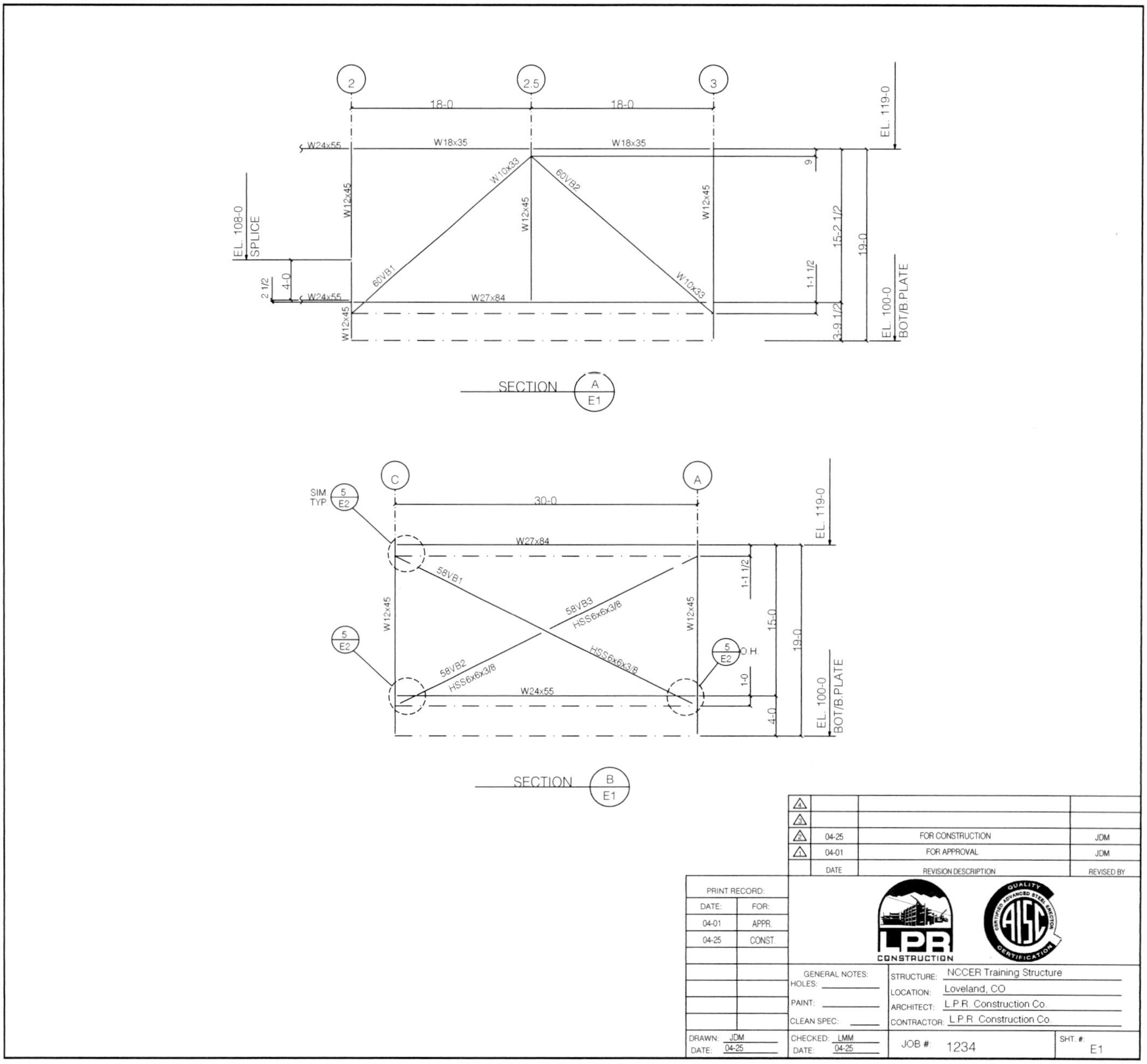

108F13.EPS

Figure 13 Framing elevations.

Detail 3 shows the joist end detail, which depicts the form deck with the concrete, showing why the T.O.S on the east and west end beams is higher than the T.O.S on the north and south beams. A 4 × 4 × 5⁄16 angle is used to retain the concrete when it is poured, and there is no significant change in height at the edge of the concrete, if it has been spread correctly.

Detail 2 shows how the inside beams are attached to the larger perimeter beams. A ½"-thick connection plate is welded to the web and flange of the larger beam, and punched with slotted holes so that the web of the smaller beam can be bolted against it. The flange of one side of the smaller beam has to be coped out so that the web can fit against the connection plate on that side.

Detail 4 shows the column attached to the beam at the first floor middle wall. The column (8C2) is mounted on a support beam to support the center of the interior wall in that section on the first floor framing plan.

Detail 5 shows how the hollow section brace is attached on section B, the elevation of the western end of the building. Two pieces of angle and a gusset plate retain the end of the HSS. The angle and gusset are welded to the HSS with ¼" fillet welds.

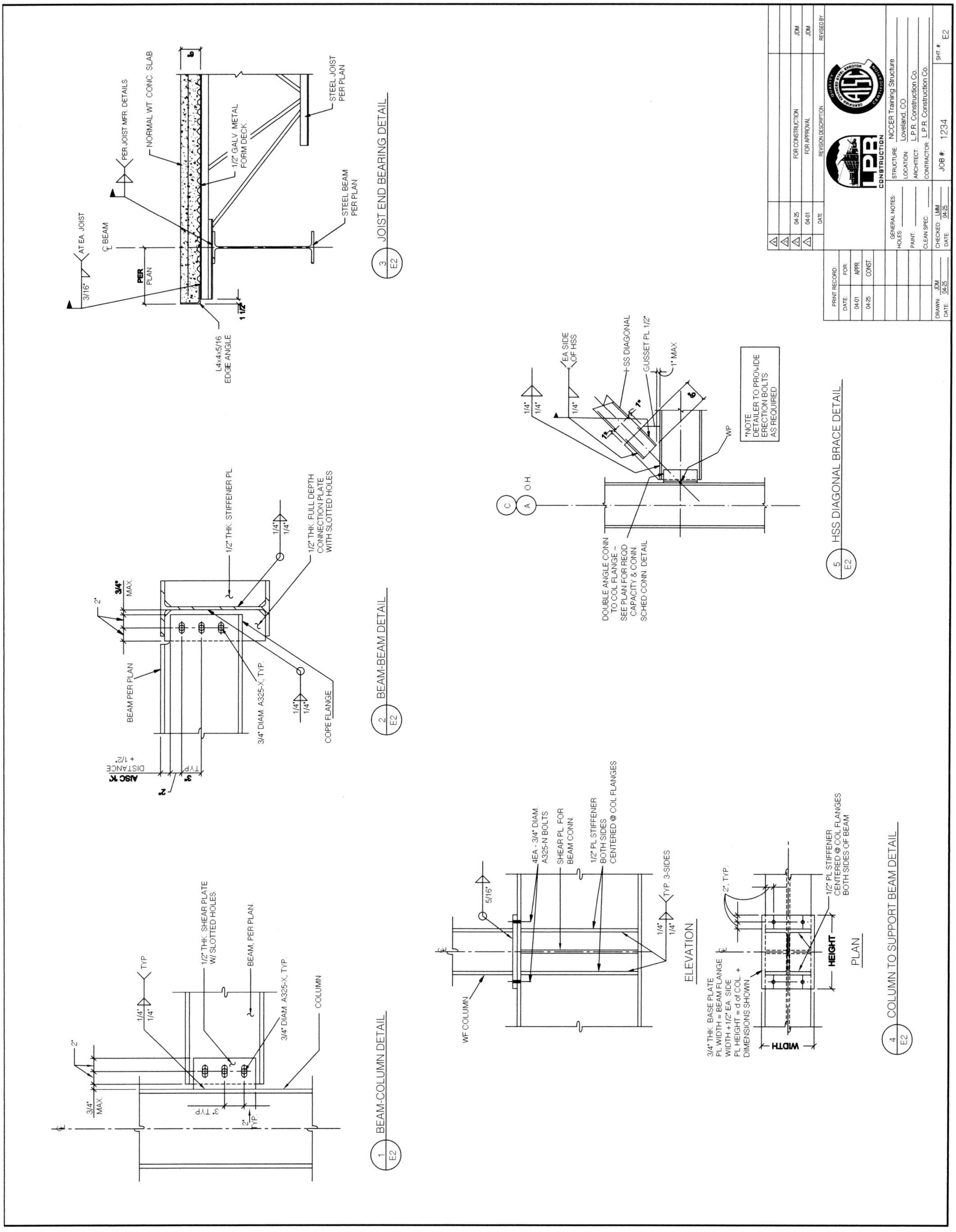

108F14.EPS

Figure 14 Framing connection details.

3.3.0 Fabrication Shop Drawings

Fabrication shop drawings are drawn to show a piece for assembly as it would be delivered from the shop. They are a full set of directions for the fabricator. It is easier to get the holes and plates aligned in a shop than in the field, and the available equipment is much quicker. Usually, a single fitter is assigned to take a unit from raw beam and angle to a completed piece. *Figures 15* and *16* are examples of ordinary pieces that a shopfitter or fabricator might make.

3.3.1 A Vertical Column

Figure 15 is Sheet 4 of the drawings. This is one of the columns shown on AB1. The column is a W12 × 45, with a beam length (not counting the base plate and top plate) of 22'-10½". The base plate is 1" thick, and the top plate is ½" thick. The overall length, as fabricated, is to be 23'. This must be one of the columns in the western half of the building structure, since one of the flanges is to face north. It has to be an exterior column, since the base plate matches Detail 1 from *Figure 11*, as shown by the hole spacing on the base plate. The holes in the web are usually punched with a beam punch at the fabrication shop, although some holes may be drilled. The beam punch is much faster, and very accurate. The plate on the north flange of the beam tells that this is one of the beams on the south end of the building, since the plate would be on the south flange if it were a northern column. Note that the plate is marked 1-fp11, an identifier that would be marked when the plate is fabricated. Examining Beam 4C3 on the first level framing plan, we see that this is the southwestern corner beam.

The holes in the web of the beam are for the mounting of the vertical supports running east and west. The plate (1-fp11) on the flange is designed for mounting the vertical and horizontal structure running north and south, as are the holes in the upper part of the north flange. The holes and the plate are gauged at 5½". In a large shop, these holes would probably be drilled with an overhead rotary drill press.

3.3.2 A Horizontal Column

The horizontal beam in *Figure 16* is Beam 30B. This beam is on the roof's framing plan, running from column 4C3 to column 9C. It is a W24 × 55, and is designed to fit into a 24' space. Because of the thickness of the flanges into which it is to be attached, it is 23'-11⁹⁄₁₆" from outside of angle clip to outside of angle clip. The angle clips are designed to bolt into the webs of the columns (see *Figure 15*) and are bolted to the beam. They are labeled 2a6, and there will be four of them made the same. Once again, these will probably be made by one fitter in the fabrication shop, along with the beam, so that the responsibility is clear. Notice that the beam is not drawn to scale. The first set of connection holes for beam 27B is shown as centered at 5'-11¹³⁄₁₆" from the outside of the left-hand angle clip, while the other holes, for 34B and 29B, will be centered 6' apart.

4.0.0 Ornamental Ironwork Drawings

An ornamental ironworker is a trade worker who is involved in the installation of metal stairways, catwalks, floor gratings, ladders, metal cabinets, and window frames. These workers also install fences and decorative ironwork. They fasten or weld prefabricated aluminum, brass, and bronze frames and panels to buildings.

Ornamental ironworkers spend most of their time fitting, aligning, and assembling preshaped and cut assemblies on the job. They must make sure that the ornamental pieces fit correctly and hold firmly. They must know how to read working drawings and how to drill, tap, fasten, and weld ornamental metals. They must be adept at reading elevation and detail drawings, plus elements of machine drawings. For example, a set of stairs will be taken from an elevation and a plan view. The handrail, however, will always be shown as a detail drawing by itself. *Figure 17* shows an example of a drawing the ironworker might be given for constructing a set of stairs leading to a platform.

5.0.0 Welding Symbols

Welding is often the method used to attach the components of a structure regardless of the construction material. Steel beams are often aligned using drift pins and bolts, and then welded. Pre-engineered structure frames are often welded. The construction drawings for each project specify the type and location of each weld with a welding symbol. These symbols include all of the information the ironworker needs to produce the proper weld. *Figure 18* shows a welding symbol base and the numerous elements of a welding symbol.

All welding symbols include one or more individual graphic characters called weld symbols. These represent specific welds. Weld symbols can be placed above or below the reference line. When the weld symbol is placed below the reference line, the weld must be made on the arrow side of the joint. Weld symbols placed above the reference line require the weld to be made on the other side of the joint. *Figure 19* shows what is meant by the arrow side and the other side of a joint.

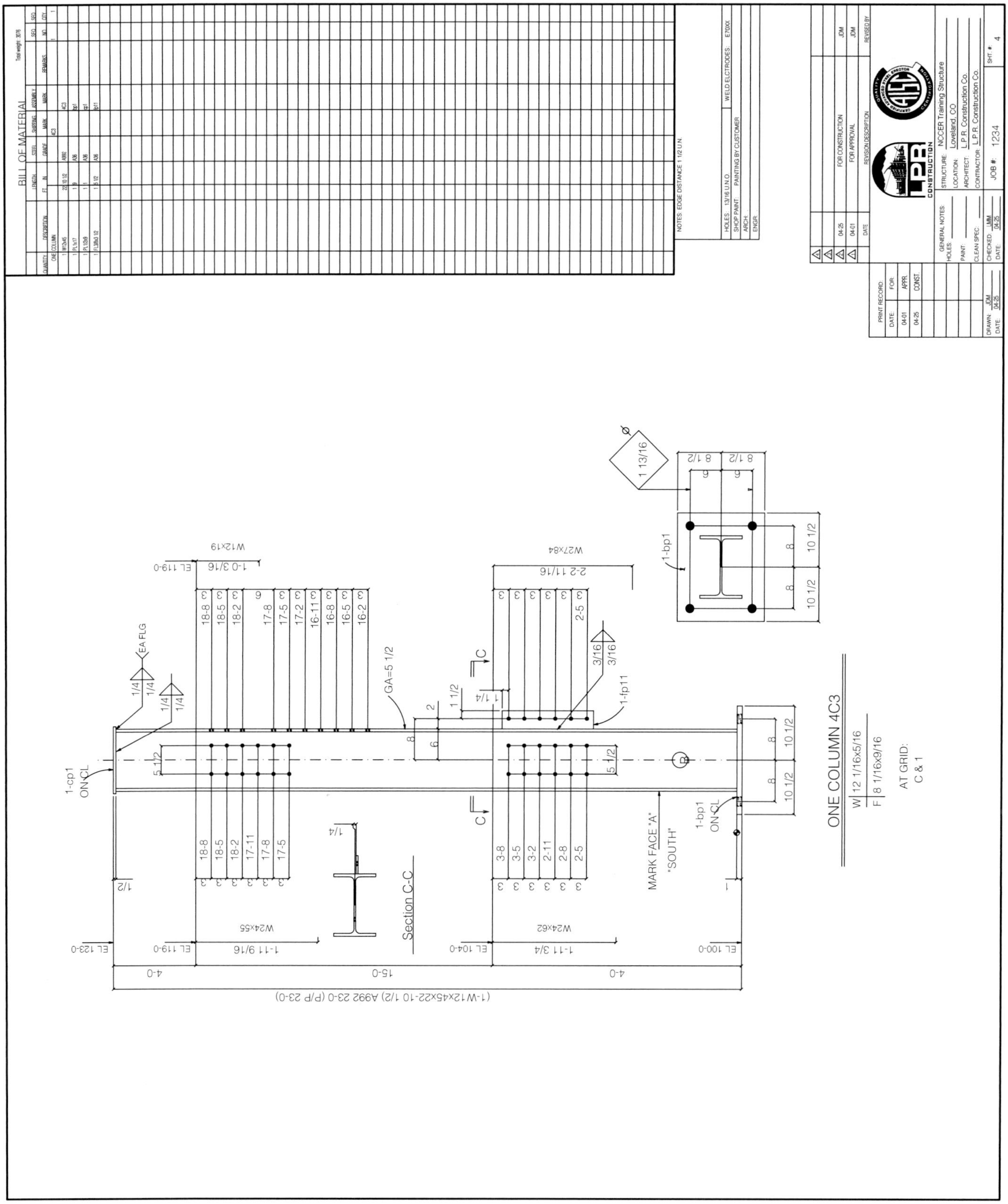

108F15.EPS

Figure 15 Column 4C3 fabrication drawing.

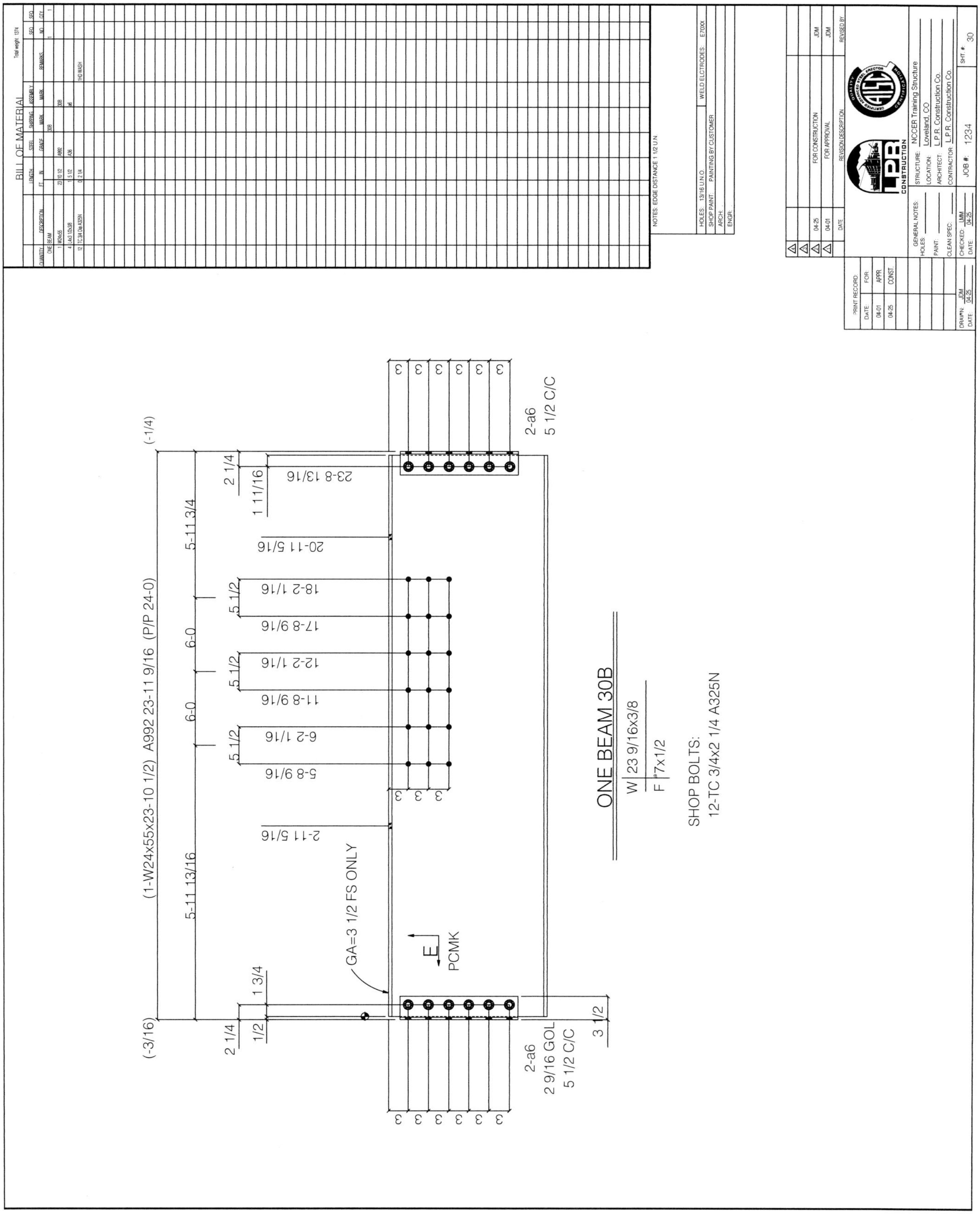

108F16.EPS

Figure 16 Beam 30B fabrication drawing.

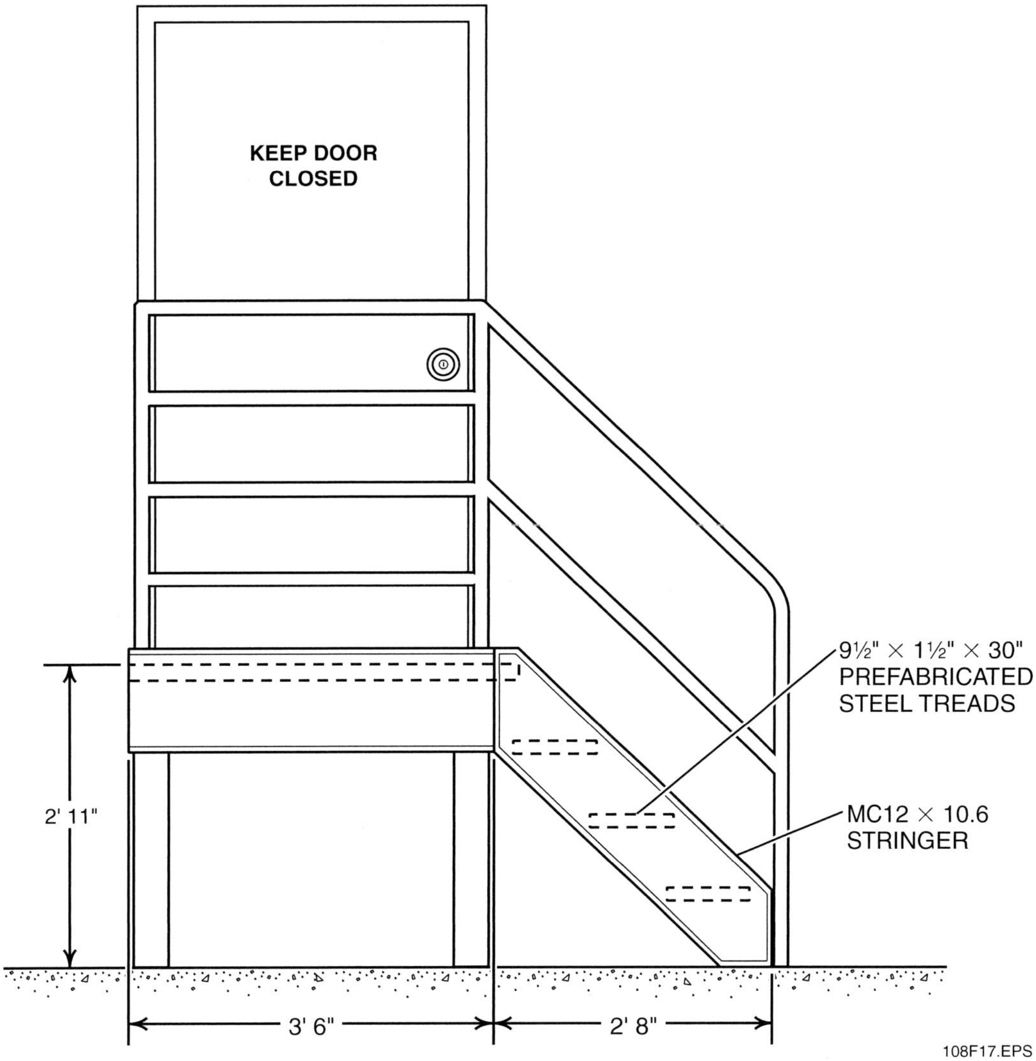

Figure 17 Drawing for a set of stairs.

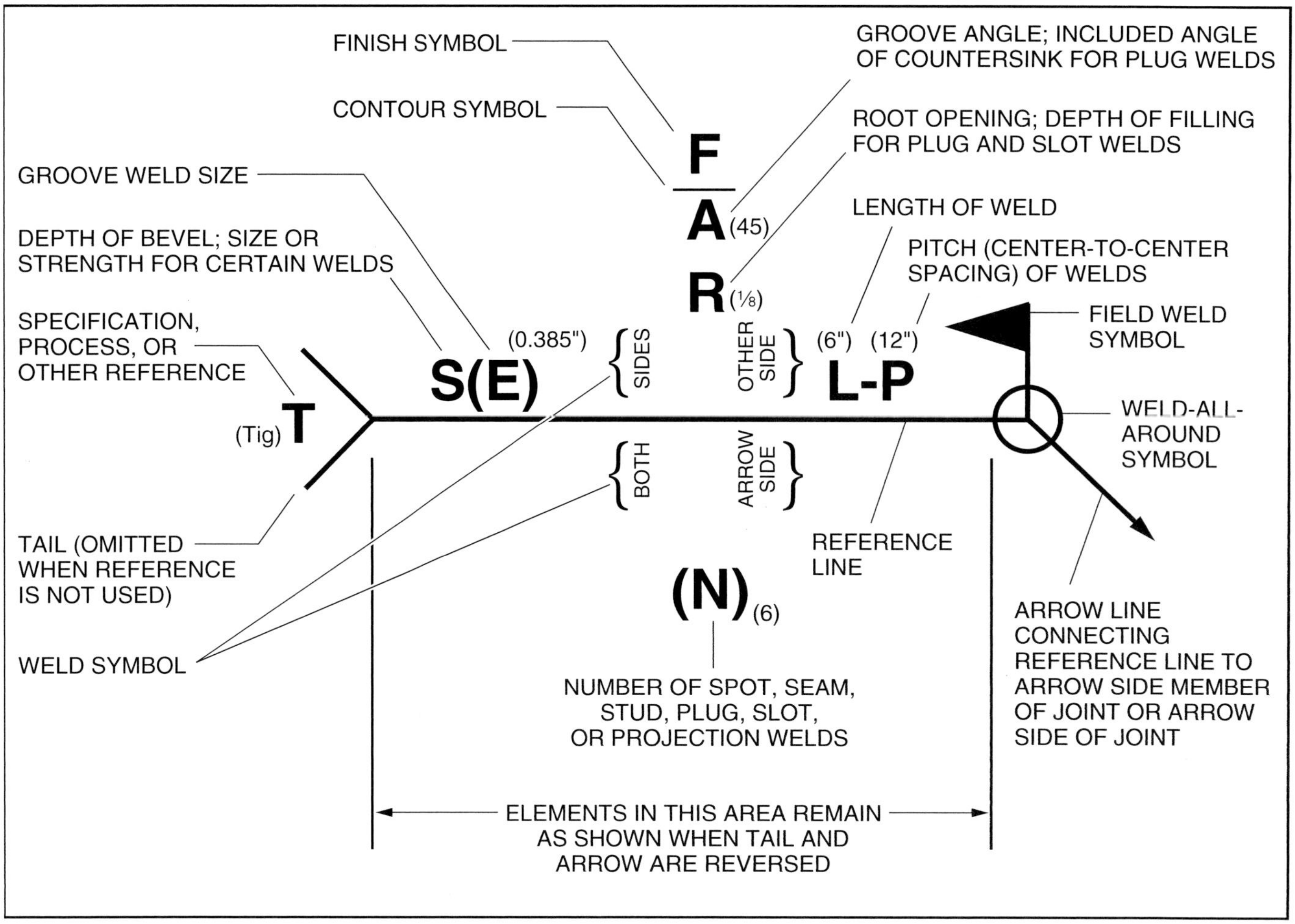

Figure 18 Welding symbol information.

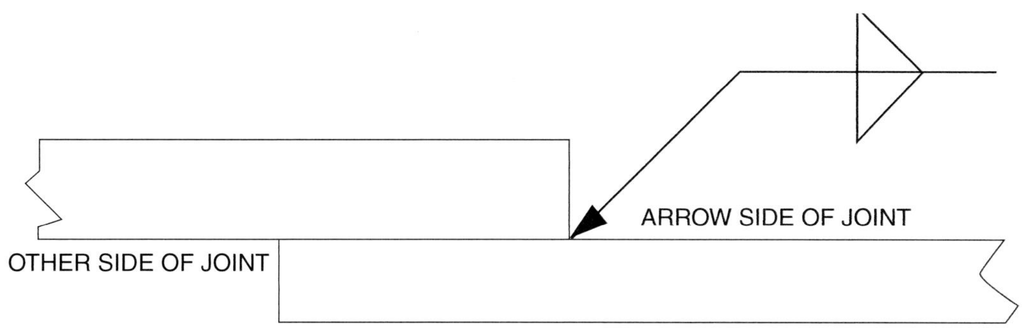

Figure 19 Arrow side and other side of joint.

Weld symbols placed on both sides of a reference line indicate the weld is to be made on both sides of the joint. The both sides designation does not apply to all weld symbols. Some symbols have no arrow side or other side significance, although supplementary symbols used in conjunction with them may vary. *Figure 20* shows basic weld symbols and their location significance.

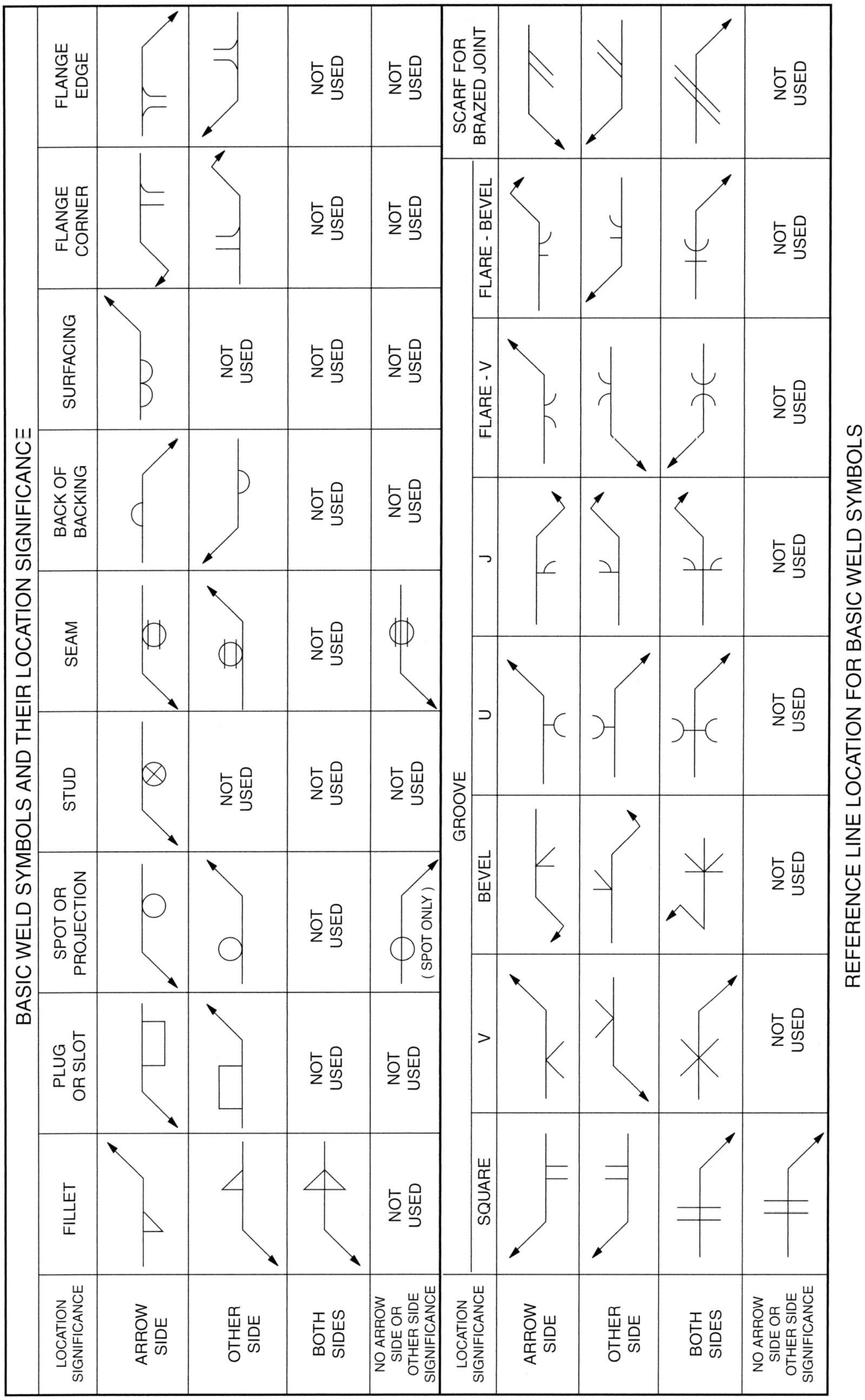

108F20.EPS

Figure 20 Significance of basic weld symbols locations.

SUMMARY

All structural materials have special methods or systems for fastening and erection that must be considered in the drafting process. The basic structural materials used in contemporary structures are steel, wood, concrete, structural clay products, and stone masonry.

Blueprints for the ironworker describe the materials and processes for construction with steel. The blueprints serve as a language and a contract for those involved in any phase of the construction project.

Review Questions

1. Large industrial and commercial buildings are constructed around a framework of structural steel or reinforced _____.
   a. wood
   b. concrete
   c. aluminum
   d. vinyl

2. In residential construction, except in multistory apartment buildings, little use is made of _____.
   a. steel
   b. concrete
   c. wood
   d. aluminum

3. Wide-flanged shapes have flanges with _____.
   a. tapered surfaces
   b. convex surfaces
   c. concave surfaces
   d. parallel surfaces

4. American Standard channels have tapered flanges and are known as _____.
   a. M-shapes
   b. R-shapes
   c. C-shapes
   d. T-shapes

5. The three common bar classifications are flat, round, and _____.
   a. tapered
   b. convex
   c. concave
   d. square

6. The frame element that gives a structure its rigidity is called a _____ and may be one or more stories in height.
   a. bent
   b. girt
   c. purlin
   d. girder

7. The main roof framing consists of the rafters or trusses in any _____.
   a. deck
   b. bent
   c. square
   d. floor

8. The roof deck rests on _____ and may be precast concrete slabs, wood planking, or plywood sheets.
   a. bays
   b. purlins
   c. girts
   d. spans

9. The drawings that show where the columns are placed with respect to each other are the _____.
   a. erection drawings
   b. detail drawings
   c. shop drawings
   d. fabrication drawings

10. The process of reading construction drawings moves back and forth between the erection drawing and the _____.
    a. shop drawings
    b. mechanical section
    c. site drawings
    d. detail drawings

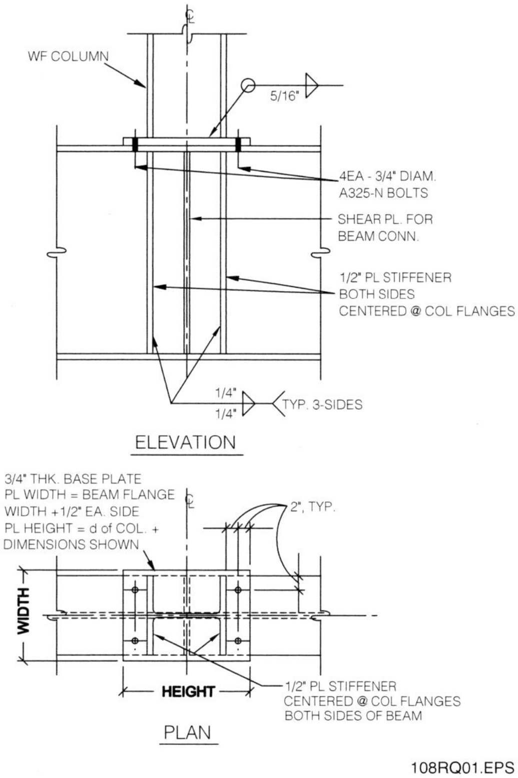

Figure 1

11. *Figure 1* shows a _____.
    a. beam to beam connection
    b. beam to support column connection
    c. column to support beam connection
    d. deck to support column connection

12. Fabrication shop drawings are drawn to show a(n) _____.
    a. completed structure
    b. piece for assembly
    c. piece of raw material
    d. detail of assembly

13. Ornamental ironworkers spend most of their time fitting and assembling _____.
    a. raw channel iron
    b. staircases only
    c. structural members
    d. preshaped pieces

14. When the weld symbol is placed below the reference line, the weld must be made on the _____.
    a. arrow side of the joint
    b. outside edge of the joint
    c. other side of the joint, opposite the arrow
    d. inside edge of the joint

15. When the weld symbol is placed above the reference line, the weld must be made on the _____.
    a. arrow side of the joint
    b. outside edge of the joint
    c. other side of the joint, opposite the arrow
    d. inside edge of the joint

Ernest Garcia

Miller and Long Concrete Construction
Sr. Foreman

How did you get started in the construction industry?
I worked on a bridge crew at Camp 26 (road gangs). In 1977, I started construction work with Hercules Construction, then went to work with Hi-Rise in 1978. I started working for Miller and Long Concrete Construction in 1982, and became a foreman in 1987. I became an ABC Metro Instructor in 2006.

Who inspired you to enter the industry? Why?
No one inspired me, it just happened. I started with the road gangs and was given the opportunity to work at a concrete company. After the concrete company learned I had worked with steel, I was moved up.

What do you enjoy most about your job?
I can see the results of my labor, and it offers a change of scenery. It also offers the ability to advance and to help others enjoy the trade.

Do you think training and education are important in construction? If so, why?
Yes, as in life. I was taught throughout life that training and education makes work and life easier and better.

Are NCCER credentials important to your career?
Yes, and they are getting more important every day, because more employers want certified people.

How has training/construction impacted your life and your career?
It has given me everything I have, including the ability to visit my folks two times a year, a house, a truck, and motorcycles. I have made friends through this work. I respect my work and have earned the respect of others.

Would you suggest construction as a career to others? If so, why?
Yes, it opens the door to learning about yourself in ways other jobs might not. I am proud to be able to point to buildings that I helped to build.

How do you define craftsmanship?
Craftsmanship is finding a way to help create something (with others) that people can admire and respect.

Trade Terms Introduced in This Module

Bent: All frames (whether rigid, truss-on-column, rafters-on-column) or other members used to span between columns in the principal plane.

Connections: Devices used to attach units to each other or to the structure.

Purlin: The horizontal roof beam used to span the bay distance and on which the roof deck rests.

Shop drawings: All drawings prepared by the engineer, divided into erection drawings, anchor setting drawings, and production drawings.

Spandrel: The beams around the outside of the floor.

Truss: Fabricated supports used to span between bents.

Additional Resources

This module presents thorough resources for task training. The following resource material is suggested for further study.

AISC 303-05, Code of Standard Practice for Steel Buildings and Bridges.

AISC 360-05, Specification for Structural Steel Buildings Allowable Stress Design and Plastic Design.

AISC S342L, Load and Resistance Factor Design Specification for Structural Steel Buildings.

AASHTO HB-17, Standard Specifications for Highway Bridges.

ASTM A6, General Requirements for Delivery of Rolled Steel Plates, Shapes, Sheet Piling, and Bars for Structural Use.

Figure Credits

Topaz Publications, Inc., Module opener

LPR Construction, Figures 10-16

AWS A2.4:2007, Figure 2, reproduced with permission from the American Welding Society (AWS), Miami, FL, and Figure 18

NCCER CURRICULA — USER UPDATE

NCCER makes every effort to keep its textbooks up-to-date and free of technical errors. We appreciate your help in this process. If you find an error, a typographical mistake, or an inaccuracy in NCCER's curricula, please fill out this form (or a photocopy), or complete the online form at **www.nccer.org/olf**. Be sure to include the exact module ID number, page number, a detailed description, and your recommended correction. Your input will be brought to the attention of the Authoring Team. Thank you for your assistance.

Instructors – If you have an idea for improving this textbook, or have found that additional materials were necessary to teach this module effectively, please let us know so that we may present your suggestions to the Authoring Team.

NCCER Product Development and Revision

13614 Progress Blvd., Alachua, FL 32615

Email: curriculum@nccer.org
Online: www.nccer.org/olf

❑ Trainee Guide ❑ Lesson Plans ❑ Exam ❑ PowerPoints Other ________________________

Craft / Level: Copyright Date:

Module ID Number / Title:

Section Number(s):

Description:

Recommended Correction:

Your Name:

Address:

Email: Phone:

Structural Ironworking One

30109-11

Trainees with successful module completions may be eligible for credentialing through NCCER's National Registry. To learn more, go to **www.nccer.org** or contact us at **1.888.622.3720**. Our website has information on the latest product releases and training, as well as online versions of our *Cornerstone* newsletter and Pearson's Contren® product catalog.

Your feedback is welcome. You may email your comments to **curriculum@nccer.org,** send general comments and inquiries to **info@nccer.org**, or use the User Update form at the back of this module.

V.1 1/11

30109-11
STRUCTURAL IRONWORKING ONE

Objectives

When you have completed this module, you will be able to do the following:

1. Identify the types of construction that use structural steel.
2. Describe the steel structure erection process.
3. State the principles of structural stresses.
4. Identify the components of common steel structures.
5. Explain the requirements of bolted connections.
6. List the advantages of pre-engineered structures.

Performance Tasks

Under the supervision of the instructor, the trainee should be able to do the following:

1. Describe different uses for structural steel.
2. Identify selected types, shapes, and grades of structural steel.
3. Identify different types of structural-steel beams.
4. Make bolted connections on structural steel.

Trade Terms

Beam
Bracing
Column
Deck
Load
Reaction
Stress

Contents

Topics to be presented in this module include:

Figures and Tables

1.0.0 INTRODUCTION

Structural ironwork involves erecting, aligning, and fastening the steel framework of bridges, buildings, and storage tanks. It also involves installing decking and doors, directing crane operators in hoisting structural components into position, and helping maneuver **beams** and girders into place.

2.0.0 STEEL

Steel is one of the most important and versatile structural materials. It has high strength and ductility, is widely available, and is very durable with minimal weather protection.

The fabricator works from the engineering and architectural drawings to produce shop detail drawings that provide the necessary dimensions. The design drawings and specifications also indicate the necessary surface finishes.

2.1.0 Steel Products

Steel lends itself to construction because the shapes can be formed at the mill. Steel ingots are rolled into shapes of varying widths and thicknesses. Most of the rolling is done on hot steel and this is called hot-rolled steel. Sometimes thinner plates are further rolled, or are formed after cooling into cold-rolled products.

Some basic steel shapes include the wide flanged W-shape used as a beam or **column**; the S-shape, also known as an I-beam; the C-shape or channel; and the L-shape, also known as the angle. *Table 1* shows the common steel shapes.

2.2.0 Steel Shapes

The wide-flange or W-shape (*Figure 1*) is the most commonly used structural shape. It is a doubly symmetrical shape (symmetrical about both the x and y axis) consisting of two rectangular-shaped flanges connected by a rectangular web plate. The flange faces are parallel with the inner flange distance.

The designation W16 × 50 indicates that the nominal overall section depth is 16 inches with a weight of 50 pounds per foot (lb/ft). S-shapes are doubly symmetrical shapes produced in accordance with dimensions adopted in the late 1800s. They are called I-beams and American Standard beams. The three essential differences between S- and W-shapes are as follows:

- The flange width of the S-shape unit is less than the flange width of the W-shape.
- The inner face of the S-shape flange has a slope of approximately 16.7 degrees.
- The theoretical depth of the S-shape is the same as the nominal depth.

M-shapes are also doubly symmetrical shapes which are not classified as W- or S-shapes. There are about 20 lightweight shapes in the M-classification. An M14 × 18 is the largest M-shape. It is important to learn the common sizes used at the job site.

C-shapes are channel shapes produced in accordance with dimensional standards adopted in 1896. The inner flange slope is the same as for S-shapes. The theoretical and nominal depths are identical. A C6 × 13 is a standard channel shape, having a depth of 6 inches and a nominal weight of 13 pounds per foot.

MC-shapes are channel shapes that are not classified as C-shapes. They are called shipbuilding channels. It is important to understand the common sizes used at a site.

L-shapes are either equal or unequal leg angles. All angles have parallel flange faces.

T-shapes, or structural tees, are made by splitting W-, S-, or M-shapes. Generally, the splitting produces a shape with half the area of the original shape section. Offset splitting, however, may be used if a deeper tee section is required. Published tables of T-shapes are based on symmetrical splitting, with no allowance made for material loss from the sawing or flame-cutting process used in splitting.

NOTE

In steel design, major factors are the yield strength and the tensile strength of the material. Yield strength is the load limit to which a steel component can be deformed and still return to its original form. Tensile strength is the ability of a component to resist pulling forces.

2.3.0 Grades

Steel making is in a constant state of progress. Research in the industry is continuously developing new steels for specific purposes and improving the versatility of the older types of steel. Unless another grade is specified in the documentation, W-shapes and WT-shapes are required to be material grade ASTM A992. Plates and bars, M, S, MT, ST, HP, L, C, and MC must be ASTM A36. HSS is required to conform to ASTM A500, grade B, and steel pipe is held to ASTM A53, grade B.

Table 1 Common Steel Shapes

DESCRIPTION	SYMBOL	EXAMPLE USE	PICTORIAL
GAUGE PITCH	G P	G = 3½ P = 2½	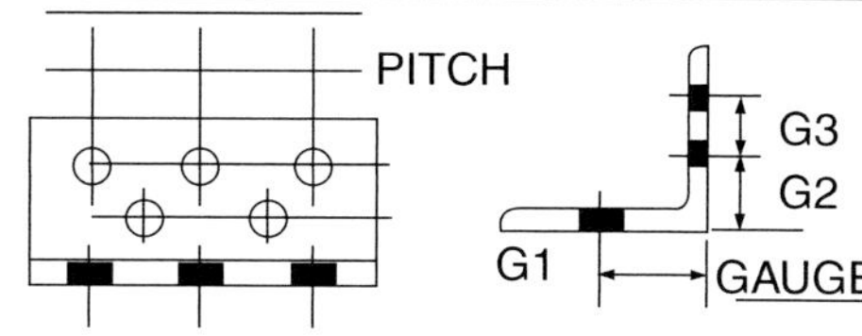
SHAPES WIDE - FLANGE AMERICAN STANDARD JOISTS AND JUNIOR BEARING PILE	 W S M HP	 W24 × 76 S15 × 42.9 M12 × 36.6 HP10 × 57	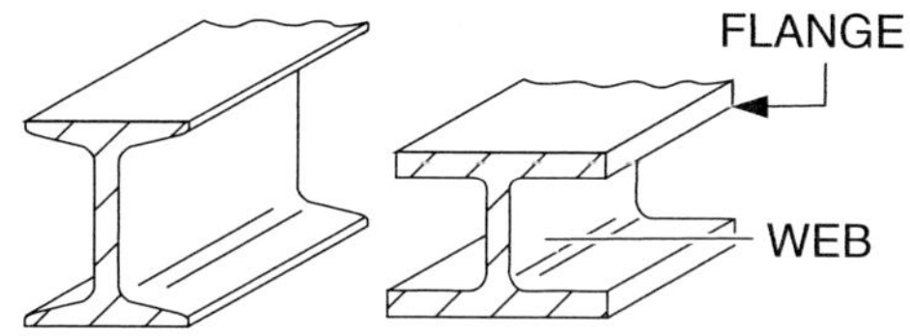
CHANNELS AMERICAN STANDARD	C	C8 × 11.5	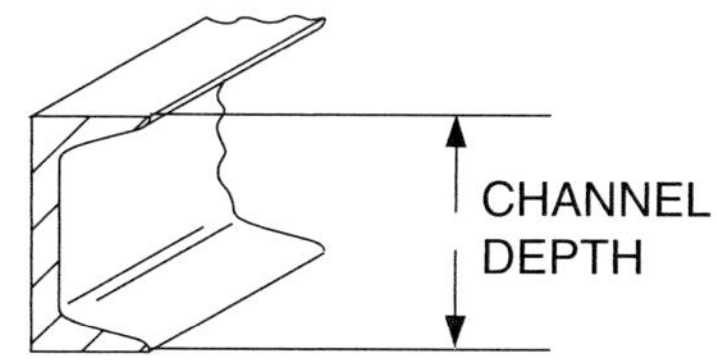
ANGLES EQUAL LEG UNEQUAL LEG	 L L	 L3 × 3 × ¼ L7 × 4 × ½	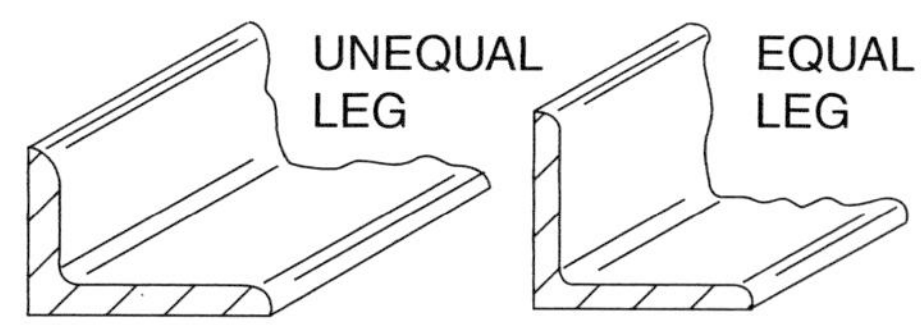
TEES	WT MT ST	WT8 × 15.5 MT5 × 4.5 ST5 × 12.7	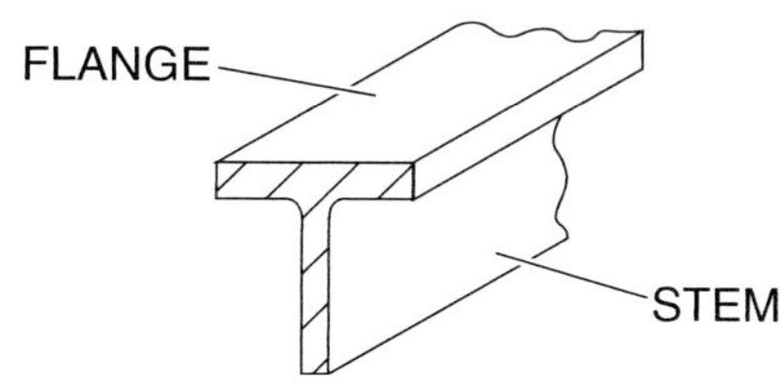
PLATE CHECKERED PLATE	PL CK PL	PL ½ × 18 × 2'-6 CK PL ½	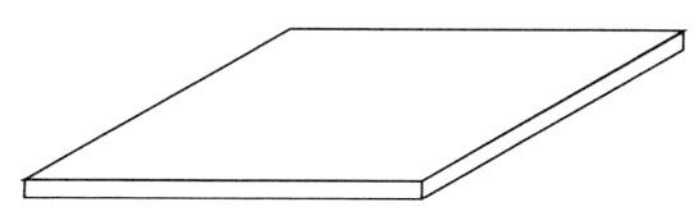
BARS FLAT SQUARE ROUND	 BAR BAR ϕBAR	 BAR 2½ × ½ BAR 2½ × 1 BAR 2½ ϕ	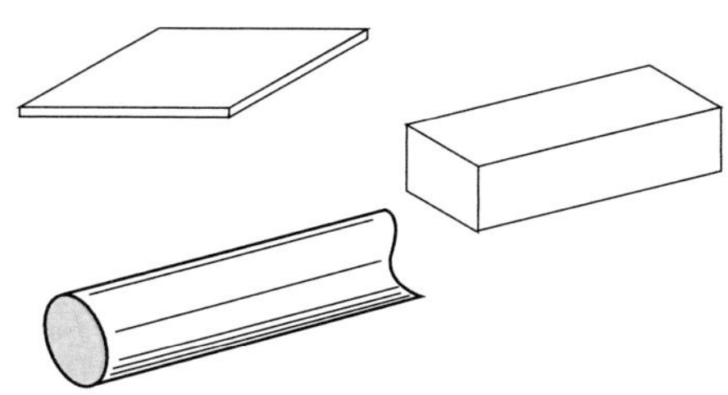
PIPE	PIPE	PIPE 4 STD	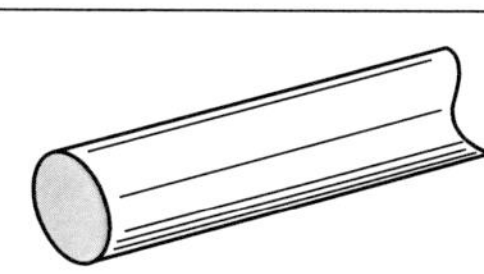

109T01.EPS

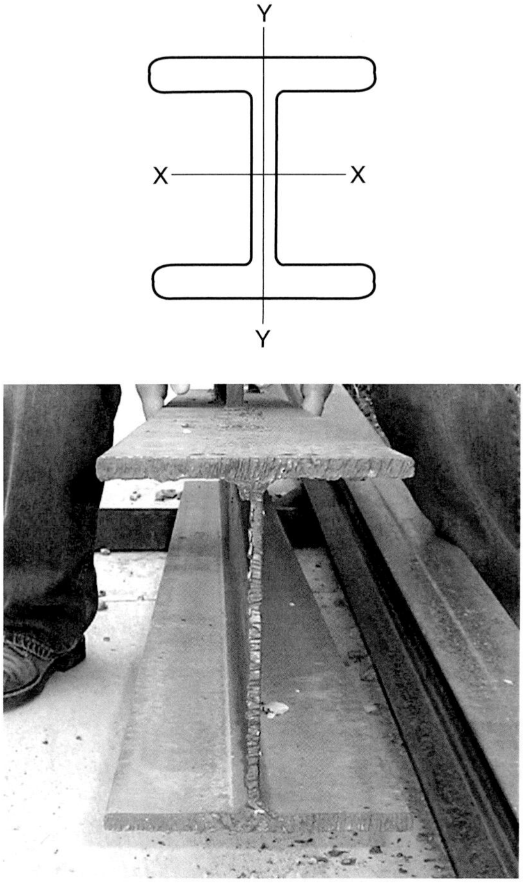

109F01.EPS

Figure 1 Wide flange shape.

3.0.0 Steel Beams

Any horizontal or inclined component that carries a **load** and is supported at two or more points is called a beam. The forces acting at the supports of these members are called **reactions**. These reactions resist the beam loads and hold the beam in equilibrium. This means that they prevent the beam from moving in any direction.

If the ends of a loaded beam are free to rotate at the points of support, the beam is called a simple beam (*Figure 2*).

If the connections at the supports are rigid and end rotation is prevented, the beam is called a fixed beam (*Figure 3*).

A cantilever beam (*Figure 4*) has one end fixed at a support and the other end unsupported. A cantilever beam may also have one end overhanging a support which is free to deflect.

A continuous beam (*Figure 5*) is a beam which spans continuously over a number of intermediate supports.

Design engineers calculate the loads that a steel structure must support to determine the size of each structural member. Component sizes, types

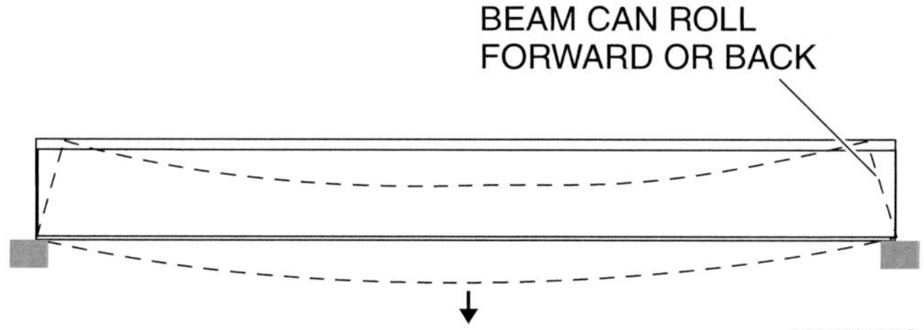

109F02.EPS

Figure 2 Simple beam.

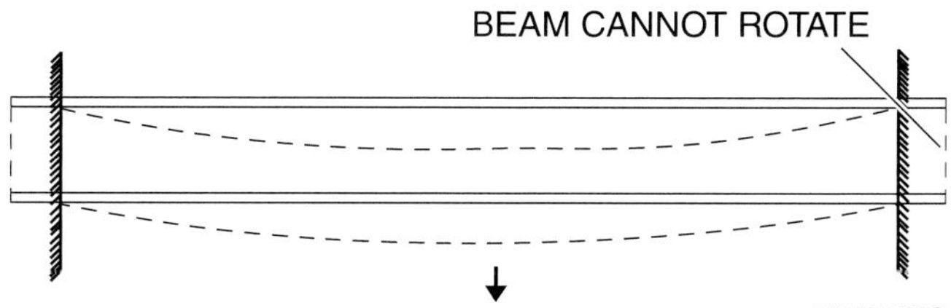

109F03.EPS

Figure 3 Fixed beam.

109F04.EPS

Figure 4 Cantilever beams.

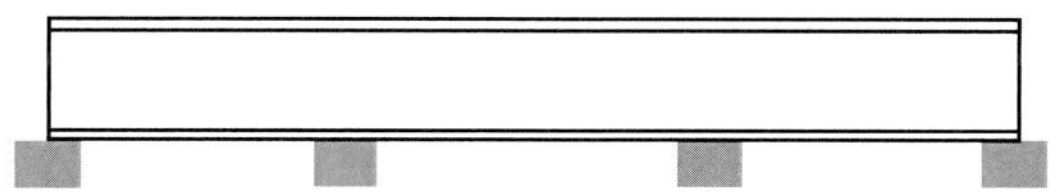

109F05.EPS

Figure 5 Continuous beam.

of connections, special reaction data, and similar design information are used by the architect in preparing drawings.

The choice of design method will decide the kinds of connections used. If there is a conflict between the drawings and the specifications, the drawings rule. However, if a difference is found, tell the supervisor so the conflict can be checked.

4.0.0 Use of Structural Steel

Structures made of steel include bridges, buildings, transmission towers, storage tanks, sign supports, and, in some cases, art. Bridges and buildings are the most common projects involving structural ironworkers.

4.1.0 Pre-Engineered Buildings

Pre-engineered buildings may include a steel frame or have a roof supported by steel members resting on loadbearing walls.

The steel skeleton of this type of building may be rigid or pinned; it may be a two- or three-hinged arch; or it may be a truss-on-column system. The truss also may be either rigid or pinned.

A building frame (*Figure 6*) is a three-dimensional skeleton but it is usually considered to be rigid in only one dimension. The frame resulting from considering only the main frame elements is called a bent and may be one or more stories in height. A bent is a framework transverse to the length of a structure designed to carry lateral as well as vertical loads. It is a self-supporting frame having at least two legs and is usually placed at right angles to the length of the structure it supports. It is used for all frames whether rigid, truss-on-column, or whatever is used to span between columns in the principal plane. The spacing between bents in the third dimension is the bay spacing. Spandrel and floor beams are used to span the bays in multistory buildings. Girders span between columns of bents. Lateral **bracing** is used in selected bays.

Pre-engineered or modular building systems use tubular steel columns with open-web or steel trusses, or hot-rolled light beam sections and standard long beam open-web steel joists. Structural frames that weigh less than four pounds per square foot of building area have been developed for single-story buildings. The frames are laid out on a uniform module that uses a maximum percentage of standard parts.

The steel frame can be erected quickly since it is assembled with standard bolt connections. Joists can be attached to the girders or trusses by field welding or fastening. A light-gauge steel roof **deck** is attached to the framework and the structural frame is enclosed with preassembled exterior wall panel systems. The exterior wall joints are usually of the batten type and use neoprene gaskets to ensure weather tightness and to accommodate expansion and contraction.

Foundations are reduced with a framing system weighing less than four pounds per square foot, and a wall system weighing six pounds or less per square foot. This allows these types of buildings to be used in many areas where conventional materials are too heavy for the available bearing surface.

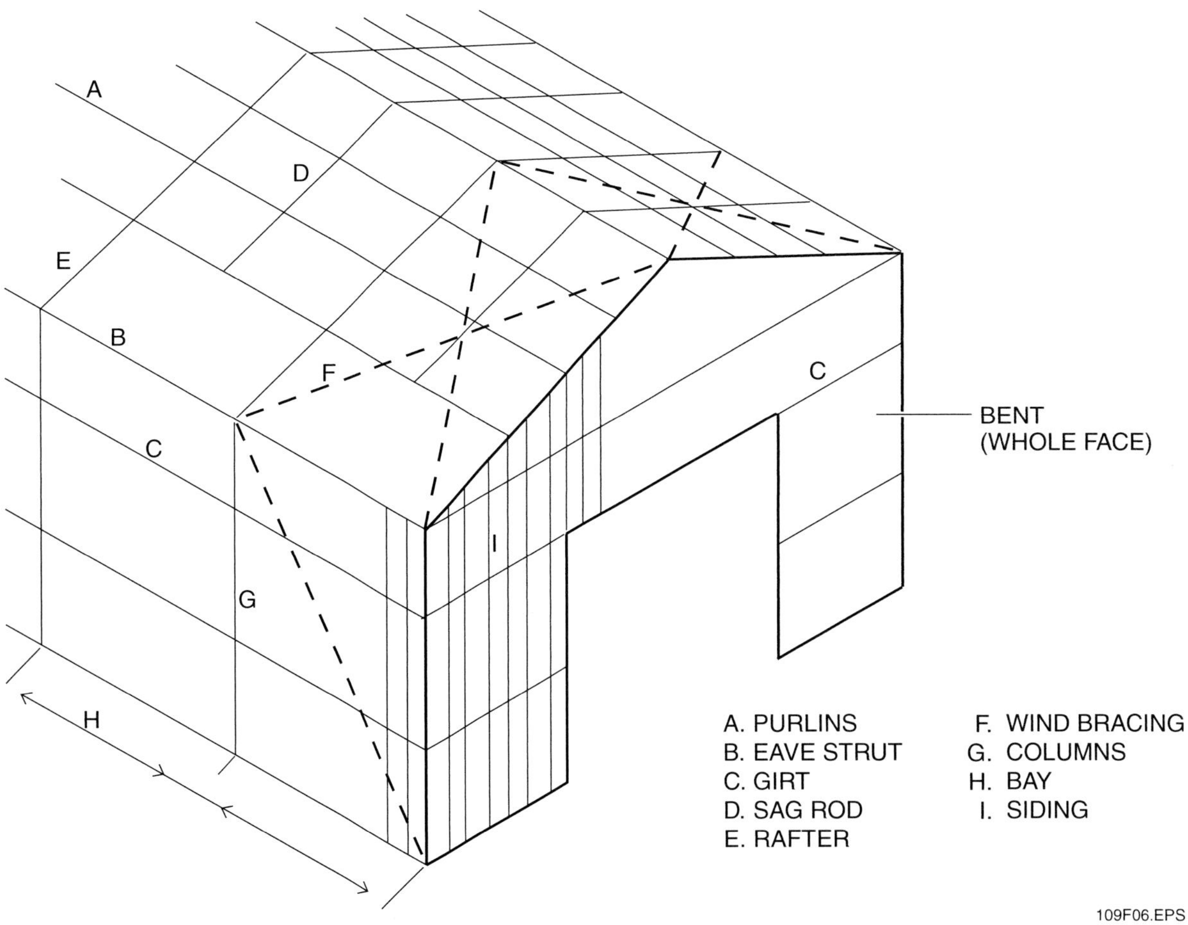

Figure 6 Typical building frame.

Manufacturers of pre-engineered buildings usually deliver all materials, components, and accessories to contractors for on-site assembly. Most drawings for this type of construction are erection manuals and contain a step-by-step procedure for structural assembly. The manual may also contain a list of tools and equipment necessary for efficient building assembly.

Building roof systems consist of framing, decking, and a waterproof covering. The main roof framing consists of the rafters or the truss in any bent. Purlins, spaced 2 to 6 feet on center, span the bay distance. Sag rods provide additional support for the purlins used on sloping roofs. The roof deck rests on the purlins and may be constructed of a metal deck, precast concrete slabs, wood planking, or gypsum sheets.

Metal sheets, curtain walls, brick, concrete block, tile, or precast or poured concrete may be used for siding. Lightweight siding is carried by spandrel beams supplemented with girts for high-rise buildings or by the eave strut and girts for industrial buildings.

4.2.0 Bridges

Bridges are classified as plate girder (*Figure 7*), truss (*Figure 8*), arch, or suspension types. The truss and plate girder bridges are the most common. The truss types include both traffic on top of the truss (deck) and the through type, where traffic passes between the trusses. Truss bridges very often combine both deck- and through-type construction. Contemporary bridge designers prefer to use girder structures, which require less fabrication over trusses.

A truss, like a joist, is a structural unit usually made from triangles. The simplest form of truss is, in fact, a triangle. There are a number of special designs for a truss, such as the Pratt truss (*Figure 9*), the queenpost truss (*Figure 10*), and the multiple kingpost truss (*Figure 11*).

Some bridge construction units are similar to those used in buildings and serve similar purposes. These units include footings, piers, caissons, walls, and slabs. A beam bridge is commonly used for short spans. If a single span is used, the end supports are called abutments or

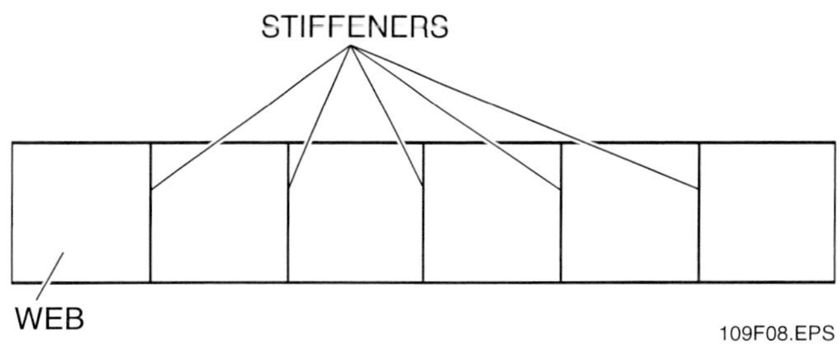

Figure 8 Truss bridge.

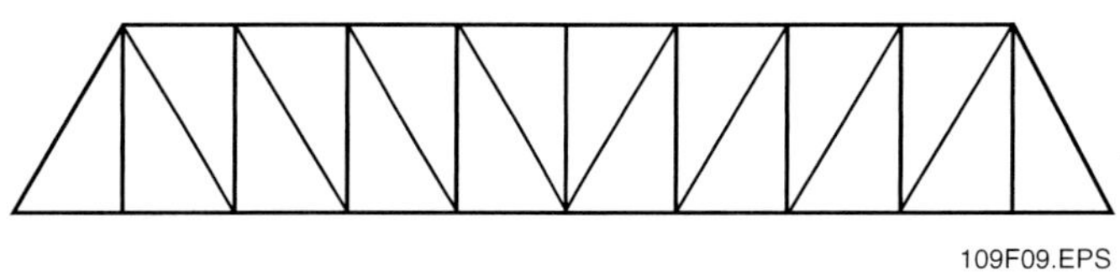

Figure 9 Pratt truss.

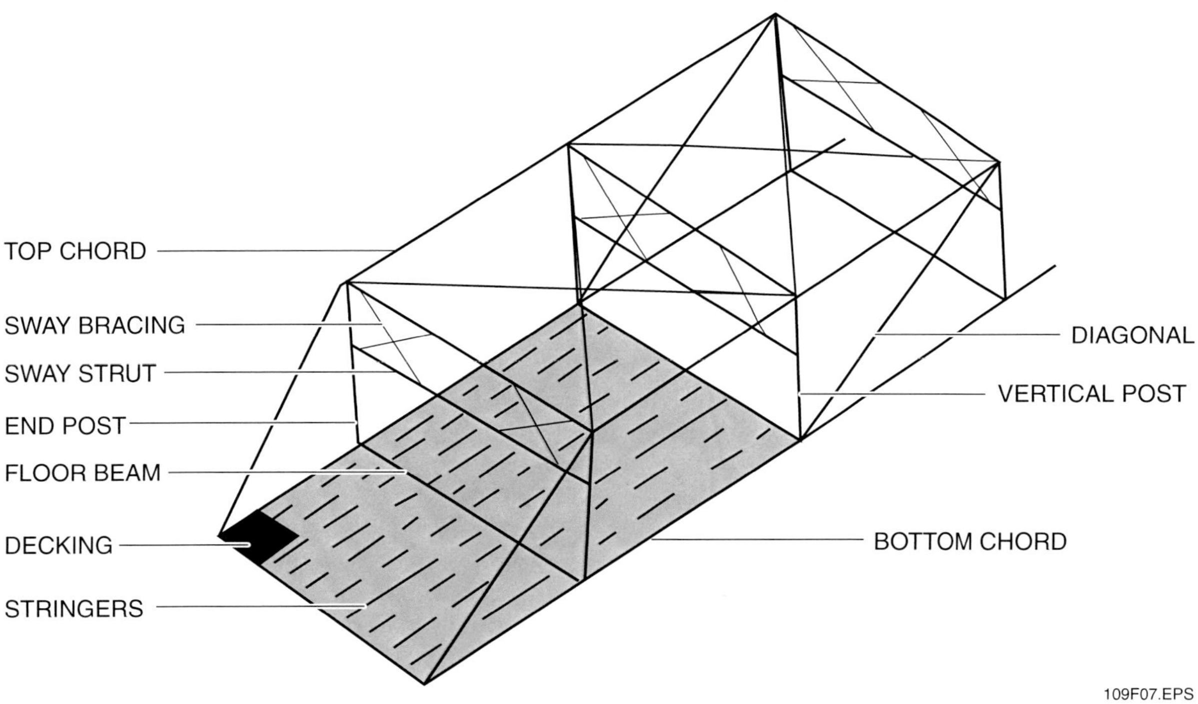

Figure 7 Plate girder.

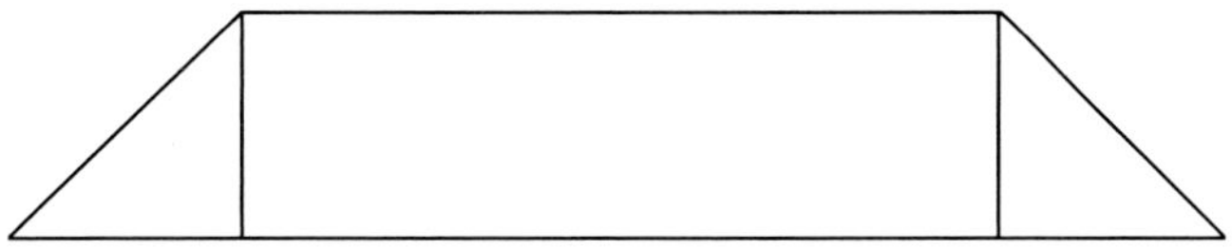

Figure 10 Queenpost truss.

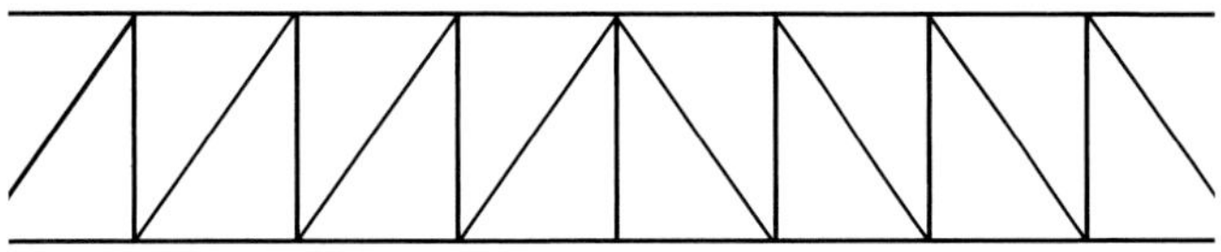

Figure 11 Kingpost truss.

end bents. When two or more spans are used, the intermediate supports are called piers or intermediate bents.

A rigid-frame bridge generally consists of footings, walls, and a deck slab cast as a unit which forms a U-shaped element.

The arch bridge has many variations. One is the multiple-span, open-spandrel bridge configuration.

> **NOTE**
>
> The American Institute of Steel Construction (AISC) is the national authority on steel construction. AISC performs research and engineering studies and provides information and standards on all the aspects of ironwork, including weights per foot of various structural members, properties of various grades of steel, and other specifications. The information is published in the *AISC Manual of Steel Construction*, and every ironworker should be familiar with this manual.

5.0.0 Work Processes

The actual erection of steel for building a structure is performed by a team having a very distinct separation of functions. The supervisor is responsible for interpreting the erection plans, assigning tasks, directing the sequence of hoisting, and making sure that the structure is plumb, fastened, and completed. The supervisor is also responsible for taking care of the equipment, tools, supplies, and materials furnished for construction. On large jobs, the supervisor may be assisted by one or more foremen.

When erection of the steel begins, one journeyman and a helper act as hook-on persons or riggers. Their job is to prepare the steel and make sure it is safe to hoist and easy to connect. They may also dress up main members by bolting connecting members together on the ground when this is possible. After the member is connected to the crane or other hoisting equipment, the supervisor or raising gang foreman usually directs the placement, indicates where the component is to go, and what direction it faces.

They work through an identified signal person to direct the hoisting operation. After the supervisor has determined that the member is placed properly, they may continue locating the next component to be erected. The only time anyone other than the authorized signal person may issue instructions to the crane operator is when an emergency stop is necessary.

6.0.0 Erection of Structural Members

The erection of structural members includes erecting the following components:

- Columns
- Girders/joists
- Trusses
- Bracing
- Deck

The following section discusses the general erection procedures for various types of structural members.

6.1.0 Columns

Columns are usually transported to the erecting equipment in a horizontal position. They are then raised upright by the erecting equipment, carried to their proper location, positioned, and secured. A hole may be provided at the top end of the shaft so that the crane can use a clevis to hoist the column and deliver it in a vertical position. Heavy column fabricators may provide lifting lugs or special holes for hoisting.

Column splices can vary according to design. The center lines of column baseplates should be scribed in both directions so the column can be centered on the scribe marks. The splice plates are shipped loose or attached to the lower shaft in the shop. As the column is lowered, either the spud end of a structural wrench or a drift pin is inserted through the connection holes to guide them. The connection is temporarily secured with erection bolts in each plate and with one or two drift pins. Drift pins have the same diameter as the holes and prevent the column from rocking with the wind until the beams and girders have been attached. The pins are removed after the column has been plumbed.

6.2.0 Girders/Joists

Girders and beams are erected using wire-rope slings because they need to be lifted into their location in the framework in a horizontal position and with the webs vertical. Sometimes, the columns may have to be spread in order to fit the girder into place.

Drift pins or the spud end of construction wrenches are used to align the holes of girders to the columns when the girders are in place. The girders are then temporarily secured with a minimum of two bolts in each end, according to OSHA and engineering specifications.

6.3.0 Trusses

Small trusses are usually shipped to the job site in one piece. Larger trusses arrive in sections and must be assembled in the field. This process requires bolting or welding, depending upon the design and available equipment. In some cases, two half-trusses, a bottom chord, and a hanger may have to be field-bolted together. Trusses are assembled immediately before erection whenever possible. For a one-story building, this is performed on the ground. Roof trusses for a multistory building could be assembled or spliced together on the uppermost floor.

The web members of trusses are often flimsy, and the truss as a whole has a tendency to twist. Sometimes, a strongback is lashed to the sides of the truss to give it lateral stability for raising. After the first truss has been raised and bolted to the columns, it is necessary to prevent it from overturning. Guys are attached to the top and bottom chords to prevent this. The second truss is erected in the same manner and the bracing is then placed. After all of the trusses and the horizontal bracings have been installed, the purlins are installed.

6.4.0 Bracing

Bottom horizontal cross-bracing for trusses consists either of angles, beams, or tees. These components are lightweight and can be easily bolted into place.

The horizontal cross-bracing of the top chord strengthens the structure and helps to square the frames. The vertical cross-bracing helps to plumb the frames.

Many tall buildings are cross-braced from the top of one column to the base of the column next to it. As soon as it has been plumbed, the structure is usually bolted. Plumbing will be discussed in more detail later in your training.

7.0.0 CONNECTIONS

The beams, girders, trusses, columns, and other components which form a completed structure are designed to support certain loads. Each of these members must transmit its load through structural joints to supporting members. For example, beams transmit loads to girders; girders transmit beam loads to trusses or columns; and columns transmit loads to footings or piers in the ground. Each of these components must be connected securely to safely transfer its assigned load to other members.

7.1.0 Bolted Connections

There are two classes of bolts used in structural building: general use and high strength. General-use A-307 bolts are sometimes called soft, machine, or unfinished bolts. These bolts have a rough shank and bearing surface. The A-307 bolt is made of steel with an ultimate tensile strength of approximately 60 (A grade) to 100 (B grade) pounds per square inch (psi). A-307 bolts are available with several head and nut configurations. The square and hexagonal head bolts are the most commonly used. A-307 bolts are cheaper than A-325 and A-490 bolts and are generally used in static load structures. A-307 bolts are used in girts, platforms, stairways, and non-structural parts.

Hardened washers are required under the turned element of high-strength bolts when using oversized bolt holes and the flange slope is greater than 1:20.

Always use two washers when the fastened material has a **stress** factor of less than 40 kilopounds per square inch (ksi). Slotted bolts, called tension control bolts, are available so that a single installer using a power driver can torque the bolt without help. Ribbed washers are available which flatten at the specified torque for visual tension checks. These are referred to as direct tension indicators (DTI). Self-locking nuts are available for installation in connections subjected to dynamic loadings, such as overhead crane rails.

Bolts for structural applications range from ½" to 1½" in diameter. Occasionally, larger diameters are used. The most common bolt diameters are ¾" and ⅞" (20 to 22 mm).

7.2.0 Welded Connections

Welding is the joining of pieces of metal by heating the points of contact to a nearly fluid state with or without the application of pressure. Nearly all structural welding is electrical and various types of electrical welding systems are used.

Most welding processes use an electric current. The current heats an electrode to a molten state, which is then deposited as a filler material between the two pieces of metal being joined. This process simultaneously melts a portion of the base metal at the joint so that the electrode metal mixes with the base metal and develops a strong joint after cooling takes place.

The following welding processes are the most common processes used for structural applications:

- *Shielded metal arc welding* – This is the most common method and uses coated stick electrodes.
- *Gas metal arc welding* – This welding method is most commonly used in shop welding. Uncoated electrodes are used in a mechanical welding unit. The unit controls electrode spacing and welding speed and has an inert gas source which shields the weld against the atmosphere.
- *Flux core welding* – This welding process uses a hollow electrode. The flux is contained in the tubular electrode.

8.0.0 Tools

The selection of items for a standard tool list depends on the needs of the contractor, the type of structure to be erected, and any special considerations. Generally speaking, the structural ironworker's tool list should include erection wrenches, pliers, bull pins, drift pins, tool holders, bolt bags, hard hats, industrial belts, full body harnesses, lanyards, and sleever bars.

Wedges are sometimes used for driving under column supports, for leveling or moving members into place. Sledges and mauls are used to drive steel into position, level columns, and force pins through holes.

Summary

Structural ironwork involves the erection of the steel skeleton that forms the framework of all large buildings and structures. Ironworkers perform a variety of tasks and need to be trained in the many skills needed in this field. Such skills include recognizing the sizes and shapes of materials, the theories behind the selection and use of materials, and the procedures used in the erection process.

Review Questions

1. I-beams are also known as _____.
   a. W-shapes
   b. S-shapes
   c. C-shapes
   d. L-shapes

2. The overall section depth of a W16 × 45 shape is _____ inches.
   a. 16
   b. 45
   c. 32
   d. 90

3. Equal or unequal leg angles are designated _____.
   a. W-shapes
   b. S-shapes
   c. C-shapes
   d. L-shapes

4. W shapes and WT shapes, unless otherwise designated, must be ASTM _____.
   a. A36
   b. A500
   c. A53
   d. A992

5. A beam that is rigidly supported on each end so that the beam ends cannot rotate is referred to as a _____.
   a. simple beam
   b. fixed beam
   c. cantilever beam
   d. continuous beam

6. The framework of a building transverse to the length of the building is referred to as the _____.
   a. purlins
   b. girt
   c. bent
   d. bay

7. Of the following types of bridges, the most common type of bridge is the _____.
   a. arch
   b. concrete
   c. truss
   d. suspension

8. A drift pin's diameter should be _____ the holes in the flanges.
   a. the same as
   b. one half of
   c. slightly larger than
   d. slightly smaller than

9. The bolts generally used in girts, platforms, stairways, and non-structural parts are _____.
   a. A-500 bolts
   b. A-490 bolts
   c. A-325 bolts
   d. A-307 bolts

10. Bolts for structural applications range in diameter from _____.
    a. ½" to ⅞"
    b. ½" to 1½"
    c. ¾" to ⅞"
    d. 1" to 2"

Stacey Cobb

Ironworker Rigger Supervisor
KBR/BE&K

How did you get started in the construction industry?
I am a sixth generation ironworker rigger. My family has always worked in construction. I started in the iron trade as a green helper and worked my way up with lots of hard work and determination, and with the help of mentors that spent the extra time to teach me to do the job right the first time. This taught me to have pride in my work. As I grew and advanced, I also took the time to pass along the knowledge that was passed on to me.

Who inspired you to enter the industry? Why?
My grandfather and father inspired me to enter the construction field from watching them as I grew up, listening to them talk and tell stories of jobs, and riding by buildings and hearing them tell how they built things and how things come together. I can remember the look of pride and sense of accomplishment on their faces. To me, construction work (ironwork) is the all-American trade. The feelings of pride and accomplishment as I tell stories to my children and show them what I have worked on and built are the same feelings my father and grandfather had.

What do you enjoy most about your job?
I most enjoy career satisfaction in seeing something that I have done become a part of a building.

Do you think training and education are important in construction? If so, why?
Yes, I think training and education are important. The industry is seeing lots of new people that have no construction background.

How important are NCCER credentials to your career?
NCCER credentials are very important in my career.

How has training/construction impacted your life and your career?
Credentials have given me the opportunity for career advancement.

Would you suggest construction as a career to others? If so, why?
Yes. I have traveled all over the United States working and enjoying all the things that construction workers have built, and making money that office workers can't dream of. The construction industry is like a big family that helps you and pushes you to be your best.

How do you define craftsmanship?
Craftsmanship is taking pride in your work, and doing the job right the first time.

Trade Terms Introduced in This Module

Beam: A single piece that is a structural element mounted more or less horizontally.

Bracing: Diagonal or horizontal supports for joists or beams. Bracing helps to keep joists or beams from twisting or rolling.

Column: A single vertical structural element.

Deck: Material used as a horizontal surface on a building level. Decking may be used to hold a concrete floor, or may serve as a roof.

Load: The total forces acting on a structural member.

Reaction: Forces acting on horizontal members of a structure that support the load.

Stress: Forces acting on each structural member.

Additional Resources

This module presents thorough resources for task training. The following resource material is suggested for further study.

Steel Construction Manual, 13th edition. Chicago, IL: American Institute of Steel Construction.

www.seaonc.org is a structural engineering website.

Figure Credits

Topaz Publications, Inc., Module opener and Figure 1 (photo)

NCCER CURRICULA — USER UPDATE

NCCER makes every effort to keep its textbooks up-to-date and free of technical errors. We appreciate your help in this process. If you find an error, a typographical mistake, or an inaccuracy in NCCER's curricula, please fill out this form (or a photocopy), or complete the online form at **www.nccer.org/olf**. Be sure to include the exact module ID number, page number, a detailed description, and your recommended correction. Your input will be brought to the attention of the Authoring Team. Thank you for your assistance.

Instructors – If you have an idea for improving this textbook, or have found that additional materials were necessary to teach this module effectively, please let us know so that we may present your suggestions to the Authoring Team.

NCCER Product Development and Revision

13614 Progress Blvd., Alachua, FL 32615

Email: curriculum@nccer.org

Online: www.nccer.org/olf

❑ Trainee Guide ❑ Lesson Plans ❑ Exam ❑ PowerPoints Other ______________________

Craft / Level: Copyright Date:

Module ID Number / Title:

Section Number(s):

Description:

Recommended Correction:

Your Name:

Address:

Email: Phone:

Plumbing, Aligning, and Guying

30110-11

Trainees with successful module completions may be eligible for credentialing through NCCER's National Registry. To learn more, go to **www.nccer.org** or contact us at **1.888.622.3720**. Our website has information on the latest product releases and training, as well as online versions of the *Cornerstone* newsletter and Pearson's Contren® product catalog.

Your feedback is welcome. You may email your comments to **curriculum@nccer.org,** send general comments and inquiries to **info@nccer.org**, or send us the User Update form at the back of this module.

V.1 1/11

30110-11

Plumbing, Aligning, and Guying

Objectives

When you have completed this module, you will be able to do the following:

1. Describe the purpose and function of aligning and plumbing steel structures.
2. Identify the tools and equipment used for aligning and plumbing steel structures.
3. Identify the components of column bases, base plate, and foundation failures.
4. Explain selected plumbing and aligning tasks.

Performance Tasks

Under the supervision of the instructor, the trainee should be able to do the following:

1. Identify selected alignment tools.
2. Demonstrate alignment methods.
3. Demonstrate plumbing a structure.

Trade Terms

Anchor bolts (anchor rods)
Come-along
Guys
Leveling nuts
Plumb bob
Push-pull jack
Spools
Steamboat ratchet
Turnbuckle
Wire rope clip

Contents

Topics to be presented in this module include:

Figures and Tables

1.0.0 INTRODUCTION

To plan the construction of a multistory, steel-framed building, it is important to understand the erection process and its timing in order to coordinate it with the other building operations. Once the sequence of erection has been developed, the job can be split up into a number of erection operations. An important process that occurs immediately after connecting the components with temporary connections is the plumbing and leveling of the structure. To be able to meet the erection tolerances, it is necessary to be able to make precise measurements based on a system of triangulation and leveling. In high-rise buildings, optical instruments can be used with greater accuracy than **plumb bobs**.

1.1.0 Current State-of-the-Art

In high-rise and bridge construction, the need for erection procedures and temporary support systems has long been established. Low-rise construction does not demand the same attention because low heights have relatively simple framing. Often, the structures are lightly loaded and the framing members are relatively light. This has led to a number of misconceptions:

- *Low-rise frames do not need bracing* – All steel frames need bracing. This is probably a carry-over from the era when steel frames were primarily used in heavy framing, and were connected in substantial ways, such as by using riveted connections.
- *Once the deck is in place, the structure is stable* – The steel-deck diaphragm is only one component of a complete system. This misconception is the result of a misunderstanding of the function of horizontal diaphragms versus vertical bracing. This may have resulted in the usefulness of diaphragms being oversold.
- *Anchor rods and footings are adequate for erection loads without evaluation* – There are many cases in which the loads on **anchor rods** and footings may be greater during erection than the loads imposed by the completed structure.
- *Bracing can be removed at any time* – The temporary supports are an integral part of the framework until it is completed and self-supporting. This may not even occur after the erection work is complete, as in the case of non-self-supporting structures.
- *The beams and tie joists are adequate as struts without evaluation* – During erection, strut forces are applied to many members which are laterally braced flexural members in the completed construction. Their axially loaded, unbraced condition must be independently evaluated.
- *Plumbing-up cables are adequate as bracing cables* – Such cables may be used as part of temporary lateral supports. Additional temporary support cables will be needed in most situations. Plumbing involves adjustment using diagonal cables. The size and number of cables for each purpose are determined by different means. For example, the lateral support cables would have a symmetrical pattern while the plumbing-up cables may all go in one direction to draw the frame into plumb.
- *Welding joist bottom-chord extensions produces full bracing* – The joist bottom chords may be a component of a bracing system and welding them would be appropriate. Other components may be lacking and temporary supports would be needed to complete the system. If the joists have not been designed for continuity, then the bottom chords must not be welded.
- *Column bases may be grouted at any convenient time in the construction process* – Until the column bases are grouted, the weight of the framework and any loads on it must be borne by the anchor rods and **leveling nuts** or shims. These elements have limited strength. The timing of grouting must be coordinated between the erector and the general contractor.

2.0.0 PURPOSES AND FUNCTIONS OF ALIGNMENT

Anchor bolts (anchor rods), grillage beams, and other basic parts may be set by others, but they must be checked by the erector before the work proceeds. The structural framework must be plumbed, leveled, and properly located before being permanently fastened. Elevator-shaft openings must be accurately plumbed from top to bottom to ensure tolerance limitations. Long members, fastened at a number of points, must be correctly placed and straight.

Fabricating structural steel to very fine limits is normally difficult. For example, the overall measurement of a long building may not be equal to that given on the working drawings. The overall dimensions could be slightly greater. When columns creep, the result is columns out of plumb. This creep of erected steelwork is noticeable on a line of columns where all the lengthwise beam connections are butt joints.

According to the American Institute of Steel Construction (AISC) Code, the following rules apply to the anchor rods on a structure:

- The variation in dimensions between centers of anchor bolts in a single anchor rod group cannot be more than ⅛ of an inch.

- The variation in dimensions between adjacent anchor rod groups must be less than or equal to ¼ of an inch.
- The variation in elevation of the tops of anchor rods cannot be greater than plus or minus ½ inch.
- The accumulated variation between centers of anchor-rod groups cannot be greater than ¼ inch per 100 feet, and not exceeding a total of 1 inch.
- And finally, the dimensional deviation of any anchor-rod group from the center line of the column line cannot be greater than ¼ inch.

Specified tolerances may be difficult to obtain due to the various vendors and crafts that are involved in any structural-steel erection project. Since a tolerance is allowed by most specifications, it may be impossible to locate the columns exactly where they should be. There must also be some tolerance for the erector in plumbing the columns. Ordinarily, the tolerances in manufacturing, fabrication, and erection balance each other out instead of adding up.

The contract specifications usually stipulate the tolerance permitted or require compliance with a code such as the *AISC Code of Standard Practice*. This code permits a specific deviation from plumbness for the columns, and out-of-level for the beams and girders. This depends on whether the columns are outside face, elevator shaft, or intermediate columns. The important columns to plumb accurately are those on the face of the building and those around the elevator shafts.

The *Code of Standard Practice* calls for "the angular variation of the working line from a plumb line to be equal to or less than ¹⁄₅₀₀ of the distance between working points..." for vertical columns. But in the case of column shipping pieces next to the elevator shaft, the displacement from the established column line cannot be more than 1 inch for the first 20 stories, and above that point ¹⁄₃₂ inch per story, to a maximum of 2 inches.

In the case of exterior columns, the allowable variation from the established column line for the first 20 stories is 1 inch toward and 2 inches away from the building line. Above the 20-story line, the tolerance is ¹⁄₁₆ inch for each story, to a maximum of 2 inches toward and 3 inches away from the building line.

When a metallic skin is to be installed on the face of a building, the plumbness of the outer columns becomes more critical, and the owner, governed by the AISC, must determine the tolerance that will be permitted in plumbing those columns.

3.0.0 Tools and Equipment

When the first panel of steel is erected, the plumbing-up **guys** are installed to plumb-up the structure. The tools and equipment required by plumbing-up should include guys ready for use and other ancillary tools and pieces of equipment. If guys are not ready for use, they must be made in the field. Guys are made up of wire rope that is ⅜", ½", ⅝", ⅞", or 1" in diameter. A loop is formed on one end around a special hook. **Wire-rope clips** are attached to hold the loop. *Figure 1* shows the initial braced box on a structure.

Similar hooks are inserted in one end of eye-and-eye **turnbuckles** in a size that matches the rope guys. The wire rope used for the guys can be new or old rope, as long as the old rope is strong enough for the purpose. Wire rope is cut to the necessary length.

3.1.0 Come-Alongs

Ratchet-lever hoists and **come-alongs** (*Figure 2*) are used for short pulls on heavy loads. The term *come-along* is widely used to identify both tools, but they are not the same thing. A come-along uses a cable, whereas a ratchet-lever hoist uses a chain. Ratchet-lever hoists can be used for vertical lifts, but cable-type come-alongs can only be used for horizontal pulls. Come-alongs and ratchet-lever hoists are portable, easy to store, and easy to use. They are available in a wide range of sizes.

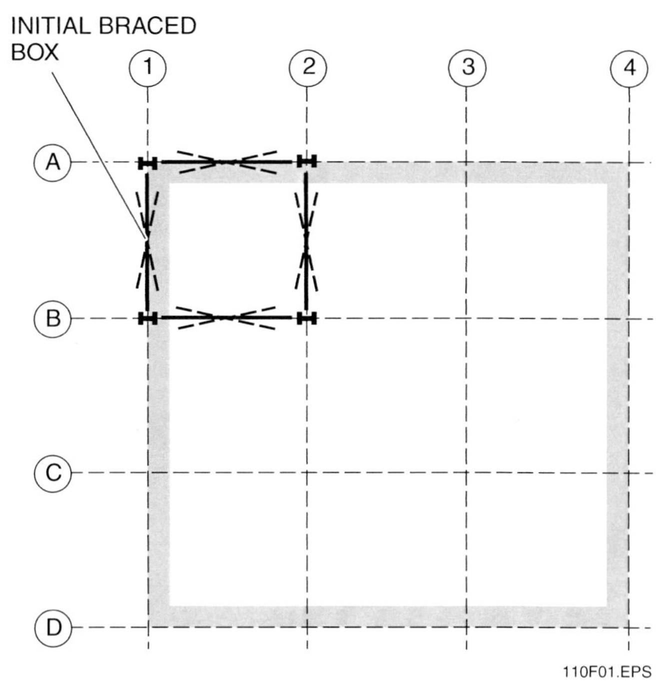

Figure 1 Initial braced box.

Figure 2 Come-along and ratchet-lever hoist.

WARNING

Never use a come-along for vertical lifts. They do not have the same safety braking mechanisms as the ratchet-lever hoist to prevent the load from slipping.

A ratchet-lever hoist consists of the following parts:

- *Suspension hook* – A steel hook used to hang the hoist.
- *Load hook* – A steel hook that attaches to the load.
- *Load chain* – The chain that attaches to the load hook and lifts the load.
- *Ratchet handle* – A handle that operates the ratchet that takes up the chain and lifts or lowers the load.
- *Fast wind handle* – A handle that takes up or lets out the chain without using the ratchet handle. It cannot be used to raise or lower the load.
- *Ratchet release* – A device that releases the ratchet so the chain can be pulled out. It also switches the device to the up or down position.

Follow these steps to select, inspect, use, and maintain a ratchet-lever hoist:

Step 1 Select a device of adequate capacity to handle the load.

Step 2 Inspect the chain and hooks to ensure that they are not excessively worn, bent, or deformed in any way.

Step 3 Inspect the device to ensure that it is not damaged in any way.

Step 4 Hang the device on a suitable support.

Step 5 Turn the ratchet release to the middle position.

Step 6 Pull the chain out enough to attach it to the load.

Step 7 Attach the load hook to the load, using the proper rigging.

Step 8 Turn the fast wind handle to take the slack out of the chain.

Step 9 Turn the ratchet release to the Up position.

Step 10 Pump the ratchet handle to raise and position the load.

Step 11 Turn the ratchet release to the Down position.

Step 12 Pump the ratchet handle to lower the load into position and until there is slack in the chain.

Step 13 Disconnect the load hook from the load.

Step 14 Store the device in its proper place.

3.2.0 Steamboat Ratchets

The **steamboat ratchet** (*Figure 3*) is used for pulling two pieces together. The ratchets are available in a wide range of capacities.

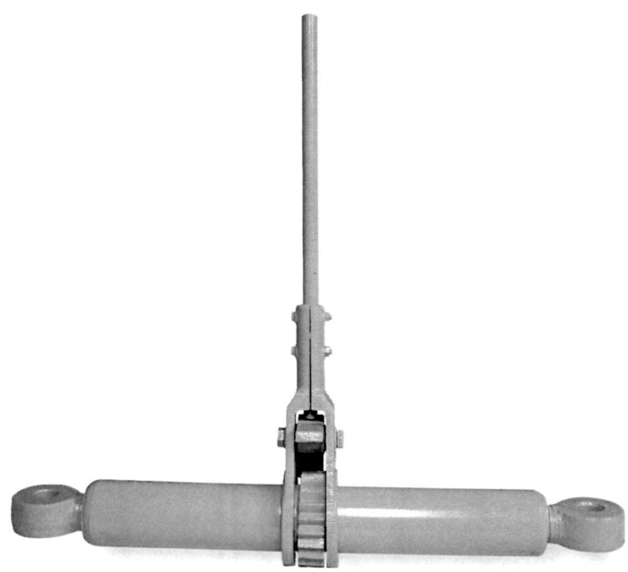

Figure 3 Steamboat ratchet.

3.3.0 Push-Pull Jack

The **push-pull jack** is similar to the steamboat ratchet except that it not only pulls pieces together but can also push them apart.

3.4.0 Mauls or Beaters

The maul (*Figure 4*) is used for driving drift pins and bull pins into connection holes, driving steel members into place, straightening bent material, and wherever force that can be achieved by a beating action is required.

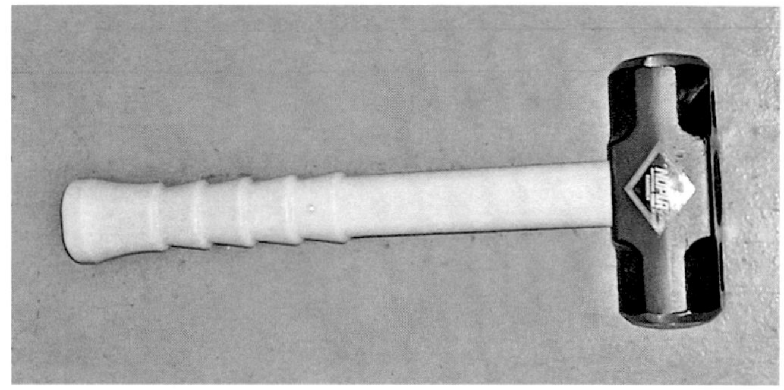

110F04.EPS

Figure 4 Maul.

3.5.0 Shackles

Screw pin and standard pin shackles (*Figure 5*) are used for placing the guys into position where the cable can be wrapped around a column.

3.6.0 Spools and Thimbles

Spools are designed to protect a wire rope guy where it is attached to a pin or bolt. They are used where a guy is bent over an attachment and a temporary loop is made in the guy using wire rope clips. When an eye at the end of a single-part guy is secured to a pin or bolt, a thimble is preferable. When the thimble is used, it is installed when splicing the eye so that it will not drop out.

3.7.0 Turnbuckles

Turnbuckles (*Figure 6*), steamboat ratchets, and come-alongs are used to provide tension on most plumbing tasks. Selection of the proper size and ends of this equipment depends on the use and capacity required for the plumbing-up guys. *Table 1* shows capacities of selected sizes of turnbuckles.

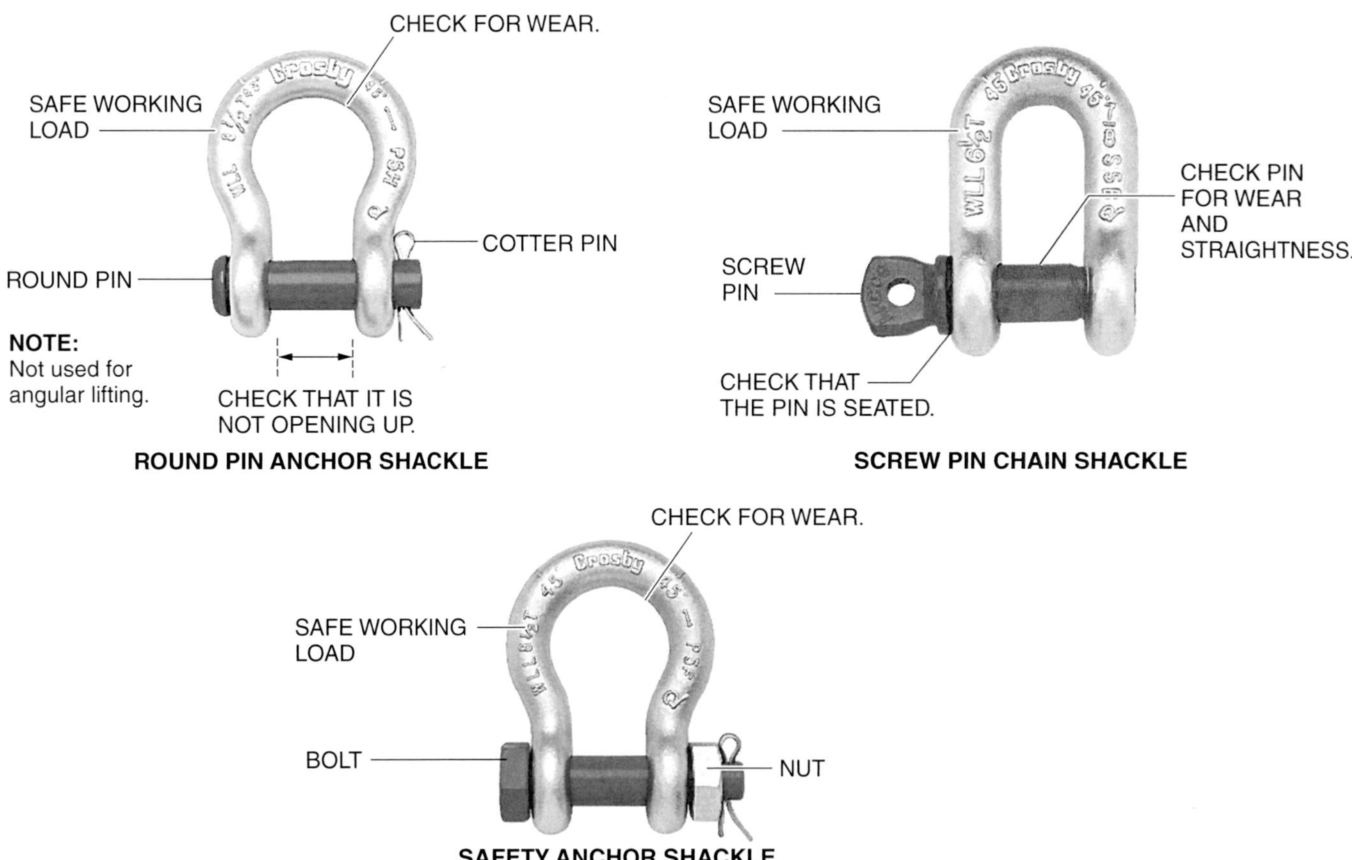

THE CROSBY GROUP, INC.

110F05.EPS

Figure 5 Shackles.

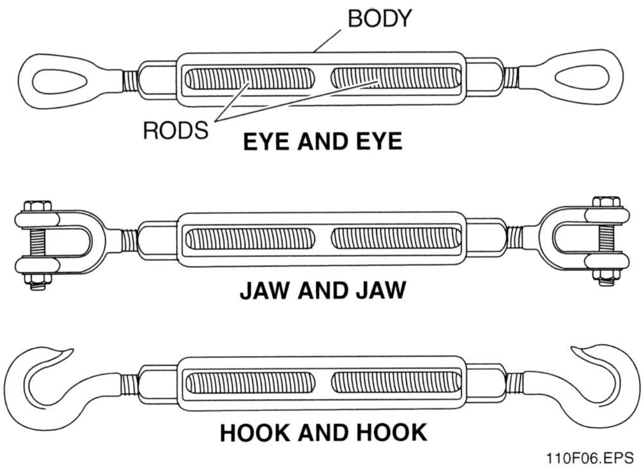

Figure 6 Turnbuckles.

3.8.0 Wire-Rope Clips

Wire-rope clips (*Figure 7*) are used for attaching turnbuckles to one end of the guy. Some ironworkers use a splice instead of clips. This involves, essentially, remaking the rope in a different shape, called an eye splice or a Molly Hogan (or Flemish eye) splice (*Figure 8*). The splice may or may not be permitted.

3.9.0 Wedges

Wedges (*Figure 9*) can be used for alignment and spacing of base plates.

Table 1 Turnbuckle Capacities

Size	Hook	Eye/Jaw
¼	400	500
5⁄16	700	800
⅜	1,000	1,200
½	1,500	2,200
⅝	2,250	3,500
¾	3,000	5,200
⅞	4,000	7,200
1	5,000	10,000
1¼	5,000	15,200

3.10.0 Drift Pins

Drift pins (*Figure 10*) are made in various diameters with the barrel being the same diameter as the holes in the steel that will be connected and pinned. The point is driven through the pieces of steel being assembled in a connection to help drift pieces into alignment.

3.11.0 Transits and Levels

Transits and levels are optical instruments that are used for precise measurements of steel structures. Transits are used for precise vertical measurements, while optical levels are used for precise horizontal measurements. A plumb bob is generally suspended from the transit or level. *Figure 11* shows a level.

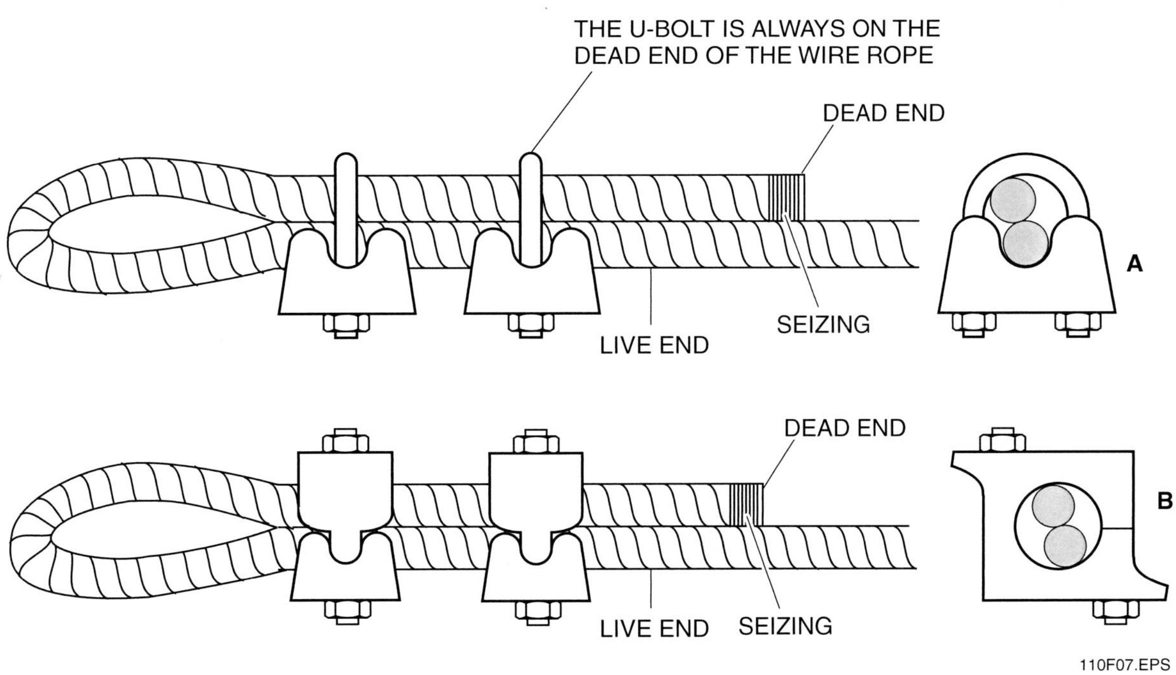

Figure 7 Wire-rope clips.

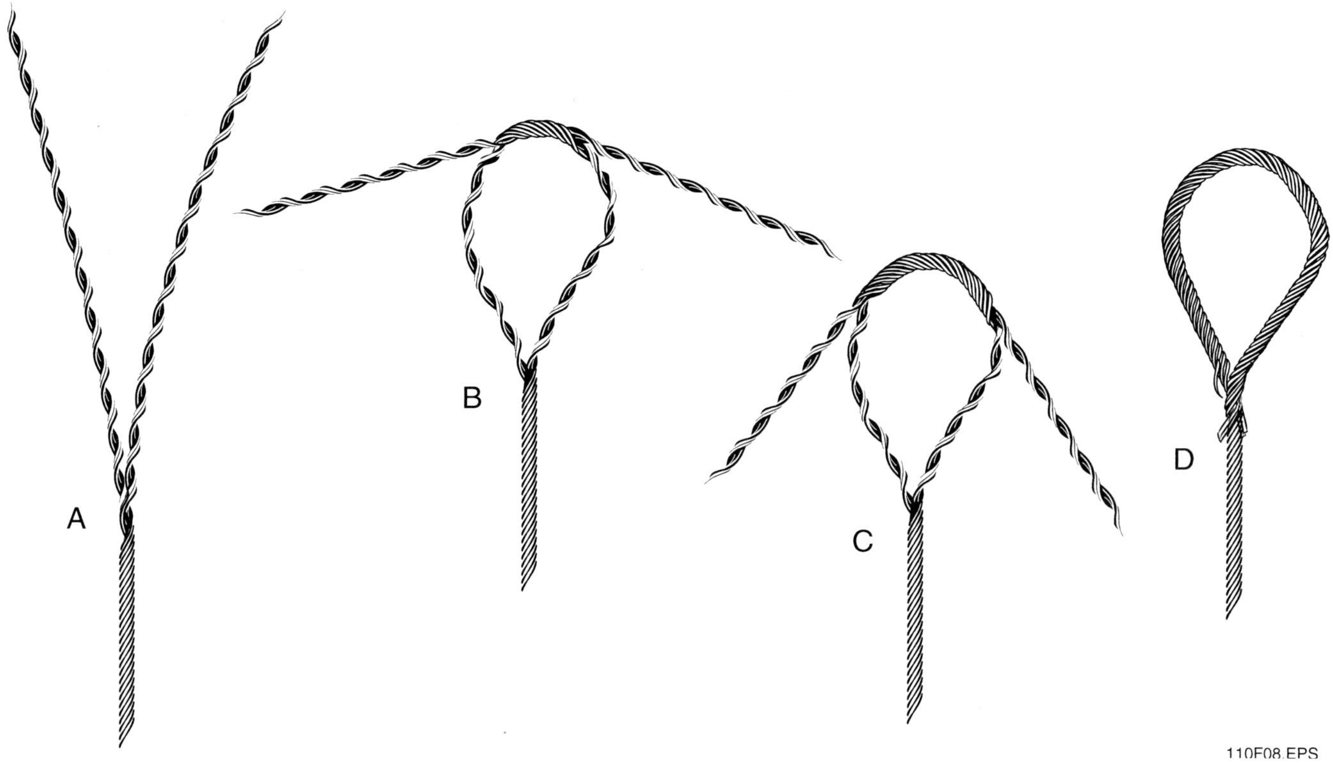

Figure 8 Steps for making a Molly Hogan splice.

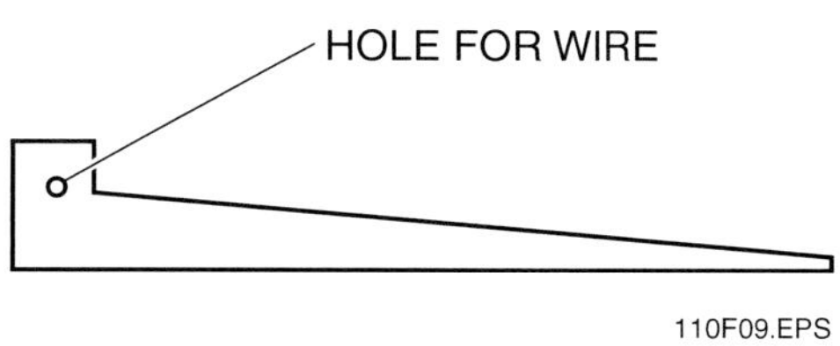

Figure 9 Wedge.

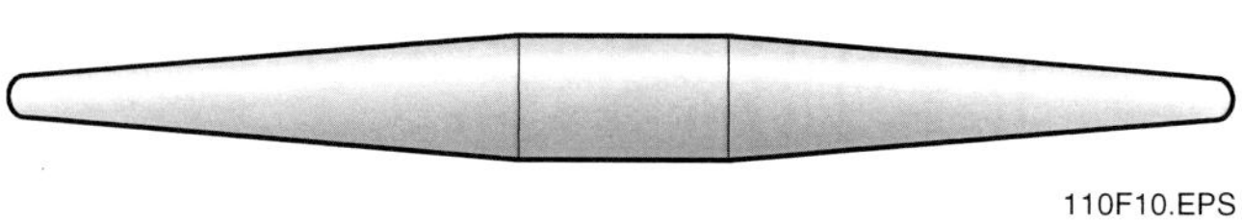

Figure 10 Drift pin.

3.12.0 Plumb Bobs/Bazooka Bob

A plumb bob is used to check vertical alignment. It is suspended by a wire or string parallel with the face or web of the column. A Bazooka Plumb Bob (*Figure 12*) is an aluminum-encased magnetic plumb bob, designed to be attached directly to steel and/or concrete work surfaces to check vertical leveling.

3.13.0 Spirit Levels

Spirit levels (*Figure 13*) are used for checking both plumbness and levelness during roughing in.

To check a spirit level (sometimes called a whiskey stick) to be sure it is reading correctly, put the level against a wall or column known to be true plumb or level. Note the position of the bubble in relation to the marks on the vial (*Figure 14*). Flip the level around and put the other side of the level against the wall or column. The bubble should be the same in relation to the marks (the way it was on the other side). If the bubble is the same for each side, the spirit level is accurate. If the bubble varies from one side to the other, then the level is inaccurate.

Turn the level over and check the other end. One end may be accurate and the other end may be bad. (Check the horizontal vial in the same manner.) It is a good idea to check a new level out of the box in this manner because even a new level can be inaccurate.

3.14.0 Havens Clamp

The havens clamp (*Figure 15*), also called a cable dog, is used to temporarily tighten the wire-rope guy for the initial attachment of a turnbuckle.

4.0.0 Column Bases

The column bases of the structure are an essential part of both the final structure and of the temporary support system. The column bases transfer the vertical and horizontal loads from the structural steel to the ground. The following components make up a column base:

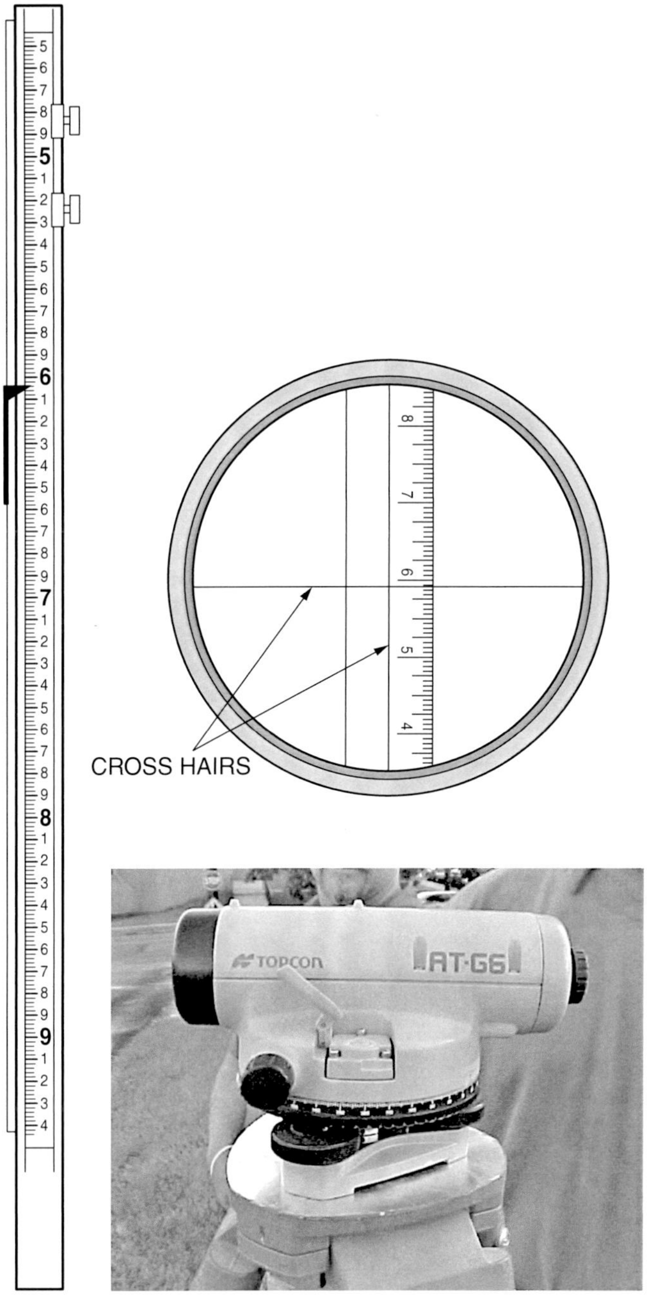

110F11.EPS

Figure 11 Taking a reading using an optical level.

110F12.EPS

Figure 12 Plumb bob/Bazooka Bob.

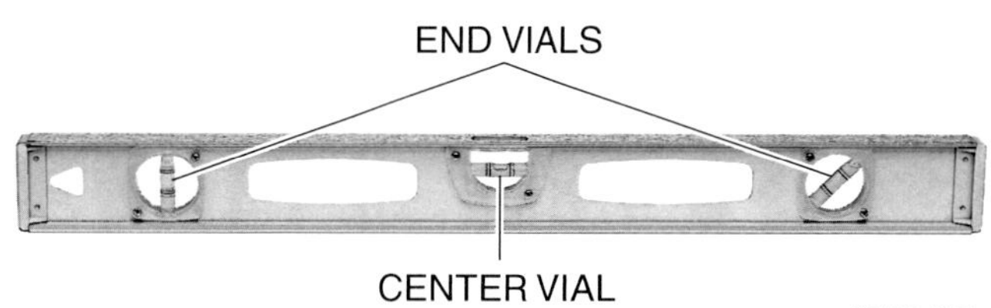

110F13.EPS

Figure 13 Spirit level.

- *Base plate* – Column base plates are square or rectangular plates which transfer loads from the column shaft to the foundation. In cases of high loading, large column bases are sometimes shipped separately from the column shafts. In the case of low-rise and one-story buildings, the base plates are usually shipped attached to the column shafts.
- *Anchor bolts* – Anchor bolts have been called anchor rods. Anchor rods may be headed similar to bolts, a threaded rod fitted with a washer and nut, or bent at a right angle on one end. In the completed construction (with base plates grouted), anchor rods carry tension forces from the beam to the foundation. During erection operations and prior to base plate grouting, anchor rods may also resist compression and shear forces depending on the temporary lateral support system. Anchor rods are embedded in the cast-in-place foundation and are terminated with either a hook or headed end, such as a heavy hex nut tack welded to prevent turning.
- *Base plate grout* – High-strength, non-shrink grout is placed between the column base plate and the supporting foundation. Where base plates are shipped loose, the base plate is usually grouted after the plate has been aligned and leveled. When plates are shipped attached to the column, one of the following methods of column support is selected:

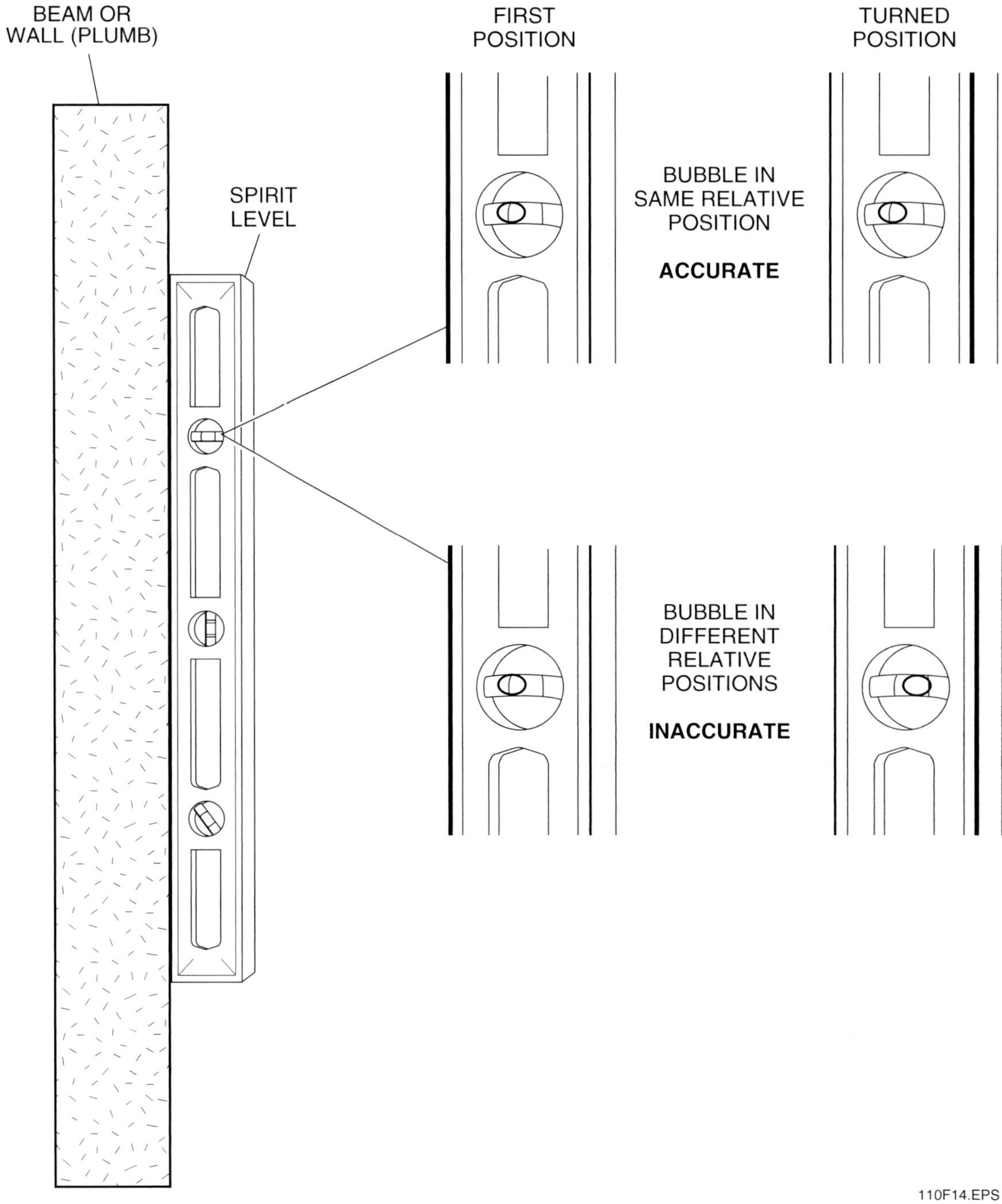

Figure 14 How to check a spirit level.

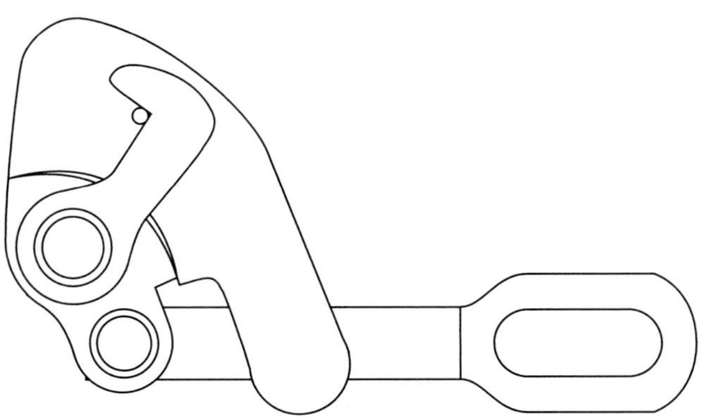

Figure 15 Havens clamp.

- The use of leveling nuts, and in some cases, washers on the anchor rods beneath the base plates.
- The use of shim stacks between the base plate bottoms and the top of the foundation.
- The use of ¼-inch steel leveling plates which are set to elevation and grouted prior to setting the columns.

Leveling nuts and shim stacks are used to transfer the column base forces to the foundation before the installation of grout. When leveling nuts are used, all forces are transferred to the foundation by the anchor rods. When shims are used, the compressive components of the column base reaction are carried by the shims, and the tension and shear components are carried by the anchor rods.

Leveling nuts support the weight of the frame until the bases are grouted. Because the anchor rod, nut, and washers have limited strength, grouting must be done before this limit is reached. For example, the strength limitation may be reached with only one tier of framing or before siding or concrete is added to metal decking. If leveling nuts are used under the base plate lever arm, the tipping force is the distance between the anchor rods. If shim stacks are used, then the tipping force is the distance from the center of the shim stack to the center of the other shim stack. *Figure 16* shows the relative moment arms.

The erection operation, sequence of work, reactions from temporary supports, and the timing of grouting may cause forces in the anchor rods and the foundation to exceed those for which the structure has been designed. One condition of loading the column base and foundation occurs when the column shaft is set on the anchor rods and the nuts tightened. Unless there is guying provided, the column is a cantilever from the base and stability is provided by the column base mounting. Diagonal cables for temporary lateral support also induce tensions and shears in the column base which must be transferred through the anchor rods to the foundation. The decked structural frame may be subjected to the wind uplift which is not counterbalanced by the final dead weight load. A net uplift in the column base may exceed the limitations of the anchor rods, base plates, or welds.

The most likely time for the collapse of a steel frame occurs during the erection sequence when the first series of columns is erected. After the crane hook is released from a column and before the column is braced, its resistance to overturning depends on the strength of the column base and the foundation system. Once the column is braced by tie members and bracing cables, it is much more stable. *Figure 17* shows base plate and foundation failures.

5.0.0 Plumbing Procedures

When steel erection begins, the anchor bolts are already in place. Always verify that the anchor bolts are placed with proper alignment and spacing in all directions. If this check is not made, the steel could be installed and have to be removed later to correct an anchor problem. Once the steel is in place, a cursory check can be made with a spirit level to make sure that a column is fairly plumb. Anchor bolts are usually tightened when a column is set. They may be loosened when plumbing and guying and then retightened when the operation has been completed.

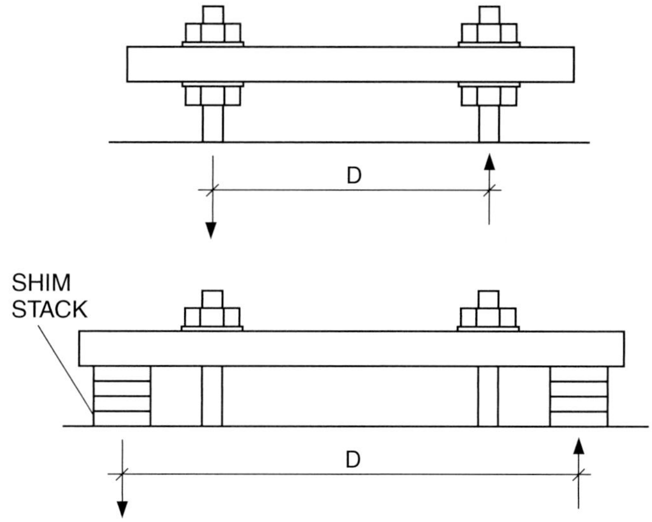

Figure 16 Moment arms.

Temporary supports, such as temporary guys, braces, falsework, cribbing, or other elements required for the erection operation is determined, furnished, and installed by the erector. These temporary supports secure the steel framing, or any partly assembled steel framing, against loads comparable in intensity to those for which the structure was designed. These loads include those resulting from wind, seismic forces, and erection operations, but not the loads resulting from the performance of work or acts by others, nor such unpredictable loads as those due to tornado, explosion, or collision.

The guys are used to plumb the column, and can be connected to the framework in a variety of ways. If a projecting plate is provided, then a spelter socket, wedge socket, or a clevis end fitting can be used. Another method of attaching to a base plate is to use a bent plate attached to one of the column anchor bolts. If no attachments are provided, the cable and hook may be wrapped around the beam. The hook on the guy is either placed over a beam flange on an upper floor, or the guy can be wrapped around a column at the upper floor and the bent end of the hook can be laid over the guy. *Figure 18* shows the proper way to position the guys.

With the turnbuckle extended almost its full length, the free end of the guy is slipped through the free eye of the turnbuckle. The guy is then pulled tight by hand or by cable dogs and a come-along. The wire-rope cables are installed in the opposite direction to form a diagonal cross. The upper end of the second guy is attached to

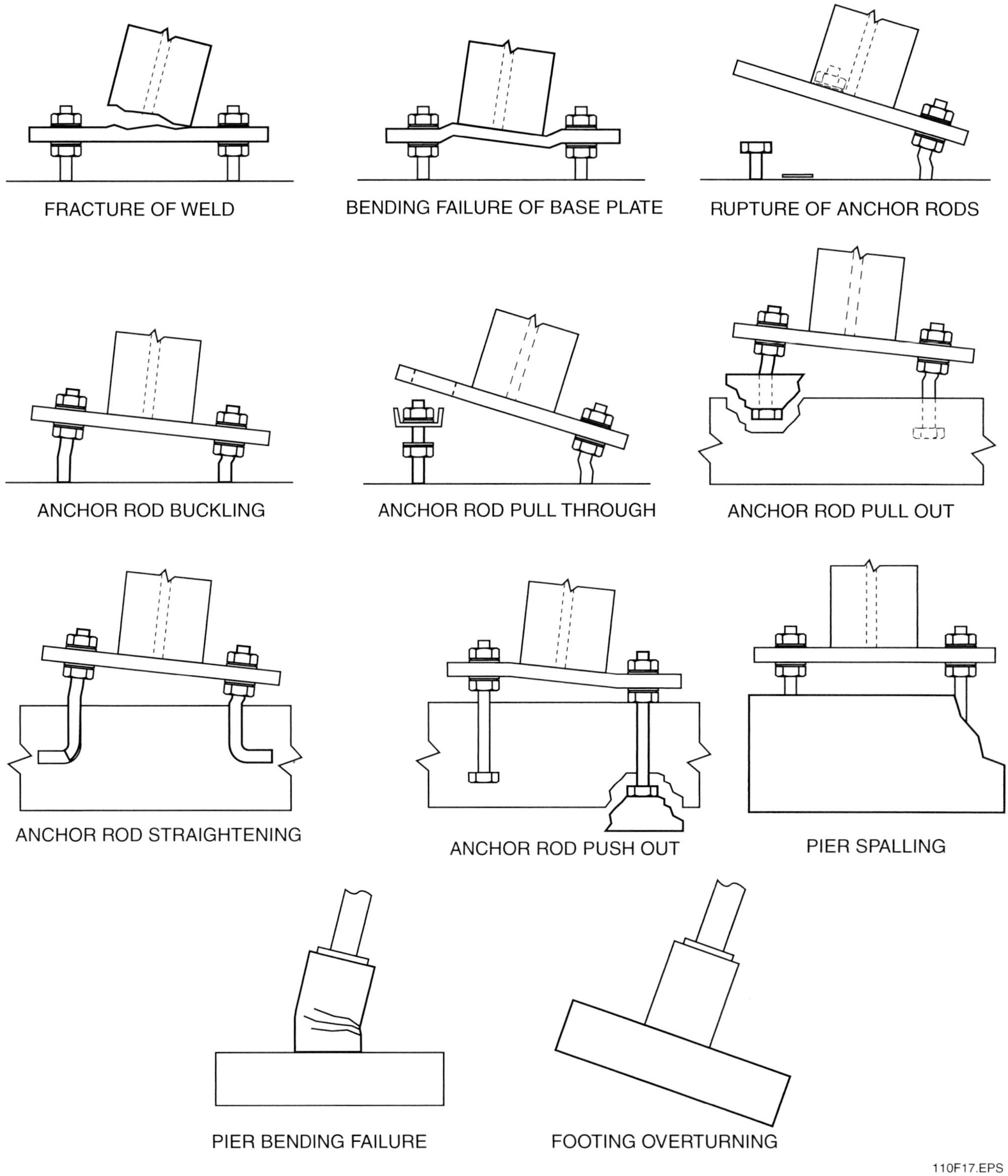

Figure 17 Baseplate and foundation failures.

an adjacent column at the same upper level, but directly over the lower end of the first guy. The second turnbuckle is hooked in place at the lower level directly below the upper end of the first guy.

By tightening one turnbuckle and loosening the other, the tops of the columns can be moved one way or the other after the connections between them have been pinned (with drift pins) and the erection bolts installed. For example, taking in on guy A (*Figure 18*) and slackening guy B will move the tops of columns one and two toward the right. Taking in on guy B and slackening guy A will move the tops of the columns toward the left. Keep a close watch on the hook end of the turnbuckle, since any strain will show there first.

Figure 18 Proper way to position the guys.

6.0.0 Aligning Methods

The plumb bob method of alignment requires a lightweight plumb bob on the end of a chalkline, or a light, flexible wire (such as a piano wire) with a heavy bob lowered from the upper level to the lower level. The plumb bob is held in place away from the face or web of the column at the upper end by a rule that the plumbing-up worker has notched at a fixed distance from the end. As the end of the rule is held against the column with the string in the notch (*Figure 19*), the worker on the lower floor reads the distance from the line to the same face of the column. If it is the same as the notch distance in both directions, such as east-west and north-south, the column is considered to be plumb. If the distance is more or less, the turnbuckles are either tightened or loosened to pull the tops of the columns until they are plumb. If the columns were originally placed correctly on their base slabs, then both columns using this set of guys would theoretically be plumb and in their correct locations. Since this is not always the case, each column must also be checked separately.

When windy conditions cause the bob and chalkline or wire to sway back and forth, a pail of water can be used around the bob to dampen the sway. In some cases, it is easier to set up a transit and sight on a rule held against the column. If the rule is held against the face, the distance to the column center must be taken into account. If the rule is held against the web of the column, half the web thickness must be added. If all the columns being plumbed have the same depth between column faces or the same thickness of web, the direct reading on the rule by the transit can be used.

The transit must be set at a known distance from the building and be on a parallel line referenced to a backsight set at the same distance from the column. Based on this distance, the reading on the tape rule is calculated as if the columns were in their correct positions. With the transit method, it is necessary to check the corner columns in two directions and then use a steel tape between the intermediate columns. The check by transit only ensures their plumbness in one direction—either away, toward the building line, or at right angles to the direction of sight of the transit. *Figure 20* shows plumbing by transit.

Checking the plumb of a selected line of columns can be performed with the transit and level. The vertical crosshair is lined up with the edge of the column, and then the instrument is rotated in a vertical plane so that the crosshair travels over the entire height of the column. If the crosshair coincides with the edge of the column throughout its vertical plane motion, the column is plumb. Two adjacent faces of the column should be checked in this manner by setting the level and transit in two different positions. If the column is out of plumb, the guys can be tightened or slackened accordingly to bring the columns into plumb. Plumb should be checked at approximately the same time of day each time because heating may change the readings.

As soon as the floor steel and the column splices have been permanently fastened, the plumbing-up guys can be removed. However, because some structural frames are so light, (and heavy winds may arise) removal of the guys before the permanent floors are installed may rack the structure out of plumbness. In these cases,

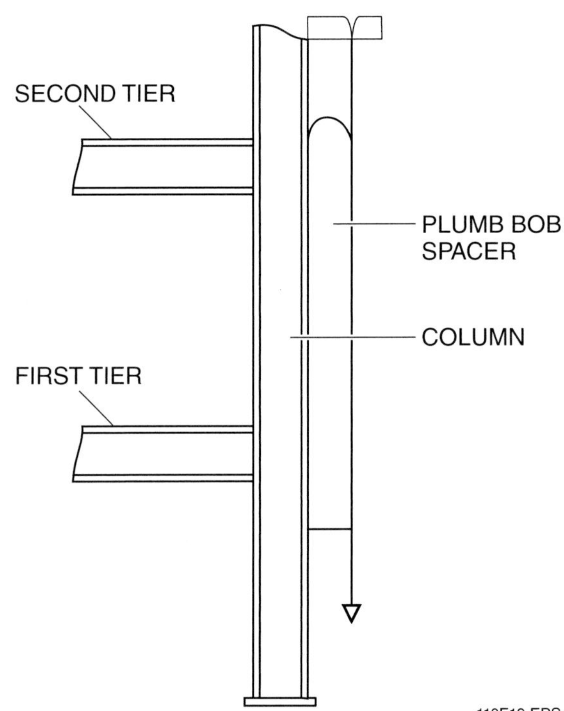

Figure 19 Use of a plumb bob.

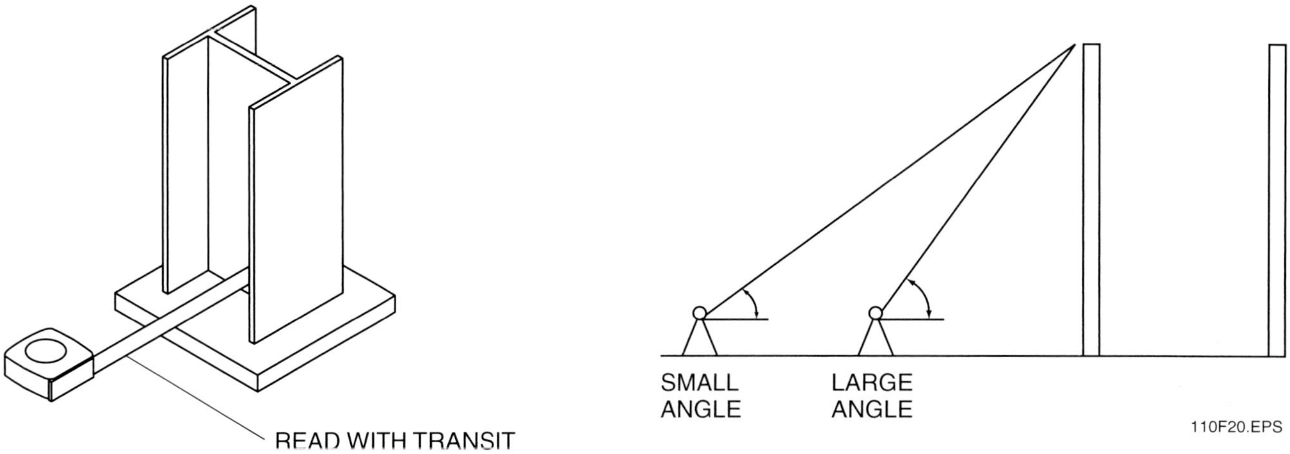

Figure 20 Plumbing by transit.

and also when the stiffness of the structure will be secured only after the permanent floors are fastened down, the guys should be left in place.

> **WARNING**
>
> Removal of the plumbing-up guys must not be done without the approval of a competent person.

On some buildings, contracts are awarded at different times so that the steel erectors may complete their work before any floors or walls are even started. Consequently, it may be advisable to leave all guys in place, especially if the structure may be unstable without the guys. Tampering with them too soon may cause the structure to rack and move out of its correct position.

After completing the plumbing of each tier, it is usually advisable to record the exact location of the tops of the columns on the outer faces of the structure and around the elevator shafts. These records are then submitted to the owner to establish the locations of the columns as left by the erector. Use shims to correct any fabricating tolerances.

With derrick or climbing crane erection, plumbing-up should be done immediately so that permanent fastening can proceed before the rig is jumped. With mobile crane erection, plumbing up should also proceed promptly so that the permanent fastenings can be installed to hold the structure correctly and safely in place.

Summary

Plumbing and leveling of a steel structure may take place before the final connecting and fastening of structural components or at the same time as the final fastening process. This depends on the type of structure and method of lifting the structural members. Guys can be made up in the tool house or fabricated on the job site from properly sized cables and selected rigging hardware. Guys are attached to structural members in a crossed pattern and are used to adjust the columns by regulated tightening or slackening of the crossed guys. Guys may be left in place for structural stability under certain conditions or may be removed immediately after final fastening of the structural members takes place. Tolerances may be specified in the contract or may be stipulated as compliance with a code. In either case, the erector is expected to fulfill the obligation within the permitted tolerances.

Review Questions

1. The important columns to plumb accurately are those on the outside of the building and those _____.
   a. on the second floor
   b. around elevator shafts
   c. beside windows
   d. in the very center of the building

2. To pull two pieces together when a stronger force is required, use a push-pull jack, turnbuckle, or a _____.
   a. come-along
   b. spool
   c. steamboat ratchet
   d. beater

3. The U-bolt of a wire rope clip is always placed on the _____ of the rope.
   a. dead end
   b. live end
   c. center
   d. very end

4. To make precise vertical measurements of steel structures, use a _____.
   a. spirit level
   b. transit
   c. plumb bob
   d. torpedo level

5. The part of the column that transfers the loads from the column shaft to the foundation is called the _____.
   a. web
   b. flange
   c. toe
   d. base plate

6. When using turnbuckles to plumb a structure, keep a close watch on the _____ since any strain will show there first.
   a. hook
   b. eye
   c. clevis
   d. clip

7. Anchor bolts may be loosened during _____, then retightened when the operation is completed.
   a. erection
   b. installation of joists
   c. plumbing and guying
   d. roofing

8. A check by transit ensures plumbness _____.
   a. in one direction
   b. in two directions
   c. in the direction of the building line
   d. at right angles to the building line

9. If the crosshair of the transit coincides with the edge of the column throughout its vertical plane motion, the column is _____.
   a. vertical
   b. level
   c. plumb
   d. cantilevered

10. Removal of plumbing-up guys must *not* be done without the approval of _____.
    a. the supervisor
    b. the architect
    c. the installer
    d. a competent person

Steve Miller

President
Miller Safety Consulting, Inc.

How did you get started in the construction industry?
I got started in the family business in Connecticut, where we worked with steel, commercial properties, and demolition. After ten years of working there, I moved to Colorado.

Who inspired you to enter the industry? Why?
My father inspired me to work in construction through watching him work with the family business.

What do you enjoy most about your job?
It is rewarding to know I am helping to prevent injuries. I also enjoy training other craftspeople and keeping the industry up-to-date with current safety practices.

Do you think training and education are important in construction? If so, why?
Yes. Training and education are essential and they keep the industry alive. Every craftworker needs training and refreshers, because lack of training/education can affect the quality of work as well as the worker's safety. It is good to have a trained workforce to keep the industry thriving.

How important are NCCER credentials to your career?
In my particular position, they are not important, because we consult on safety practices.

How has training/construction impacted your life and your career?
It has impacted my entire career. I got to this point through training and mentors.

Would you suggest construction as a career to others? If so, why?
Yes. There is always a need for new construction workers. It is a career that is always growing. Even with massive reductions in other industries, construction is constant.

How do you define craftsmanship?
Craftsmanship is the reflection of the finished product and the ability to follow directions well and precisely.

Trade Terms Introduced in This Module

Anchor bolts (anchor rods): Threaded rods with a nut on the end used to transfer tension forces to the foundation. They are usually embedded into the concrete foundation.

Come-along: Portable hoist used with guys to pull frames into plumb.

Guys: Cables and fixtures attached to a building frame used to rack the building into plumb.

Leveling nuts: Leveling nuts are used to bear the weight of the frame until the bases are grouted.

Plumb bob: A cone-shaped weight suspended at the end of a string or wire used to check the plumb of a vertical member.

Push-pull jack: A fixture similar to the steamboat ratchet but capable of pulling or pushing.

Spools: Round fixtures attached to the end of a guy to reduce bending strain.

Steamboat ratchet: A fixture used to pull two construction pieces together when more force is required than can be achieved with a turnbuckle.

Turnbuckle: A fixture used with guys to pull framing into plumb.

Wire rope clips: U-bolt or clamp clips that clamp wire ropes to form loops for attachments.

Additional Resources

This module presents thorough resources for task training. The following resource material is suggested for further study.

American Institute of Steel Construction 303-05: Code of Standard Practice for Steel Buildings and Bridges.

Figure Credits

Topaz Publications, Inc., Module opener and Figure 4

Lug-All Corporation, Figure 2 (come-along)

JET brand of WHM Tool Group, Figure 2 (ratchet-lever hoist)

Photo supplied by Simplex, a division of Actuant Corporation., Figure 3

The Crosby Group, Inc., Figure 5

John Hoerlein, Figure 11 (photo)

Bazooka Plumb Bob Company, Figure 12

The Stanley Works, Figure 13

Bryan McClure, Figure 18

NCCER CURRICULA — USER UPDATE

NCCER makes every effort to keep its textbooks up-to-date and free of technical errors. We appreciate your help in this process. If you find an error, a typographical mistake, or an inaccuracy in NCCER's curricula, please fill out this form (or a photocopy), or complete the online form at **www.nccer.org/olf**. Be sure to include the exact module ID number, page number, a detailed description, and your recommended correction. Your input will be brought to the attention of the Authoring Team. Thank you for your assistance.

Instructors – If you have an idea for improving this textbook, or have found that additional materials were necessary to teach this module effectively, please let us know so that we may present your suggestions to the Authoring Team.

NCCER Product Development and Revision

13614 Progress Blvd., Alachua, FL 32615

Email: curriculum@nccer.org
Online: www.nccer.org/olf

❑ Trainee Guide ❑ Lesson Plans ❑ Exam ❑ PowerPoints Other ______________________

Craft / Level: Copyright Date:

Module ID Number / Title:

Section Number(s):

Description:

Recommended Correction:

Your Name:

Address:

Email: Phone:

Oxyfuel Cutting

29102-09

Trainees with successful module completions may be eligible for credentialing through NCCER's National Registry. To learn more, go to **www.nccer.org** or contact us at **1.888.622.3720**. Our website has information on the latest product releases and training, as well as online versions of our *Cornerstone* newsletter and Pearson's Contren® product catalog.

Your feedback is welcome. You may email your comments to **curriculum@nccer.org,** send general comments and inquiries to **info@nccer.org**, or use the User Update form at the back of this module.

BV.1 1/11

29102-09
Oxyfuel Cutting

Objectives

When you have completed this module, you will be able to do the following:

1. Identify and explain the use of oxyfuel cutting equipment.
2. Set up oxyfuel equipment.
3. Light and adjust an oxyfuel torch.
4. Shut down oxyfuel cutting equipment.
5. Disassemble oxyfuel equipment.
6. Change cylinders.
7. Perform oxyfuel cutting:
   - Straight line and square shapes
   - Piercing and slot cutting
   - Bevels
   - Washing
   - Gouging
8. Operate a motorized, portable oxyfuel gas cutting machine.

Performance Tasks

1. Set up oxyfuel equipment.
2. Light and adjust an oxyfuel torch.
3. Shut down oxyfuel cutting equipment.
4. Disassemble oxyfuel equipment.
5. Change empty cylinders.
6. Cut shapes from various thicknesses of steel, emphasizing:
   - Straight line
   - Square shape
   - Piercing
   - Bevel
   - Slot
7. Perform washing.
8. Perform gouging.

Trade Terms

Backfire
Carburizing flame
Drag lines
Dross
Ferrous metals
Flashback
Kerf
Neutral flame
Oxidizing flame
Pierce
Soapstone

Contents

Topics to be presented in this module include:

Contents (continued)

Figures and Tables

Figures and Tables (continued)

1.0.0 INTRODUCTION

Oxyfuel cutting (OFC), also called flame cutting or burning, is a process that uses the flame and oxygen from a cutting torch to cut **ferrous metals**. The flame is produced by burning a fuel gas mixed with pure oxygen. The flame heats the metal to be cut to the kindling temperature (a cherry-red color); then a stream of high-pressure pure oxygen is directed from the torch at the metal's surface. This causes the metal to instantaneously oxidize or burn. The cutting process results in oxides that mix with molten iron and produce **dross**, which is blown from the cut by the jet of cutting oxygen. This oxidation process, which takes place during the cutting operation, is similar to a greatly sped-up rusting process. *Figure 1* shows oxyfuel cutting.

The oxyfuel cutting process is usually used only on ferrous metals such as straight carbon steels, which oxidize rapidly. This process can be used to quickly cut, trim, and shape ferrous metals, including the hardest steel.

Oxyfuel cutting can be used for certain metal alloys, such as stainless steel; however, the process requires higher preheat temperatures (white heat) and about 20 percent more oxygen for cutting. In addition, sacrificial steel plate or rod may have to be placed on top of the cut to help maintain the burning process. Other methods, such as carbon arc cutting, powder cutting, inert gas cutting, and plasma arc cutting, are much more practical for cutting steel alloys and nonferrous metals. These methods are covered in other modules.

102F01.EPS

Figure 1 Oxyfuel cutting.

2.0.0 Oxyfuel Safety Summary

The proper safety equipment and precautions must be used when working with oxyfuel equipment because of the potential danger from the high-pressure flammable gases and high temperatures used. The following is a summary of safety procedures and practices that must be observed while cutting or welding. Keep in mind that this is just a summary. Complete safety coverage is provided in the *Welding Safety* module. If you have not completed that module, it is recommended that you do so before continuing. Above all, be sure to wear appropriate protective clothing and equipment when welding or cutting.

2.1.0 Protective Clothing and Equipment

- Always use safety glasses with a full face shield or a helmet (*Figure 2*). The glasses, face shield, or helmet lens must have the proper light-reducing tint for the type of welding or cutting to be performed.
- Wear proper protective leather and/or flame-retardant clothing along with welding gloves that will protect you from flying sparks and molten metal, as well as heat.
- Wear 8" or taller high-top safety shoes or boots. Make sure that the tongue and lace area of the footwear will be covered by a pant leg. If the tongue and lace area is exposed or the footwear must be protected from burn marks, wear leather spats under the pants or chaps and over the top of the footwear.

Figure 2 Personal protective equipment for oxyfuel cutting.

- Wear a solid material (nonmesh) hat with a bill pointing to the rear or, if much overhead cutting or welding is required, a full leather hood with a welding face plate and the correctly tinted lens. If a hard hat is required, use a hard hat that allows the attachment of rear deflector material and a face shield.
- If a full leather hood is not worn, wear a face shield over safety glasses for gas welding or cutting. Either the face shield or the lenses of the safety glasses must be an approved shade. A shade 3 to 6 filter is recommended, depending on the thickness of the metal being cut, as required by *ANSI Z49.1*.
- If a full leather hood is not worn, wear earmuffs, or at least earplugs, to protect your ear canals from sparks.

2.2.0 Fire/Explosion Prevention

- Never carry matches or gas-filled lighters in your pockets. Sparks can cause the matches to ignite or the lighter to explode, resulting in serious injury.
- Never perform any type of heating, cutting, or welding until a hot work permit is obtained and an approved fire watch is established. Most work site fires caused by these types of operations are started by cutting torches.
- Never use oxygen to blow off clothing. The oxygen can remain trapped in the fabric for a time. If a spark hits the clothing during this time, the clothing can burn rapidly and violently out of control.
- Make sure that any flammable material in the work area is moved or shielded by a fire-resistant covering. Approved fire extinguishers must be available before attempting any heating, welding, or cutting operations.
- Always comply with any site requirement for a hot work permit and/or a fire watch.
- Never release a large amount of oxygen or use oxygen as compressed air. Its presence around flammable materials or sparks can cause rapid and uncontrolled combustion. Keep oxygen away from oil, grease, and other petroleum products.
- Never release a large amount of fuel gas, especially acetylene. Methane and propane tend to concentrate in and along low areas and can ignite at a considerable distance from the release point. Acetylene is lighter than air but is even more dangerous. When mixed with air or oxygen, it will explode at much lower concentrations than any other fuel.
- To prevent fires, maintain a neat and clean work area and make sure that any metal scrap or dross is cold before disposal.
- Before cutting containers such as tanks or barrels, check to see if they have contained any explosive, hazardous, or flammable materials, including petroleum products, citrus products, or chemicals that decompose into toxic fumes when heated. As a standard practice, always clean and then fill any tanks or barrels with water, or purge them with a flow of inert gas to displace any oxygen (*Figure 3*).

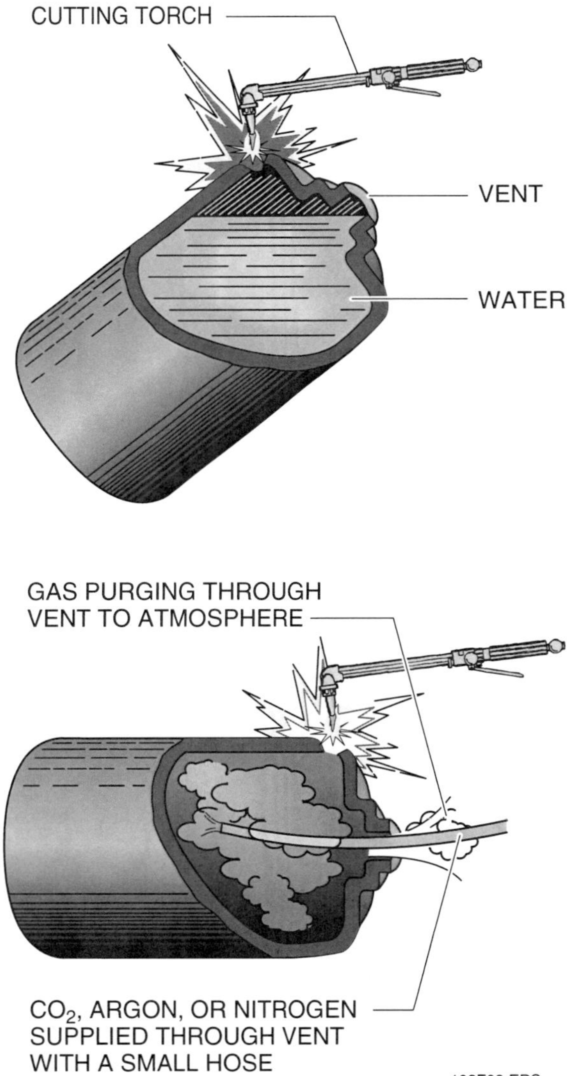

Figure 3 Eliminating/minimizing oxygen in a container.

2.3.0 Work Area Ventilation

- Make sure confined-space procedures are followed before conducting any welding or cutting in the confined space.
- Always perform cutting operations in a well-ventilated area. Cutting operations involving materials or coatings containing cadmium, mercury, lead, zinc, chromium, and beryllium result in toxic fumes. For long-term cutting of such materials, always wear an approved full face, supplied-air respirator (SAR) that uses breathing air supplied externally of the work area. For occasional, very short-term exposure to zinc or copper fumes, a high-efficiency particulate arresting (HEPA)-rated or metal-fume filter may be used on a standard respirator.
- Make sure confined spaces are ventilated properly for cutting or welding purposes.
- Never use oxygen in confined spaces for ventilation purposes.

WARNING

An accumulation of an oxygen and fuel-gas mixture can result in a dangerous explosion when ignited. A 2" balloon filled with an oxygen and acetylene mixture has the explosive power of an M80 firecracker (one-quarter stick of dynamite). The mixed gases from 100' of ¼" twin hose will fill a 9" balloon. A 9" balloon filled with mixed oxygen and acetylene gases has an explosive power that is 90 times more powerful than an M80 firecracker.

On Site

Oxygen Consumption

You should be aware of the fact that two-thirds of the oxygen in a neutral flame comes from the breathing air inside the work space. This could be an issue in a tight work space.

3.0.0 Oxyfuel Cutting Equipment

The equipment used to perform oxyfuel cutting includes oxygen and fuel gas cylinders, oxygen and fuel gas regulators, hoses, and a cutting torch. A typical movable oxyfuel (oxyacetylene) cutting outfit is shown in *Figure 4*.

3.1.0 Oxygen

Oxygen (O_2) is a colorless, odorless, tasteless gas that supports combustion. Combined with burning material, pure oxygen causes a fire to flare and burn out of control. When mixed with fuel gases, oxygen produces the high-temperature flame required in order to flame cut metals.

3.1.1 Oxygen Cylinders

Oxygen is stored at more than 2,000 psi in hollow steel cylinders. The cylinders come in a variety of sizes based on the cubic feet they hold. The smallest standard cylinder holds about 85 cubic feet of oxygen, and the largest ultra-high-pressure cylinder holds about 485 cubic feet. The most common size oxygen cylinder used for welding and cutting operations is the 227 cubic foot cylinder. It is more than 4' tall and 9" in diameter. *Figure 5* shows standard high-pressure oxygen cylinder markings and sizes. The cylinders must be tested every 10 years.

Oxygen cylinders have bronze cylinder valves on top (*Figure 6*). Turning the cylinder valve

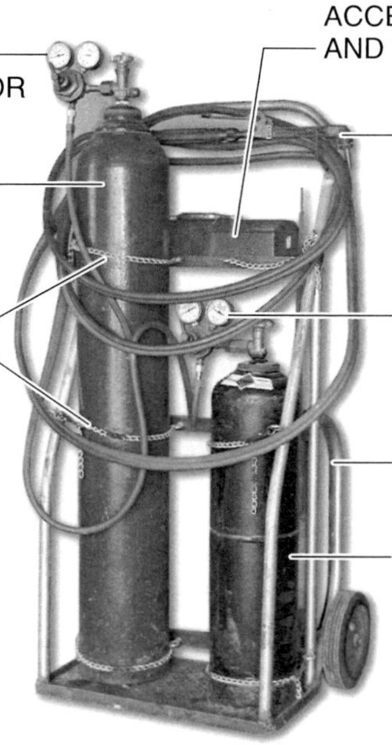

Figure 4 Typical oxyacetylene welding/cutting outfit.

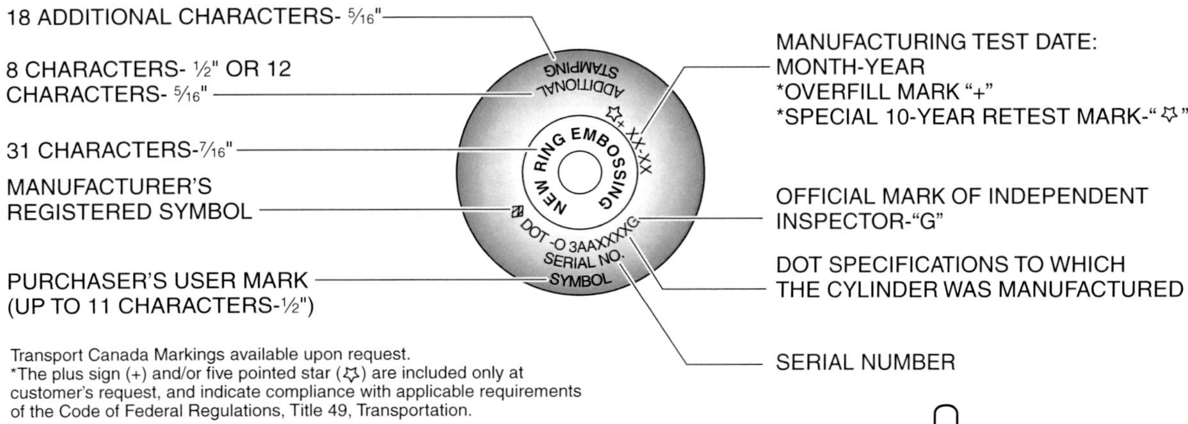

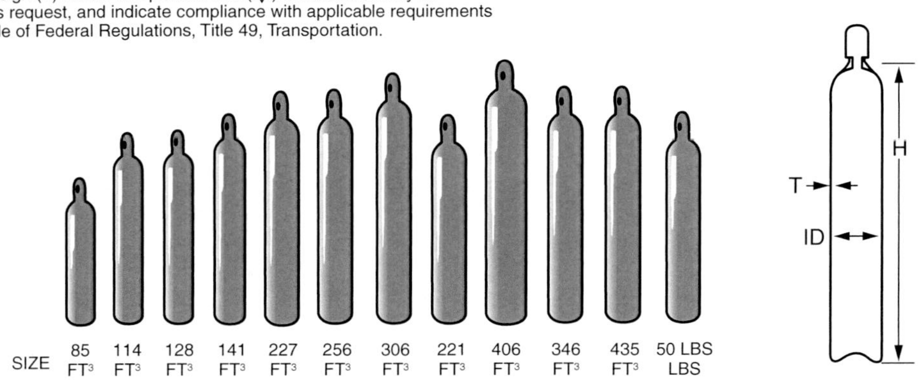

HIGH PRESSURE CYLINDER MARKINGS

DOT SPECIFICATIONS	O_2 CAPACITY (FT^3)		WATER CAPACITY (IN^3)		NOMINAL DIMENSIONS (IN)			NOMINAL WEIGHT (LB)	PRESSURE (PSI)	
	AT RATED SERVICE PRESSURE	AT 10% OVERCHARGE	MINIMUM	MAXIMUM	AVG. INSIDE DIAMETER "ID"	HEIGHT "H"	MINIMUM WALL "T"		SERVICE	TEST
STANDARD HIGH PRESSURE CYLINDERS[1]										
3AA2015	85	93	960	1040	6.625	32.50	0.144	48	2015	3360
3AA2015	114	125	1320	1355	6.625	43.00	0.144	61	2015	3360
3AA2265	128	140	1320	1355	6.625	43.00	0.162	62	2265	3775
3AA2015	141	155	1630	1690	7.000	46.00	0.150	70	2015	3360
3AA2015	227	250	2640	2710	8.625	51.00	0.184	116	2015	3360
3AA2265	256	281	2640	2710	8.625	51.00	0.208	117	2265	3775
3AA2400	306	336	2995	3060	8.813	55.00	0.226	140	2400	4000
3AA2400	405	444	3960	4040	10.060	56.00	0.258	181	2400	4000
ULTRALIGHT® HIGH PRESSURE CYLINDERS[1]										
E-9370-3280	365	NA	2640	2710	8.625	51.00	0.211	122	3280	4920
E-9370-3330	442	NA	3181	3220	8.813	57.50	0.219	147	3330	4995
ULTRA HIGH PRESSURE CYLINDERS[2]										
3AA3600	347[3]	374	2640	2690	8.500	51.00	0.336	170	3600	6000
3AA6000	434[3]	458	2285	2360	8.147	51.00	0.568	267	6000	10000
E-10869-4500	435[3]	NA	2750	2890	8.813	51.00	0.260	148	4500	6750
E-10869-4500	485[3]	NA	3058	3210	8.813	56.00	0.260	158	4500	6750

1. Regulators normally permit filling these cylinders with 10% overcharge, provided certain other requirements are met.
2. Under no circumstances are these cylinders to be filled to a pressure exceeding the marked service pressure at 70°F.
3. Nitrogen capacity at 70°F.

All cylinders normally furnished with 3/4" NGT internal threads, unless otherwise specified.
Nominal weights include neck ring but exclude valve and cap, add 2 lbs. (.91 kg) for cap and 1 1/2 lb. (.8 kg) for valve.
Cap adds approximately 5 in. (127 mm) to height.
Cylinder capacities are approximately 5 in. (127 mm) to height.
Cylinder capacities are approximately at 70°F. (21°C).

102F05.EPS

Figure 5 High-pressure oxygen cylinder markings and sizes.

Figure 6 Oxygen cylinder valve.

handwheel controls the flow of oxygen out of the cylinder. A safety plug on the side of the cylinder valve allows oxygen in the cylinder to escape if the pressure in the cylinder rises too high. Oxygen cylinders are usually equipped with Compressed Gas Association (CGA) 540 valves for service up to 3,000 per square inch gauge (psig). Some cylinders are equipped with CGA 577 valves for up to 4,000 psig service or CGA 701 valves for up to 5,500 psig service. Use care when handling oxygen cylinders because oxygen is stored at such high pressures. When it is not in use, always cover the cylinder valve with the protective steel safety cap and tighten it securely (*Figure 7*).

WARNING

Do not remove the protective cap unless the cylinder is secured. If the cylinder falls over and the nozzle breaks off, the cylinder will be propelled like a rocket, causing severe injury or death to anyone in its way.

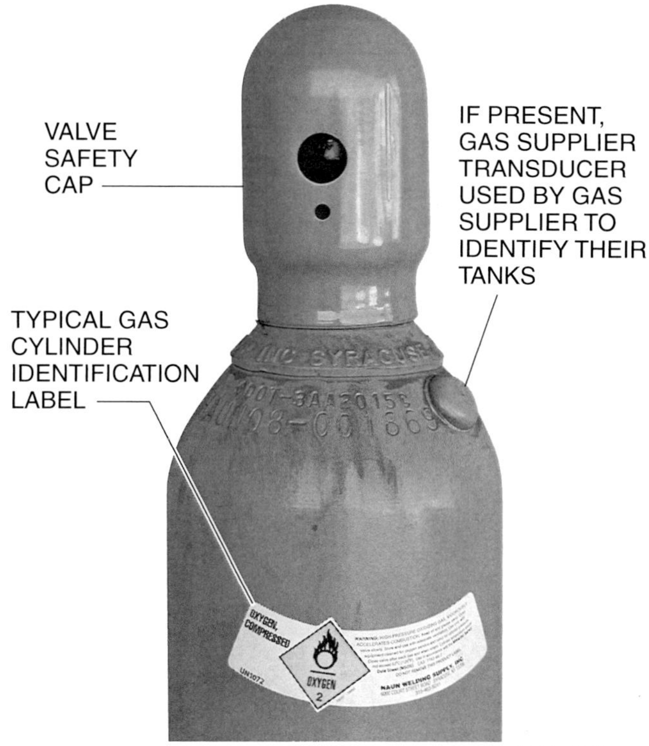

Figure 7 Oxygen cylinder with standard safety cap.

On Site

Tank Identification Transducers

Many gas suppliers mount transducers on their tanks so that they can readily identify the tanks. The transducer is electronically scanned, and a coded number is matched against the supplier's records for identification. This aids in quickly determining the purchaser or user of the tank as well as the required retesting date.

Alternate High-Pressure Cylinder Valve Cap

High-pressure cylinders can also be equipped with a clamshell cap that can be closed to protect the cylinder valve with or without a regulator installed on the valve. This enables safe movement of the cylinder after the cylinder valve is closed. This type of cap is usually secured to the cylinder body cap threads when it is installed so that it cannot be removed. When the clamshell is closed, it can also be padlocked to prevent unauthorized operation of the cylinder valve.

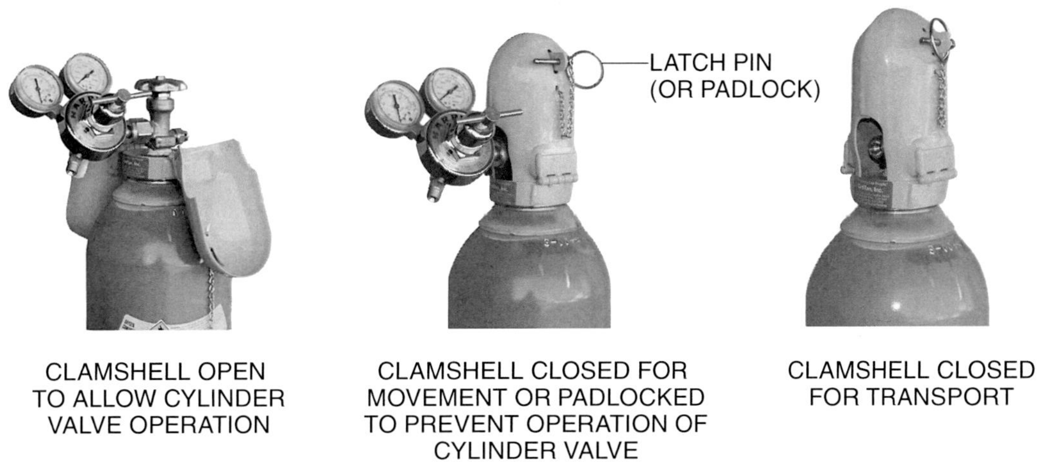

102SA01.EPS

3.2.0 Acetylene

Acetylene gas (C_2H_2), a compound of carbon and hydrogen, is lighter than air. It is formed by dissolving calcium carbide in water. It has a strong, distinctive, garlic-like odor, which is added to the gas so that it can be detected. In its gaseous form, acetylene is extremely unstable and explodes easily. Because of this instability, it cannot be compressed at pressures of more than 15 psi when in its gaseous form. At higher pressures, acetylene gas breaks down chemically, producing heat and pressure that could result in a violent explosion. When combined with oxygen, acetylene creates a flame that burns hotter than 5,500°F, one of the hottest gas flames. Acetylene can be used for flame cutting, welding, heating, flame hardening, and stress relieving.

3.2.1 Acetylene Cylinders

Because of the explosive nature of acetylene gas, it cannot be stored above 15 psi in a hollow cylinder. To solve this problem, acetylene cylinders are specially constructed to store acetylene at higher pressures. The acetylene cylinder is filled with a porous material that creates a solid, instead of a hollow, cylinder. The porous material is soaked with acetone, which absorbs the acetylene, stabilizing it for storage at higher pressures. Because of the liquid acetone inside the cylinder, acetylene cylinders must always be used in an upright position. If the cylinder is tipped over, stand the cylinder upright and wait at least one hour before using it. If liquid acetone is withdrawn from a cylinder, it will gum up the safety check valves and regulators. Always take care to withdraw acetylene gas from a cylinder at pressures less than 15 psig and at hourly rates that do not exceed one-seventh of the cylinder capacity. Higher rates may cause liquid acetone to be withdrawn along with the acetylene.

Acetylene cylinders have safety fuse plugs in the top and bottom of the cylinder that melt at 220°F (*Figure 8*). In the event of a fire, the fuse plugs will release the acetylene gas, preventing the cylinder from exploding.

Acetylene cylinders are available in a variety of sizes based on the cubic feet of acetylene that they can hold. The smallest standard cylinder holds about 10 cubic feet of gas. The largest standard

Figure 8 Standard acetylene cylinder valve and safety plugs.

cylinder holds about 420 cubic feet of gas. A cylinder that holds about 850 cubic feet is also available. *Figure 9* shows standard acetylene cylinder markings and sizes. Like oxygen cylinders, acetylene cylinders must be tested every 10 years.

Acetylene cylinders are usually equipped with a standard CGA 510 brass cylinder valve (see *Figure 8*). The handwheel of the valve controls the flow of acetylene from the cylinder to a regulator. Some acetylene cylinders are equipped with an alternate standard CGA 300 valve. Some obsolete valves still in use require a special long-handled wrench with a square socket end to operate the valve.

On Site

Alternate Acetylene Cylinder Safety Cap

Acetylene cylinders can be equipped with a ring guard cap that protects the cylinder valve with or without a regulator installed on the valve. This enables safe movement of the cylinder after the cylinder valve is closed. This type of cap is usually secured to the cylinder body cap threads when it is installed so that it cannot be removed.

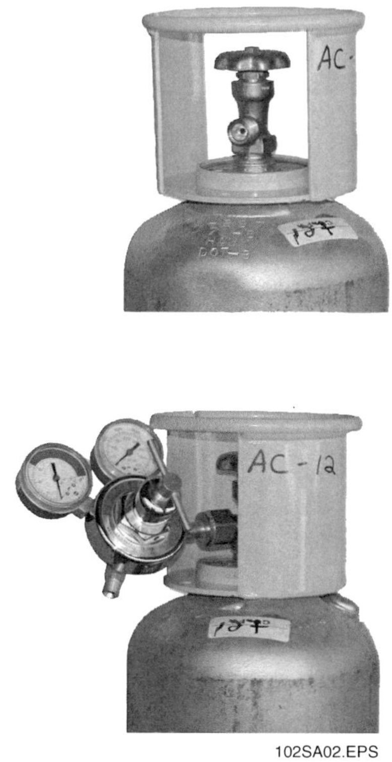

The smallest standard acetylene cylinder, which holds 10 cubic feet, is equipped with a CGA 200 small series valve, and 40-cubic-foot cylinders use a CGA 520 small series valve. As with oxygen cylinders, place a protective valve cap on the acetylene cylinders during transport (*Figure 10*).

3.3.0 Liquefied Fuel Gases

WARNING

Do not remove the protective cap unless the cylinder is secured. If the cylinder falls over and the nozzle breaks off, the cylinder will release highly explosive gas.

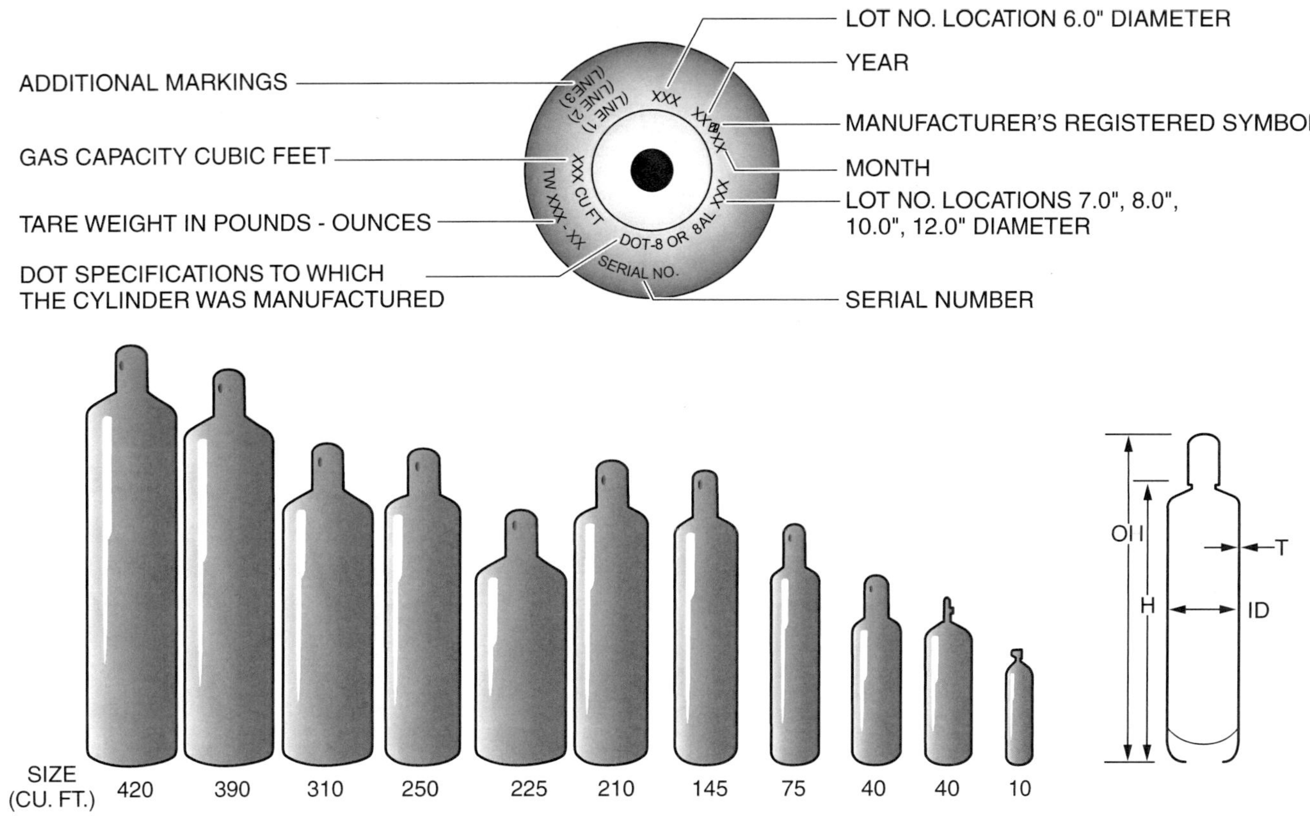

ACETYLENE CYLINDER MARKINGS

DOT SPECIFICATIONS	CAPACITY: ACETYLENE (FT³)	CAPACITY: MIN. WATER (IN³)	CAPACITY: MIN. WATER (LB.)	NOMINAL DIMENSIONS (IN): AVG. INSIDE DIAMETER "ID"	NOMINAL DIMENSIONS (IN): HEIGHT W/OUT VALVE OR CAP "H"	NOMINAL DIMENSIONS (IN): HEIGHT W/VALVE AND CAP "OH"	NOMINAL DIMENSIONS (IN): MINIMUM WALL "T"	ACETONE (LB - OZ)	APPROXIATE TARE WEIGHT WITH VALVE WITHOUT CAP (LB)
8 AL[1]	10	125	4.5	3.83	13.1375	14.75	0.0650	1-6	8
8[1]	40	466	16.8	6.00	19.8000	23.31	0.0870	5-7	25
8[2]	40	466	16.8	6.00	19.8000	28.30	0.0870	5-7	28
8[3]	75	855	30.8	7.00	25.5000	31.25	0.0890	9-8	45
8	100	1055	38.0	7.00	30.7500	36.50	0.0890	12-2	55
8	145	1527	55.0	8.00	34.2500	40.00	0.1020	18-10	76
8	210	2194	79.0	10.00	32.2500	38.00	0.0940	25-13	105
8AL	225	2630	94.7	12.00	27.5000	32.75	0.1280	29-6	110
8	250	2606	93.8	10.00	38.0000	43.75	0.0940	30-12	115
8AL	310	3240	116.7	12.00	32.7500	38.50	0.1120	39-5	140
8AL	390	4151	150.0	12.00	41.0000	46.75	0.1120	49-14	170
8AL	420	4375	157.5	12.00	43.2500	49.00	0.1120	51-14	187
8	60	666	24.0	7.00	25.79 OH		0.0890	7-11	40
8	130	1480	53.3	8.00	36.00 OH		0.1020	17-2	75
8AL	390	4215	151.8	12.00	46.00 OH		0.1120	49-14	180

1. Tapped for 3/8" valve but are not equipped with valve protection caps.
2. Includes valve protection cap.
3. Can be tared to hold 60 ft³ (1.7 m³) of acetylene gas.

Standard tapping (except cylinders tapped for 3/8") 3/4"-14 NGT.

Weight includes saturation gas, filler, paint, solvent, valve, fuse plugs.
Does not include cap of 2 lb. (.91 kg.)
Cylinder capacities are based upon commercially pure acetylene gas at 250 psi (17.5 kg/cm²), and 70°F (15°C).

102F09.EPS

Figure 9 Acetylene cylinder markings and sizes.

102F10.EPS

Figure 10 Acetylene cylinder with standard valve safety cap.

Many fuel gases other than acetylene are used for cutting. They include natural gas and liquefied fuel gases such as methylacetylene propadiene (MAPP®), propylene, and propane. Their flames are not as hot as acetylene, but they have higher British thermal unit (Btu) ratings and are cheaper and safer to use. The supervisor at your job site will determine which fuel gas to use. *Table 1* compares the flame temperatures of oxygen mixed with various fuel gases.

MAPP® is a Dow Chemical Company product that is a chemical combination of acetylene and propane gases. MAPP® gas burns at temperatures almost as high as acetylene and has the stability of propane. Because of this stability, it can be used at pressures over 15 psi and is not as likely as acetylene to **backfire** or **flashback**. MAPP® also has an offensive odor that can be detected easily. MAPP® gas can be used for flame cutting, heating, stress relieving, brazing, soldering, and scarfing (cleaning cutting dross or other material from the workpiece).

Propylene mixtures are hydrocarbon-based gases that are stable and shock-resistant, making them relatively safe to use. They are purchased under trade names such as High Purity Gas (HPG™), Apachi™, and Prestolene™. These gases and others have distinctive odors to make leak detection easier. They burn at temperatures around 5,193°F, hotter than natural gas and propane. Propylene gases are used for flame cutting, scarfing, heating, stress relieving, brazing, and soldering.

Propane is also known as liquefied petroleum (LP) gas. It is stable and shock-resistant, and it has a distinctive odor for easy leak detection. It burns at 4,580°F, which is the lowest temperature of any fuel gas. It has a slight tendency toward backfire and flashback and is used quite extensively for cutting procedures.

Natural gas is delivered by pipeline rather than by cylinders. It burns at about 4,600°F. Natural gas is relatively stable and shock-resistant and has a slight tendency toward backfire and flashback. Because of its recognizable odor, leaks are easily detectable. Natural gas is used primarily for cutting on job sites with permanent cutting stations.

Table 1 Flame Temperatures of Oxygen with Various Fuel Gases

Type of Gas	Flame Temperature
Acetylene	More than 5,500°F
MAPP®	5,300°F
Propylene	5,130°F
Natural gas	4,600°F
Propane	4,580°F

3.3.1 Liquefied Fuel Gas Cylinders

Liquefied fuel gases are shipped in hollow steel cylinders (*Figure 11*). When empty, they are much lighter than acetylene cylinders.

The liquefied fuel gas is stored in hollow steel cylinders of various sizes. They can hold from 30 to 225 pounds of fuel gas. As the cylinder valve is opened, the vaporized gas is withdrawn from the cylinder. The remaining liquefied gas absorbs heat and releases additional vaporized gas. The pressure of the vaporized gas varies with the outside temperature. The colder the outside temperature, the lower the vaporized gas pressure will be. If high volumes of gas are removed from a liquefied fuel gas cylinder, the pressure drops, and the temperature of the cylinder will also drop. A ring of frost can form around the base of the cylinder. If high withdrawal rates continue,

On Site

MAPP® Gas

Unlike acetylene, which must be used at a pressure less than 15 psi, MAPP® gas can be used at the full cylinder pressure of 95 psi. For that reason, it can be used for underwater cutting in deep water. Because it is stored as a liquid, MAPP® is sold by the pound, rather than the cubic foot like acetylene. MAPP® cylinders, unlike acetylene cylinders, do not require acetone or other packing. Therefore, the same size cylinder can hold much more MAPP® than acetylene.

Source: www.free-ed.net

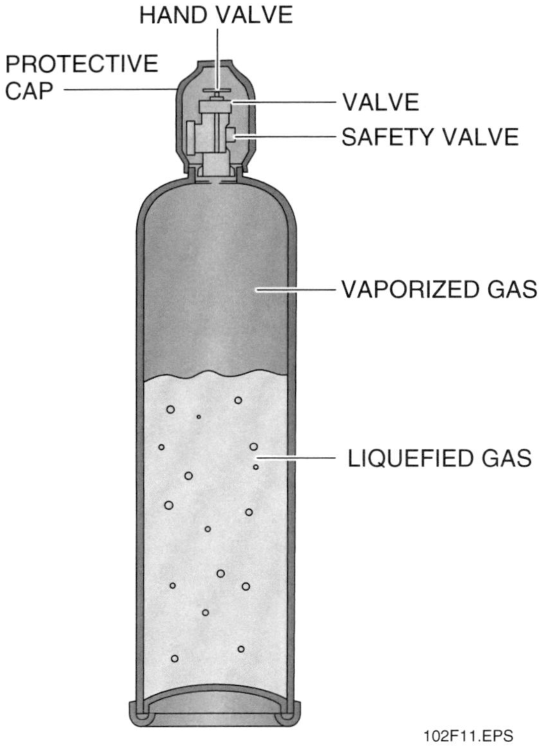

Figure 11 Liquefied fuel gas cylinder.

the regulator may also start to ice up. If high withdrawal rates are required, special regulators with electric heaters should be used.

> **WARNING**
>
> Never apply heat directly to a cylinder or regulator. This can cause excessive pressures, resulting in an explosion.

On Site

Handling and Storing Liquefied Gas Cylinders

Liquefied fuel gas cylinders have a safety valve built into the valve at the top of the cylinder. The safety valve releases gas if the pressure begins to rise. Use care when handling fuel gas cylinders because the gas in cylinders is stored at such high pressures. Cylinders should never be dropped or hit with heavy objects, and they should always be stored in an upright position. When not in use, the cylinder valve must always be covered with the protective steel cap.

The pressure inside a liquefied fuel gas cylinder is not an indicator of how full or empty the cylinder is. The weight of a cylinder determines how much liquefied gas is left. Liquefied fuel gas cylinders are equipped with CGA 510, 350, or 695 valves, depending on the fuel and storage pressures.

> **WARNING**
>
> Do not remove the protective cap on liquefied fuel gas cylinders unless the cylinder is secured. If the cylinder falls over and the nozzle breaks off, the cylinder will release highly explosive gas. Cylinders containing a liquid such as propane must be kept in an upright position. If the valve is broken or is opened with the cylinder horizontal, the fuel can emerge as a liquid that will shoot a long distance before it vaporizes. If it is ignited, it will act like a flamethrower.

3.4.0 Regulators

Regulators (*Figure 12*) are attached to the cylinder valve. They reduce the high cylinder pressures to the required lower working pressures and maintain a steady flow of gas from the cylinder.

The pressure-adjusting screw controls the gas pressure. Turned clockwise, it increases the pressure of gas. Turned counterclockwise, it reduces the pressure of gas. When turned counterclockwise until loose (released), it stops the flow of gas.

Most regulators contain two gauges. The high-pressure or cylinder-pressure gauge indicates the actual cylinder pressure; the low-pressure or working-pressure gauge indicates the pressure of the gas leaving the regulator.

Oxygen regulators differ from fuel gas regulators. Oxygen regulators may have green markings and always have right-hand threads on all connections. The oxygen regulator's high-pressure gauge generally reads up to 3,000 psi and includes a second scale that shows the amount of oxygen in the cylinder in terms of cubic feet. The low-pressure- or working-pressure gauge may read 100 psi or higher.

Fuel gas regulators may have red markings and usually have left-hand threads on all the connections. As a reminder that the regulator has left-hand threads, a V-notch may be cut around the nut. The fuel gas regulator's high-pressure gauge usually reads up to 400 psi. The low-pressure or working-pressure gauge may read up to 40 psi. Acetylene gauges, however, are always red-lined at 15 psi as a reminder that acetylene pressure should not be increased over 15 psi.

There are two types of regulators: single-stage and two-stage.

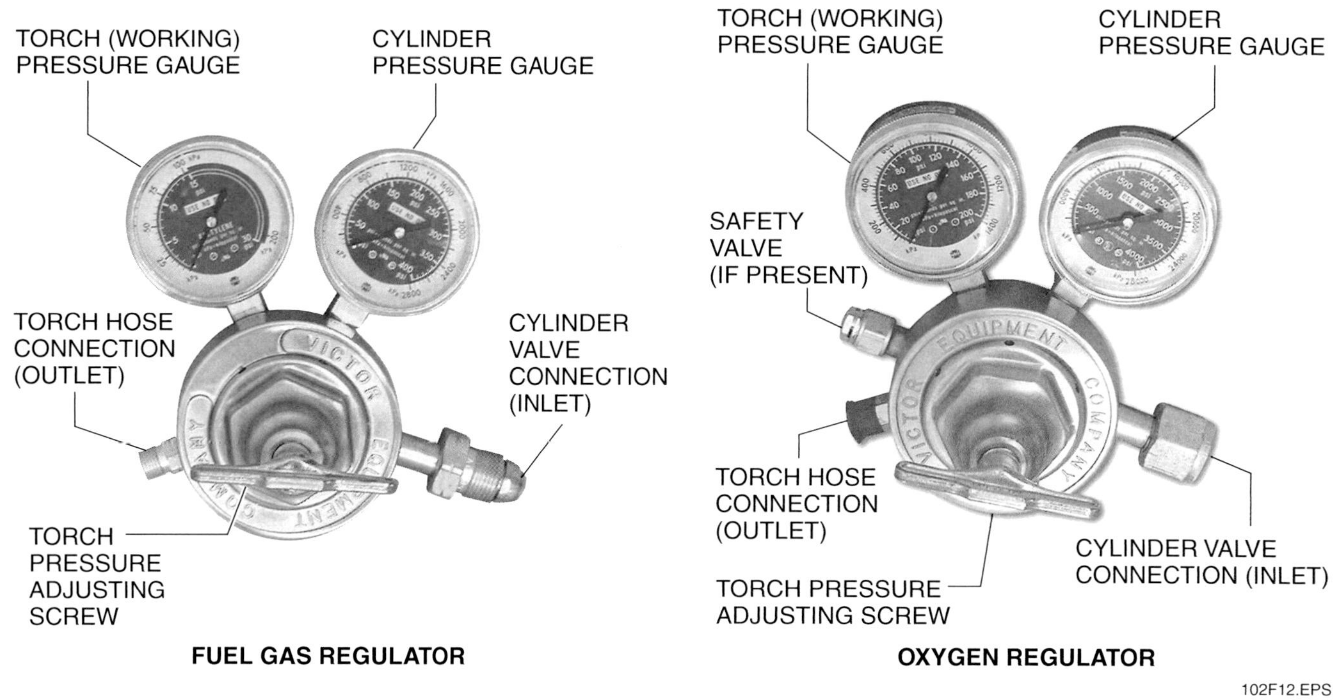

Figure 12 Oxygen and acetylene regulators.

WARNING

To prevent injury and damage to regulators, always follow these guidelines:

- Never subject regulators to jarring or shaking, as this can damage the equipment beyond repair.
- Always check that the adjusting screw is fully released before the cylinder valve is turned on and when the welding has been completed.
- Always open cylinder valves slowly and stand with the valve stem between you and the regulator.
- Never use oil to lubricate a regulator. This can result in an explosion when the regulator is in use.
- Never use fuel gas regulators on oxygen cylinders or oxygen regulators on fuel gas cylinders.
- Never work with a defective regulator. If it is not working properly, shut off the gas supply and have the regulator repaired by someone who is qualified to work on it.
- Never use large wrenches, pipe wrenches, pliers, or slipjoint pliers to install or remove regulators.

3.4.1 Single-Stage Regulators

Single-stage, spring-compensated regulators reduce pressure in one step. As gas is drawn from the cylinder, the internal pressure of the cylinder decreases. A single-stage, spring-compensated regulator is unable to automatically adjust for this decrease in internal cylinder pressure. Therefore, it becomes necessary to adjust the spring pressure to periodically adjust the output gas pressure as the gas in the cylinder is consumed. These regulators are the most commonly used because of their low cost and high flow rates.

3.4.2 Two-Stage Regulators

The two-stage, pressure-compensated regulator reduces pressure in two steps. It first reduces the input pressure from the cylinder to a predetermined intermediate pressure. The intermediate pressure is then adjusted by the pressure-adjusting screw. With this type of regulator, the delivery pressure to the torch remains constant, and no readjustment is necessary as the gas in the cylinder is consumed. Standard two-stage regulators are more expensive than single-stage regulators and have lower flow rates. There are also heavy-duty types with higher flow rates that are usually preferred for thick material and/or continuous-duty cutting operations.

3.4.3 Check Valves and Flashback Arrestors

Check valves and flashback arrestors (*Figure 13*) are safety devices for regulators, hoses, and torches. Check valves allow gas to flow in one direction only. Flashback arrestors stop fire.

Check valves consist of a ball and spring that open inside a cylinder. The valve allows gas to move in one direction but closes if the gas attempts to flow in the opposite direction. When a torch is first pressurized or when it is being shut off, back-pressure check valves prevent the entry and mixing of acetylene with oxygen in the oxygen hose or the entry and mixing of oxygen with acetylene in the acetylene hose.

Flashback arrestors prevent flashbacks from reaching the hoses and/or regulator. They have a flame-retarding filter that will allow heat, but not flames, to pass through. Most flashback arrestors also contain a check valve.

Add-on check valves and flashback arrestors are designed to be attached to the torch handle connections and to the regulator outlets. As a minimum, flashback arrestors with check valves should be attached to the torch handle connections. Both devices have arrows on them to indicate flow direction. When installing add-on check valves and flashback arrestors, be sure the arrow matches the desired gas flow direction.

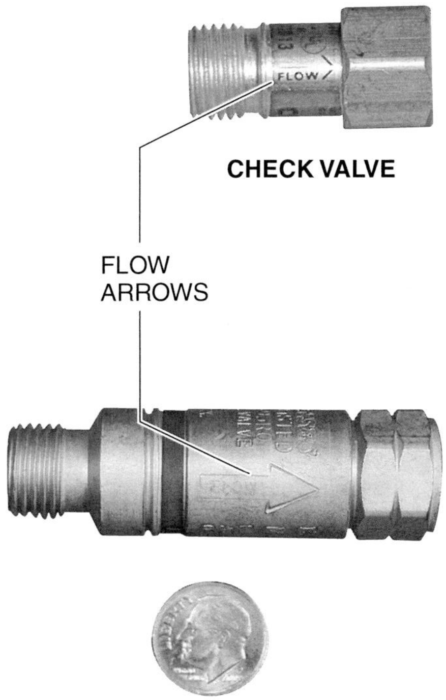

Figure 13 Torch handle add-on check valve and flashback arrestor.

On Site

Torch Wrenches

Only a torch wrench, sometimes called a gang wrench, should be used to install regulators, hose connections, check valves, flashback arrestors, torches, and torch tips. The universal torch wrench shown is equipped with various size wrench cutouts for use with a variety of equipment and standard CGA components. The fittings for oxyfuel equipment are brass or bronze, and certain components are often fitted with soft, flexible, O-ring seals. The seal surfaces of the fittings or O-rings can be easily damaged by overtightening with standard wrenches. The length of a torch wrench is limited to reduce the chances of damage to fittings because of excessive torque. In some cases, manufacturers specify only hand-tightening for certain fitting connections of a torch set (tips or cutting/welding attachments). In any event, follow the manufacturer's specific instructions when connecting the components of a torch set.

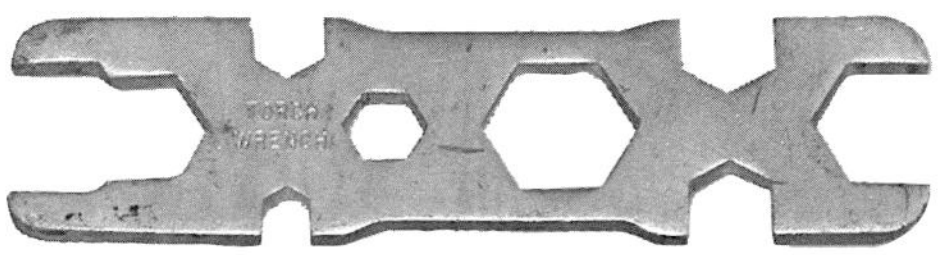

102SA03.EPS

On Site

Flashback Arrestors

It is highly recommended that add-on flashback arrestors with check valves, rather than just check valves, be installed at both the torch handle and regulator connections to prevent serious injury or property damage. If the flame front of a flashback, sometimes called the backburn, gets by the check valves into the hoses or regulators, the hoses could burn through, and a large, very dangerous, uncontrolled fire could start. Newer torches are available with check valves and flashback arrestors built into the torch handle.

102SA04.EPS

3.5.0 Hoses

Hoses transport gases from the regulators to the torch. Oxygen hoses are usually green with right-hand threaded connections. Hoses for fuel gas are usually red and have left-hand threaded connections. The fuel gas connections may also be grooved as a reminder that they have left-hand threads.

Proper care and maintenance of the hose is important for maintaining a safe, efficient work area. Remember the following guidelines for hoses:

- Protect the hose from molten dross or sparks, which will burn the exterior. Although some hoses are flame retardant, they will burn.
- Remove the hoses from under the metal being cut. If the hot metal falls on the hose, the hose will be damaged.
- Frequently inspect and replace hoses that show signs of cuts, burns, worn areas, cracks, or damaged fittings.
- Never use pipe-fitting compounds or lubricants around hose connections. These compounds often contain oil or grease, which ignite and burn or explode in the presence of oxygen.

> **NOTE**
> MAPP®, propane, and propylene require hoses designed for the mixture of hydrocarbons present in these fuels. Check with a hose manufacturer to obtain their recommendations for hoses suitable for use with these gases.

3.6.0 Cutting Torches

Cutting torches mix oxygen and fuel gas for the torch flame and control the stream of oxygen necessary for the cutting jet. Depending on the job site, you may use either a one-piece or a combination cutting torch.

3.6.1 One-Piece Hand Cutting Torch

The one-piece hand cutting torch, sometimes called a demolition torch, contains the fuel gas and oxygen valves that allow the gases to enter the chambers and then flow into the part of the torch where they are mixed. The main body of the torch is called the handle. The torch valves control the fuel gas and oxygen used for preheating the metal to be cut. The cutting oxygen lever, which is spring-loaded, controls the jet of cutting oxygen. Hose connections are located at the end of the torch body behind the valves. *Figure 14* shows a three-tube one-piece positive-pressure hand cutting torch in which the preheat fuel and oxygen are mixed in the tip. These torches are designed for heavy-duty cutting. They have long supply tubes from the torch handle to the torch head to reduce radiated heat to the operator's hands. The torches are generally available in capacities for cutting steel up to 12" thick. Larger-capacity torches, with the ability to cut steel up to 36" thick, can also be obtained.

Two different types of oxyfuel cutting torches are in general use. The positive-pressure torch is designed for use with fuel supplied through a regulator from pressurized fuel storage cylinders. The injector torch is designed to use a vacuum created by oxygen flow to draw the necessary amount of fuel from a very-low-pressure fuel source, such as a natural gas line or acetylene generator. The injector torch, when used, is most often found in continuous-duty high-volume manufacturing applications. Both types may employ one of two different fuel-mixing methods:

- Torch-handle or supply-tube mixing
- Torch-head or tip mixing

The two methods can normally be distinguished by the number of supply tubes from the torch handle to the torch head. Torches that use three tubes (see *Figure 14*) from the handle to the head mix the preheat fuel and oxygen at the torch head or tip. This method tends to help eliminate any flashback damage to the torch head supply tubes and torch handle. Torches with two tubes usually mix the preheat fuel and oxygen in a mixing chamber in the torch body or in one of the supply tubes. Injector torches usually have the injector located in one of the supply tubes, and the mixing occurs in that tube from the injector to the torch head. Some

On Site

Built-In Flashback Arrestors

Newer one-piece hand cutting torches and combination torches with built-in check valves and/or flashback arrestors are available from most manufacturers.

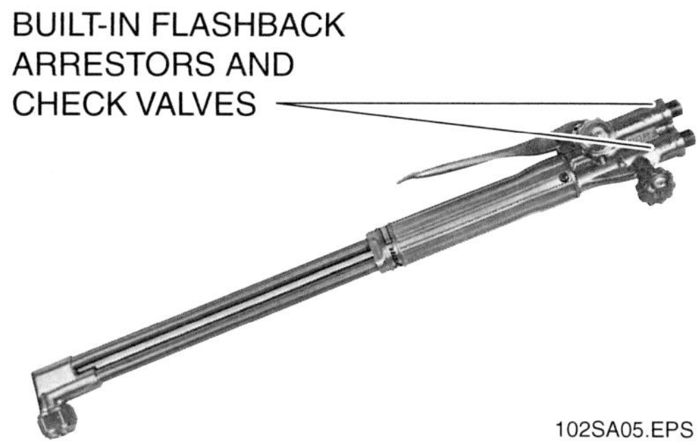

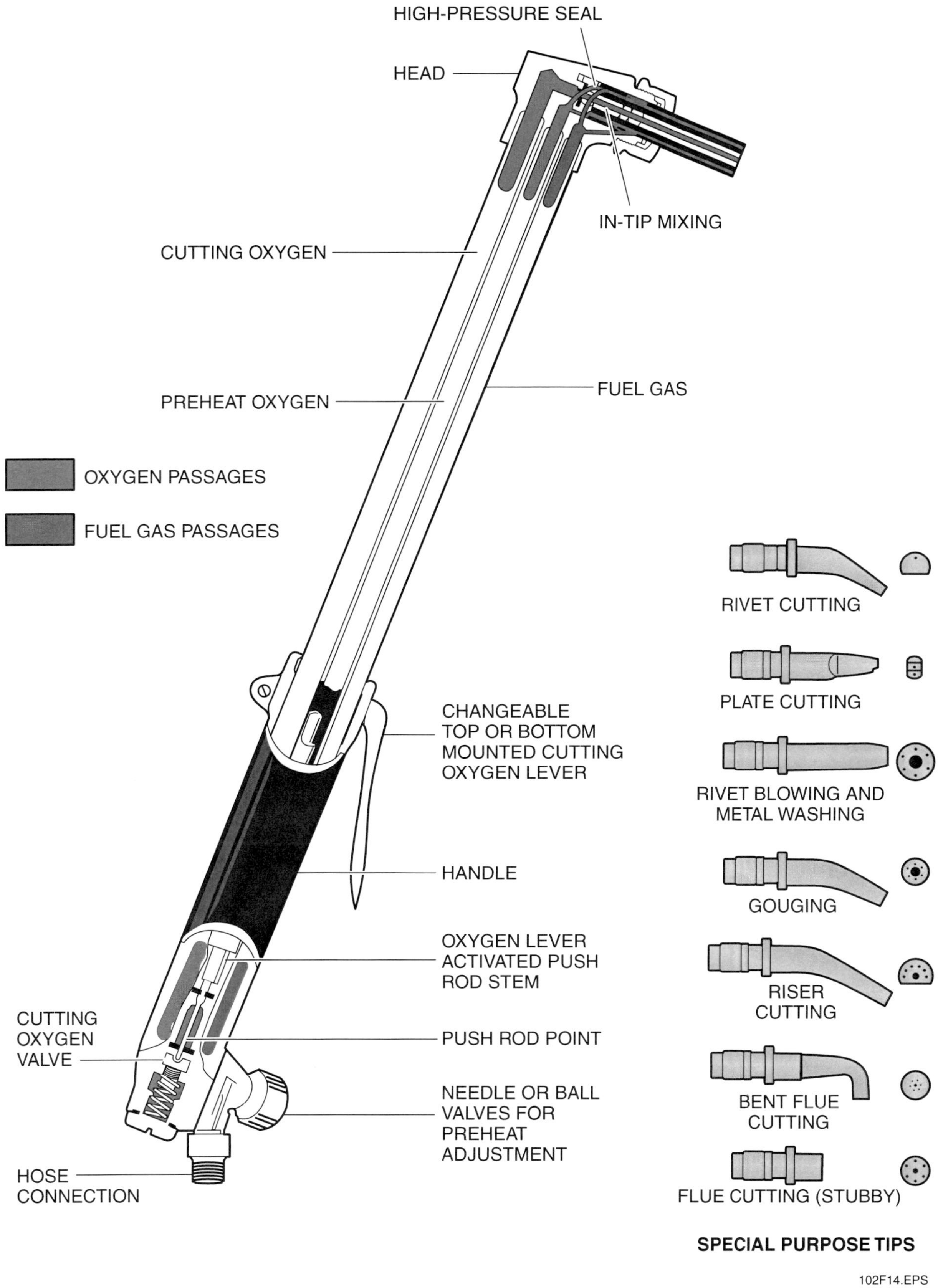

Figure 14 Heavy-duty three-tube one-piece positive-pressure hand cutting torch.

older torches that have only two visible tubes are actually three-tube torches that mix the preheat fuel and oxygen in the torch head or tip. This is accomplished by using a separate preheat fuel tube inside a larger preheat oxygen tube.

3.6.2 Combination Torch

The combination torch consists of a cutting torch attachment that fits onto a welding torch handle. These torches are normally used in light-duty or medium-duty applications. Fuel gas and oxygen valves are on the torch handle. The cutting attachment has a cutting oxygen lever and another oxygen valve to control the preheat flame. When the cutting attachment is screwed onto the torch handle, the torch handle oxygen valve is opened all the way, and the preheat oxygen is controlled by an oxygen valve on the cutting attachment. When the cutting attachment is removed, welding and heating tips can be screwed onto the torch handle. *Figure 15* shows a two-tube combination torch in which the preheat mixing is accomplished in a supply tube. These torches are usually positive-pressure torches with mixing occurring in the attachment body, supply tube, head, or tip. These torches may be equipped with built-in flashback arrestors and check valves.

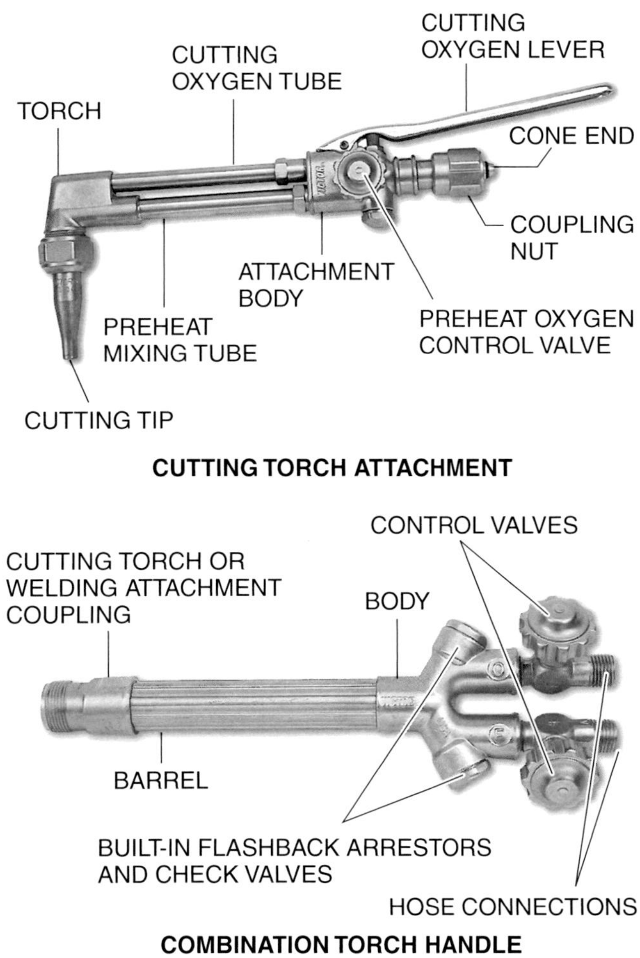

Figure 15 Combination cutting torch.

3.7.0 Cutting Torch Tips

Cutting torch tips, or nozzles, fit into the cutting torch and are either screwed in or secured with a tip nut. There are one- and two-piece cutting tips (*Figure 16*).

One-piece cutting tips are made from a solid piece of copper. Two-piece cutting tips have a separate external sleeve and internal section.

Torch manufacturers supply literature explaining the appropriate torch tips and gas pressures to be used for various applications. *Table 2* shows a sample cutting tip chart that lists recommended tip sizes and gas pressures for use with acetylene fuel gas and a specific manufacturer's torch and tips.

WARNING

Do not use the cutting tip chart from one manufacturer for the cutting tips of another manufacturer. The gas flow rate of the tips may be different, resulting in excessive flow rates. Different gas pressures may also be required. The cutting torch tip to be used depends on the base metal thickness and fuel gas being used. Special-purpose tips are also available for use in such operations as gouging and grooving.

The cutting torch tip to be used depends on the base metal thickness and fuel gas being used. Special-purpose tips are also available for use in such operations as gouging and grooving.

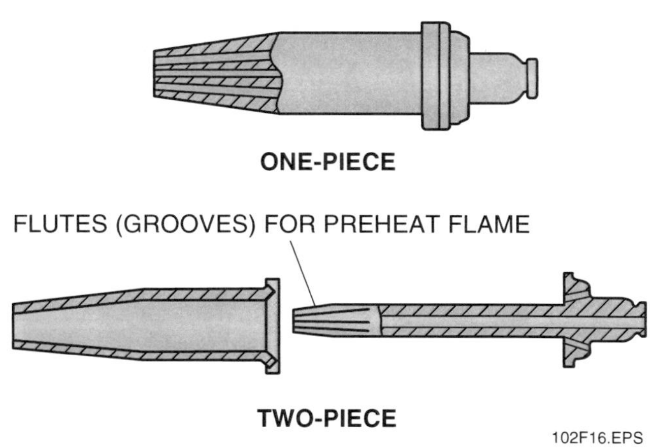

Figure 16 One- and two-piece cutting tips.

Table 2 Sample Acetylene Cutting Tip Chart

Cutting Tip Series 1-101, 3-101, and 5-101						
Metal Thickness	Tip Size	Cutting Oxygen Pressure (psig)*	Preheat Oxygen (psig)*	Acetylene Pressure (psig)	Speed (in./min.)	Kerf Width
⅛"	000	20/25	3/5	3/5	20/30	0.04
¼"	00	20/25	3/5	3/5	20/28	0.05
⅜"	0	25/30	3/5	3/5	18/26	0.06
½"	0	30/35	3/6	3/5	16/22	0.06
¾"	1	30/35	4/7	3/5	15/20	0.07
1"	2	35/40	4/8	3/6	13/18	0.09
2"	3	40/45	5/10	4/8	10/12	0.11
3"	4	40/50	5/10	5/11	8/10	0.12
4"	5	45/55	6/12	6/13	6/9	0.15
6"	6	45/55	6/15	8/14	4/7	0.15
10"	7	45/55	6/20	10/15	3/5	0.34
12"	8	45/55	7/25	10/15	3/4	0.41

*The lower side of the pressure listing is for hand cutting and the higher side is for machine cutting.

On Site

Torch Tip Styles

Nearly all manufacturers use different tip-to-torch mounting designs, sealing surfaces, and diameters. In addition, tip sizes and flow rates are usually not the same between manufacturers even though the number designations may be the same. This makes it impossible to safely interchange cutting tips between torches from different manufacturers. Even though some tips from different manufacturers may appear to be the same, do not interchange them. The sealing surfaces are very precise, and serious leaks may occur that could result in a dangerous fire or flashback.

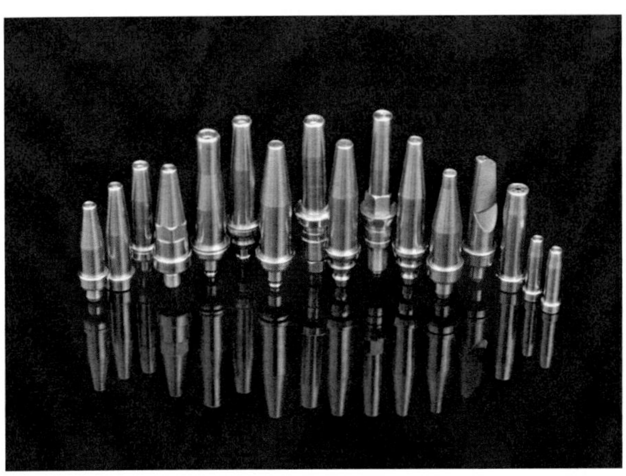

102SA06.EPS

3.7.1 Cutting Tips for Acetylene

One-piece torch tips are usually used with acetylene cutting because of the high temperatures involved. They can have four, six, or eight preheat holes in addition to the single cutting hole. *Figure 17* shows the arrangement of typical acetylene torch cutting tips.

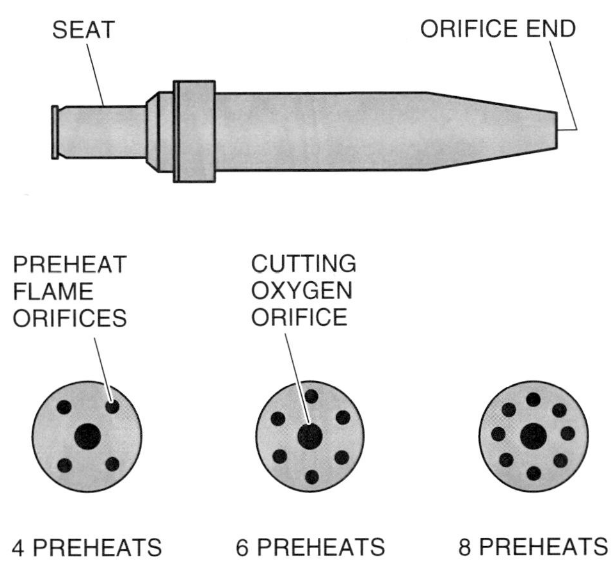

102F17.EPS

Figure 17 Orifice end views of one-piece acetylene torch cutting tips.

On Site

Acetylene Flow Rates

Manufacturers provide listings of the maximum fuel flow rate for each acetylene tip size in addition to recommended acetylene pressures. When selecting a tip, make sure that its maximum flow rate (in cubic feet per hour) does not exceed one-seventh of the total fuel capacity (in cubic feet) for the acetylene cylinder in use. Multiple cylinders must be manifolded together if the flow rate exceeds the cylinder(s) in use in order to prevent withdrawal of acetone along with acetylene.

3.7.2 Cutting Tips for Liquefied Fuel Gases

Tips used with liquefied fuel gases must have at least six preheat holes. Because fuel gases burn at lower temperatures than acetylene, more holes are necessary for preheating. Tips used with liquefied fuel gases can be one- or two-piece cutting tips. *Figure 18* shows typical cutting tips used with liquefied fuel gases.

3.7.3 Special-Purpose Cutting Tips

Special-purpose tips are available for special cutting jobs. These jobs include cutting sheet metal, rivets, risers, and flues, as well as washing and gouging. *Figure 19* shows special-purpose torch cutting tips.

- The sheet metal cutting tip has only one preheat hole. This minimizes the heat and prevents distortion in the sheet metal. These tips are normally used with a motorized carriage but can also be used for hand cutting.
- Rivet cutting tips are used to cut off rivet heads, bolt heads, and nuts.
- Riser cutting tips are similar to rivet cutting tips and can also be used to cut off rivet heads, bolt heads, and nuts. They have extra preheat holes to cut risers, flanges, or angle legs faster. They can be used for any operation that requires a cut close to and parallel to another surface, such as in removing a metal backing.
- Rivet blowing and metal washing tips are heavy-duty tips designed to withstand high heat. They are used for coarse cutting and for removing such items as clips, angles, and brackets.
- Gouging tips are used to groove metal in preparation for welding.
- Flue cutting tips are designed to cut flues inside boilers. They also can be used for any cutting operation in tight quarters where it is difficult to get a conventional tip into position.

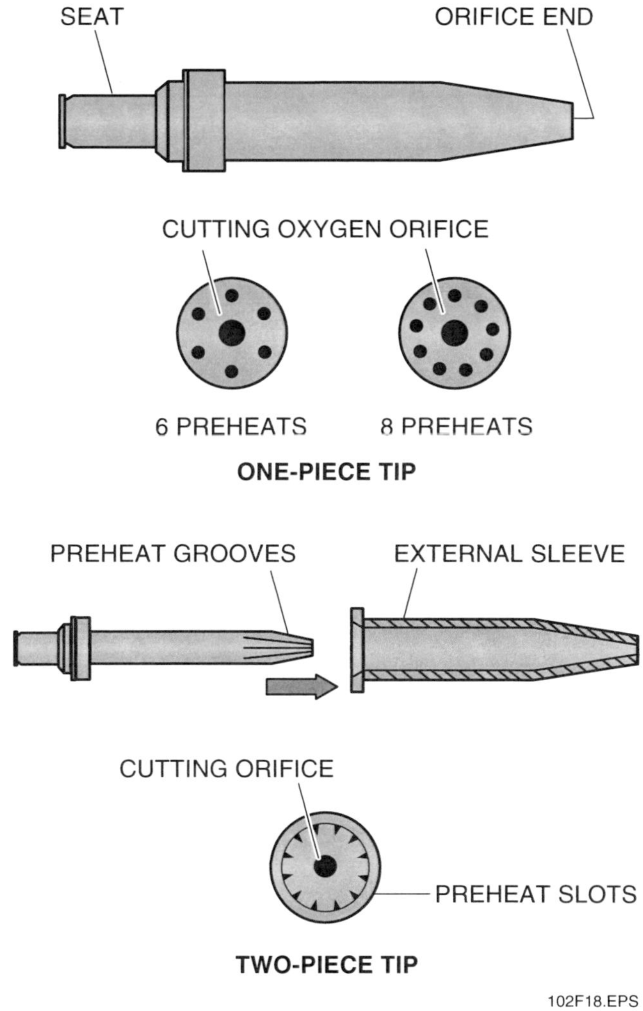

Figure 18 Orifice end views of cutting tips for liquefied fuel gases.

3.8.0 Tip Cleaners and Tip Drills

With use, cutting tips become dirty. Carbon and other impurities build up inside the holes, and molten metal often sprays and sticks onto the surface of the tip. A dirty tip will result in a poor-quality cut with an uneven **kerf** and excessive dross buildup. To ensure good cuts with straight kerfs and minimal dross buildup, clean cutting tips with tip cleaners or tip drills (*Figure 20*).

Tip cleaners are small round files. They usually come in a set with files to match the diameters of the various tip holes. In addition, each set usually includes a file that can be used to lightly recondition the face of the cutting tip. Tip cleaners are inserted into the tip hole and moved back and forth a few times to remove deposits from the hole.

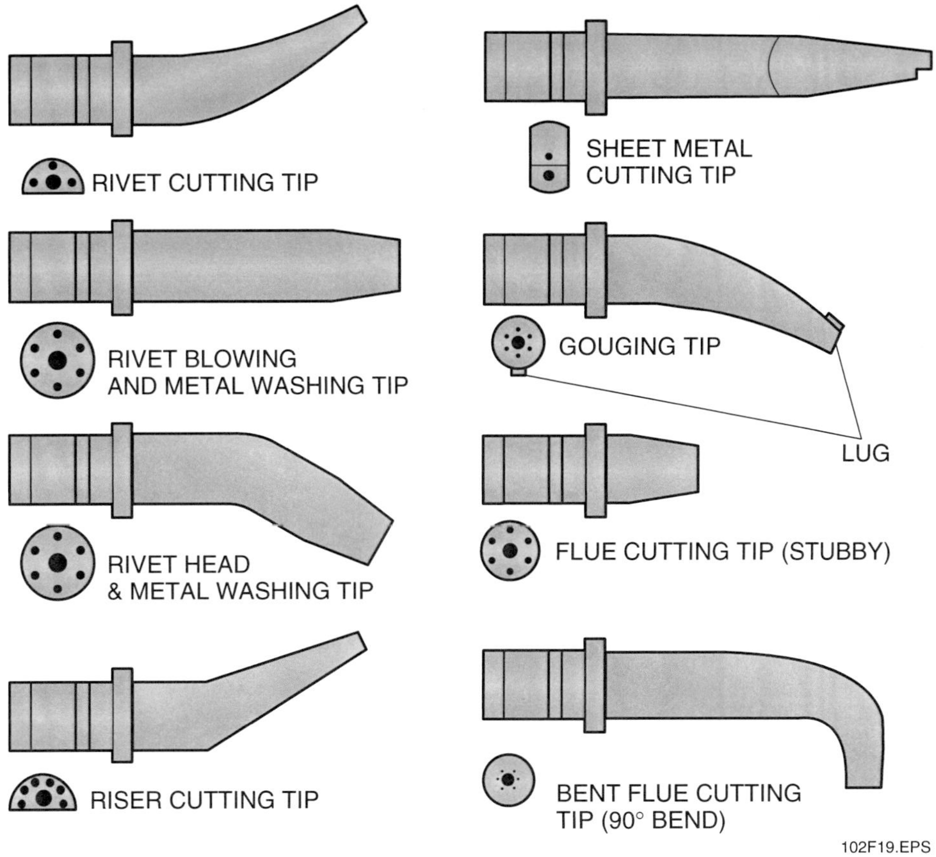

Figure 19 Special-purpose torch cutting tips.

On Site

Alternate Design Solid Copper Tip

A manufacturer of replacement torch tips for various brands of torches has a two-piece design for use in place of a standard solid copper tip. It consists of a single tip-to-head sealing adapter and a number of small interchangeable tip sizes that screw onto the adapter. The small screw-on tips are faster to change and somewhat less costly than a full standard tip.

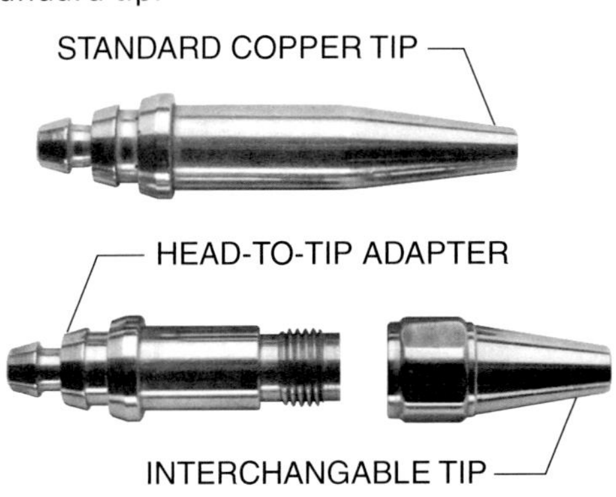

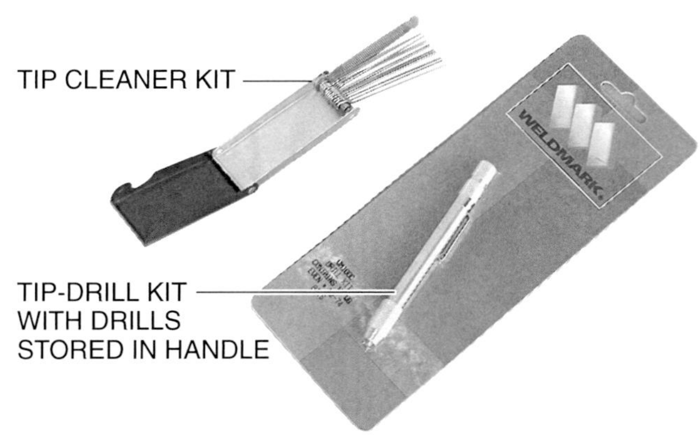

Figure 20 Tip cleaner and drill kits.

Tip drills are used for major cleaning and for holes that are plugged. Tip drills are tiny drill bits that are sized to match the diameters of tip holes. The drill fits into a drill handle for use. The handle is held, and the drill bit is turned carefully inside the hole to remove debris. They are more brittle than tip cleaners, making them more difficult to use.

CAUTION

Tip cleaners and tip drills are brittle. If you are not careful, they may break off inside a hole. Broken tip cleaners are difficult to remove. Improper use of tip cleaners or tip drills can enlarge the tip, causing improper burning of gases. If this occurs, tips must be discarded. If the end of the tip has been partially melted or deeply gouged, do not attempt to cut it off or file it flat. The tip should be discarded and replaced with a new tip. This is because some tips have tapered preheat holes, and if a significant amount of metal is removed from the end of the tip, the preheat holes will become too large.

3.9.0 Friction Lighters

Always use a friction lighter (*Figure 21*), also known as a striker or spark-lighter, to ignite the cutting torch. The friction lighter works by rubbing a piece of flint on a steel surface to create sparks.

WARNING

Do not use a match or a gas-filled lighter to light a torch. This could result in severe burns and/or could cause the lighter to explode.

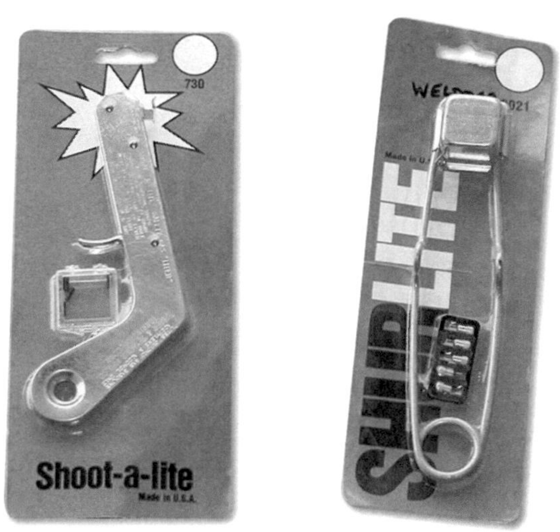

TRIGGER OPERATED STRIKER

COMMON CUP-TYPE STRIKER

102F21.EPS

Figure 21 Friction lighters.

On Site

Cup-Type Striker

When using a cup-type striker to ignite a welding torch, hold the cup of the striker slightly below and to the side of the tip, parallel with the fuel gas stream from the tip. This prevents the ignited gas from deflecting back toward you from the cup and reduces the amount of carbon soot in the cup. Note that the flint in a striker can be replaced.

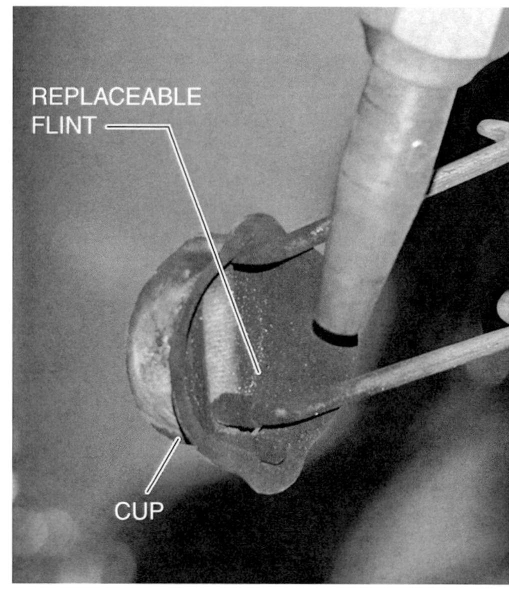

102SA08.EPS

3.10.0 Cylinder Cart

The cylinder cart, or bottle cart, is a modified hand truck that has been equipped with seats and chains to hold cylinders firmly in place. Bottle carts help ensure the safe transportation of gas cylinders. *Figure 22* shows a typical cylinder cart for a cutting rig. Some carts are equipped with tool/accessory trays or boxes as well as rod holders.

3.11.0 Soapstone Markers

Because of the heat involved in cutting operations, along with the tinted lenses that are required, ordinary pen or pencil marking for cutting lines or welding locations is not effective. The oldest and most common material used for marking is **soapstone** in the form of sticks or cylinders (*Figure 23*). Soapstone is soft and feels greasy and slippery. It is actually steatite, a dense, impure form of talc that is heat resistant. It also shows up well through a tinted lens under the illumination of

Figure 22 Cylinder cart.

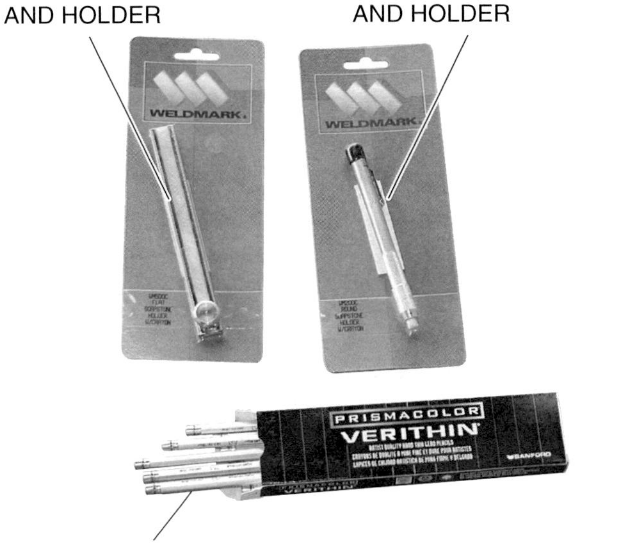

Figure 23 Soapstone and graphite markers.

an electric arc or gas cutting flame. Some welders prefer to use silver-graphite pencils for marking dark materials and red-graphite pencils for aluminum or other bright metals. Graphite is also highly heat resistant (*Figure 23*). A few manufacturers also market heat-resistant paint/dye markers for cutting and welding.

On Site

Sharpening Soapstone Sticks

The most effective way to sharpen a soapstone stick marker is to shave it on one side with a file. By leaving one side flat, accurate lines can be drawn very close to a straightedge or a pattern.

3.12.0 Specialized Cutting Equipment

In addition to the common hand-cutting torches, other types of equipment are used in oxyfuel cutting applications. This equipment includes mechanical guides used with a hand cutting torch, various types of motorized cutting machines, and oxygen lances. All of the motorized units use special straight body machine cutting or welding torches with a gear rack attached to the torch body to set the tip distance from the work.

3.12.1 Mechanical Guides

On long, circular, or irregular cuts, it is very difficult to control and maintain an even kerf with a hand-cutting torch. Mechanical guides can help maintain an accurate and smooth kerf along the cutting line. For straight line or curved cuts, use a one- or two-wheeled accessory that clamps on the torch tip in a fixed position. The wheeled accessory maintains the proper tip distance while the tip is guided by hand along the cutting line.

Perform arc or circular cuts with the circle cutting accessory shown in *Figure 24*. The torch tip fits through and is secured to a rotating mount between the two small metal wheels. The wheel heights are adjustable so that the tip distance from the work can be set. The radius of the circle is set by moving the pivot point on a radius bar. After a starting hole is cut (if needed), the torch tip is

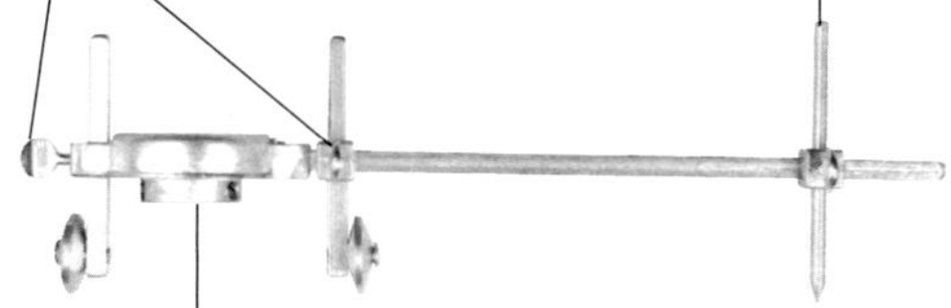

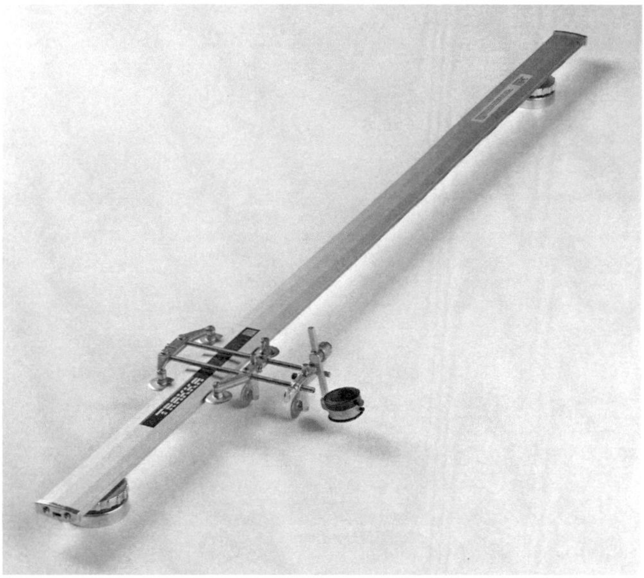

102F24.EPS

Figure 24 Circle cutting guide.

placed through and secured to the circle cutter rotating mount. Then the pivot point is placed in a drilled hole or a magnetic holder at the center of the circle. When the cut is restarted, the torch can be moved in a circle around the cut, guided by the circle cutter. The magnetic guide shown in *Figure 24* is used for straight line cuts. The magnets hold it securely in place during cutting operations.

When large work with an irregular pattern must be cut, a template is often used. The torch is drawn around the edges of the template to trace the pattern as the cut is made. If multiple copies must be cut, a metal pattern, held in place and spaced for tip distance from the work by stacked magnets, is usually used. For a one- or two-time copy, a heavily weighted Masonite or aluminum template that is spaced off the workpiece could be carefully used and discarded.

3.12.2 Motor-Driven Equipment

A variety of fixed and portable motorized cutting equipment is available for straight and curved cutting/welding. The computer-controlled gantry cutting machine (*Figure 25*) and the optical pattern-tracing machine (*Figure 26*) are fixed-location machines used in industrial manufacturing applications. The computer-controlled machine can be programmed to **pierce** and then cut any pattern from flat metal stock. The optical pattern-tracing machine follows lines on a drawing using a light beam and an optical detector. They both can be rigged to cut multiple items using multiple torches operated in parallel. Both units have a motor-driven gantry that travels the length of a table and a transverse motor-driven beam or torch head that moves back and forth across the table. Both units

On Site

Shop-Made Straight-Line Cutting Guide

A simple solution for straight-line cutting is to clamp a piece of angle iron to the work and use a band clamp around the cutting torch tip to maintain the cutting tip distance from the work. When the cut is started, the band clamp rests on the top of the vertical leg of the angle iron, and the torch is drawn along the length of the angle iron at the correct cutting speed.

102SA10.EPS

102F25.EPS

Figure 25 Computer-controlled gantry cutting machine.

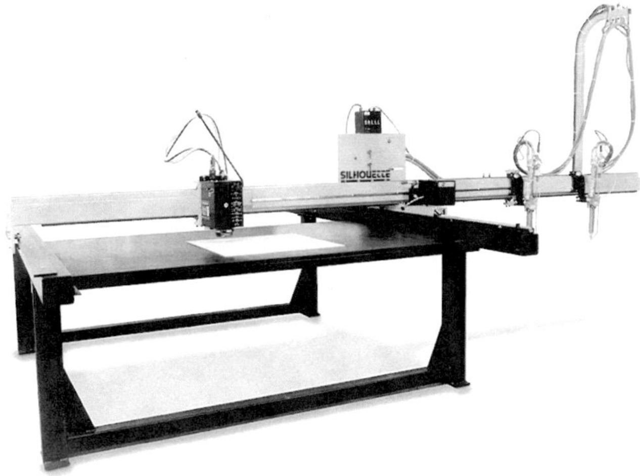

102F26.EPS

Figure 26 Optical pattern-tracing machine.

102F28.EPS

Figure 28 Track burner with a plasma machine torch.

are also equipped to use both oxyfuel cutting and plasma cutting torches. The size of the patterns can also be adjusted.

Other types of pattern-tracing machines use metal templates that are clamped in the machine. A follower wheel traces the pattern from the template. The pattern size can be increased or decreased by electrical or mechanical linkage to a moveable arm holding one or more cutting torches that cut the pattern from flat metal stock.

Portable track cutting machines (track burners) can be used in the field for straight or curved cutting and beveling. *Figures 27* and *28* show units driven by a variable-speed motor. The unit shown in *Figure 27* is available with track extensions for any length of straight cutting or beveling, along with a circle cutting attachment. The unit shown in *Figure 28* uses a somewhat flexible magnetic track for both flat straight-line or large-diameter object cutting or beveling. It is shown equipped with an optional plasma machine torch. Both units can be adapted to metal inert gas (MIG) or tungsten inert gas (TIG) welding.

A portable, motor-driven band track or hand-cranked ring gear cutter/beveler can be set up in the field for cutting and beveling pipe with oxyfuel or plasma machine torches (*Figures 29* and *30*). The stainless steel band track cutter uses a chain and motor sprocket drive to rotate the machine cutting torch around the pipe a full 360 degrees. The all-aluminum ring gear type of cutter/ beveler is positioned on the pipe, and then the saddle is clamped in place. In operation, the ring gear and the cutting torch rotate at different rates around the saddle for a full 360-degree cut.

A hand-guided oxyfuel cutting torch with an integral-precision, variable-speed motor drive

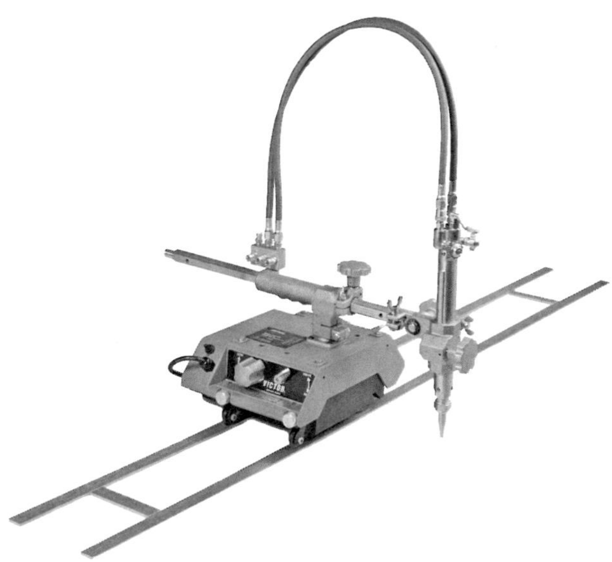

102F27.EPS

Figure 27 Track burner with oxyfuel machine torch.

102F29.EPS

Figure 29 Band-track pipe cutter/beveler.

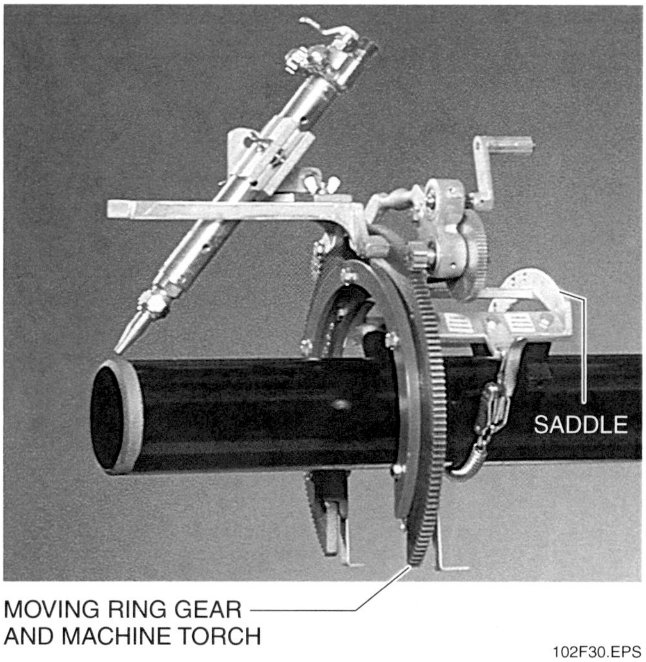

Figure 30 Ring gear pipe cutter/beveler.

(*Figure 31*), which can be used for straight line and curved cutting to achieve machine-quality cuts, is available. A circle cutting accessory is also available for the unit.

3.12.3 Exothermic Oxygen Lances

Exothermic (combustible) oxygen lances are a special oxyfuel cutting tool usually used in heavy industrial applications and demolition work. The lance is a steel pipe that contains magnesium- and aluminum-cored powder or rods (fuel). In operation, the lance is clamped into a holder (*Figure 32*) that seals the lance to a fitting that supplies oxygen to the lance through a hose at pressures of 75 to 80 psi. With the oxygen turned on, the end of the lance is ignited with an acetylene torch or flare. As long as the oxygen is applied, the lance will burn and consume itself. The oxygen-fed flame of the burning magnesium, aluminum, and steel pipe creates temperatures approaching 10,000°F. At this temperature, the lance will rapidly cut or pierce any material, including steel, metal alloys, and cast iron, even under water. The lances for the holder shown in *Figure 32* are 10' long and range in size from ⅜" to 1" in diameter. The larger sizes can be coupled to obtain a longer lance. A small pistol-grip heat-shielded unit that can, if desired, be used with an electric welder is also available. The arc, combined with an oxygen lance, can create temperatures exceeding 10,000°F. This small unit uses lances from ¼" to ⅜" in diameter that are 22" to 36" long and that cut very rapidly at a maximum burning time of 60 to 70 seconds. The small unit is primarily used to burn out large frozen pins and frozen headless bolts or rivets. Like a large lance, it can be used to

Figure 31 Hand-guided motorized oxyfuel cutting torch.

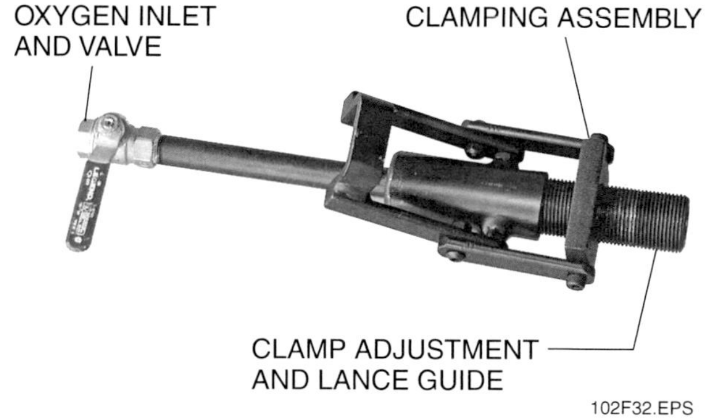

Figure 32 Oxygen lance holder.

On Site

Cutting through Difficult Materials

Fire brick, reinforced concrete, and large, thick iron castings or steel objects are some of the many materials that can quickly be cut or pierced with an exothermic oxygen lance. With the larger lances, use specially insulated heat-reflective suits and full head shields because of the amount of heat generated.

On Site

New Cutting Technologies

Manufacturers of welding and cutting equipment are constantly working on new products and methods. Two new cutting technologies were emerging at the time this module was being written.

One of these technologies uses water as the heat source. A special generator separates the water into hydrogen and oxygen using electrolysis. The two gases are then recombined in the cutting torch in the same manner as oxygen and acetylene. This generator can be used in welding, cutting, soldering, and brazing applications. It also has potential for use as a fuel source for automobiles. The product is called Brown's Gas in honor of Yull Brown, the man who devoted his professional life and fortune to the development of this process (www.brownsgas.com).

Another emerging technology uses regular gasoline in place of acetylene. Like an oxyacetylene setup, the gasoline and oxygen are combined in the torch to produce the cutting flame (www.petrogen.com).

cut any material, including concrete-lined pipe. It also can be used to remove hard-surfacing material that has been applied to wear surfaces. Both units are relatively inexpensive and can be set up in the field with only an oxygen cylinder, hose, and ignition device.

4.0.0 Setting Up Oxyfuel Equipment

When setting up oxyfuel equipment, you must follow certain procedures to ensure that the equipment operates properly and safely. The following sections explain the procedures for setting up oxyfuel equipment.

4.1.0 Transporting and Securing Cylinders

Follow these steps to transport and secure cylinders:

WARNING

Always handle cylinders with care. They are under high pressure and should never be dropped, knocked over, rolled, or exposed to heat in excess of 140°F. When moving cylinders, always be certain that the valve caps are in place. Use a cylinder cage to lift cylinders. Never use a sling or electromagnet.

Step 1 Transport cylinders to the workstation in the upright position on a hand truck or bottle cart (*Figure 33*).

Step 2 Secure the cylinders at the workstation.

Step 3 Remove the protective cap from each cylinder and inspect the outlet nozzles to make sure that the seat and threads are not damaged. Place the protective caps where they will not be lost and where they will be available when the cylinders are empty.

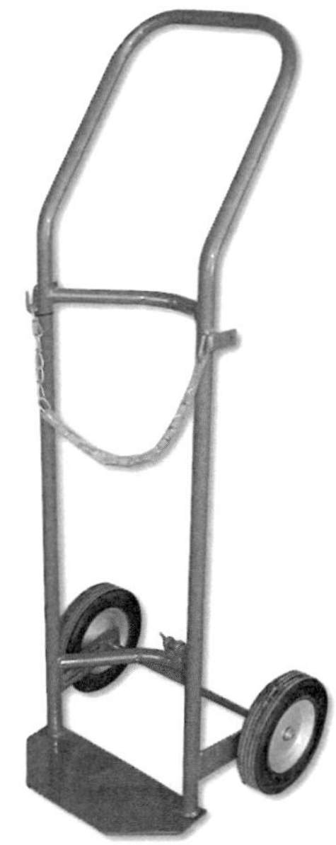

SINGLE-CYLINDER HAND CART

HEAVY-DUTY TWIN-CYLINDER HAND CART

102F33.EPS

Figure 33 Carts for transporting cylinders.

On Site

Hoisting Cylinders

Never attempt to lift a cylinder using the holes in a safety cap. Always use a lifting cage. Make sure that the cylinder is secured in the cage. Various size cages are available for high-pressure cylinders and cylinders containing liquids.

102SA12.EPS

CAUTION

Do not transport or immediately use an acetylene cylinder found resting on its side. Stand it upright and wait at least an hour to allow the acetone to settle before using it.

On Site

Fuel and Oxygen Cylinder Separation for Fixed Installations

For fixed installations involving one or more cylinders coupled to a manifold, fuel and oxygen cylinders must be separated by at least 20' or be divided by a wall 5' or more high with a 30-minute burn rating (*American National Standards Institute Z49.1*).

4.2.0 Cracking Cylinder Valves

Follow these steps to crack cylinder valves:

Step 1 Crack open the cylinder valve momentarily to remove any dirt from the valves (*Figure 34*).

WARNING

Always stand with the valve stem between you and the regulator when opening valves to avoid injury from dirt that may be lodged in the valve.

Step 2 Wipe out the connection seat of the valves with a clean cloth. Dirt frequently collects in the outlet nozzle of a cylinder valve and must be cleaned out to keep it from entering the regulator when pressure is turned on.

WARNING

Be sure the cloth used does not have any oil or grease on it. Oil or grease mixed with compressed oxygen will cause an explosion.

102F34.EPS

Figure 34 Cracking a cylinder valve.

4.3.0 Attaching Regulators

Follow these steps to attach the regulators:

Step 1 Check that the regulator is closed (adjustment screw loose).

> **CAUTION**
>
> Use care not to overtighten connections. The brass connections used will strip if overtightened. Repair or replace equipment that does not seal properly. If the torch valves leak, try lightly tightening the packing nut at the valves. If a cutting attachment or torch tip leaks, disassemble and check the sealing surfaces and/or O-rings for damage. If the equipment or torch tip is new and the sealing surfaces do not appear damaged, overtighten the connection slightly to seat the sealing surfaces; then loosen and retighten normally. If the leaks persist or the equipment is visibly damaged, replace the torch, cutting attachment (if used), or cutting tip as necessary.

Step 2 Check the regulator fittings to ensure that they are free of oil and grease (*Figure 35*).

Step 3 Connect and tighten the oxygen regulator to the oxygen cylinder using a torch wrench (*Figure 36*).

> **WARNING**
>
> Do not work with a regulator that shows signs of damage, such as cracked gauges, bent thumbscrews, or worn threads. Set it aside for repairs. Never attempt to repair regulators yourself. When tightening the connections, always use a torch wrench.

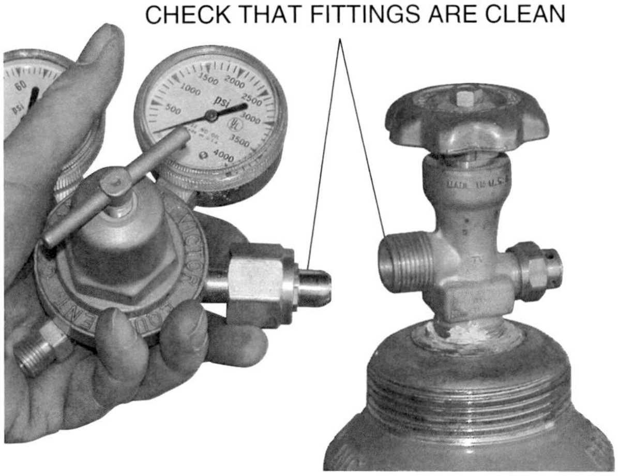

102F35.EPS

Figure 35 Checking connection fittings.

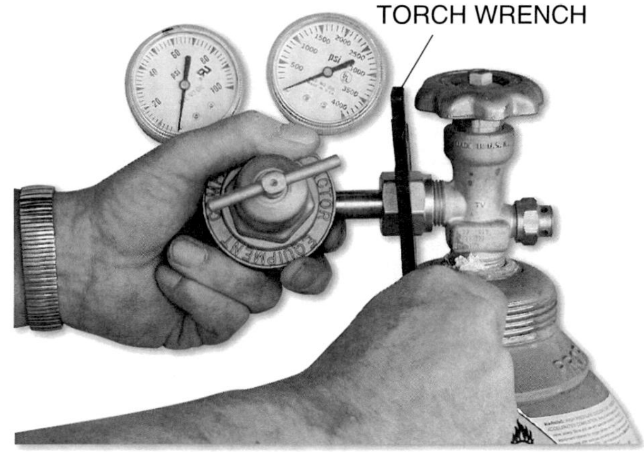

102F36.EPS

Figure 36 Tightening regulator connection

Step 4 Connect and tighten the fuel gas regulator to the fuel gas cylinder. Remember that most fuel gas fittings have left-hand threads.

Step 5 Crack the cylinder valve slightly and open the regulator to expel any debris from the outlet. Shut the cylinder valve and close the regulator (*Figure 37*).

4.4.0 Installing Flashback Arrestors or Check Valves

Follow these steps to install flashback arrestors or check valves:

> **WARNING**
>
> At least one flashback arrestor must be used with each hose. The absence of a flashback arrestor could result in flashback. Flashback arrestors can be attached either to the regulator, the torch, or both; however, flashback arrestors installed at the torch handle are preferred if only one is being used.

102F37.EPS

Figure 37 Cleaning the regulator.

Step 1 Attach a flashback arrestor or check valve to the hose connection on the oxygen regulator (*Figure 38*) and tighten with a torch wrench.

Step 2 Attach and tighten a flashback arrestor or check valve to the hose connection on the fuel gas regulator. Keep in mind that most fuel gas fittings have left-hand threads.

4.5.0 Connecting Hoses to Regulators

New hoses contain talc and loose bits of rubber. These materials must be blown out of the hoses before the torch is connected. If they are not blown out, they will clog the torch needle valves.

Follow these steps to connect the hoses to the regulators:

Step 1 Inspect both the oxygen and fuel gas hoses for any damage, burns, cuts, or fraying.

Step 2 Replace any damaged hoses.

Step 3 Connect the oxygen hose to the oxygen regulator flashback arrestor or check valve (*Figure 39*).

Step 4 Connect the fuel gas hose to the fuel gas regulator flashback arrestor or check valve. Keep in mind that most fuel gas fittings have left-hand threads.

Step 5 Open the regulator and purge the hose until it is clear.

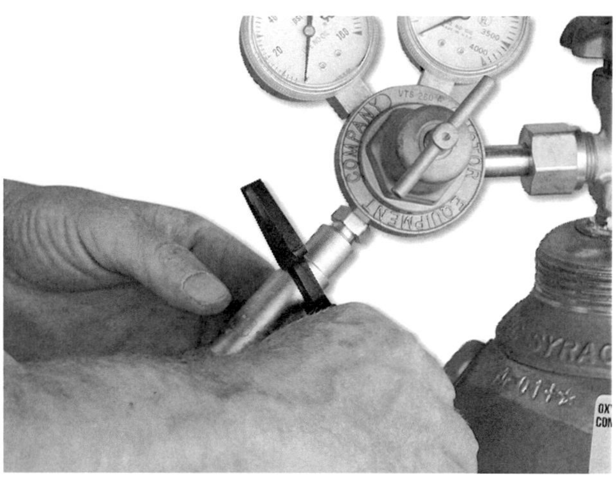

102F38.EPS

Figure 38 Attaching a flashback arrestor.

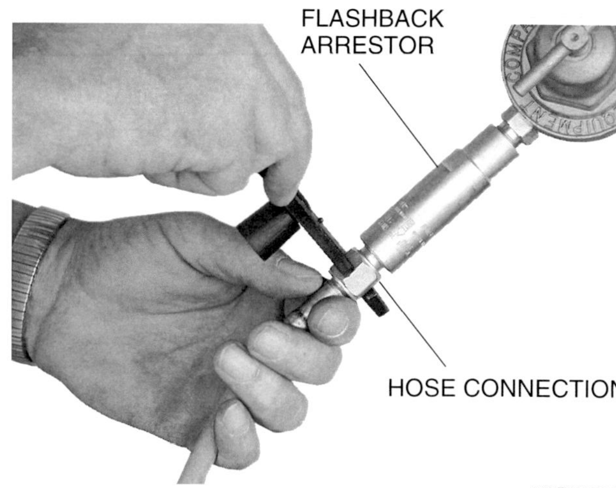

102F39.EPS

Figure 39 Connecting hose to regulator flashback arrestor.

4.6.0 Attaching Hoses to the Torch

Follow these steps to attach the hoses to the torch:

Step 1 Attach flashback arrestors to the oxygen and fuel gas hose connections on the torch body unless the torch has built-in flashback arrestors and check valves. Keep in mind that most fuel gas fittings have left-hand threads.

Step 2 Attach and tighten the oxygen hose to the oxygen fitting on the flashback arrestor or torch (*Figure 40*).

Step 3 Attach and tighten the hose to the fuel gas fitting on the flashback arrestor or torch. Remember that all hose fittings have left-hand threads.

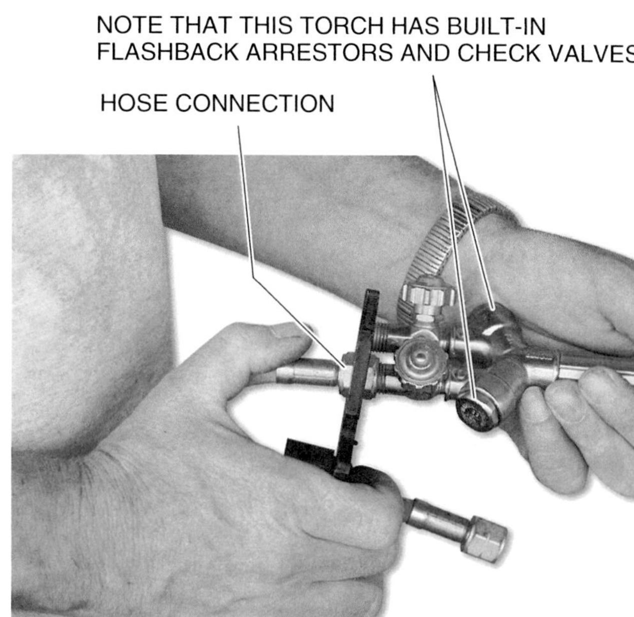

102F40.EPS

Figure 40 Connecting hoses to torch body.

4.7.0 Connecting Cutting Attachments (Combination Torch Only)

If a combination torch is being used, connect cutting attachments as follows:

Step 1 Check the torch manufacturer's instructions for the correct method of installing the attachment.

Step 2 Connect the attachment and tighten by hand as required.

4.8.0 Installing Cutting Tips

Follow these steps to install a cutting tip in the cutting torch:

Step 1 Identify the thickness of the material to be cut.

> **WARNING**
>
> If acetylene fuel is being used, make sure that the maximum fuel flow rate per hour of the tip does not exceed one-seventh of the fuel cylinder capacity. If a purplish flame is observed when the torch is operating, the fuel rate is too high and acetone is being withdrawn from the acetylene cylinder along with the acetylene gas.

Step 2 Identify the proper size cutting tip from the manufacturer's recommended tip size chart for the fuel being used.

Step 3 Inspect the cutting tip sealing surfaces and orifices for damage or plugged holes. If the sealing surfaces are damaged, discard the tip. If the orifices are plugged, clean them with a tip cleaner or drill.

Step 4 Check the torch manufacturer's instructions for the correct method of installing cutting tips.

Step 5 Install the cutting tip, securing it with a torch wrench or by hand as required (*Figure 41*).

4.9.0 Closing Torch Valves and Loosening Regulator Adjusting Screws

Follow these steps to close the torch valves and loosen the regulator adjusting screws (*Figure 42*):

Step 1 Check the fuel and oxygen valves on the torch to be sure they are closed. Closing the torch gas valves prevents gases from backing up inside the torch.

102F41.EPS

Figure 41 Installing a cutting tip.

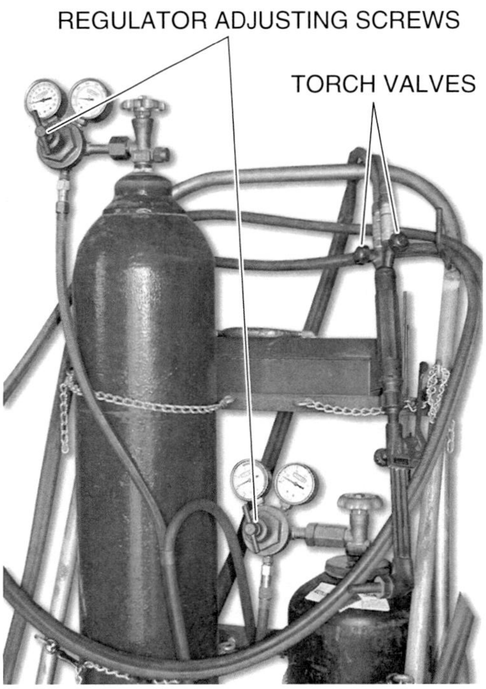

102F42.EPS

Figure 42 Torch valves and regulator adjusting screws.

> **CAUTION**
>
> Loosening regulator adjusting screws closes the regulators and prevents damage to the regulator diaphragms when the cylinder valves are opened.

Step 2 Check the oxygen regulator adjusting screw to be sure it is loose (backed out).

Step 3 Check the fuel gas regulator adjusting screw to be sure it is loose.

4.10.0 Opening Cylinder Valves

Follow these steps to open cylinder valves (*Figure 43*):

> **WARNING**
>
> Never stand directly in front of or behind a regulator. The regulator adjusting screw can blow out, causing serious injury. Always open the cylinder valve gradually. Quick openings can damage a regulator or gauge or even cause a gauge to explode.
>
> Oxygen cylinder valves must be opened all the way until the valve seats at the top. Seating the valve at the fully open position prevents high-pressure leaks at the valve stem.

Step 1 Standing on the cylinder valve side of the oxygen regulator, slowly open the oxygen cylinder valve until the desired pressure is reached. Allow the pressure in the cylinder pressure gauge to rise gradually until the gauge indicates the oxygen cylinder pressure, then open it all the way.

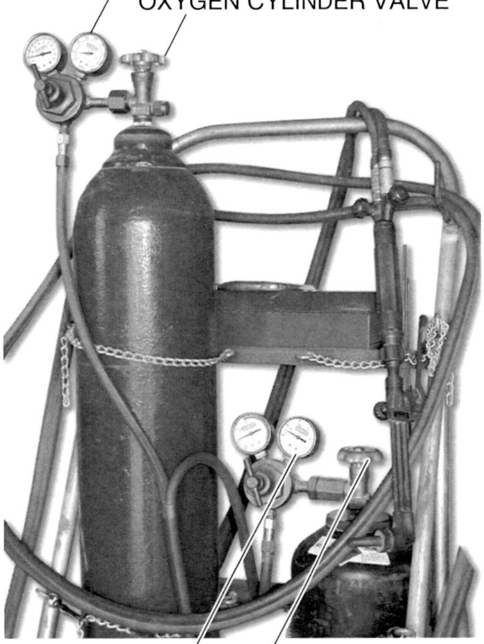

Figure 43 Cylinder valves and gauges.

> **On Site**
>
> ### Portable Oxyacetylene or Oxyfuel Equipment
>
> The equipment shown in *Figure 43* is mounted on a hand truck and is very heavy. This type of equipment is typically used in a shop or on a job site when extensive cutting must be accomplished. For small jobs, portable equipment that can be hand carried by one person may be used.

> **WARNING**
>
> When opening a cylinder valve, keep the valve stem between you and the regulator.

Step 2 Slowly open the fuel gas cylinder valve until the cylinder pressure gauge indicates the cylinder pressure, but no more than 1½ turns. This allows it to be quickly closed in case of a fire.

4.11.0 Purging the Torch and Setting the Working Pressures

Follow these steps to purge the torch and set the working pressures (*Figure 44*):

> **WARNING**
>
> The working pressure gauge readings on single-stage regulators will rise after the torch valves are turned off. This is normal. However, if acetylene is being used as the fuel gas, make sure that the static pressure does not rise above 15 psig. Make sure that equipment is purged and leak tested in a well-ventilated area to avoid creating an explosive concentration of gases.

Step 1 Fully open the oxygen valve on the torch. Then depress and hold or lock open the cutting oxygen lever.

Step 2 Tighten the oxygen regulator adjusting screw until the working pressure gauge shows the correct oxygen gas working pressure with the gas flowing. Allow the gas to flow for five to ten seconds to purge the torch and hoses of air or fuel gas.

Step 3 At the torch, release the cutting lever and close the oxygen valve.

Step 4 Open the fuel valve on the torch about ⅛ of a turn.

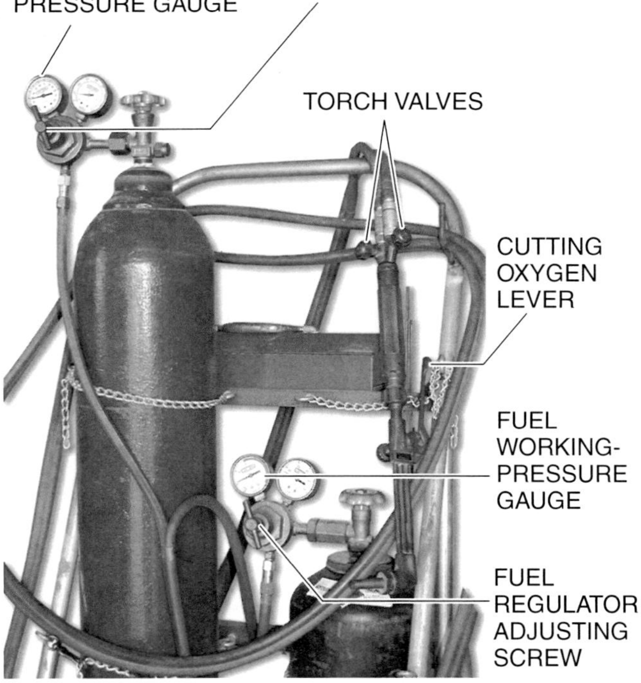

Figure 44 Typical points for purging the equipment and setting the working pressures.

On Site

Fuel Cylinder Wrench

If the fuel cylinder is equipped with a valve requiring a T-wrench, always leave the wrench in place on the valve so that the fuel can be quickly turned off. This type of valve is obsolete but still in use.

Step 5 Tighten the fuel regulator adjusting screw until the working pressure gauge shows the correct fuel gas working pressure with the gas flowing. Allow the gas to flow for five to ten seconds to purge the hoses and torch of air.

Step 6 At the torch, close the fuel valve. If acetylene is used, check that the acetylene static pressure does not rise above 15 psig. If it does, immediately open the torch fuel valve and reduce the regulator output pressure as required.

4.12.0 Testing for Leaks

Equipment must be tested for leaks immediately after it is set up and periodically thereafter. The torch should be checked for leaks each time before use. Leaks could cause a fire or explosion if undetected. To test for leaks, brush a commercially prepared leak-testing formula or a solution of detergent and water on the following points. If bubbles form, a leak is present.

> **WARNING**
>
> Make sure the detergent contains no oil. In the presence of oxygen, oil can cause fires or explosions.

Leak points include:

- Oxygen cylinder valve
- Fuel gas cylinder valve
- Oxygen regulator and regulator inlet and outlet connections
- Fuel gas regulator and regulator inlet and outlet connections
- Hose connections at the regulators, check valves/flashback arrestors, and torch
- Torch valves and cutting oxygen lever valve
- Cutting attachment connection (if used)
- Cutting tip

If there is a leak at the fuel gas cylinder valve stem, attempt to stop it by tightening the packing gland. If this does not stop the leak, mark and remove the cylinder and notify the supplier. For other leaks, tighten the connections slightly with a wrench. If this does not stop the leak, turn off the gas pressure, open all connections, and inspect the screw threads.

> **WARNING**
>
> Do not use Teflon® tape or pipe dope on these fittings because you will not get a good seal. Make sure that equipment is purged and leak tested in a well-ventilated area to avoid creating an explosive concentration of gases.

4.12.1 Initial and Periodic Leak Testing

Perform the following steps at initial equipment setup and periodically thereafter:

Step 1 Set the equipment to the correct working pressures with the torch valves turned off.

Step 2 Using a leak-test solution, check for leaks at the cylinder valves, regulator relief ports, and regulator gauge connections (*Figure 45*). Also, check for leaks at hose connections, regulator connections, and check valve/flame arrestor connections up to the torch.

4.12.2 Leak-Down Testing of Regulators, Hoses, and Torch

Perform the following steps to quickly leak test the regulators, hoses, and torch before the torch is ignited.

Step 1 Set the equipment to the correct working pressures with the torch valves turned off. Then loosen both regulator adjusting screws. Check the working pressure gauges after a minute or two to see if the pressure drops. If the pressure drops, check the hose connection and regulators for leaks; otherwise, proceed to Step 2.

Step 2 Place a thumb over the cutting tip orifices and press to block the orifices.

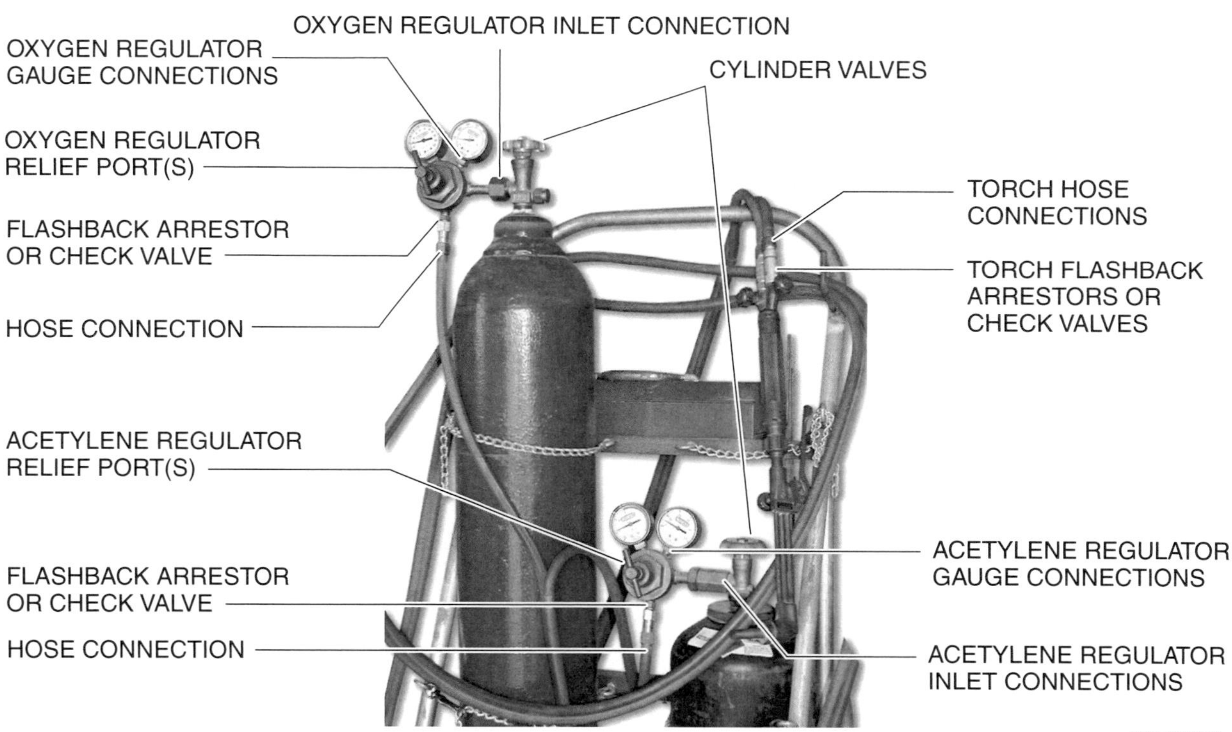

Figure 45 Typical initial and periodic leak-test points.

Step 3 Turn on the torch oxygen valve and then depress and hold the cutting oxygen lever down.

Step 4 After the gauge pressure drops slightly, observe the oxygen working pressure gauge for a minute to see if the pressure continues to drop. If the pressure keeps dropping, perform the leak test described in the following section to determine the source of the leak. If the pressure does not change, close the torch oxygen valve and release the pressure at the cutting tip.

Step 5 With the tip blocked, turn on the torch fuel valve. After the gauge pressure drops slightly, carefully observe the fuel working pressure gauge for a minute. If the pressure continues to drop, perform the leak test described in the following section to determine the source of the leak. If the pressure does not change, close the torch fuel valve and release the pressure at the cutting tip.

Step 6 If no leaks are apparent, set the equipment to the correct working pressures.

4.12.3 Full Leak Testing of a Torch

Perform the following steps to test for and isolate torch leaks (*Figure 46*).

Step 1 Set the equipment to the correct working pressures with the torch valves turned off.

Step 2 Place a thumb over the cutting tip orifices and press to block the orifices.

Step 3 Turn on the torch oxygen valve and then depress and lock the cutting oxygen lever down.

Step 4 With the cutting tip blocked, check for leaks using a leak-test solution at the torch oxygen valve, cutting oxygen lever valve, cutting attachment connection (if used), preheat oxygen valve (if used), and cutting tip seal at the torch head.

Step 5 Release the cutting oxygen lever and close the torch oxygen valve. Release the pressure at the cutting tip.

Step 6 With the cutting tip blocked, open the torch fuel valve.

Step 7 Using a leak-test solution, check for leaks at the torch fuel valve, cutting attachment (if used), and cutting tip seal at the torch head.

Step 8 Close the torch fuel valve and remove thumb from the cutting tip.

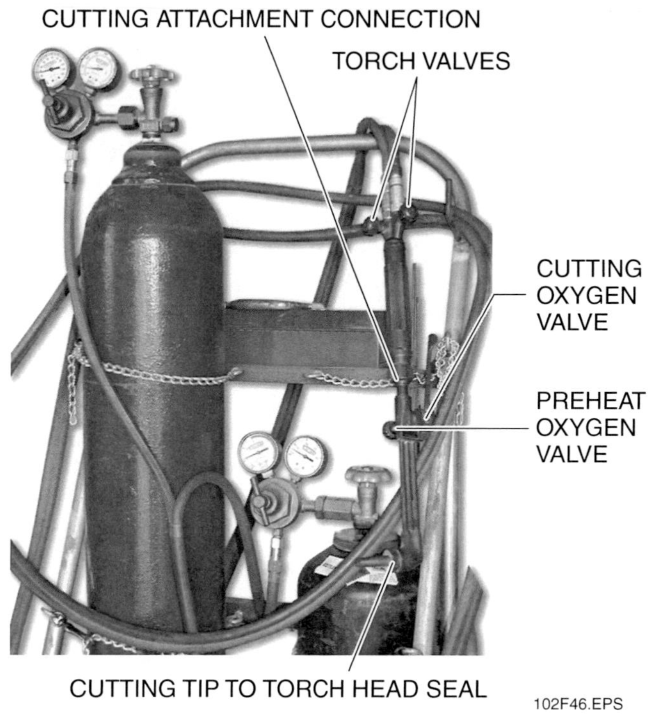

Figure 46 Torch leak-test points.

On Site

Always Leak Test a Cutting/Welding Torch

Always take the time to perform a leak test on regulators, hoses, and the torch before you use them the first time that day. Always perform a leak test on a torch after you change tips or if you convert the torch from welding to cutting or vice versa. Leaks, especially fuel leaks, can cause a fire or explosion after a cutting or welding operation begins. A dangerous fire occurring at or near the torch may not be immediately noticed by the welder due to limited visibility caused by the tinted lenses in the welding goggles or face shield.

5.0.0 Controlling the Oxyfuel Torch Flame

To be able to safely use a cutting torch, the operator must understand the flame and be able to adjust it and react to unsafe conditions. The following sections will explain the oxyfuel flame and how to control it safely.

5.1.0 Oxyfuel Flames

There are three types of oxyfuel flames: **neutral flame**, **carburizing flame**, and **oxidizing flame**.

- *Neutral flame* – A neutral flame burns proper proportions of oxygen and fuel gas. The inner cones will be light blue in color, surrounded by a darker blue outer flame envelope that results when the oxygen in the air combines with the super-heated gases from the inner cone. A neutral flame is used for all but special cutting applications.
- *Carburizing flame* – A carburizing flame has a white feather created by excess fuel. The length of the feather depends on the amount of excess fuel present in the flame. The outer flame envelope is longer than that of the neutral flame, and it is much brighter in color. The excess fuel in the carburizing flame (especially acetylene) produces large amounts of carbon. The carbon will combine with red-hot or molten metal, making the metal hard and brittle. The carburizing flame is cooler than a neutral flame and is never used for cutting. It is used for some special heating applications.
- *Oxidizing flame* – An oxidizing flame has an excess of oxygen. The inner cones are shorter, much bluer in color, and more pointed than a neutral flame. The outer flame envelope is very short and often fans out at the ends. An oxidizing flame is the hottest flame. A slightly oxidizing flame is recommended with some special fuel gases, but in most cases it is not used. The excess oxygen in the flame can combine with many metals, forming a hard, brittle, low-strength oxide. However, the preheat flames of a properly adjusted cutting torch will be slightly oxidizing when the cutting oxygen is shut off.

Figure 47 shows the various flames that occur at a cutting tip for both acetylene and LP gas.

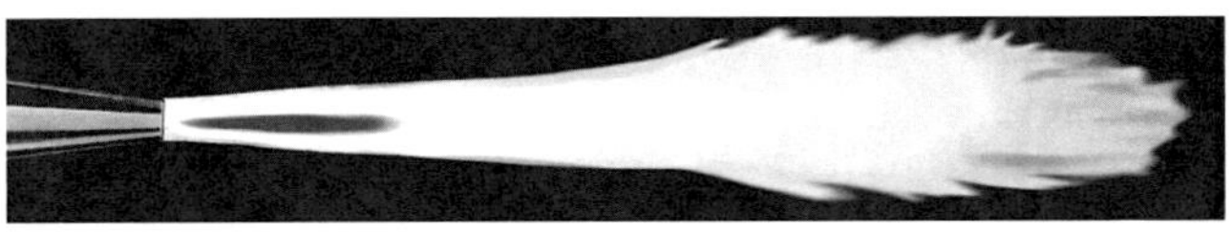

Acetylene Burning in Atmosphere
Open fuel gas valve until smoke clears from flame.

Carburizing Flame
(Excess acetylene with oxygen)
Preheat flames require more oxygen.

Neutral Flame
(Acetylene with oxygen) Temperature 5589°F (3087°C).
Proper preheat adjustment when cutting.

Neutral Flame with Cutting Jet Open
Cutting jet must be straight and clean.
If it flares, the pressure is too high for the tip size.

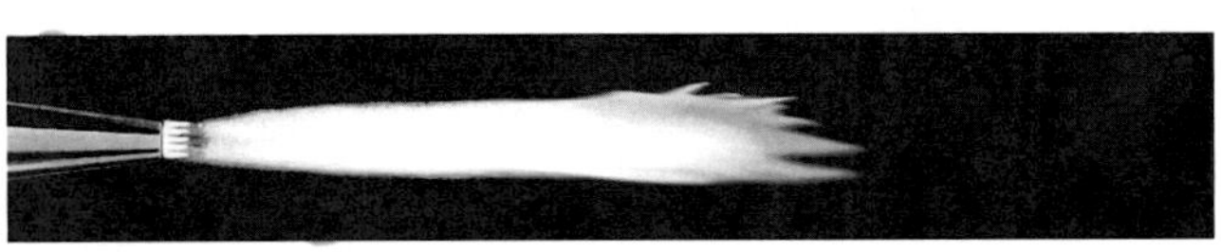

Oxidizing Flame
(Acetylene with excess oxygen) Not recommended for average cutting. However, if the preheat flame is adjusted for neutral with the cutting oxygen on, then this flame is normal after the cutting oxygen is off.

102F47A.EPS

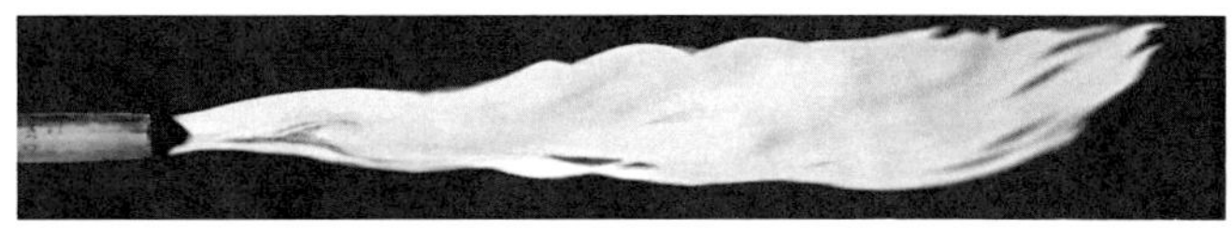

LP Gas Burning in Atmosphere
Open fuel gas valve until flame begins to leave tip end.

Reducing Flame
(Excess LP-gas with oxygen) Not hot enough for cutting.

Neutral Flame
(LP-gas with oxygen) For preheating prior to cutting.

Oxidizing Flame with Cutting Jet Open
Cutting jet stream must be straight and clean.

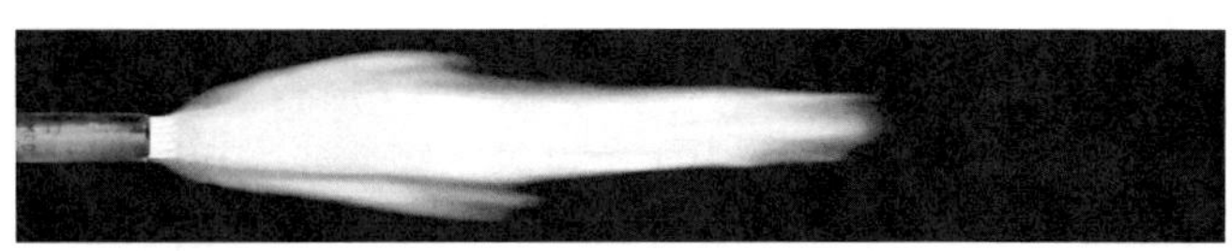

Oxidizing Flame without Cutting Jet Open
(LP-gas with excess oxygen) The highest temperature flame for fast starts and high cutting speeds.

102F47B.EPS

Figure 47 Acetylene and LP gas flames.

5.2.0 Backfires and Flashbacks

When the torch flame goes out with a loud pop or snap, a backfire has occurred. Backfires are usually caused when the tip or nozzle touches the work surface or when a bit of hot dross briefly interrupts the flame. When a backfire occurs, you can relight the torch immediately. Sometimes the torch even relights itself. If a backfire recurs without the tip making contact with the base metal, shut off the torch and find the cause. Possible causes are:

- Improper operating pressures
- A loose torch tip
- Dirt in the torch tip seat or a bad seat

When the flame goes out and burns back inside the torch with a hissing or whistling sound, a flashback is occurring. Immediately shut off the oxygen valve on the torch; the flame is burning inside the torch. If the flame is not extinguished quickly, the end of the torch will melt off. The flashback will stop as soon as the oxygen valve is closed. Therefore, quick action is crucial. Flashbacks can cause fires and explosions within the cutting rig and, therefore, are very dangerous. Flashbacks can be caused by:

- Equipment failure
- Overheated torch tip
- Dross or spatter hitting and sticking to the torch tip
- Oversized tip (tip is too large for the gas flow rate being used)

After a flashback has occurred, wait until the torch has cooled. Then, blow oxygen (not fuel gas) through the torch for several seconds to remove soot that may have built up in the torch during the flashback before relighting it. If you hear the hissing or whistling after the torch is reignited or if the flame does not appear normal, shut off the torch immediately and have the torch serviced by a qualified technician. If the torch or tip has been damaged, replace the flashback arrestors.

5.3.0 Igniting the Torch and Adjusting the Flame

After the cutting equipment has been properly set up and purged, the torch can be ignited and the flame adjusted for cutting. Follow these steps to ignite the torch:

Step 1 Choose the appropriate cutting torch tip according to the base metal thickness you will be cutting and fuel gas you are using.

> **NOTE**
> Refer to the manufacturer's charts. You may have to readjust the oxygen and fuel gas pressure depending on the tip selected.

Step 2 Inspect the tip sealing surfaces and orifices. Attach the tip to the cutting torch or cutting attachment by placing it on the end of the torch and tightening the nut.

> **NOTE**
> Some manufacturers recommend tightening the nut with a torch wrench. Others recommend tightening the nut by hand. Check the tip manual to see if the manufacturer recommends that the nut be tightened manually or with a torch wrench.

Step 3 Put on proper protective clothing, gloves, and eye/face protection.

Step 4 Release the oxygen cutting lever. If present, close the preheat oxygen valve and open the torch oxygen valve fully.

Step 5 Open the fuel gas valve on the torch handle about one-quarter turn.

Step 6 Holding the friction lighter near the side and to the front of the torch tip, ignite the torch.

> **WARNING**
> Hold the friction lighter near the side of the tip when igniting the torch to prevent deflecting the ignited gas back toward you. Always use a friction lighter. Never use matches or cigarette lighters to light the torch because this could result in severe burns and/or could cause the lighter to explode. Always point the torch away from yourself, other people, equipment, and flammable material.

Step 7 Once the torch is lit, adjust the torch fuel gas flame by adjusting the flow of fuel gas with the fuel gas valve. Increase the flow of fuel gas until the flame stops smoking or pulls slightly away from the tip. Decrease the flow until the flame returns to the tip.

Step 8 Open the preheat oxygen valve (if present) or the oxygen torch valve very slowly and adjust the torch flame to a neutral flame.

On Site

Obtaining Maximum Fuel Flow

Increasing the fuel flow until the flame pulls away from the tip and then decreasing the flow until the flame returns to the tip sets the maximum fuel flow for the tip size in use.

Step 9 Press the cutting oxygen lever all the way down and observe the flame. It should have a long, thin, high-pressure oxygen cutting jet up to 8" long extending from the cutting oxygen hole in the center of the tip. If it does not:

- Check that the working pressures are set as recommended on the manufacturer's chart.
- Clean the cutting tip. If this does not clear up the problem, change the cutting tip.

Step 10 With the cutting oxygen on, observe the preheat flame. If it has changed slightly to a carburizing flame, increase the preheat oxygen until the flame is neutral. After this adjustment, the preheat flame will change slightly to an acceptable oxidizing flame when the cutting oxygen is shut off.

5.4.0 Shutting Off the Torch

Follow these steps to shut off the torch after a cutting operation is completed:

WARNING

Always turn off the oxygen flow first to prevent a possible flashback into the torch.

Step 1 Release the cutting oxygen lever. Then close the torch or preheat oxygen valves.

Step 2 Close the torch fuel gas valve quickly to extinguish the flame.

6.0.0 Shutting Down Oxyfuel Cutting Equipment

When a cutting job is completed and the oxyfuel equipment is no longer needed, it must be shut down. Follow these steps to shut down oxyfuel cutting equipment (*Figure 48*):

Step 1 Close the fuel gas and oxygen cylinder valves.

Step 2 Open the fuel gas valve to allow gas to escape, then close it. Open the oxygen valve to allow gas to escape, then close it. Do not proceed to the next step until all pressure is released and all regulator gauges read zero.

Step 3 Back out the fuel gas and oxygen regulator adjusting screws until they are loose.

Step 4 Coil the hose and secure the torch to prevent damage.

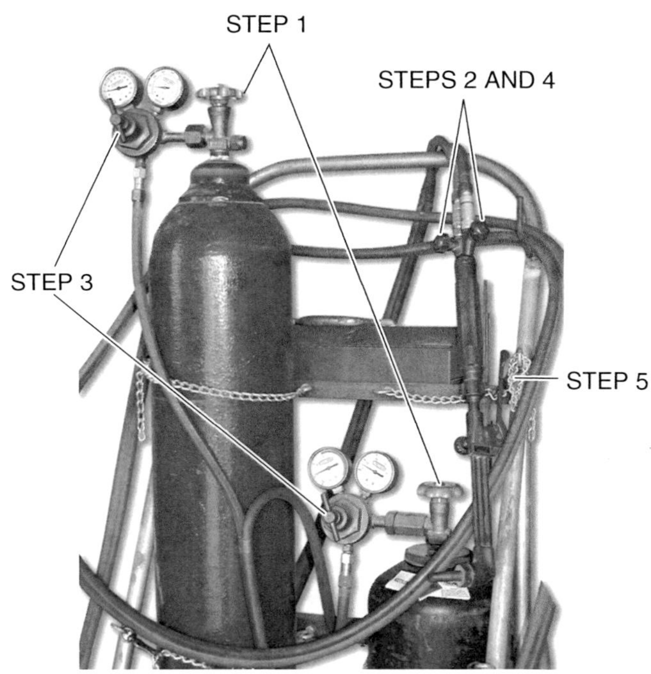

102F48.EPS

Figure 48 Shutting down oxyfuel cutting equipment.

7.0.0 DISASSEMBLING OXYFUEL EQUIPMENT

Follow these steps if the oxyfuel equipment must be disassembled after use:

Step 1 Check to be sure the equipment has been properly shut down. This includes checking that the cylinder valves are closed and all pressure gauges read zero.

Step 2 Remove both hoses from the torch.

Step 3 Remove both hoses from the regulators.

Step 4 Remove both regulators from the cylinder valves.

Step 5 Replace the protective caps on the cylinders.

Step 6 Return the oxygen cylinder to its proper storage place.

> **WARNING**
> Always transport and store gas cylinders in the upright position. Be sure they are properly secured (chained) and capped.

Step 7 Return the fuel gas cylinder to its proper storage place.

> **WARNING**
> Regardless if the cylinders are empty or full, never store fuel gas cylinders and oxygen cylinders together. Storing cylinders together is a violation of OSHA and local fire regulations and could result in a fire and explosion.

On Site

Marking and Tagging Cylinders

Do not use permanent markers on cylinders; use soapstone or another temporary marker. If a cylinder is defective, place a warning tag on it.

8.0.0 CHANGING CYLINDERS

Empty is a relative term when discussing gas cylinders. These cylinders should never be completely emptied because reverse flow could occur. Oxygen tanks should never get below the required working pressure or 25 psi. Follow these procedures to change a cylinder:

> **WARNING**
> When moving cylinders, always be certain that they are in the upright position and the valve caps are secured in place. Never use a sling or electromagnet to lift cylinders. To lift cylinders, use a cylinder cage.

Step 1 Check to be sure the equipment has been properly shut down. This includes checking that the cylinder valves are closed and all pressure gauges read zero.

Step 2 Remove the regulator from the empty cylinder.

Step 3 Replace the protective cap on the empty cylinder.

Step 4 Transport the empty cylinder from the workstation to the storage area.

Step 5 Mark MT (empty) and the date (or the accepted site notation for indicating an empty cylinder) near the top of the cylinder using soapstone (*Figure 49*).

Step 6 Place the empty cylinder in the empty cylinder section of the cylinder storage area for the type of gas in the cylinder.

102F49.EPS

Figure 49 Typical empty cylinder marking.

9.0.0 Performing Cutting Procedures

The following sections explain how to recognize good and bad cuts, how to prepare for cutting operations, and how to perform straight-line cutting, piercing, bevel cutting, washing, and gouging.

9.1.0 Inspecting the Cut

Before attempting to make a cut, you must be able to recognize good and bad cuts and know what causes bad cuts. This is explained in the following list and illustrated in *Figure 50*:

- A good cut features a square top edge that is sharp and straight, not ragged. The bottom edge can have some dross adhering to it but not an excessive amount. What dross there is should be easily removable with a chipping hammer. The **drag lines** should be near vertical and not very pronounced.
- When preheat is insufficient, bad gouging results at the bottom of the cut because of slow travel speed.
- Too much preheat will result in the top surface melting over the cut, an irregular cut edge, and an excessive amount of dross.
- When the cutting oxygen pressure is too low, the top edge will melt over because of the resulting slow cutting speed.
- Using cutting oxygen pressure that is too high will cause the operator to lose control of the cut, resulting in an uneven kerf.
- A travel speed that is too slow results in bad gouging at the bottom of the cut and irregular drag lines.
- When the travel speed is too fast, there will be gouging at the bottom of the cut, a pronounced break in the drag line, and an irregular kerf.

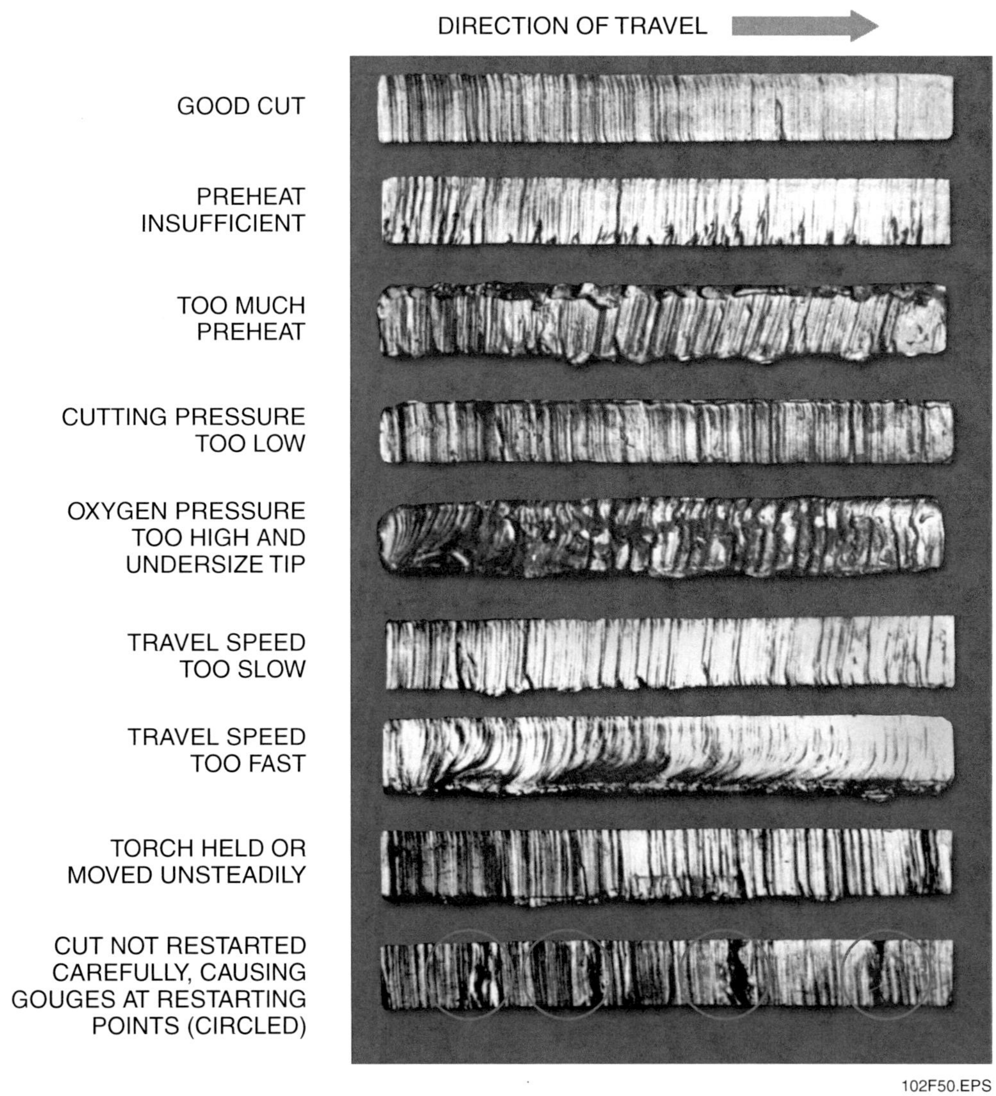

Figure 50 Examples of good and bad cuts.

- A torch that is held or moved unsteadily across the metal being cut can result in a wavy and irregular kerf.
- When a cut is lost and then not restarted carefully, bad gouges will result at the point where the cut is restarted.

NOTE

The following tasks are designed to develop your skills with a cutting torch. Practice each task until you are thoroughly familiar with the procedure. As you complete each task, take it to your instructor for evaluation. Do not proceed to the next task until your instructor tells you to continue.

9.2.0 Preparing for Oxyfuel Cutting with a Hand Cutting Torch

Before metal can be cut, the equipment must be set up and the metal prepared. One important step is to properly lay out the cut by marking it with soapstone or punch marks. The few minutes this takes will result in a quality job, reflecting craftsmanship and pride in your work. Follow these steps to prepare to make a cut:

Step 1 Prepare the metal to be cut by cleaning any rust, scale, or other foreign matter from the surface.

Step 2 If possible, position the work so you will be comfortable when cutting.

Step 3 Mark the lines to be cut with soapstone or a punch.

Step 4 Select the correct cutting torch tip according to the thickness of the metal to be cut, the type of cut to be made, the amount of preheat needed, and the type of fuel gas to be used.

Step 5 Ignite the torch.

Step 6 Use the procedures outlined in the following sections for performing particular types of cutting operations.

9.3.0 Cutting Thin Steel

Thin steel is ³⁄₁₆" thick or less. A major concern when cutting thin steel is distortion caused by the heat of the torch and the cutting process. To minimize distortion, move as quickly as you can without losing the cut. Follow these steps to cut thin steel:

Step 1 Prepare the metal surface.

Step 2 Light the torch.

Step 3 Hold the torch so that the tip is pointing in the direction the torch is traveling at a 15- to 20-degree angle. Make sure that a preheat orifice and the cutting orifice are centered on the line of travel next to the metal (*Figure 51*).

CAUTION

Holding the tip upright when cutting thin steel will overheat the metal, causing distortion.

Step 4 Preheat the metal to a dull red. Use care not to overheat thin steel because this will cause distortion.

NOTE

The edge of the tip can be rested on the surface of the metal being cut and then slid along the surface when making the cut.

Step 5 Press the cutting oxygen lever to start the cut, and then move quickly along the line to be cut. To minimize distortion, move as quickly as you can without losing the cut.

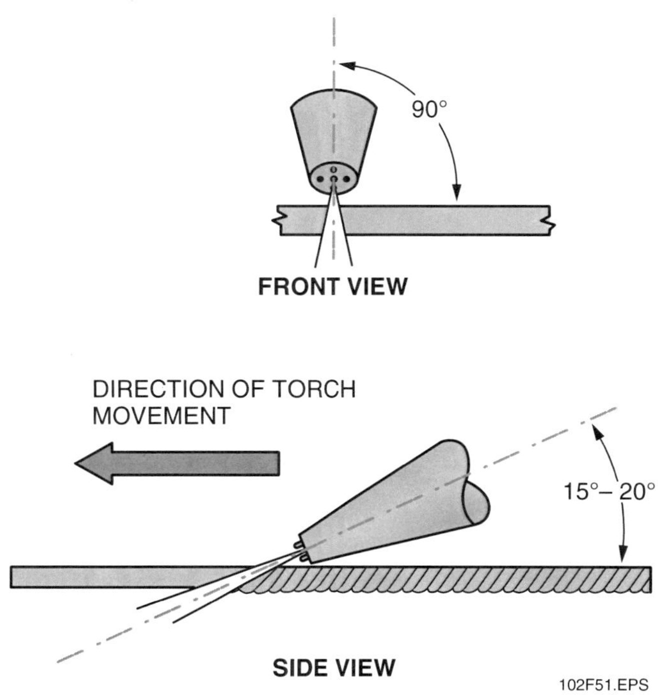

Figure 51 Cutting thin steel.

9.4.0 Cutting Thick Steel

Most oxyfuel cutting will be on steel that is more than 3⁄16" thick. Whenever heat is applied to metal, distortion is a problem, but as the steel gets thicker, it becomes less of a problem. Follow these steps to cut thick steel with a hand cutting torch:

Step 1 Prepare the metal surface and torch. Light the torch.

Step 2 Ignite and adjust the torch flame.

> **NOTE**
> The torch can be moved from either right to left or left to right. Choose the direction that allows the best visibility of the cut.

Step 3 Follow the number sequence shown in *Figure 52* to perform the cut.

> **NOTE**
> When cutting begins, the tips of the preheat flame should be held 1⁄16" to 1⁄8" above the workpiece. For steel up to 3⁄8" thick, the first and third procedures can usually be omitted.

9.5.0 Piercing a Plate

Before holes or slots can be cut in a plate, the plate must be pierced. Piercing puts a small hole through the metal where the cut can be started. Because more preheat is necessary on the surface of a plate than at the edge, choose the next-larger cutting tip than is recommended for the thickness to be pierced. When piercing steel that is more than 3" thick, it may help to preheat the bottom side of the plate directly under the spot to be pierced. Follow these steps to pierce a plate for cutting:

Step 1 Prepare the metal surface and torch.

Step 2 Ignite the torch and adjust the flame.

Step 3 Hold the torch tip 1⁄4" to 5⁄16" above the spot to be pierced until the surface is a bright cherry red (*Figure 53*).

Step 4 Slowly press the cutting oxygen lever. As the cut starts, raise the tip about 1⁄2" above the metal surface and tilt the torch slightly so that molten metal does not blow back onto the tip. The tip should be raised and tipped before the cutting oxygen lever is fully depressed.

Step 5 Maintain the tipped position until a hole burns through the plate. Then rotate the tip vertically.

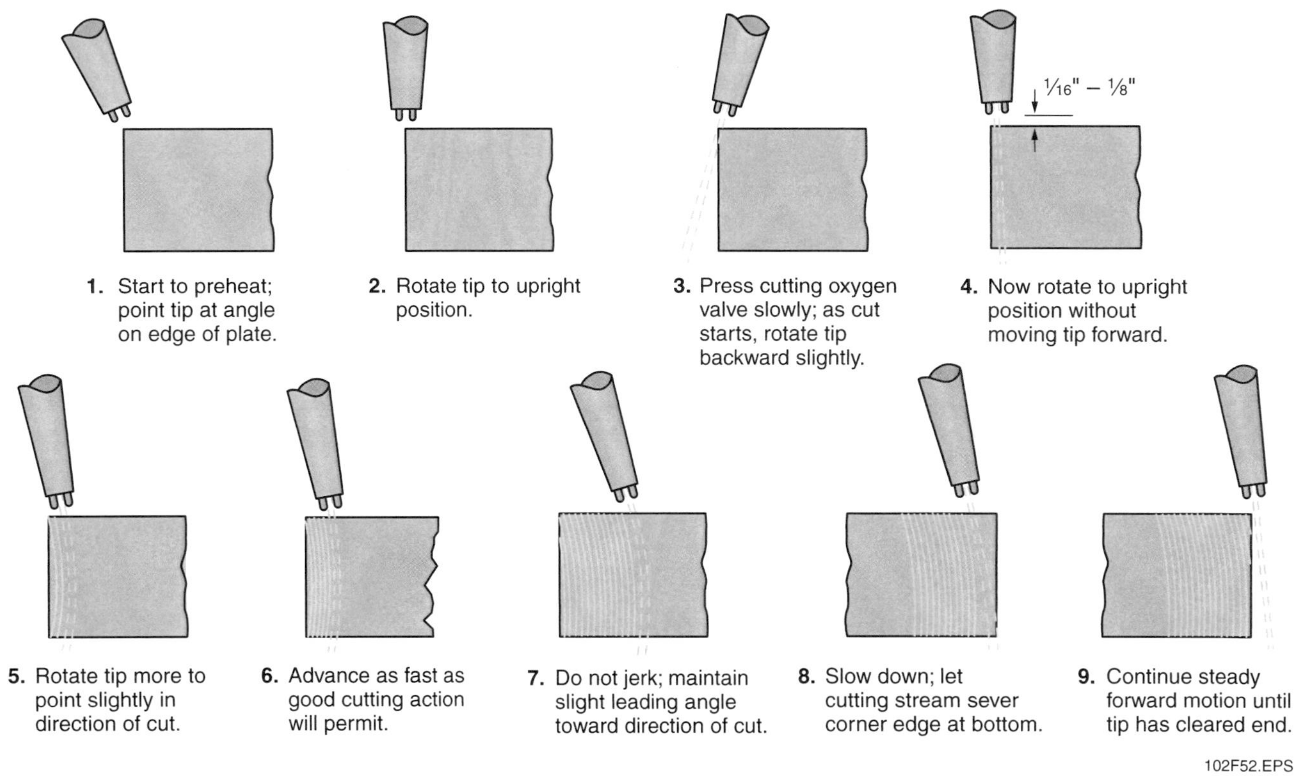

Figure 52 Flame cutting with a hand torch.

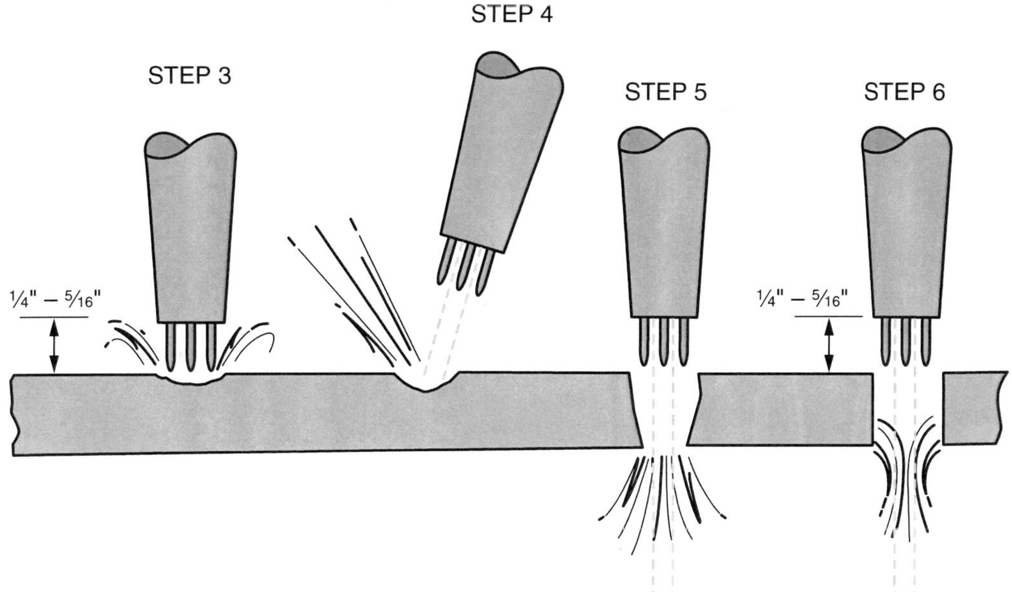

Figure 53 Steps for piercing steel.

Step 6 Lower the torch tip to about 3⁄16" above the metal surface and continue to cut outward from the original hole to the edge of the line to be cut.

9.6.0 Cutting Bevels

Bevel cutting is often performed to prepare the edge of steel plate for welding. Follow these steps to perform bevel cutting (*Figure 54*):

Step 1 Prepare the metal surface and the torch.

Step 2 Ignite the torch and adjust the flame.

Step 3 Hold the torch so that the tip faces the metal at the desired bevel angle.

> **NOTE**
> An angle iron can be used as a guide.

Step 4 Preheat the edge to a bright cherry red.

Step 5 Press the cutting oxygen lever to start the cut.

Step 6 As cutting begins, move the torch tip at a steady rate along the line to be cut. Pay particular attention to the torch angle to ensure it is uniform along the entire length of the cut.

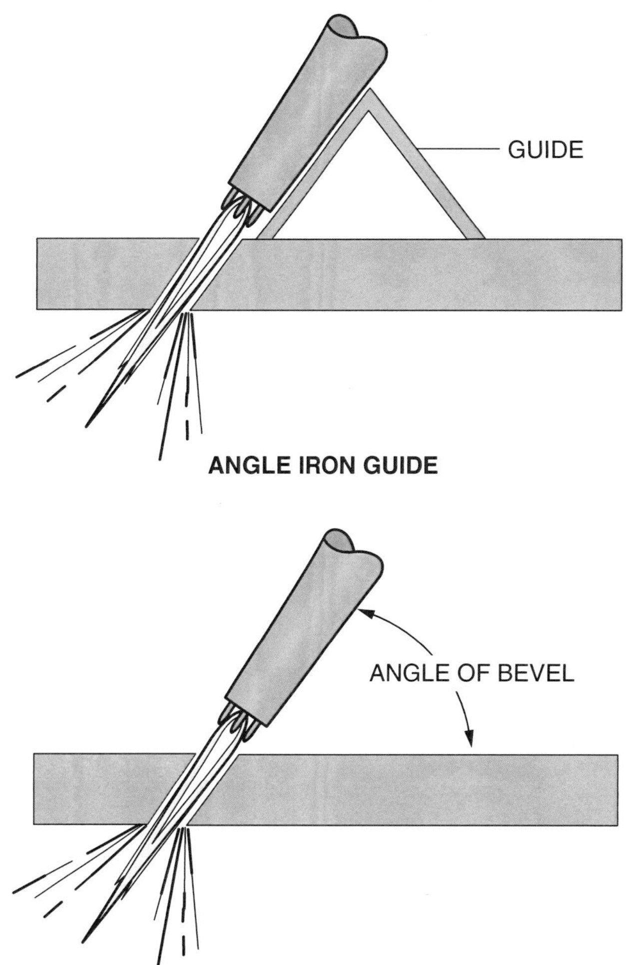

Figure 54 Cutting a bevel.

9.7.0 Washing

Washing is a term used to describe the process of cutting out bolts or rivets. Washing operations use a special tip with a large cutting hole that produces a low-velocity stream of oxygen. The low-velocity oxygen stream helps prevent cutting into the surrounding base metal. Washing tips can also be used to remove items such as blocks, angles, or channels that are welded onto a surface. Follow these steps to perform washing (*Figure 55*):

Step 1 Prepare the metal surface and torch.

Step 2 Ignite the torch and adjust the flame.

Step 3 Preheat the metal to be cut until it is a bright cherry red.

Step 4 Move the cutting torch at a 55-degree angle to the metal surface.

Step 5 At the top of the material, press the cutting oxygen lever to cut the material to be removed. Continue moving back and forth across the material while rotating the tip to a position parallel with the material. Move the tip back and forth and down to the surrounding metal. Use care not to cut into the surrounding metal.

> **CAUTION**
>
> As the surrounding metal heats up, there is a greater danger of cutting into it. Try to complete the washing operation as quickly as possible. If the surrounding metal gets too hot, stop and let it cool down.

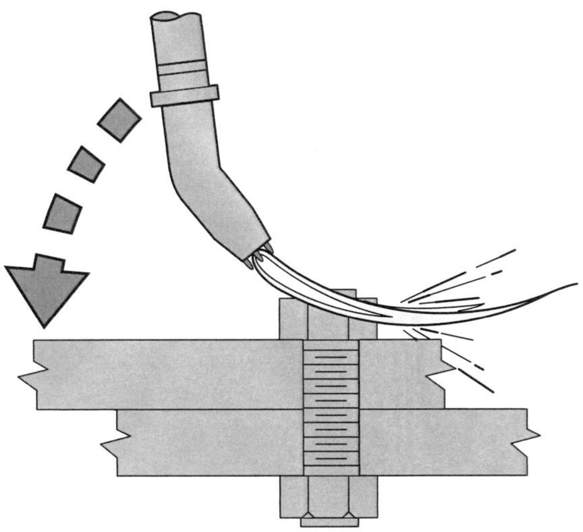

Figure 55 Washing.

9.8.0 Gouging

Gouging is the process of cutting a groove into a surface. Gouging operations use a special curved tip that produces a low-velocity stream of oxygen that curves up, allowing the operator to control the depth and width of the groove. It is an effective means to gouge out cracks or weld defects for welding. Gouging tips can also be used to remove steel backing from welds or to wash off bolt or rivet heads. Gouging tips are not as effective as washing tips for removing the shank of the bolt or rivet. Follow these steps to gouge (*Figure 56*):

Step 1 Prepare the metal surface and torch.

Step 2 Ignite the torch and adjust the flame.

Step 3 Holding the torch so that the preheat holes are pointed directly at the metal, preheat the surface until it becomes a bright cherry red.

Step 4 When the steel has been heated to a bright cherry red, slowly roll the torch away from the metal so that the holes are at an angle that will enable you to cut the gouge to the correct depth. As you roll the torch away, press the cutting oxygen lever gradually.

Step 5 Move the cutting torch along the line of the gouge. As you move the torch, rock it back and forth as necessary to create a gouge of the required depth and width.

> **NOTE**
>
> The travel speed and torch angle are very important when gouging. If the travel speed or torch angle is incorrect, the gouge will be irregular and there will be a buildup of dross inside the gouge. Practice until the gouge is clean and even.

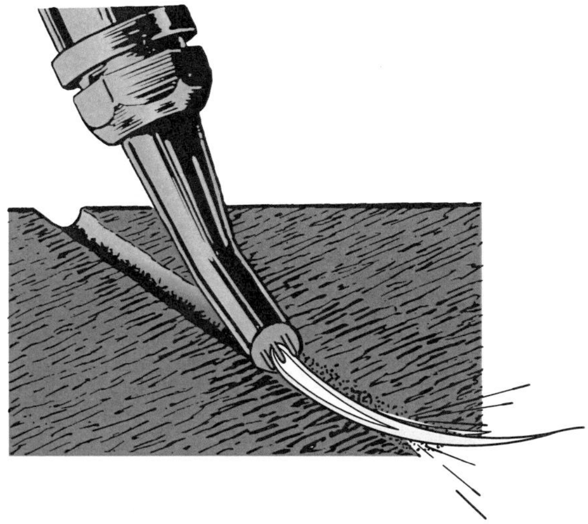

Figure 56 Gouging.

10.0.0 Portable Oxyfuel Cutting Machine Operation

As explained previously, machine oxyfuel gas cutters or track burners are basic guidance systems driven by a variable speed electric motor to enable the operator to cut or bevel straight lines at any desired speed. The device (*Figure 57*) is usually mounted on a track or used with a circle-cutting attachment to enable the operator to cut various diameters from 4" to 96". It consists of a heavy-duty tractor unit fitted with an adjustable torch mount and gas hose attachments. It is also equipped with an ON/OFF switch, a reversing switch, a clutch, and a speed-adjusting dial calibrated in feet/meters per minute.

The device shown in *Figure 57* offers the following operational features:

- Makes straight-line cuts of any length
- Makes circle cuts up to 96" in diameter
- Makes bevel or chamfer cuts
- Has an infinitely variable cutting speed from 1" to 60″ per minute
- Has dual speed and clutch controls to enable operation of the machine from either end

Figure 58 shows the location of the following controls:

- *Directional control* – Set the machine direction by toggling the FWD-OFF-REV toggle switch located next to the power cord.
- *Speed control*– Turn the large knob on either end of the machine to position the speed indicator at the desired cutting speed.
- *Clutch operation*– Engage the clutch by rotating one of the two clutch levers, located on either end of the machine, to the Drive position. Place the clutch lever in the Free position to permit easy manual positioning of the machine prior to or after the actual cutting operation.

10.1.0 Torch Adjustment

The rack assembly permits the torch holder assembly to move toward or away from the tractor unit. The torch holder allows vertical positioning of the torch. The torch bevel adjustment allows torch positioning at any angle from 190 to 290 degrees in a plane perpendicular to the track. After adjusting the torch to the desired position, tighten all clamping screws to prevent the torch from making any unexpected movements.

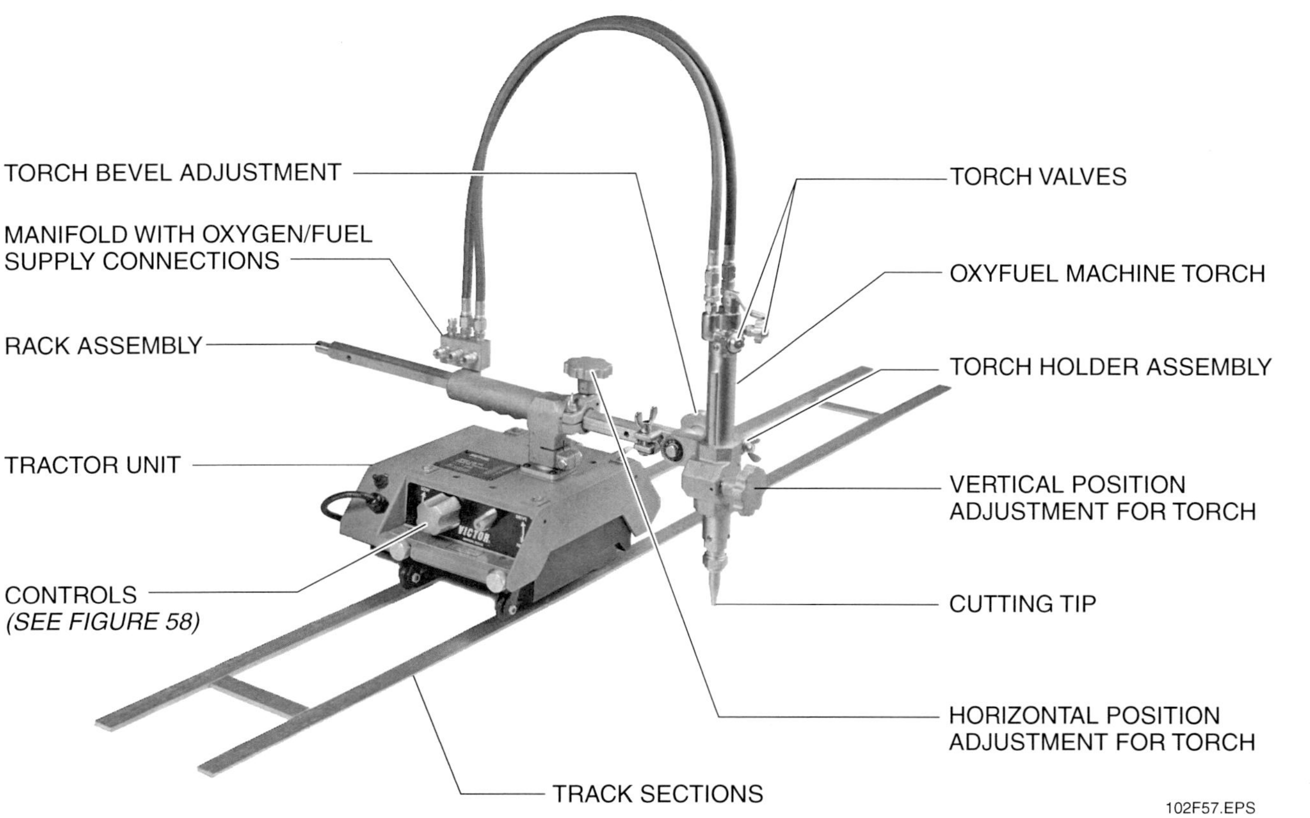

Figure 57 Victor track burner.

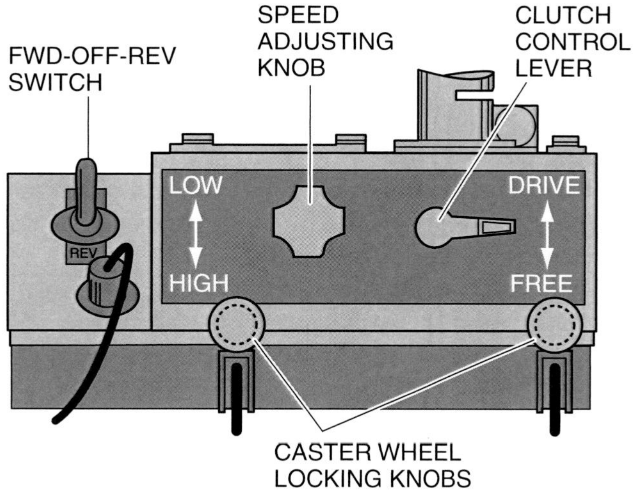

Figure 58 Victor track burner controls.

10.2.0 Straight-Line Cutting

Cut straight lines using the following procedure:

> **WARNING**
> Most cutting machines are not designed to detect the end of their track or workpiece. Take care that an unattended machine does not fall from an elevated workpiece.

Step 1 Place the machine track on the workpiece and line it up before placing the machine on the track.

Step 2 Be sure the track is long enough for the cut to be made. If not, install additional track. Connect track sections carefully. Extend the track on both sides of the cut and support the track. When properly connected, the machine should travel smoothly from one track section to the next. If the cut is long, the track may have to be clamped at both ends beyond the cut to keep the track from moving during the cut.

Step 3 Place the machine on the track. Place the clutch lever in the Free position. Be sure that the supply gas hoses and the power lines are long enough and free to move with the machine so that it can complete the cut properly.

Step 4 Move the machine to the approximate point where the cut will start. Set the drive speed control to the desired cutting speed. Set the FWD-OFF-REV switch to the Off position. Plug the power cord into a 115 alternating current (AC), 60 Hertz (Hz) power outlet.

Step 5 Ensure that all clamping screws are properly tightened. Ignite and properly adjust the torch, then preheat the start of the cut. Set the FWD-OFF-REV switch to the desired direction of travel. Simultaneously turn on the cutting oxygen and set the clutch lever to the Drive position.

Step 6 When the cut is completed, stop the machine and shut off the torch.

10.3.0 Bevel Cutting

Perform bevel cutting operations using the following procedure:

> **WARNING**
> Most cutting machines are not designed to detect the end of their track or workpiece. Take care that an unattended machine does not fall from an elevated workpiece.

Step 1 Place the machine track on the workpiece and line it up before placing the machine on the track.

Step 2 Be sure the track is long enough for the cut to be made. If not, install additional track. Connect track sections carefully. Extend the track on both sides of the cut and support the track. When properly connected, the machine should travel smoothly from one track section to the next. If the cut is long, the track may have to be clamped at both ends beyond the cut to keep the track from moving during the cut.

Step 3 Place the machine on the track. Place the clutch lever in the Free position. Be sure that the supply gas hoses and the power lines are long enough and free to move with the machine so that it can complete the cut properly.

Step 4 Loosen the bevel adjusting knob, set the torch angle to the desired bevel angle, and then tighten the bevel adjusting knob.

Step 5 Move the machine to the approximate point where the cut will start. Set the drive speed control to the desired cutting speed. Set the FWD-OFF-REV switch to the Off position. Plug the power cord into a 115AC, 60Hz power outlet.

Step 6 Ensure that all clamping screws are properly tightened. Ignite and properly adjust the torch, then preheat the start of the cut. Set the FWD-OFF-REV switch to the desired direction of travel. Simultaneously turn on the cutting oxygen and set the clutch lever to the DRIVE position.

Step 7 When the cut is completed, stop the machine and shut off the torch.

SUMMARY

Oxyfuel cutting has many uses on job sites. It can be used to cut plate and shapes to size, prepare joints for welding, clean metals or welds, and disassemble structures. Because of the high pressures and flammable gases involved, there is a danger of fire and explosion when using oxyfuel equipment. However, these risks can be minimized when the oxyfuel cutting operator is well trained and knowledgeable. Be sure you know and understand the safety precautions and equipment presented in this module before using oxyfuel equipment.

Review Questions

1. For gas welding, either the face shield or the lenses of the safety glasses must be an approved shade _____ filter.

   a. 2
   b. 3 to 6
   c. 7 to 8
   d. 9 to 10

2. Most worksite fires caused by welding or cutting are started by _____.

   a. improper disposal of dross
   b. escape of acetylene
   c. welding torches
   d. cutting torches

3. When mixed with air or oxygen, _____ will explode at much lower concentrations than any other fuel.

   a. propylene
   b. propane
   c. acetylene
   d. MAPP®

4. When pure oxygen is combined with fuel gases, the pure oxygen produces _____.

   a. argon
   b. hydrogen
   c. a colorless, odorless, and tasteless gas
   d. a high-temperature flame needed for flame cutting

5. Oxygen has a strong garlic-like odor so leaks can be readily detected.

   a. True
   b. False

6. The smallest standard oxygen cylinder holds about _____ cubic feet of oxygen.

   a. 40
   b. 65
   c. 80
   d. 85

7. Acetylene gas must be withdrawn from a cylinder at an hourly rate that does *not* exceed _____ of the cylinder capacity.

   a. ½
   b. ⅓
   c. ⅕
   d. ⅐

8. Of the fuel gases listed, _____ burns with the hottest flame.

   a. propane
   b. propylene
   c. butane
   d. MAPP®

9. The amount of liquefied gas remaining in a cylinder is determined by the _____ of the cylinder.

   a. color
   b. heat
   c. weight
   d. pressure

10. The regulators used on fuel gas cylinders usually have red markings and usually have _____ on all connections.

    a. right-hand threads
    b. left-hand threads
    c. metric threads
    d. safety latches

11. The attachment on the top of a fuel gas cylinder that allows the gas to flow only in one direction is called a _____.

    a. single-stage regulator
    b. two-stage regulator
    c. flashback arrestor
    d. check valve

12. The cutting tips used with liquefied fuel gases must have at least _____ preheat holes.

    a. four
    b. five
    c. six
    d. seven

13. When lifting oxyfuel cutting cylinders, always use a(n) _____.

    a. sling cable
    b. sling strap
    c. cylinder cage
    d. electromagnet

14. If an acetylene cylinder is found lying on its side, you should _____ before using it.
    a. contact the supplier for instructions
    b. release a small amount to the air
    c. stand it upright and wait at least one hour
    d. roll it around to stabilize the gas

15. Most fuel gas fittings have _____ threads.
    a. left-hand
    b. right-hand

16. The first step in selecting a cutting tip is to _____.
    a. inspect the cutting tip sealing surfaces and orifices for damage
    b. determine the size of cutting tip to use
    c. determine the kind of gas being used
    d. identify the thickness of the material to be cut

17. Before opening cylinder valves, verify that the adjusting screws on the oxygen and fuel gas regulators are _____.
    a. tight
    b. fully clockwise
    c. loose
    d. fully counterclockwise

18. When a cutting flame has an excess of fuel, the flame is called a(n) _____ flame.
    a. cold
    b. neutral
    c. oxidizing
    d. carburizing

19. When disassembling oxyfuel equipment, verify that all pressure gauges read _____ before starting to take the equipment apart.
    a. 21.0
    b. 0
    c. within 0.3 of 0
    d. within 0.2 of 0

20. When inspecting a completed cut made with oxyfuel cutting equipment, the drag lines of the cut should be near _____ and not very pronounced.
    a. 30 degrees
    b. 45 degrees
    c. horizontal
    d. vertical

Trade Terms Quiz

Fill in the blank with the correct term that you learned from your study of this module.

1. The gap produced by a cutting process is called a(n) ____________.
2. A loud snap or pop as a torch flame is extinguished is called a(n) ____________.
3. The soft, white material used to mark metal is ____________.
4. ____________ contain iron.
5. A flame burning with too much fuel is called a(n) ____________.
6. A flame burning with too much oxygen is called a(n) ____________.
7. When correct portions of fuel gas and oxygen are fed to a flame, the flame is said to be a(n) ____________.
8. When a thermal process is used for cutting, the material expelled from the kerf is called ____________.
9. When a flame burning back into the tip of a torch causes a high-pitched whistling sound, the condition is called ____________.
10. When metal plate is penetrated with an oxyfuel cutting torch, the action is called a(n) ____________.
11. On the edge of a cut, the lines resulting from the cutting oxygen stream into, through, and out of the metal are called ____________.

Trade Terms

Backfire
Carburizing flame
Drag lines
Dross
Ferrous metals
Flashback
Kerf
Neutral flame
Oxidizing flame
Pierce
Soapstone

Ray French

Execution Coordinator,
Special Projects
Flint Hills Resource

As a young man, Ray French got off on the wrong foot, but eventually he found his niche in the welding trade.

How did you get your start in the welding trade?
I received a college football scholarship but flunked out of college in my freshman year. I then spent a year loading ships and barges until my father suggested that I learn a trade. My uncle had a welding shop, so I decided to try welding. I really liked it! I went to welding school for two years while working part time at my uncle's welding shop.

How did your career progress after welding school?
I was hired by an oil refinery and worked in the welding shop until I received my first-class welder classification. Then I was assigned to a hot-shot team made up of welders, pipefitters, boilermakers, pump mechanics, and electricians. During my time on that team, I had the opportunity to learn a lot about each of these trades. I supervised several shutdowns—that's where a unit is taken out of service and completely overhauled—and eventually was promoted to maintenance foreman. Today, my job title is Execution Coordinator, and I supervise large projects at a major oil refinery.

What do you like most about your job?
I like the pride that comes from doing a good job and being able to see the results of my work every day. I also like the respect I get because of my skills and my knowledge of the trade.

What do you think it takes to be a success?
You have to take pride in your work. Do a quality job every time and don't take shortcuts. Your reputation is everything. It will follow you from job to job. No matter where you go, someone will know you or know someone who knows you.

Cornerstone of Craftsmanship

Drew Fontenot

Calcasieu Parish School Systems
Welding Instructor
Lake Charles, Louisiana

The seeds of a welding career took root when Drew was just a boy as he watched his dad and uncle repair and build farm equipment. As a teenager, Drew studied welding in high school. Then he went off into the world to work in various jobs—from maintenance welder in oil refineries and the petrochemical industry, to welder for general contractors on various projects, to sharing his knowledge with young welders through teaching. Drew's philosophy for a successful career in welding is simple: have diverse skills and get as much training as you can. According to Drew, this will make you competitive in the job market and an excellent candidate for advancement.

How did you choose a career in the field?
As a youngster, I was intrigued with welding, and I liked to watch my dad and uncle repair and build farm equipment. Later on in high school, I heard about a welding program that was offered starting in the tenth grade. I enrolled and took welding for the next three years. Near the end of my senior year, I got a job as a maintenance welder for Cities Service Oil Company, which later became CITGO. After graduation, I kept working for Cities Service evenings and nights, and I studied business agriculture at college during the day. By the time graduation rolled around, jobs in my major were few and far between, so I kept on working in welding. Eventually I got a job teaching welding back at my old high school. Ironically, this was the same position that had been held by the teacher who taught me welding in the first place.

What types of training have you been through?
I got my formal welding training in a high school welding program, but I also had lots of on-the-job experience. With respect to training and certifications, I am a Certified Welding Educator and Certified Welding Inspector through the AWS, and I've completed the NCCER Craft Instructor and Master Trainer programs.

What kinds of work have you done in your career?
Over the course of my career, I've worked as a maintenance welder in oil refineries and in the petrochemical industry, worked as a welder for several general contractors on a variety of projects, and done contract welding. I'm currently teaching welding at the high school level. I've served as an officer for my local AWS section, and am presently chairman of the Educational Committee for the section. I also serve as an advisor on the National AWS Educators Committee. With everything else that's going on, I still like to find the time to do welding repair work on the family farm.

Tell us about your present job and what you like about it.
In my present job, I teach the NCCER Core Curriculum, NCCER Welding Level One and Welding Level Two, and the AWS SENSE (Schools Excelling through National Skill Education) Entry Level Welder. I am also the lead SkillsUSA advisor for our school. Something I really enjoy is talking with former students and having them tell me about how successful they are in their welding careers.

What factors have contributed most to your success?
I've made it a point to participate in professional organizations such as the AWS and SkillsUSA. I've always found a lot of value in participating in these organizations—not only in the benefits offered, but especially in the opportunity to actively promote the goals of organizations like the AWS and SkillsUSA.

What advice would you give to those new to the field?

Get as much training in as many processes as you can before you start working. Once you start to work full time, you may find it difficult to go back and get additional training. The more diverse your skills are, the easier it will be for you to find work. Diverse skills and additional training will also give you a greater opportunity for advancement.

Tell us some interesting career-related facts or accomplishments.

When I was a high school senior, I had the honor of winning first place in the Louisiana State Vocational Welding Contest. I've been recipient of the AWS Section and District Educator Awards. In 2005, I was among those chosen to attend the AWS Instructors Institute in Miami, Florida. Most recently, in 2007, the Louisiana Department of Education selected me to serve as Subject Matter Expert (SME) on NCCER's Welding Committee.

Trade Terms Introduced in This Module

Backfire: A loud snap or pop as a torch flame is extinguished.

Carburizing flame: A flame burning with an excess amount of fuel; also called a reducing flame.

Drag lines: The lines on the edge of the material that result from the travel of the cutting oxygen stream into, through, and out of the metal.

Dross: The material (oxidized and molten metal) that is expelled from the kerf when cutting using a thermal process. It is sometimes called slag.

Ferrous metals: Metals containing iron.

Flashback: The flame burning back into the tip, torch, hose, or regulator, causing a high-pitched whistling or hissing sound.

Kerf: The gap produced by a cutting process.

Neutral flame: A flame burning with correct proportions of fuel gas and oxygen.

Oxidizing flame: A flame burning with an excess amount of oxygen.

Pierce: To penetrate through metal plate with an oxyfuel cutting torch.

Soapstone: Soft, white stone used to mark metal.

Additional Resources

This module presents thorough resources for task training. The following resource material is suggested for further study.

ANSI Z49.1, Safety in Welding, Cutting, and Allied Processes, American Welding Society, Miami, FL.

Figure Credits

Lincoln Electric Company, Module opener, Figure 1 (left)
J. R. Yochum, Figure 1 (right)
Topaz Publications, Inc., Figures 2, 4, 6–8, SA01–SA04, Figures 10, 12, 13, 20–23, SA08–SA10, Figures 32–46, SA13, SA14, and Figure 48
Thermadyne, Inc., SA05, Figures 15, 24B, 27, 31, 52, 57, and 58
American Torch Tip Company, SA06 and SA07
Lenco d/b/a NLC, Inc., Figure 24A
Koike Aronson, Inc., Figure 25
ESAB Welding and Cutting Products, Figure 26
Bug-O Systems, Inc., Figure 28
H&M Pipe Beveling Machine Company, Inc., Figures 29 and 30
SA11 courtesy of American Welding Society *Welding Handbook Volume 2*, Figure 14.26, Reproduced with permission from the American Welding Society, Miami, FL, Eighth Edition.
Saf-T-Cart, SA12
Smith Equipment, Figure 47
Figure 50 courtesy of American Welding Society *Welding Handbook Volume 2*, Figure 8.16, Reproduced with permission from the American Welding Society, Miami, FL, Eighth Edition.

Performance Accreditation Tasks

The Performance Accreditation Tasks (PATs) correspond to and support the learning objectives in *AWS EG2.0, Guide for the Training and Qualification of Welding Personnel: Entry-Level Welder*.

Note that in order to satisfy all learning objectives in AWS EG2.0, the instructor must also use the PATs contained in the second level in the Welding Contren® Learning Series.

PATs provide specific acceptable criteria for performance and help to ensure a true competency-based welding program for students.

The following task is designed to test your competency with an oxyfuel cutting torch. Do not perform this cutting task until directed to do so by your instructor.

SETTING UP, IGNITING, ADJUSTING, AND SHUTTING DOWN OXYFUEL EQUIPMENT

Using oxyfuel equipment that has been completely disassembled, demonstrate how to:

- Set up oxyfuel equipment
- Ignite and adjust the flame
  - Carburizing
  - Neutral
  - Oxidizing
- Shut off the torch
- Shut down the oxyfuel equipment

Criteria for Acceptance:

- Set up the oxyfuel equipment in the correct sequence ____________
- Demonstrate that there are no leaks ____________
- Properly adjust all three flames ____________
- Shut off the torch in the correct sequence ____________
- Shut down the oxyfuel equipment ____________

CUTTING A SHAPE FROM THIN STEEL

Using a carbon steel plate, lay out and cut the shape and holes shown in the figure. If available, use a machine track cutter to straight cut the 6" dimensions.

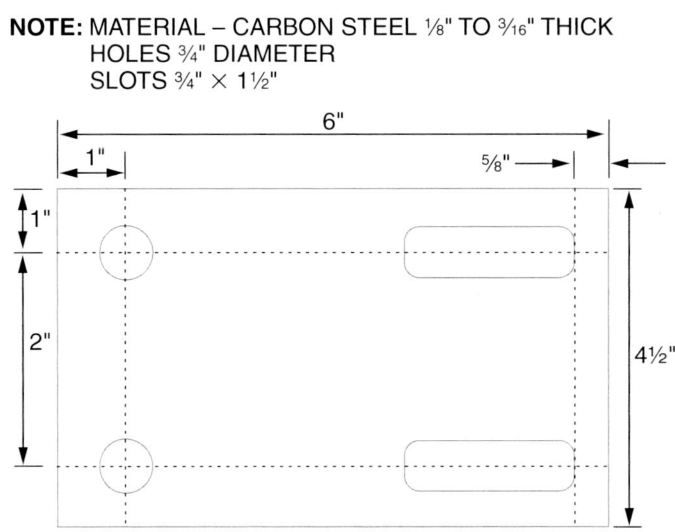

102A01.EPS

Criteria for Acceptance:

- Outside dimensions ±1/16" ______
- Inside dimensions (holes and slots) ±1/8" ______
- Square ±5° ______
- Minimal amount of dross sticking to plate which can be easily removed ______
- Square kerf face with minimal notching not exceeding 1/16" deep ______

CUTTING A SHAPE FROM THIN STEEL

Using a carbon steel plate, lay out and cut the shape and holes shown in the figure. If available, use a machine track cutter to straight cut the 6" dimensions.

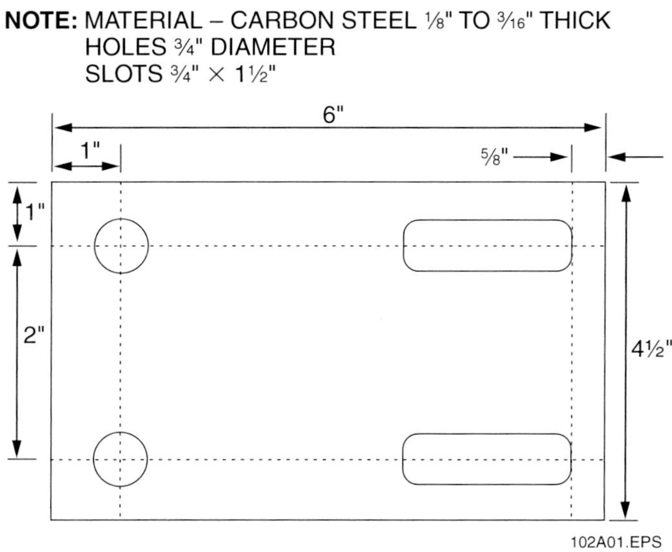

102A01.EPS

Criteria for Acceptance:

- Outside dimensions ±1/16" ________
- Inside dimensions (holes and slots) ±1/8" ________
- Square ±5° ________
- Minimal amount of dross sticking to plate which can be easily removed ________
- Square kerf face with minimal notching not exceeding 1/16" deep ________

CUTTING A SHAPE FROM THICK STEEL

Using a carbon steel plate, lay out and cut the shape and holes shown in the figure. If available, use a machine track cutter to bevel and straight cut the 6" dimensions.

NOTE: MATERIAL – CARBON STEEL OVER 1/4" THICK OR GREATER
HOLES 3/4" DIAMETER
SLOTS 3/4" × 1½"

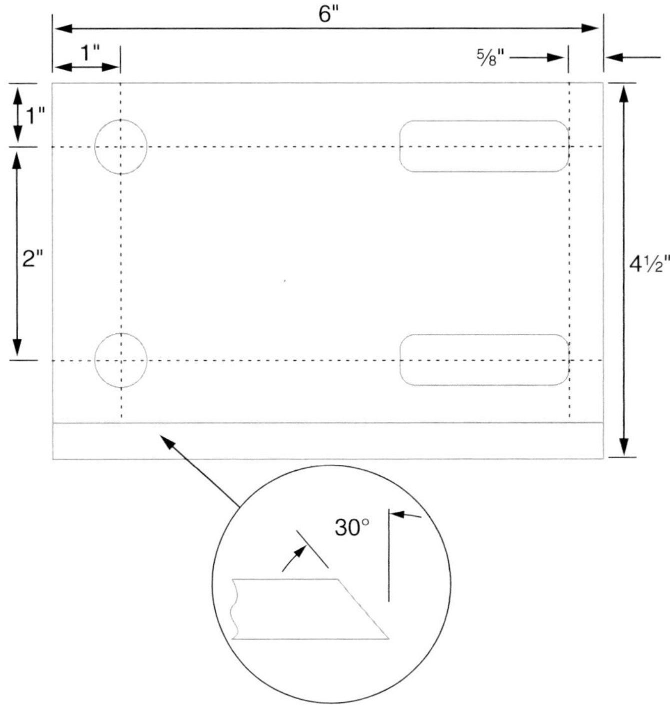

102A02.EPS

Criteria for Acceptance:

- Outside dimensions ±1/16" ________
- Inside dimensions (holes and slots) ±1/8" ________
- Square ±5° ________
- Bevel ±2° ________
- Minimal amount of dross sticking to plate which can be easily removed ________
- Square kerf face with minimal notching not exceeding 1/16" deep ________

NCCER CURRICULA — USER UPDATE

NCCER makes every effort to keep its textbooks up-to-date and free of technical errors. We appreciate your help in this process. If you find an error, a typographical mistake, or an inaccuracy in NCCER's curricula, please fill out this form (or a photocopy), or complete the online form at **www.nccer.org/olf**. Be sure to include the exact module ID number, page number, a detailed description, and your recommended correction. Your input will be brought to the attention of the Authoring Team. Thank you for your assistance.

Instructors – If you have an idea for improving this textbook, or have found that additional materials were necessary to teach this module effectively, please let us know so that we may present your suggestions to the Authoring Team.

NCCER Product Development and Revision

13614 Progress Blvd., Alachua, FL 32615

Email: curriculum@nccer.org
Online: www.nccer.org/olf

❑ Trainee Guide ❑ Lesson Plans ❑ Exam ❑ PowerPoints Other ____________________

Craft / Level: Copyright Date:

Module ID Number / Title:

Section Number(s):

Description:

Recommended Correction:

Your Name:

Address:

Email: Phone:

Introduction to Arc Welding

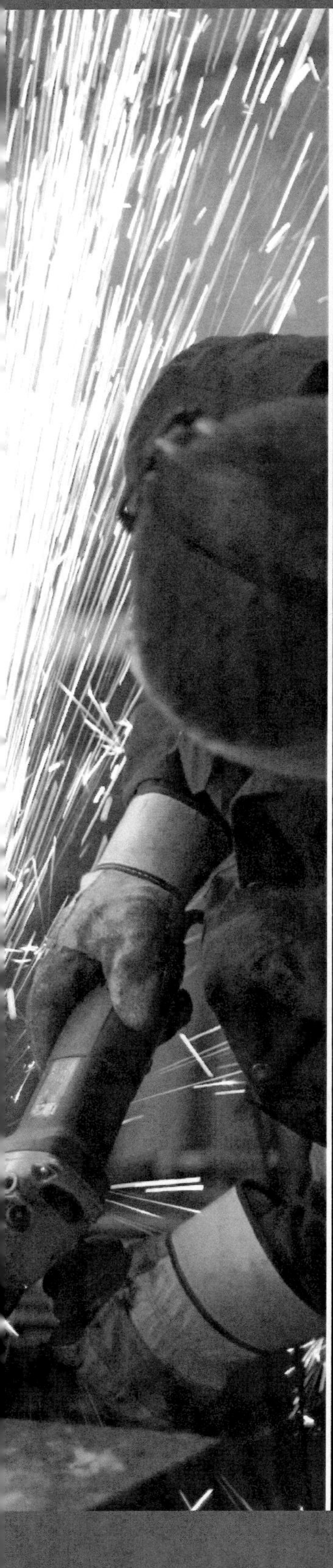

30112-11

Trainees with successful module completions may be eligible for credentialing through NCCER's National Registry. To learn more, go to **www.nccer.org** or contact us at **1.888.622.3720**. Our website has information on the latest product releases and training, as well as online versions of our *Cornerstone* newsletter and Pearson's Contren® product catalog.

Your feedback is welcome. You may email your comments to **curriculum@nccer.org,** send general comments and inquiries to **info@nccer.org**, or use the User Update form at the back of this module.

V.1 11/10

30112-11
Introduction to Arc Welding

Objectives

When you have completed this module, you will be able to do the following:

1. Identify different welding processes and welding equipment.
2. State safety precautions associated with arc welding.
3. Identify and explain shielded metal arc welding (SMAW) electrodes.
4. Identify weld joints, their dimensions, and their applications from weld symbols and drawings.
5. Explain how to set up and use SMAW equipment to weld steel.
6. Explain the different codes governing welding.

Performance Tasks

Under the supervision of the instructor, the trainee should be able to do the following:

1. Identify welding equipment.
2. Identify SMAW electrodes.
3. Identify welding joints from weld symbols and drawings.
4. Set up SMAW equipment and weld steel plate.

Trade Terms

Alloy
Alternating current (AC)
Amperage
Arc
Arc blow
Arc strike
Arrow line
Backing
Condensation
Direct current (DC)
Distortion
Drag angle
Ductile
Electrode
Flux
Ground
Heat-affected zone
Hermetically sealed
Hot work permit
Low-hydrogen electrode
Melt-through
Notch toughness
Porosity
Procedure qualification
Procedure qualification record (PQR)
Push angle
Reference line
Restart
Root face
Root opening
Root pass
Spatter
Stringer bead
Surfacing
Traceability
Vertical welding
Voltage
Weave bead
Weld axis
Weld coupon
Weld symbols
Work angle
Workpiece

Contents

Topics to be presented in this module include:

Contents (continued)

Figures and Tables

Figures and Tables (continued)

1.0.0 INTRODUCTION

Welding is a process by which intense heat is used to raise the temperature of metals above their melting point. This creates a pool of molten metal. When the heat source is removed, the pool hardens, fusing the metals. In many cases, a filler metal is added to fill the space between the metals being welded.

Many ironworkers make the effort required to become certified welders because welding is an important part of their craft and of the construction industry. During the installation of any structure, metal must be cut, shaped, and either bolted or welded. Because welds must hold and not fail, not everyone can legally weld. Government, state, local, and industry regulations specify who can legally weld on structural steels. This module only introduces the basics of welding.

2.0.0 WELDING AND CUTTING PROCESSES

Four common welding processes are used in construction and industrial work:

- Shielded metal **arc** welding (SMAW)
- Gas metal arc welding (GMAW)
- **Flux** cored arc welding (FCAW)
- Gas tungsten arc welding (GTAW)

Other methods are used to cut and prepare metal for welding. These include plasma arc cutting and carbon arc cutting, in addition to the oxyfuel cutting method covered in *Oxyfuel Cutting*.

2.1.0 Shielded Metal Arc Welding

SMAW is sometimes referred to as stick welding. It is a process that fuses two pieces of metal together using an electric arc as the heat source (*Figure 1*). As shown in *Figure 2*, one lead of a welding machine is connected to the **workpiece**. The other lead has a flux-covered metal **electrode** clamped in it. When the electrode is touched to the base metal, it short circuits. This creates an arc between the electrode and the base metal. The extreme heat of the arc melts the base metal, the electrode, and the flux coating together. Some of

112F01.EPS

Figure 1 SMAW on structural steel.

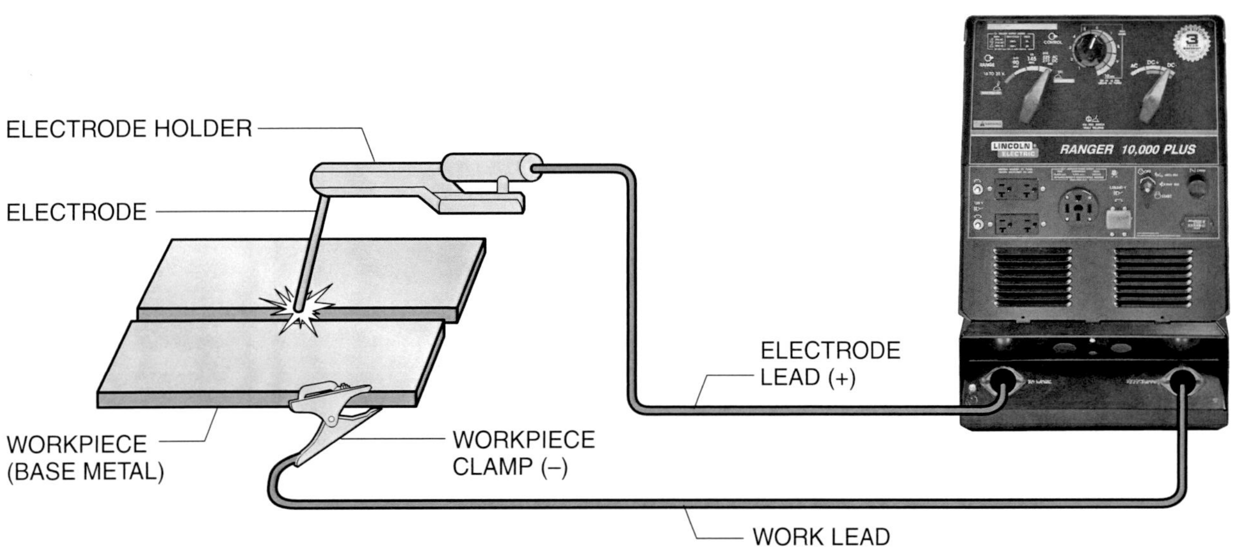

112F02.EPS

Figure 2 SMAW setup.

the flux also vaporizes, forming a gas shield that protects the arc. Melted flux floats to the surface of the weld puddle. Here it cools and forms a coating over the weld bead called slag. *Figure 3* shows the fusion process in a SMAW puddle.

2.2.0 GMAW/FCAW

GMAW (*Figure 4*) uses an electric arc to melt a consumable wire electrode and the base metal fusing them together (*Figure 5*). Note that GMAW is a constant-voltage process. The arc generates an intense heat of 6,000°F to 10,000°F. The GMAW gun nozzle dispenses shielding gas. The gas stabilizes the arc and protects the molten filler and base metals from oxidation. The wire is automatically fed into the weld puddle as it melts. The process usually operates with **direct current (DC)** electrode positive (DCEP). *Figure 6* shows equipment used in GMAW/FCAW and a typical welding setup.

FCAW (*Figure 7*) uses an electric arc to melt a flux-cored (tubular) wire electrode and fuse it with the base metal to form a weld. A wire feeder automatically feeds the flux cored wire as it is being consumed in the arc. The equipment used

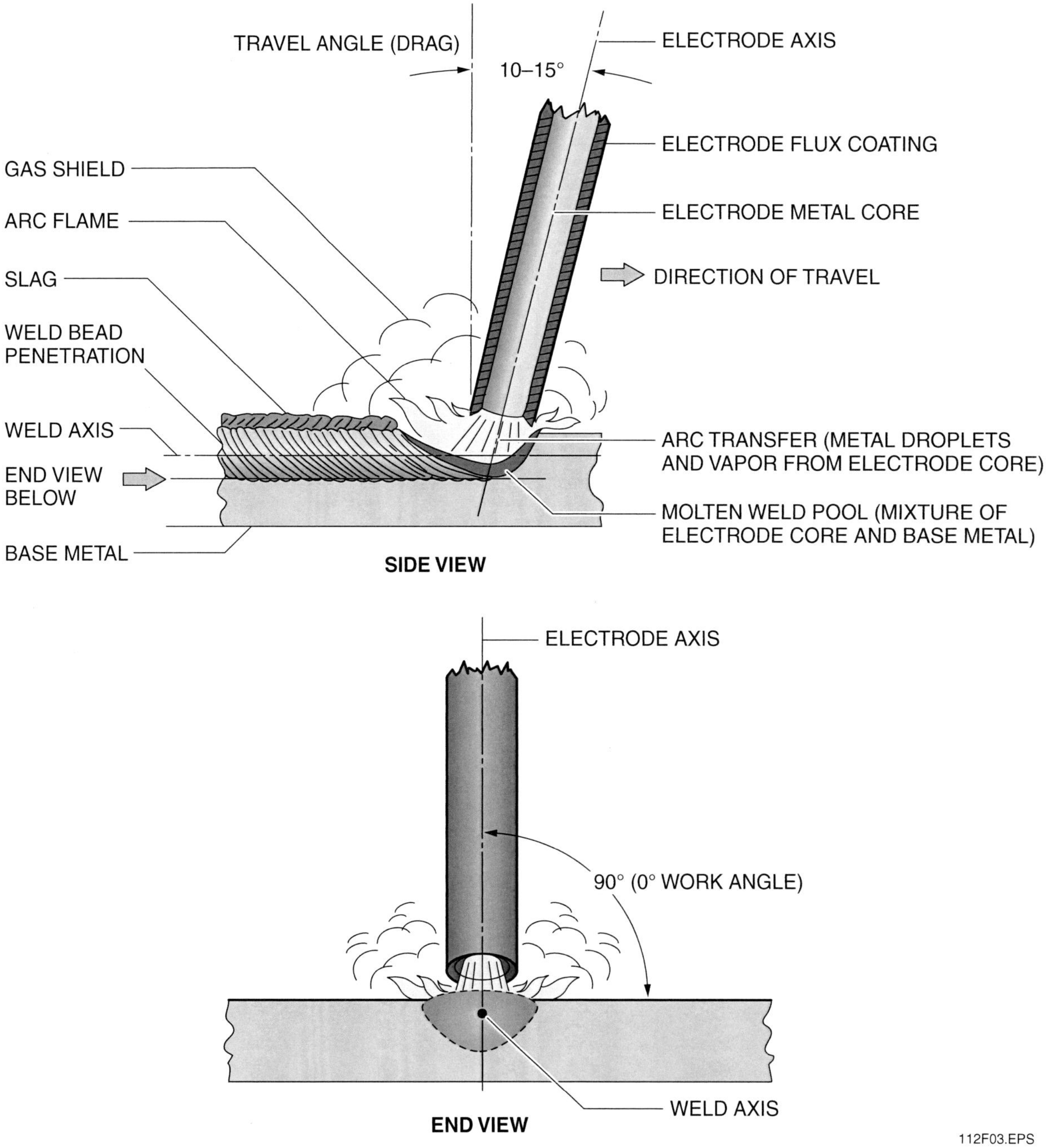

Figure 3 Fusion process in a SMAW puddle.

Figure 4 GMAW.

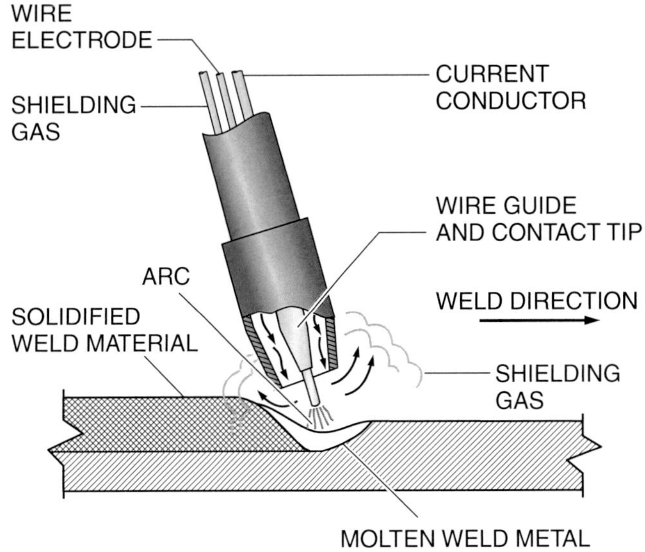

Figure 5 Electric arc.

for FCAW is basically the same as that used for GMAW. The main difference between the two processes is that GMAW uses solid wire or composite core wire electrodes within a shielding gas. FCAW, in contrast, uses a tubular flux-cored wire electrode with or without shielding gas.

There are two basic FCAW processes. One uses an external shielding gas. The other is self-shielding. The type of flux-cored wire used determines whether a shielding gas is used. When the flux-cored wire does not require an external shielding gas, the process is called self-shielding FCAW or FCAW-S. When an external shielding gas is required, the process is called gas-shielded FCAW or FCAW-G. Equipment specifically designed for self-shielding FCAW does not have provisions

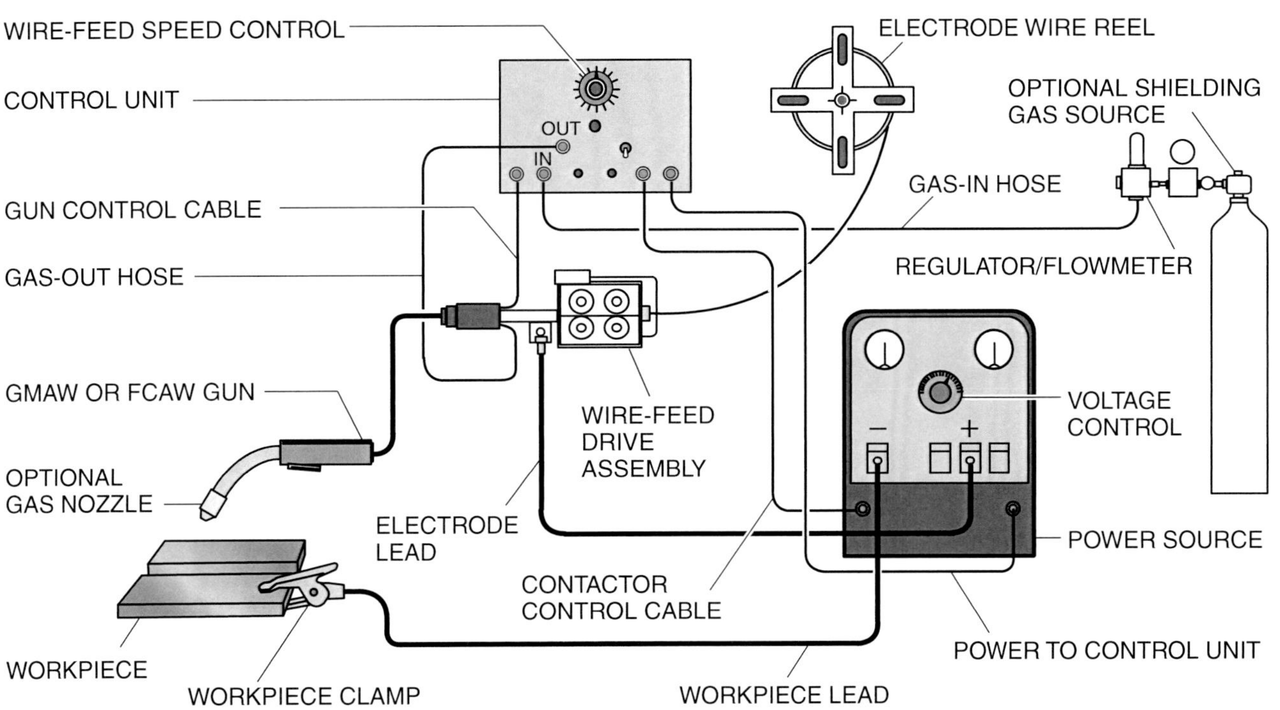

NOTE: THE POLARITY OF THE GUN AND WORKPIECE LEAD IS DETERMINED BY THE TYPE OF FILLER METAL AND APPLICATION.

112F06.EPS

Figure 6 Typical GMAW welding setup.

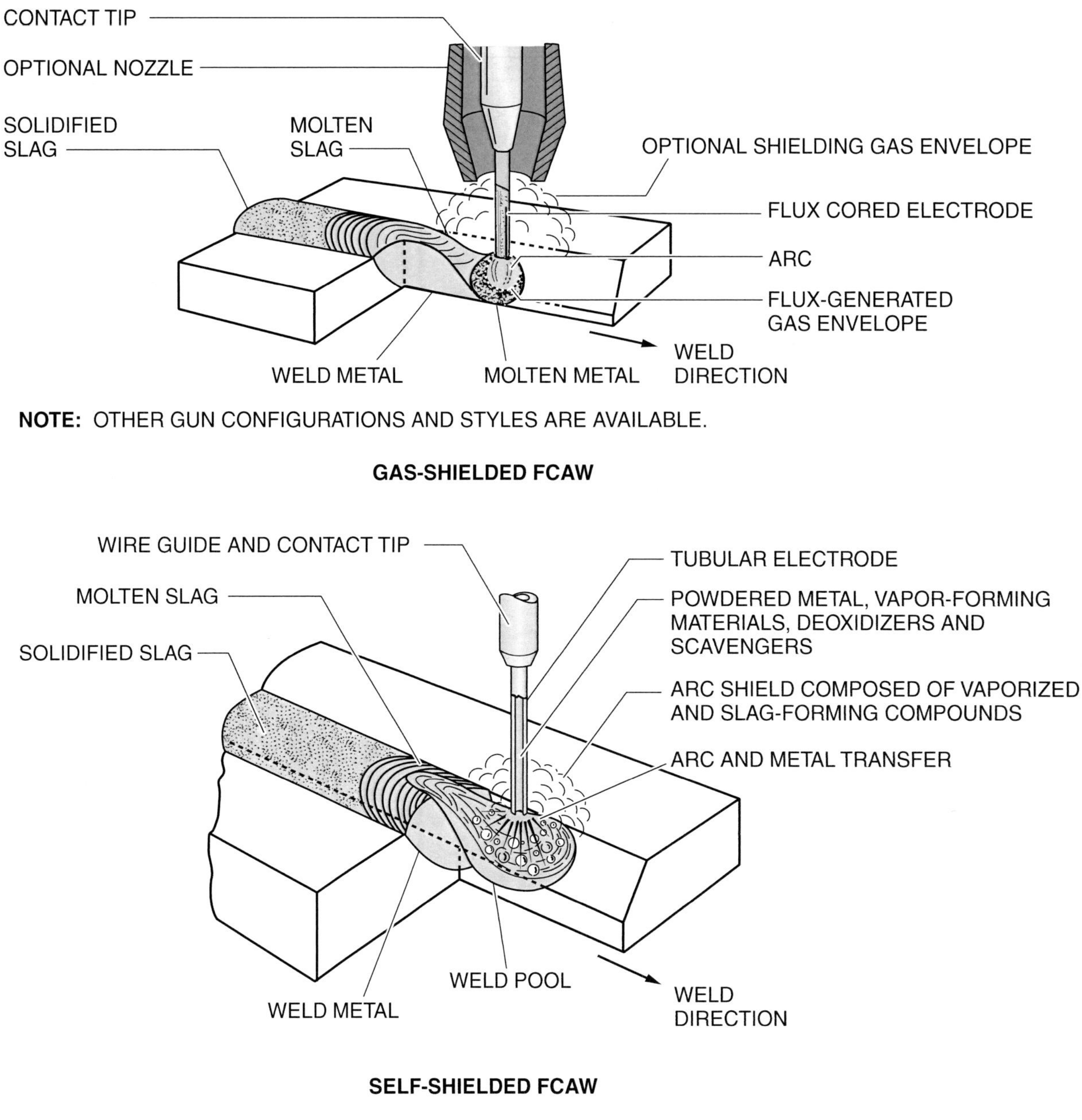

Figure 7 Basic FCAW processes.

for shielding gas. FCAW-S guns do not have gas nozzles. Equipment used for gas-shielded FCAW welding is the same as that used for GMAW except for the wire feed drive assemblies. The wire feed drive assemblies in FCAW-G have to be able to handle the core wire, which is softer and generally larger in diameter than solid wire.

2.3.0 Gas Tungsten Arc Welding

GTAW is another arc welding process. It uses a welding power supply to produce an electric arc between a relatively nonconsumable tungsten electrode and the base metal (*Figure 8*). The arc melts and fuses the base metal to form the weld. If required, a filler metal can be added. In manual welding, the filler metal is usually a handheld metal rod or wire. The wire is manually fed into the leading edge of the weld metal pool. An inert gas or gas mixture is used to shield the electrode and molten weld metals. The gas also prevents oxidation and contamination by the atmosphere. The finished weld has no slag because no flux is used. Also, since no filler metal is transported within the arc, the process produces little or no **spatter**. *Figure 9* shows the basic GTAW process.

The arc produced during GTAW is very stable. This gives a higher degree of control and mini-

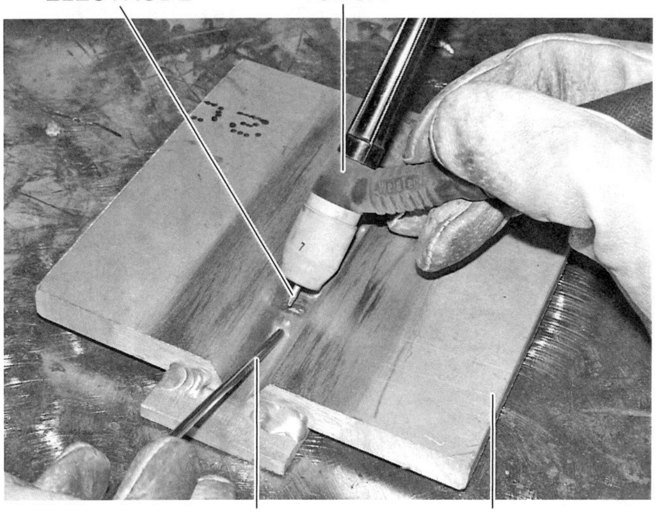

Figure 8 Performing manual GTAW.

mizes the heat-affected area. GTAW is often used on thin materials, where precise heat control is needed. Aluminum and stainless steel are often joined using GTAW.

In the past, GTAW was called tungsten inert gas (TIG) welding. Today, the nonstandard term TIG is still used, especially in the control nomenclature marked on some welding machines.

The welding arc required for GTAW is produced by electrical current from a constant-current welding power source. Electrical current is the flow of electrons along a conductor. The current flows through two welding cables. One cable is connected to the base metal (work). The other is connected to the torch, which holds a tungsten electrode. In welding, an arc is established in the gap between the end of the tungsten electrode and the work. The gap has a high resistance. When the electrical current meets resistance, heat is generated. When the arc is struck across the gap, the resistance generates intense heat of 6,000°F to 10,000°F. The heat melts the base metal and forms a weld. The area surrounding the arc is flooded with a shielding gas. The gas prevents oxidation of the filler rod and base metal surface during the welding process. If filler metal is added, the filler metal rod must be held at specific angles to produce the best weld.

NOTE

Most welding machines have a heat-activated circuit breaker that shuts off the machine automatically when it overheats. The machine cannot be turned back on until it has cooled.

Welding power sources specifically designed for GTAW have a high-frequency generator built into them (*Figure 10*). If conventional power sources are used for GTAW, a separate auxiliary high-frequency generator can be attached if required. The generators require 110VAC to operate. Some auxiliary high-frequency generators only have high-frequency controls. Others have a combination of high-frequency controls and gas and cooling water controls. Most have a torch switch that plugs into the console. The switch attaches to the torch and can be used to turn the power, shielding gas, and cooling water (if used) on or off. The auxiliary high-frequency generator is connected to the power source by the electrode cable.

2.4.0 Plasma Arc Cutting Process

Plasma arc cutting (PAC) uses a jet of plasma to pierce, cut, and gouge metal (*Figure 11*). The plasma is created by superheating gas in an electric arc. The temperature of the plasma jet is hot enough to melt any metal. The process can be set up for manual or mechanized operation. In most cases, PAC is faster and more efficient than any other cutting methods. PAC offers the following advantages:

- Cuts both ferrous and nonferrous metals
- Minimal slag
- Very thin **heat-affected zone**
- Very little or no **distortion**
- High cutting speeds

The PAC process requires an electrical power supply and a cutting gas. The cutting gas can include compressed air. Depending on the equipment, a separate shielding or cooling gas may also be required. PAC torches convert the cutting gas into plasma by passing the gas through an electric arc inside the torch. The torch has a relatively small opening (orifice) which constricts the arc and high-pressure gas flow. This results in the plasma stream exiting the torch as a supersonic jet hotter than any flame. Depending on the current flow, the plasma jet may reach temperatures of 30,000°F or higher. When the jet contacts metal, it transfers tremendous heat. This melts the metal instantly. The high-velocity jet then blasts the molten metal away, forming a hole, groove, or gouge.

2.5.0 Air Carbon Arc Cutting Process

There are three common applications of air carbon arc cutting (CAC-A):

- Cutting or gouging out defective metal

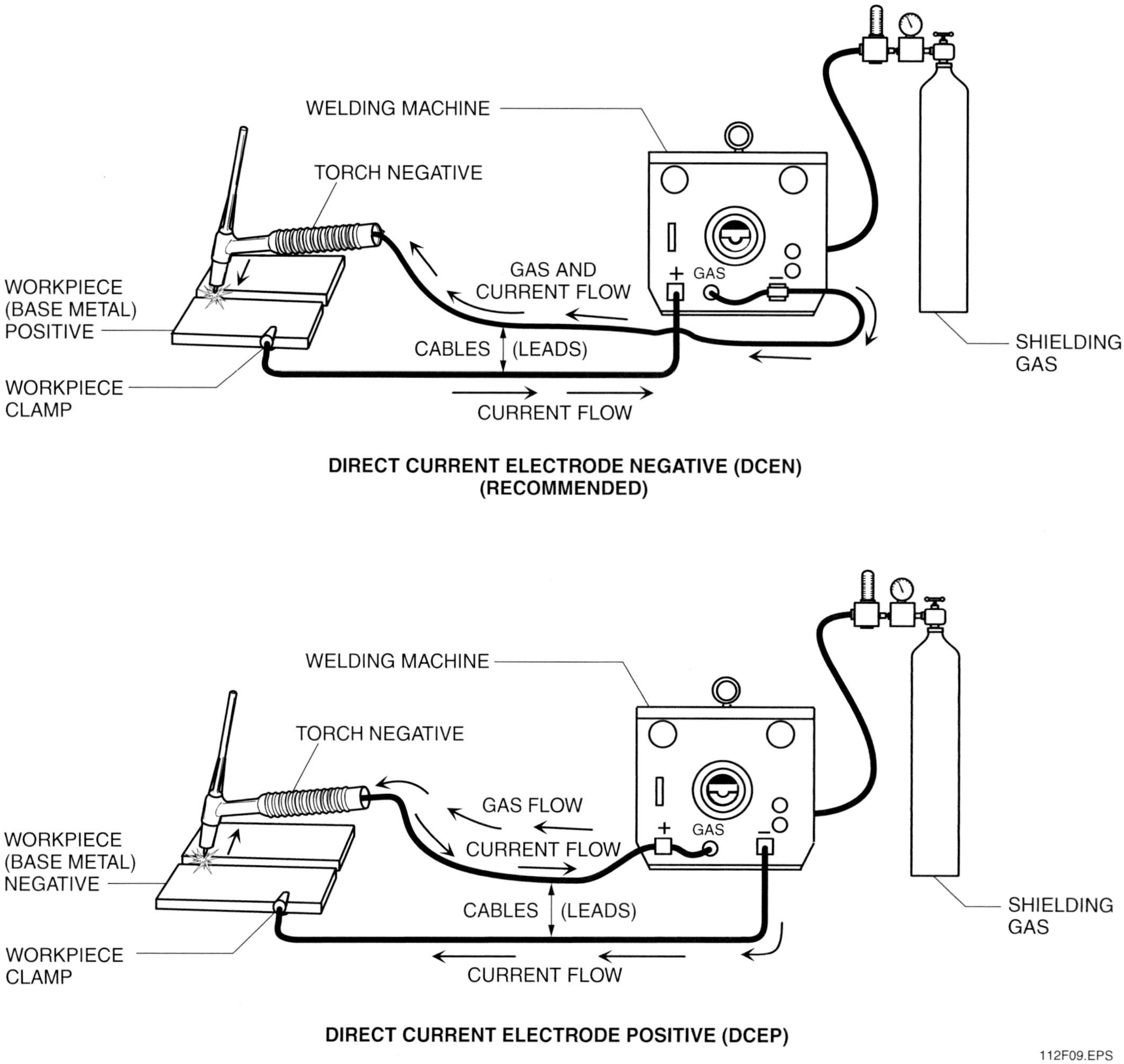

Figure 9 GTAW welding setup.

- Removing old or inferior welds
- Creating a groove for a welding pool

The CAC-A process (*Figure 12*) uses a high-current electric arc between the base metal and the end of a carbon-graphite electrode to melt the base metal. The currents used with CAC-A electrodes are much higher than those required for similarly sized SMAW electrodes. Special torch heads have air jet orifices. These orifices direct jets of compressed air along the length of the CAC-A electrode. The compressed air blows the molten metal pool away from the cut. When the electrode is moved along the surface of the base metal, a groove is made.

The depth and width of the groove are controlled by the following:

- Electrode size
- Arc **amperage**
- Electrode angle to the base metal
- Electrode advancement rate
- Electrode movement
- Regulated air pressure

2.6.0 Shielding Gas

Shielding gas provides a protective barrier around the electrode, filler metal, and weld puddle. This protective bubble prevents any slag or impurities

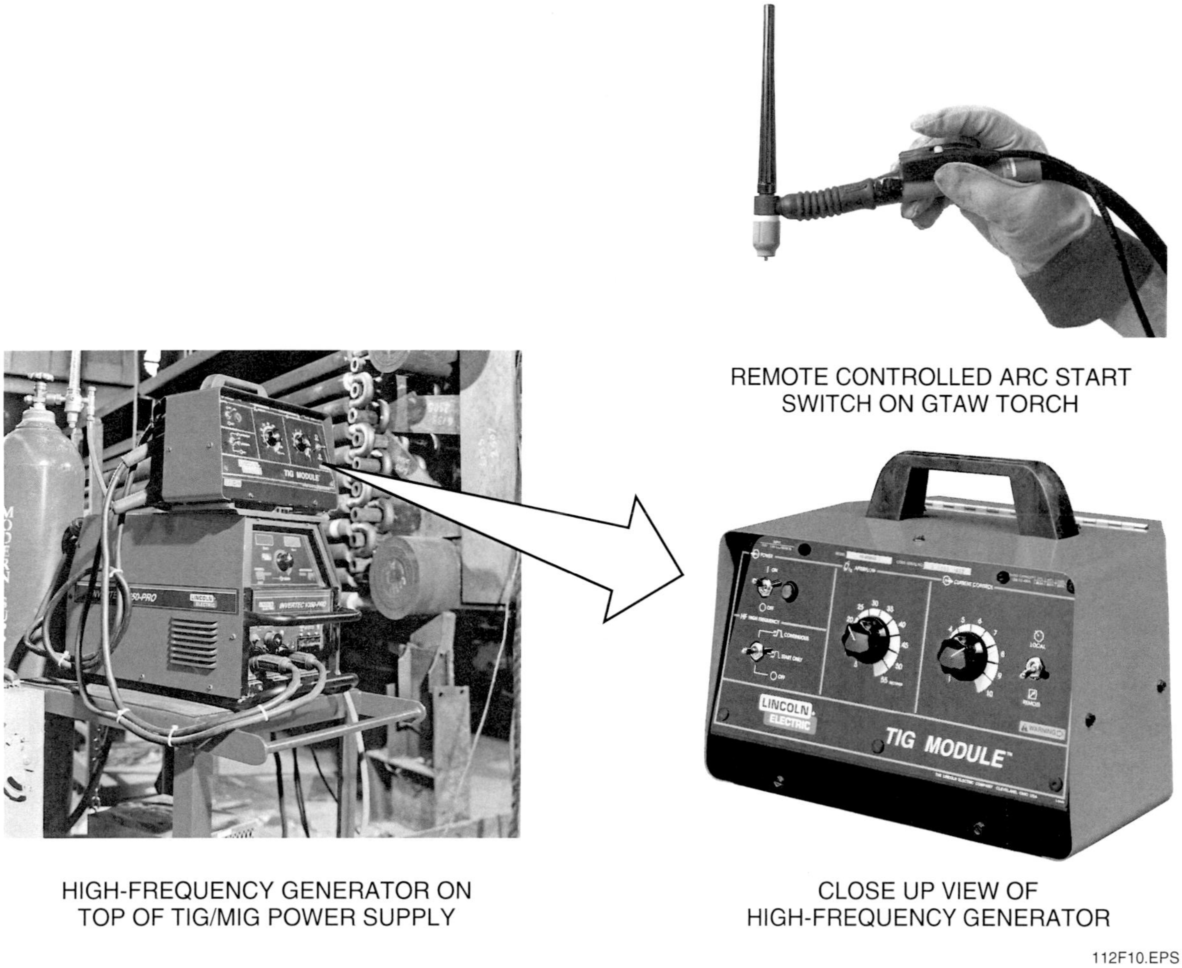

Figure 10 GTAW welding machine.

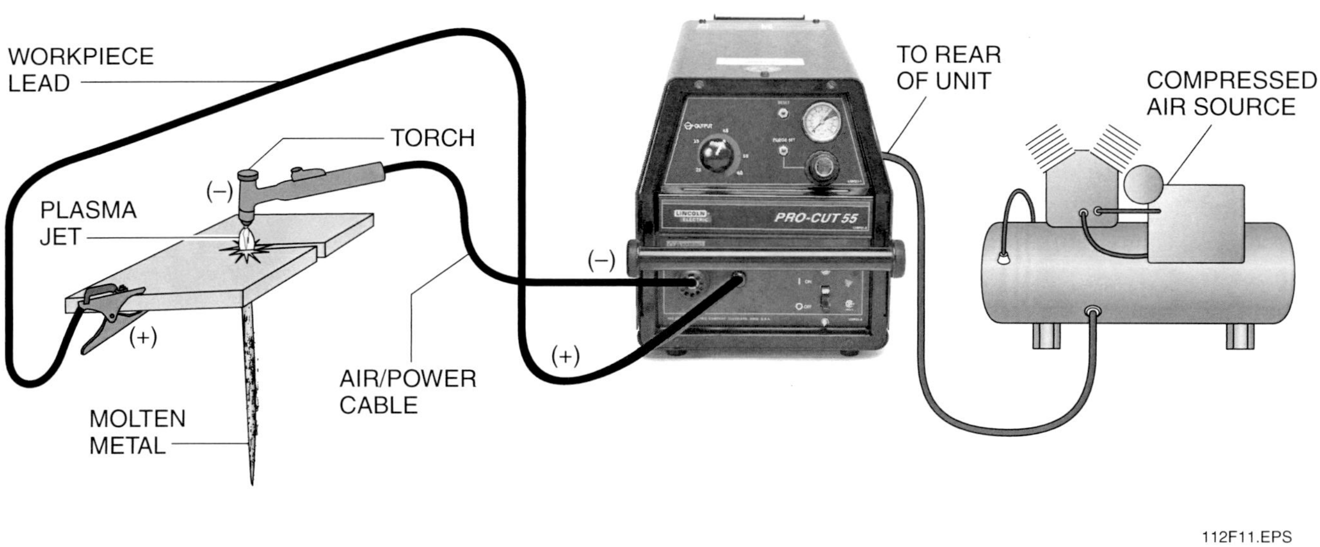

Figure 11 PAC.

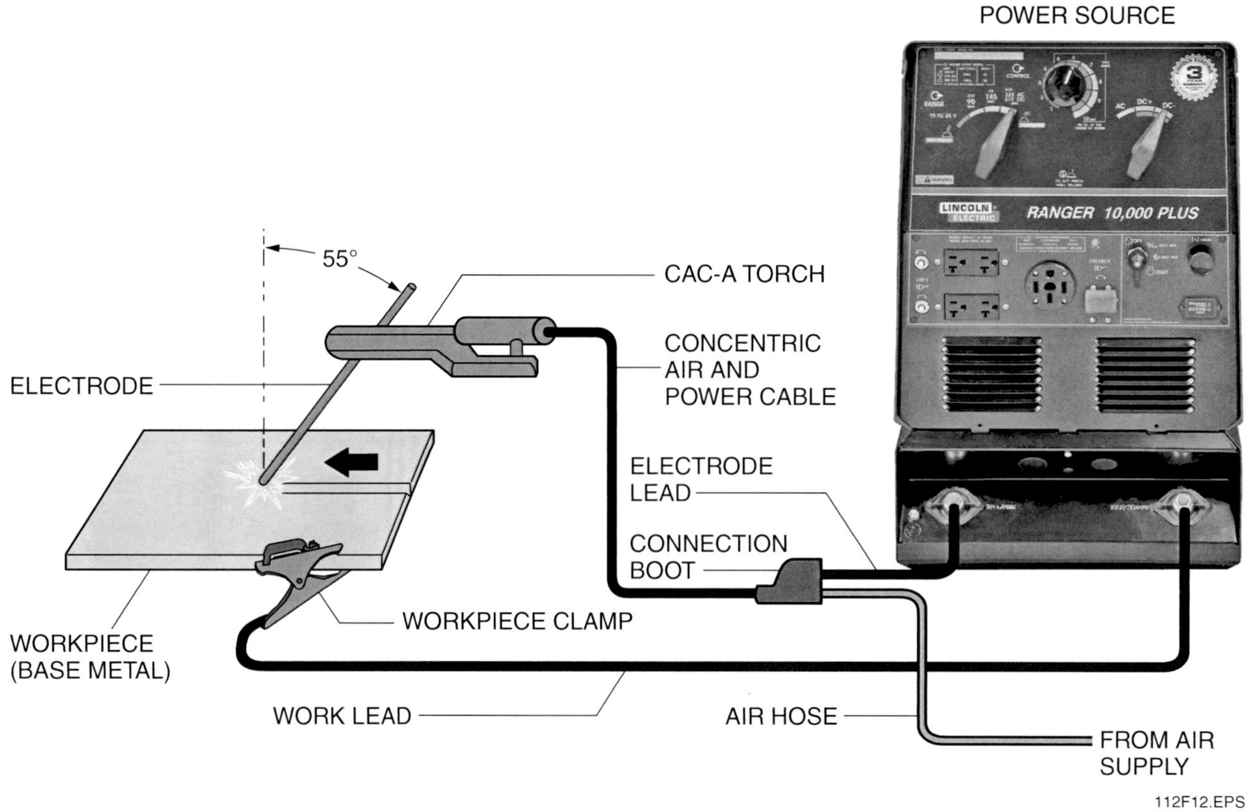

Figure 12 CAC-A.

from entering the weld puddle and contaminating the weld.

The shielding gases used are commonly known as inert gases. The term *inert* means that they will not easily react with any other substance. This protects the weld and surrounding metal during the welding process.

The GTAW and GMAW processes both use a shielding gas to protect the weld site. This gas is supplied from a storage bottle, through a pressure regulator and supply hose. It eventually exits the end of the welding torch in order to surround the weld area. The flow of gas is started and stopped by pressing the trigger on the torch or a foot pedal.

Two common shielding gases are helium and argon. Other inert gases may be used as well, but must be carefully matched to the particular weld being performed. Factors like **voltage**, base metal type, filler metal type, welding position, and base metal thickness can all affect the choice of shielding gas.

Shielding gas is also used inside of pipes, tubes, or containers being welded. The outside welds transmit heat through the walls to the inside surfaces. The trapped heat can cause the formation of scale. To prevent scale formation, openings are covered or plugged and a shielding gas is pumped in. The enclosed area is vented during the weld-

GOING GREEN

New Cutting Technologies

Manufacturers of welding and cutting equipment are constantly working on new products and methods. One such technology uses water as the heat source. A special generator separates the water into hydrogen and oxygen using electrolysis. The two gases are then recombined in the cutting torch in the same manner as oxygen and acetylene. This generator can be used in welding, cutting, soldering, and brazing applications. It also has potential for use as a fuel source for automobiles. The product is called Brown's Gas in honor of Yull Brown, the man who devoted his professional life and fortune to the development of this process (www.brownsgas.com).

Another emerging technology uses regular gasoline in place of acetylene. Like an oxyacetylene setup, the gasoline and oxygen are combined in the torch to produce the cutting flame.

ing process, but the shielding gas still covers the inside surfaces. The shielding gas used to protect the inside surfaces is often called backing gas. After the welding is completed and the weld has cooled, the covers or plugs are removed.

3.0.0 Safety

Welding and cutting can be extremely dangerous. It is important to follow proper safety practices to reduce the risk of burns, explosions, electric shock, and other hazards.

3.1.0 PPE

When performing welding or oxyfuel cutting, wear clothing that offers protection from flying sparks and heat. Shirts should be long-sleeved, have pocket flaps, and be worn with the collar buttoned. Pants should not have cuffs and should fit so they hang straight down the leg, touching the shoe tops without creases. Cuffs and creases can catch sparks, which can cause fires. Never wear polyester or other synthetic fibers. Sparks will melt these materials, causing serious burns. Wear materials treated for fire resistance or outer coverings of leather instead of synthetic fibers. *Figure 13* shows PPE.

WARNING

Never carry matches or gas-filled lighters in your pockets. Sparks could cause the matches to ignite or the lighter to explode, causing serious injury.

Never wear low-cut shoes when welding or oxyfuel cutting. Sparks and slag can enter low-cut shoes, causing burns to the feet. Wear leather work boots with high tops (at least 8"). They should fit tightly around the leg to prevent sparks from going down into the boot. A non-mesh cap, worn with the bill pointing to the back, protects the head and prevents sparks from going down the back of your shirt collar. Turning the cap so that the bill is over the ear closest to the work adds more protection to the ear and neck on that side. To protect the hands and wrists, always wear all-leather, gauntlet-type welding gloves when cutting.

For cutting, wear snug-fitting safety goggles and a face shield to protect the face and eyes from glare, heat, flying sparks, and slag. The goggles or face shield should have an approved colored filter lens. Filter lenses with shade three or six are generally recommended for most oxyfuel cutting, depending upon the thickness of the metal.

For electric arc welding, wear safety glasses and a welding hood with the correct tinted lens. A clear lens is often placed outside the colored filter plate on the welding hood to protect the tinted lens from spatter.

Earplugs should also be worn when cutting or welding. The earplugs prevent flying sparks from entering the ear and burning the ear canal or perforating the eardrum.

3.2.0 Ventilation

Always perform welding and cutting operations in a well-ventilated area, depending on the metals or welding materials used. Welding and cutting certain metals creates toxic fumes. The most common hazardous material is zinc, which is present in galvanized coatings and brass. Stainless steel or its **alloys** contain nickel and chromium 6 that can cause cancer. If doing demolition work, make sure that you know what paints are on the metals. If the demolition work is on vessels or containers, make sure you know what substances they

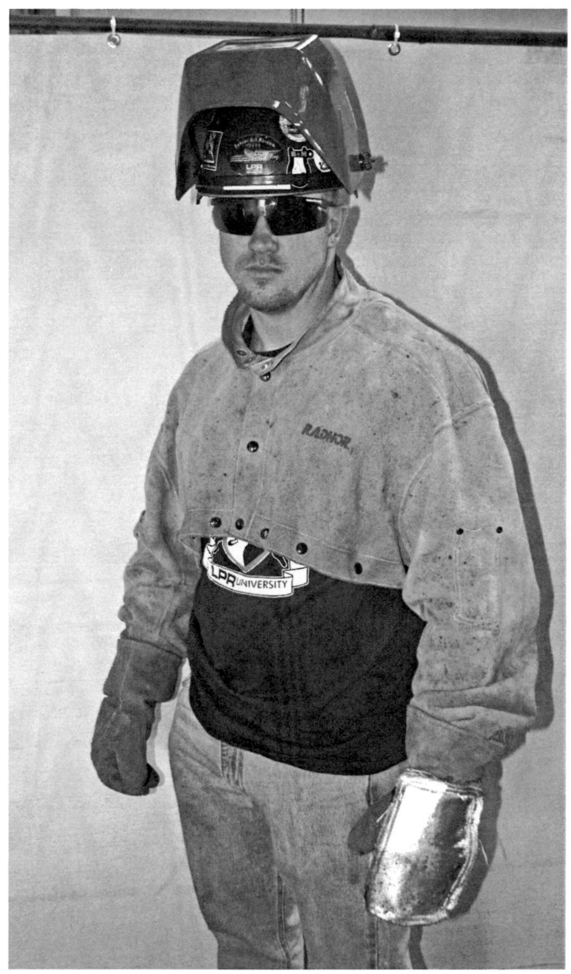

112F13.EPS

Figure 13 PPE.

once contained. Always make sure that confined spaces are properly ventilated for cutting or welding purposes. Follow all required confined space procedures. Never use oxygen in confined spaces for ventilation purposes.

> **WARNING**
> Never perform any type of heating, cutting, or welding until a hot work permit is obtained and a fire watch is established. If unsure of the procedure, check with a supervisor. Violation of the hot work permit and fire watch can result in serious injury or death.

3.3.0 Hot Work Permits and Fire Watches

Before beginning a cutting, welding, or grinding operation, check the area for fire hazards. Cutting sparks can fly 30 feet or more and can drop several floors. Any flammable material in the area must be moved or covered. In addition, have an approved fire extinguisher ready before starting any cutting operation.

Whenever oxyfuel cutting equipment is used, there is a great danger of fire. **Hot work permits** and fire watches have been developed to reduce this danger. Most sites require the use of hot work permits and fire watches. When they are violated, severe penalties are imposed. Check the site requirements regarding fire watches and hot work permits before starting any cutting operations.

A hot work permit is an official authorization from the site manager to perform work that may pose a fire hazard. The permit will include information such as the time, location, and type of work being done. The hot work permit system promotes the development of standard fire safety guidelines. Permits also help managers keep records of who is working where and at what time. This information is essential in the event of an emergency or other times when personnel need to be evacuated.

> **WARNING**
> Never cut galvanized or cadmium material without proper ventilation. The zinc oxide or cadmium fumes given off as the material is cut are hazardous. Also, use a respirator when cutting galvanized or cadmium material.

During a fire watch, one person other than the cutting operator or welder constantly scans the work area for fires. Fire watch personnel should have ready access to fire extinguishers and alarms, and know how to use them. Cutting or welding operations should never be performed without a fire watch.

3.4.0 Cutting Containers

Before welding or cutting containers such as tanks or barrels, check to see if they have contained combustible and/or hazardous materials or residues of these materials. Such materials include the following:

- Petroleum products, such as gasoline or oil
- Chemicals that give off toxic fumes when heated
- Acids that could produce hydrogen gas as a result of a chemical reaction
- Any explosive or flammable residue

To identify the contents of containers, check the label. Then refer to the material safety data sheet (MSDS). The MSDS provides information that will help you determine if the material is hazardous. If the container label is missing, do not proceed until the material is identified. Make sure that the container label matches the MSDS data. Container labels have been known to differ from the official MSDS data. Also, do not proceed if the container may have been used to hold materials other than what is on the label.

If a container has held any hazardous materials, it must be cleaned before cutting or welding takes place. Clean containers by steam cleaning, flushing with water, or washing with detergent until all traces of the material have been removed.

> **WARNING**
> Clean containers only in well-ventilated areas. Vapors can collect during cleaning, causing explosions or injury.

After cleaning the container, fill it with water or an inert gas, such as argon, nitrogen, or carbon dioxide (CO_2), for additional safety. Air, which contains oxygen, is displaced from inside the container by the water or inert gas. Without oxygen, combustion cannot take place. When using water, position the container to reduce the air space. When using an inert gas, provide a vent hole so the inert gas can purge out the air containing oxygen.

Figure 14 shows how to use water or inert gas in a container to reduce free air space.

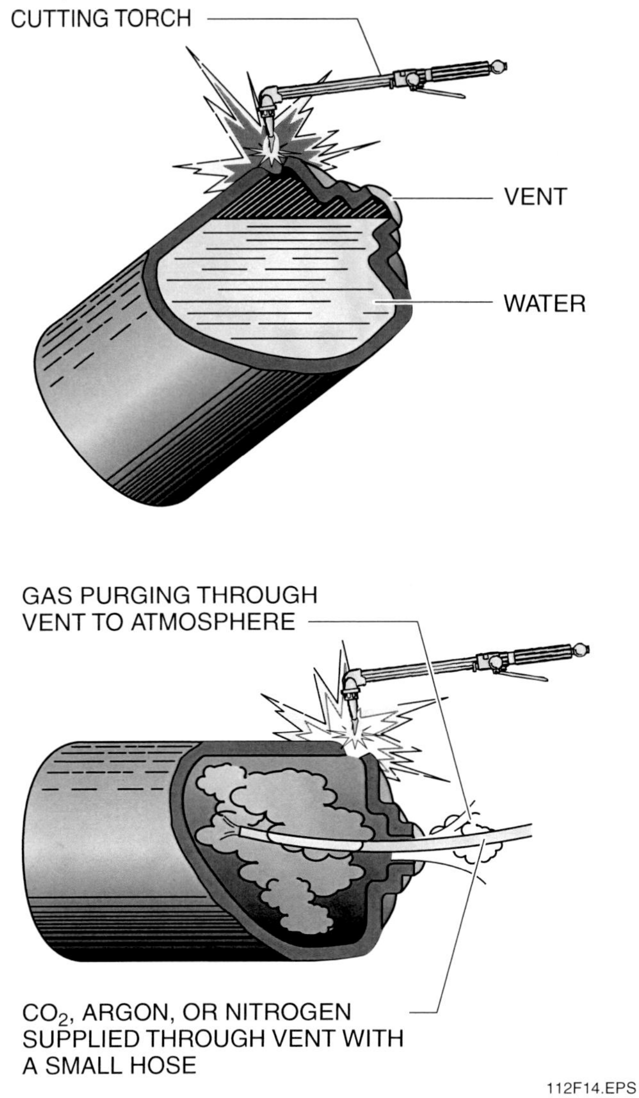

Figure 14 Minimizing free air space.

3.5.0 Oxygen Hazards

Oxygen supports combustion. Its presence around flammable materials or sparks can create conditions that encourage flames to burn out of control. Therefore, keep oxygen away from combustible materials and sources of flame. Also, keep it away from substances containing oil, grease, or other petroleum products. Never use the oxygen in cylinders as compressed air. The oxygen can react with the oil residue in air tools, causing an explosion. Do not use oxygen to blow away dust or debris. Even a small spark will cause an explosion.

WARNING

Never use oxygen to blow off clothing. If a spark hits the clothing, the oxygen trapped in the clothing will cause it to burn rapidly and out of control.

3.6.0 Electrical Safety

Electric shock from welding and cutting equipment can kill or cause severe burns to bare skin. Serious injury can also result if a fall occurs because of the shock. The amount of current that passes through the human body determines the outcome of an electrical shock. The higher the current, the greater the chance exists for a fatal shock. Electrical current follows any path to its source. Electrical equipment that is properly **grounded** provides a path for the current to flow that has less resistance than a human body.

If you come in contact with a live conductor, you become a load. *Figure 15* shows how much resistance the human body presents under various conditions and how this converts to amps or milliamps when the voltage is 110V. Note that the potential for shock increases greatly if the skin is damp. A cut also reduces your resistance and increases the chance of a shock. Currents of less than one amp can severely injure and even kill a person.

WARNING

Always follow these electrical safety precautions:

- Never operate arc welding equipment on a wet or damp floor.
- The electrode, electrode holder, and work circuit are electrically energized when the output is on. Never touch the bare metal of an electrode holder and a grounded surface at the same time.
- Never place electrode holders in contact with a grounded metal surface. Doing so could short-circuit the welding machine.
- Floors must be kept dry. Use wooden platforms or rubberized carpet/floor coverings or other insulated material.
- Only experienced electricians may work on electric arc welding machine power connections.
- Ensure that workbenches are grounded.
- With the power off, check cable connections. Look for cracked insulators, loose contacts, and worn or cut hoses.
- Never operate a machine above its rated capacity.

Another electrical area that ironworkers must be concerned with are the areas under, over, or near high-voltage lines, transformers, and switching devices. High-voltage lines and transformers in particular create electromagnetic fields. Those electrical fields can be induced into nearby steel structures. Temporary grounding devices can

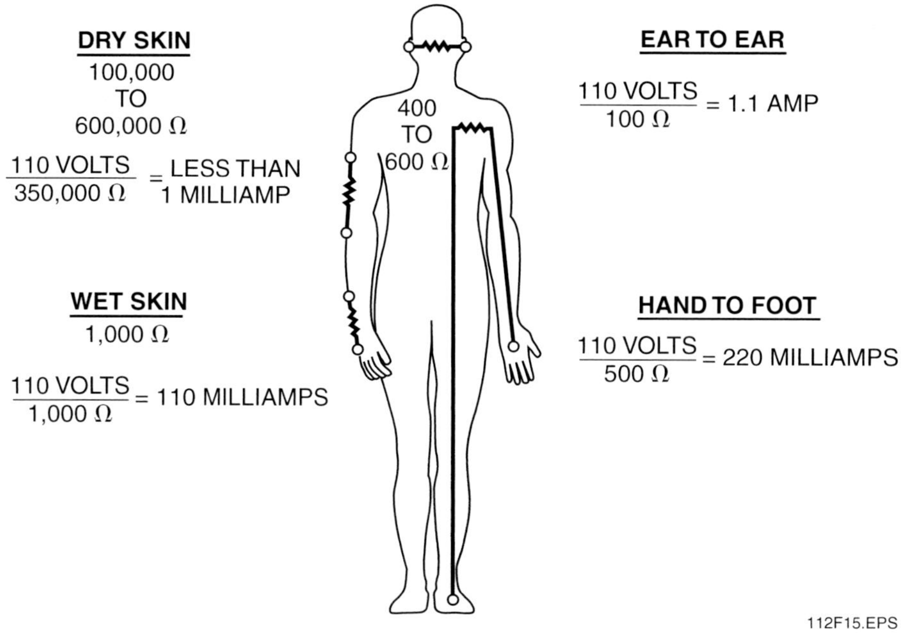

Figure 15 Typical body resistances and current.

be used to ensure that any induced voltages are shorted to ground. Ironworkers also need to be aware of potential arc flash situations near their work areas. An arc flash situation is one in which a short circuit or backfeed into an electrical circuit can cause the circuit or switching device to blow up. Such situations can create a fireball of up to 35,000°F. Under new OSHA guidelines, companies are supposed to be identifying and marking such arc flash areas so that unqualified personnel stay away from them. If in doubt, ask your supervisor(s) to make sure that the structures on which you are working are electrically safe.

Each welder and operator of the equipment must be trained to recognize the dangers associated with each type of equipment to avoid injury and death.

Because electrical shock affects the heart more than anything else in the body, it is a good idea to know first aid procedures, such as cardiopulmonary resuscitation (CPR). For electrical burns, treat them as any other type of burn by applying ice or a cold compress over the burned area. Cover with a clean, dry dressing. Seek medical attention if needed.

4.0.0 SMAW Electrodes

Shielded metal arc welding cannot take place without electrodes. The electrodes are often referred to as stick electrodes. Ironworkers must be able to tell the difference between types of electrodes and choose the right electrode for the job. In addition, ironworkers must know how to handle and store electrodes. Some electrodes require special handling.

This section explains the electrode classification system, electrodes, and the factors that determine the right electrode for the job. It also explains how to properly store electrodes in compliance with the appropriate welding codes.

An SMAW electrode has a wire core coated with flux. The wire core transfers the welding current from the electrode holder to the work. The arc at the end of the electrode melts the metal core, the flux coating, and the base metal at a temperature exceeding 6,000°F. The wire core mixes with the melted base metal, forming the weld. As the flux coating is melted, it does the following:

- Gives off a shield of carbon dioxide and carbon monoxide gas that protects the molten weld by pushing away harmful oxygen, hydrogen, ozone, and nitrogen oxide present in the air
- Acts as a fluxing agent to clean and deoxidize the molten metal
- Stabilizes the arc and reduces spatter
- Forms a slag covering over the weld to further protect it from the air and slow the cooling rate

Figure 16 shows an illustration of an electrode in use.

Some flux coatings also have powdered metal and alloying elements in them. The powdered metal adds filler metal. It also helps improve the appearance of the weld. The alloying elements change the chemical composition and strength of the deposited weld metal. Alloys are normally added to electrodes so the deposited weld metal

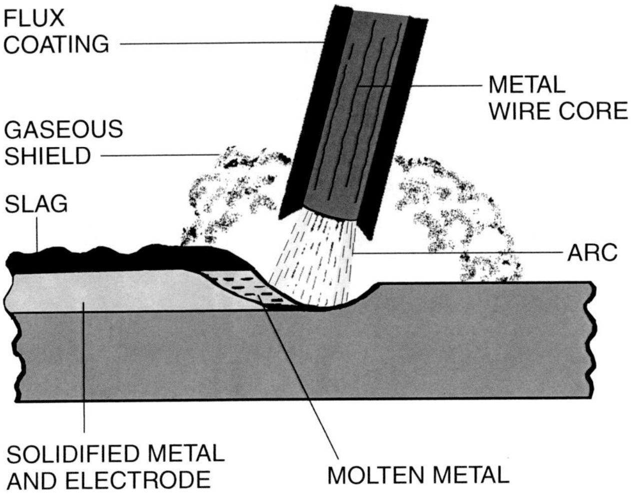

Figure 16 Electrode in use.

matches the base metal chemical composition and strength.

Obtain only the electrodes to be used for a particular welding exercise at one time. Have some type of pouch or rod holder in which to store the electrodes, to prevent them from becoming damaged. Never store electrodes loose on a table. They may fall on the floor, become damaged, or create a tripping hazard. Some type of metal container, such as a pail, must also be available to discard hot electrode stubs. Return unused electrodes to the proper storage location.

WARNING

Do not throw electrode stubs on the floor. They easily roll, and could cause someone to slip and fall.

4.1.0 AWS Filler Metal Specification System

The American Welding Society (AWS) writes specifications for welding consumables. Consumables include the electrodes, filler rods, and fluxes used with all the various welding processes. The series of specifications for consumables are identified as *AWS A5.XX*. These specifications are used in all major welding codes. They are accepted in all industries except where other approvals are specified. The exceptions include the American Bureau of Shipping (ABS), the U.S. Coast Guard, the Federal Highway Administration, Military Specifications (MIL), and Lloyds of London. The American Society of Mechanical Engineers (ASME) Boiler Code electrode specifications closely follow the AWS specifications.

The purpose of the specifications is to set standards that all manufacturers must follow when producing welding consumables. This ensures consistency for the user regardless of which company manufactured the product. The specifications set standards for the following:

- Classification system, identification, and marking
- Chemical composition of the deposited weld metal
- Mechanical properties of the deposited weld metal

The following are some examples of AWS specifications that pertain to shielded metal arc welding electrode:

- *A5.1 – Specification for Carbon Steel Electrodes for Shielded Metal Arc Welding*
- *A5.3 – Specification for Aluminum and Aluminum-Alloy Electrodes for Shielded Metal Arc Welding*
- *A5.4 – Specification for Stainless Steel Electrodes for Shielded Metal Arc Welding*
- *A5.5 – Specification for Low-Alloy Steel Electrodes for Shielded Metal Arc Welding*
- *A5.6 – Specification for Copper and Copper-Alloy Electrodes for Shielded Metal Arc Welding*
- *A5.11 – Specification for Nickel and Nickel-Alloy Welding Electrodes for Shielded Metal Arc Welding*
- *A5.15 – Specification for Welding Electrodes and Rods for Cast Iron*
- *A5.16 – Specification for Titanium and Titanium Alloy Welding Electrodes and Rods*
- *A5.21 – Specification for Bare Electrodes and Rods for Surfacing*
- *A5.24 – Specification for Zirconium and Zirconium-Alloy Welding Electrodes and Rods*

The most common electrodes are discussed in the *A5.1* and *A5.5* specifications. The electrodes described in the *A5.1* specification are used to weld carbon (mild) steel. The electrodes in the *A5.5* specification are used to weld low-alloy, high-strength steels. These two specifications are explained later in this module.

NOTE

The specification number is generally followed by the year of publication, such as *A5.1:2004* or *A5.5:2006*. The year of publication is the year that the specification was last revised. These documents are normally on a five-year revision cycle. Always use the latest version.

4.2.0 Electrode Classification System

The classification system used to identify SMAW *A5.1* carbon steel electrodes and *A5.5* low-alloy electrodes uses the prefix E followed by four or five numbers, such as E6010 or E10018. This is printed on each electrode near the end where the electrode holder grasps the electrode. *Figure 17* shows electrode markings.

The meanings of the markings are as follows (*Figure 18*):

- *E* stands for electrode to signify use with electric welding.
- The first two digits, or three if 100 or over, multiplied by 1,000 indicates the minimum tensile strength in pounds per square inch of the deposited weld metal:
  - E6010 has a minimum tensile strength of 60,000 psi.
  - E7018 has a minimum tensile strength of 70,000 psi.
  - E12018 has a minimum tensile strength of 120,000 psi.

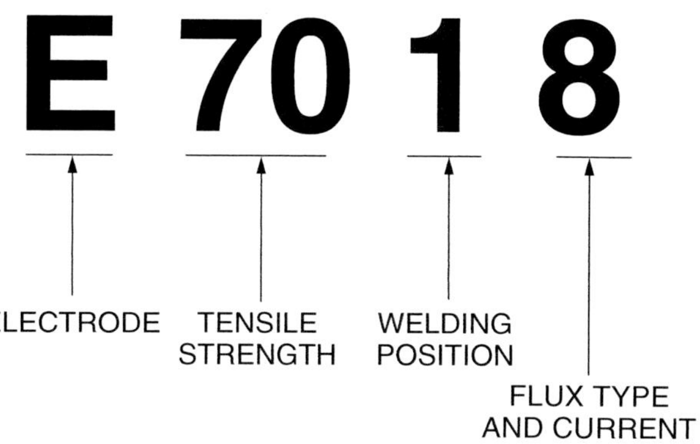

112F18.EPS

Figure 18 Electrode marking system.

112F17.EPS

Figure 17 Electrode markings.

- The third or fourth digit indicates the possible welding positions:
  - 1 can be used to weld in all positions.
  - 2 can be used to weld in flat and horizontal fillet weld positions only.
  - 4 can be used to weld in flat, horizontal, overhead, and vertical down positions.
- The fourth or fifth digit indicates special features of the flux coating. It also indicates the type of welding current required, such as **alternating current (AC)**, DCEN, or DCEP.

Table 1 shows the characteristics of mild steel electrodes covered by the *A5.1* specification.

The low-alloy steel electrodes that are covered by *A5.5* have a suffix in addition to the standard classification numbers. The suffix has a letter and a number. The letter indicates the chemical composition of the deposited weld metal. The number indicates the composition of the chemical classification. *Figure 19* shows an electrode classification marking from the *A5.5* specification.

The suffix letters indicate the following:

- *A* – carbon-molybdenum alloy steel
- *B* – chromium-molybdenum alloy steel
- *C* – nickel steel alloy
- *D* – manganese-molybdenum alloy steel
- *G* – other alloys with minimal elements

The meaning of the numbers varies by alloy type but always indicates the percentages of the following alloying elements:

- Carbon (C)
- Manganese (Mn)
- Phosphorus (P)
- Sulfur (S)
- Silicon (Si)
- Nickel (Ni)
- Chromium (Cr)
- Molybdenum (Mo)
- Vanadium (V)

Figure 20 shows the electrode marking system for low-alloy, high-strength electrodes.

CAUTION

Be very careful when selecting electrodes. For example, a 7018-B2 cannot be used in place of a 7018 electrode because the additional alloy will cause a major change in the weld's metallic properties.

Figure 21 shows examples of various electrodes available listed by alloy content.

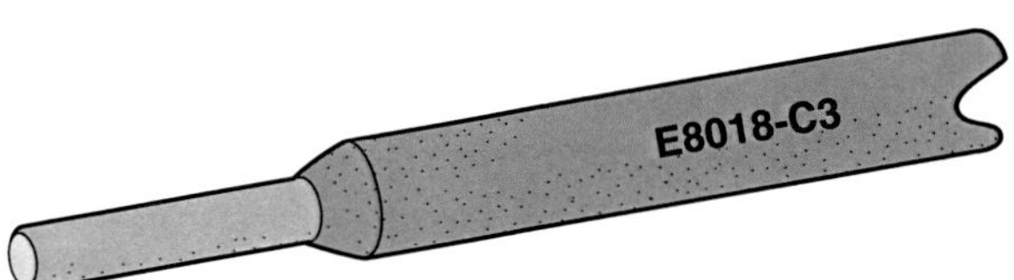

112F19.EPS

Figure 19 Marking on a low-alloy, high-strength electrode.

Table 1 Mild Steel Electrode Characteristics

AWS Class	Flux Coating	Current Requirements	Position	Characteristics
EXX 10	Cellulose Sodium	DCEP	All	Deep Penetration, Flat Beads
EXX 20			Flat, Hor. Fillet	
EXX 11	Cellulose Potassium	AC, DCEP	All	Deep Penetration, Flat Beads
EXX 12	Titania Sodium	AC, DCEN	All	Medium Penetration
EXX 13	Titania Potassium	AC, DCEP, DCEN	All	Shallow Penetration
EXX 14	Titania Iron Powder	AC, DCEP, DCEN	All	Medium Penetration, Fast Deposit
EXX 24			Flat, Hor. Fillet	
EXX 15	Low-Hydrogen Sodium	DCEP	All	Moderate Penetration
EXX 16	Low-Hydrogen Potassium	AC, DCEP	All	Moderate Penetration
EXX 27	Iron Powder, Iron Oxide	AC, DCEP, DCEN	Flat, Hor. Fillet	Medium Penetration
EXX 18	Iron Powder Low-Hydrogen	AC, DCEP	All	Shallow to Medium Penetration
EXX 28			Flat, Hor. Fillet	

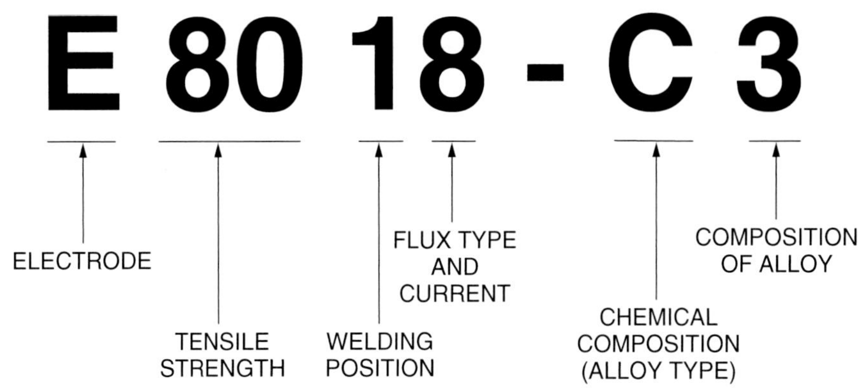

112F20.EPS

Figure 20 Example of marking system.

4.3.0 Manufacturers' Classification

All electrodes used for code work must have the AWS classification number printed on them. It also must be printed on the shipping container. In addition, electrode manufacturers generally print their own unique classification name or number on the electrode and/or on the shipping container. Manufacturers may also make more than one electrode within the same AWS classification.

For example, Lincoln Electric® manufactures two electrodes with the AWS classification E6010. One is called Fleetweld 5P, and the other is called Fleetweld 5P+. Lincoln Electric® recommends Fleetweld 5P for general fabrication and maintenance welding. They state that it is good for pipe. However, Lincoln Electric® also recommends Fleetweld 5P+ as the best choice for pipe. The major difference between the two electrodes is the mechanical properties of the deposited weld metal. Even though both fall into the E6010 classification, the 5P+ electrode decreases cracking in heat-affected zones. It is also better suited for open root joints. When 5P+ is used for the **root pass**, its greater elongation improves resistance to heat-affected zone cracking. Also, 5P+ has higher tensile and yield strength.

When there is more than one electrode in an AWS classification and you are unsure about which electrode to use, check with your supervisor.

Table 2 shows some examples of manufacturers' and AWS classifications for common mild-steel electrodes.

CARBON-MOLYBDENUM STEEL

E7010-A1
E7011-A1
E7015-A1
E7016-A1
E7018-A1
E7020-A1
E7027-A1

NICKEL STEEL

E8016-C1
E8018-C1
E8016-C2
E8018-C2
E8016-C3
E8018-C3

MANGANESE-MOLYBDENUM STEEL

E9015-D1
E9018-D1
E10015-D2
E10016-D2
E10018-D2

CHROMIUM-MOLYBDENUM STEEL

E8016-B1
E8018-B1
E8015-B2L
E8016-B2
E8018-B2
E8018-B2L
E9015-B3L
E9015-B3
E9016-B3
E9018-B3
E9018-B3L
E8015-B4L
E8016-B5

OTHER LOW-ALLOY STEEL

EXX10-G
EXX11-G
EXX13-G
EXX15-G
EXX16-G
EXX18-G
E7020-G
E9018-M
E10018-M
E11018-M
E12018-M

112F21.EPS

Figure 21 Examples of available electrodes listed by alloy.

4.4.0 Electrode Sizes

Mild steel electrodes for SMAW are available in standard sizes, usually ranging from 3⁄32" to 1⁄4". They are typically packaged in 10- and 50-pound cans. They also come in smaller containers as shown in *Figure 22*. The amperage application for which the electrodes are used is proportional to the diameter of the electrode. The number of electrodes in a 50-pound can varies with the diameter and length of the electrodes. The length of the electrode may vary from among manufacturers. *Table 3* shows a table taken from the McKay® electrode catalog. It lists some mild steel electrodes and the number of electrodes in one pound. Lincoln Electric® electrodes and electrodes from other manufacturers are very similar in terms of weight.

Table 2 Manufacturers' and AWS Classifications for Common Mild-Steel Electrodes

Manufacturer	AWS Classification				
	E6010	E6011	E6013	E7014	E7018
Hobart	Pipemaster 60	335A, 335C	447A, 447C	14A	718, 718C
Airco	Pipecraft	6011, 6011C	6013, 6013D	7014	Easy Arc 7018MR
ESAB	AP100, SW-610, SW-10P	SW-14	SW-15	SW-15P	Atom Arc 7018
Lincoln	Fleetwood 5P, 5P1	Fleetwood 35, 35LS, 180	Fleetwood 37, 57	Fleetwood 47	Jetweld LH-70
McKay	6010M	6011	6013	7014	—

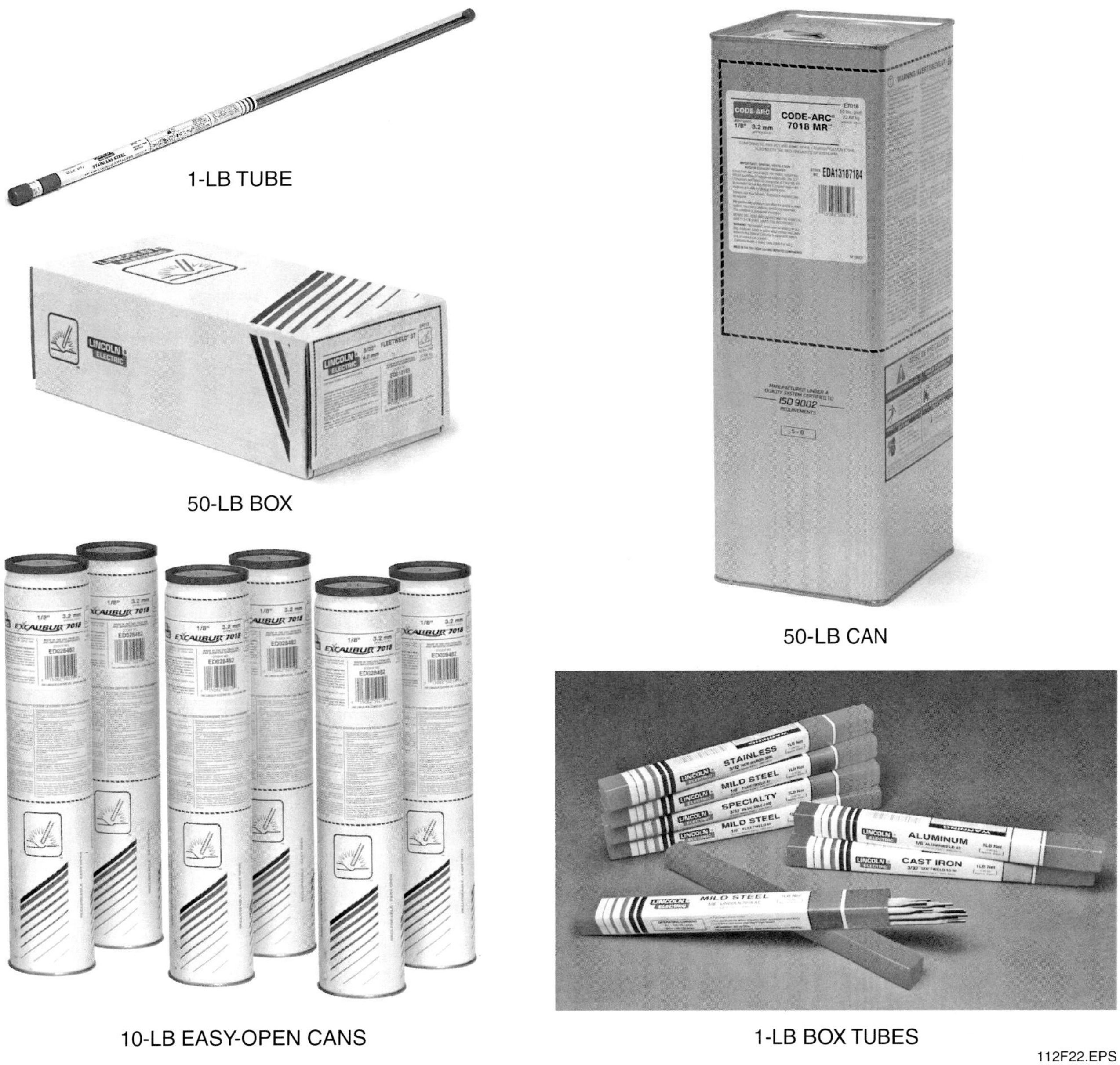

Figure 22 Various containers of electrodes.

Table 3 Number of Electrodes per Pound

McKay Type	Diameter: Length:	$\frac{3}{32}$" (2.4mm) 14" (35.5cm)	$\frac{1}{8}$" (3.2mm) 14" (35.5cm)	$\frac{5}{32}$" (4.0mm) 14" (35.5cm)	$\frac{3}{16}$" (4.8mm) 14" (35.5cm)	$\frac{7}{32}$" (5.6mm) 18" (45.7cm)	$\frac{1}{4}$" (6.4mm) 18" (45.7cm)
6010 PM		30	17	12	8	—	—
6011, Soft-Arc 6011		25	15	11	7	—	—
6013		25	15	10	7	—	—
7014		24	13	9	6	—	—
7024		—	10	7	—	4	2
7018 XLM, Soft-Arc 7018-1		32	15	10	7	—	3
7018 AC, Millennia 7018		32	15	10	7	4	3

5.0.0 Selecting SMAW Electrodes

Because so many different electrodes are available, the welder must be aware of the differences between them to choose the right electrode for the job. The following sections explain how to select electrodes.

5.1.0 Electrode Groups

Electrodes are often classified into one of four groups based on their general characteristics:

- Fast-freeze
- Fast-fill
- Fill-freeze
- **Low-hydrogen**

5.1.1 Fast-Freeze Electrodes

Fast-freeze electrodes are all-position electrodes. They provide deep penetration. However, the arc does produce a lot of fine weld spatter. The weld bead is flat with distinct ripples. It has a light slag coating that can be difficult to remove. The arc is easy to control, which makes fast-freeze electrodes good for overhead and **vertical welding**. Electrodes in the fast-freeze group include the following:

- E6010
- E7010
- E6011

Typical applications for fast-freeze electrodes include the following:

- Vertical-up and overhead plate welding
- Pipe welding (cross-country and in-plant)
- Welds made on galvanized, plated, or painted surfaces
- Joints requiring deep penetration, such as square butts
- Sheet metal welds

WARNING

Many surface coatings, such as galvanized coatings, give off very toxic fumes when heated, cut, or welded. Fumes from certain types of electrodes may also be hazardous. Breathing these fumes can cause damage to the lungs and other organs. Check the MSDS for the coatings with which you are working. The MSDS will identify any hazard for a specific coating. Also, read the MSDS contained in the electrode package. Use proper ventilation and take the necessary steps to protect yourself.

5.1.2 Fast-Fill Electrodes

Fast-fill electrodes have powdered iron in the flux. This gives them high deposition rates. It also means they can only be used for flat welds and horizontal fillet welds. These electrodes have shallow penetration. They produce an excellent weld appearance with almost no spatter. The heavy slag covering is easy to remove. Electrodes in the fast-fill group also have a 2 as the next-to-last number. They include the following:

- E6027
- E7024
- E7028 (also in the low-hydrogen group)

Typical applications for fast-fill electrodes are production welds on plate more than ¼" thick and flat and horizontal fillets and lap welds.

5.1.3 Fill-Freeze Electrodes

Fill-freeze electrodes are also called fast-follow electrodes. They are all-position electrodes commonly used for vertical and flat welding. They have medium deposition rates and penetration. Fill-freeze electrodes are excellent for sheet metal. The weld bead ranges from smooth and ripple-free to even with distinct ripples. The medium slag covering is easy to remove. Electrodes in the fill-freeze group include the following:

- E6012
- E6013
- E7014

Typical applications for fill-freeze electrodes include the following:

- Vertical fillet or lap welds
- Sheet metal lap and fillet welds
- Joints having poor fit-up in the flat position
- General purpose welding

5.1.4 Low-Hydrogen Electrodes

Low-hydrogen electrodes are designed for welding high-sulfur, phosphorus, and medium- to high-carbon steels. These metals tend to develop **porosity** and cracks under the weld bead. These problems are caused by hydrogen absorbed during the welding process. Low-hydrogen electrodes are available with fast-fill and fill-freeze features. These electrodes produce dense, X-ray-quality welds. The welds have excellent **notch toughness** and ductility. Electrodes in the low-hydrogen group include the following:

- E7015
- E7016
- E7018
- E7028
- E7048

Typical applications for low-hydrogen electrodes include the following:

- X-ray-quality welds
- Welds requiring high mechanical properties
- Crack-resistant welds in medium- to high-carbon steels
- Welds that resist hot-short cracking in phosphorus steels
- Welds that minimize porosity in sulfur steels
- Welds in thick or highly restrained mild or alloy steels to reduce the risk of weld cracking
- Welds in alloy steels requiring a strength of 70,000 psi or more
- Multiple-pass welds in all positions

If hydrogen is present in the weld zone, it creates defects. The greatest source of hydrogen is water moisture. Water is composed of oxygen and hydrogen. Low-hydrogen electrodes are carefully made to eliminate moisture in the flux coating. Low hydrogen means there is less than 16 milliliters (ml) per 100 grams of weld metal. The amount of hydrogen in these electrodes varies from 4 to 16 ml per 100 grams of weld metal.

Low-hydrogen electrodes have two unique suffixes. The M suffix is for compliance. Most military codes require an M suffix.

When the suffix H and a number are added, the numbers relate to the amount of hydrogen in the electrodes. As a result, the electrodes have suffixes such as H4, H8, and H16. These three suffixes indicate the dryness of the electrodes. An additional suffix may also be added to the dryness suffixes. H4R is an example. The R in this case indicates that the electrode was designed to meet specific low-moisture pickup limits under controlled humidification tests. *Figure 23* shows a 7018 electrode with the H4R suffix.

If low-hydrogen electrodes are exposed to the air, they start to absorb moisture and are no longer low-hydrogen. For this reason, all low-hydrogen electrodes are shipped in **hermetically sealed** metal containers. Once opened, the electrodes must be stored in a heated oven. *Figure 24* shows low-hydrogen electrodes properly stored. Guidelines for the storage and handling of low-hydrogen electrodes are covered later in this module.

CAUTION

The welding codes establish strict guidelines for the storage, handling, and redrying of low-hydrogen electrodes. These procedures will be referenced in your site quality standards and welding procedure specifications (WPSs). Severe penalties will result from violation of these standards. Improperly storing these electrodes will lead to poor-quality welds. Be sure you read and understand your site standards before working with low-hydrogen electrodes.

Figure 23 Electrode with suffix.

112F24.EPS

Figure 24 Storing low-hydrogen electrodes.

5.2.0 Electrode Selection Considerations

Many factors enter into the selection of electrodes for the job. The available welding equipment and its current ranges limit the choice of electrodes. Base metal properties play an important role in the selection. Also, design factors and the position in which the weld must be made affect the choice of electrodes for a particular application. *Table 4* is taken from Hobart's welding catalog. It indicates position, polarity, and common usages for some of the common steel electrodes. The electrodes are listed by AWS classification.

5.2.1 Welding Procedure Specification

If there is a WPS for the weld to be made, the filler metal (electrode) will be specified. It will be given as an AWS classification and/or manufacturer's standard. It will sometimes include a manufacturer's name. The filler metal specified in the WPS must be used. On some critical jobs, the electrodes and filler metals are marked with coded numbers and/or letters. Such **traceability** allows government inspectors and insurers to back track on any failed welds.

5.2.2 Base Metal Type

The filler metal should match the base metal's chemical composition and mechanical properties. The composition and mechanical properties are described in the base metal mill specifications. The mill specifications can then be compared to various tables that list base metal and filler metal acceptability. These tables are available in a number of codes, such as the *AWS Structural Welding Code – Steel D1.1, Table 3.1 Prequalified Base Metal – Filler Metal Combinations for Matching Strength*, and different sections of the *ASME Boiler and Pressure Vessel Code*. Tables are also available from the electrode or base metal manufacturers. If the filler metal and base metal are not compatible, the weld produced will be defective.

A weld made on mild steel with a high-tensile electrode such as E12018 is weaker than one made with E7018 because the high-tensile weld will not give. Instead, all of the stress is transferred to the heat-affected zone along the weld, which causes cracking. A weld made with E7018 gives along with the base metal, distributing the stress across the weld. This makes a stronger joint. The opposite occurs when mild steel electrodes are used to weld high-strength steel. All the stress occurs in the weld because the surrounding base metal is so much stronger. This causes the weld to fail prematurely.

5.2.3 Base Metal Thickness

Normally, an electrode is used that is smaller in diameter than the thickness of the metal to be welded. For sheet metal, a 3⁄32" E6013 electrode may be recommended because it has shallow penetration. For very thick sections, always use a low-hydrogen electrode for its **ductile** qualities. To minimize joint preparation, a deep penetration E6010 or E7010 electrode can be used for the root pass.

Table 4 Common Steel Electrode Selection Factors

AWS Class	Position	Polarity	Usage
E6011	All	AC, DCEP	Good for dirty, rusty steel
E6013	All	AC, DCEN, DCEP	Good for thin steel and poor fit-up
E7014	All	AC, DCEN, DCEP	Good for thin steel
E7018	All	AC, DCEP	Good for high deposition; good for medium- or high-carbon steels

DCEN – DC Electrode Negative (Straight Polarity)
DCEP – DC Electrode Positive (Reverse Polarity)

5.2.4 Base Metal Surface Condition

When possible, the base metal surface should always be cleaned before welding. Surface corrosion, plating, coatings, and paint should be removed. If this is not possible, use a deep penetrating electrode, such as an E6010.

5.2.5 Welding Position

The position of welding is important in electrode selection. Some electrodes, such as E7024 and E7028, can only be used in the flat and horizontal fillet weld positions (*Figure 25*). In addition, except for low-hydrogen electrodes, electrodes larger than 3⁄16" should never be used for welding in vertical and overhead positions. The limitation for low-hydrogen electrodes and EXX14 electrodes is 5⁄32". As a general rule, position the joint for flat welding so you can use the largest electrode possible.

5.2.6 Joint Design

The type of joint to be welded is an important consideration when selecting an electrode. When T- and lap joints (*Figure 26*) can be positioned flat, they are usually welded with E7024 or E7028 electrodes. Butt joints that need deep penetration should be welded with E6010 electrodes. If the joint has poor fit-up, use a shallow penetrating electrode such as an E6012 or E6013. If low-hydrogen electrodes are used for open root joints, back gouging may be required. To eliminate the need for back gouging and to make the root pass easier to run, use metal or ceramic **backing**. Alternately, put the root in with an E6010 electrode and then fill out with a low-hydrogen electrode.

5.2.7 Welding Current

The most common type of welding current is DC. Because all electrodes can be run on DC, there is no problem when DC current is available. Welding machines that can produce only AC current cannot use E6010 or E7010 electrodes. Although most other electrodes can run on AC, they run better using DC current.

The amount of current produced by the welding machine is also a factor. Running large diameter electrodes that require high amperages near the maximum output of the welding machine causes the welding machine to overheat and cut out. Check the welding machine current type, rated output, and duty cycle before selecting the electrode size.

6.0.0 SMAW Filler Metal Storage and Control

Electrodes used for shielded metal arc welding must be dry. The dry mineral flux coatings on electrodes start to absorb moisture as soon as they are exposed to the air. In as little as one-half hour, some electrodes can pick up enough moisture from the air to cause weld defects. Some of

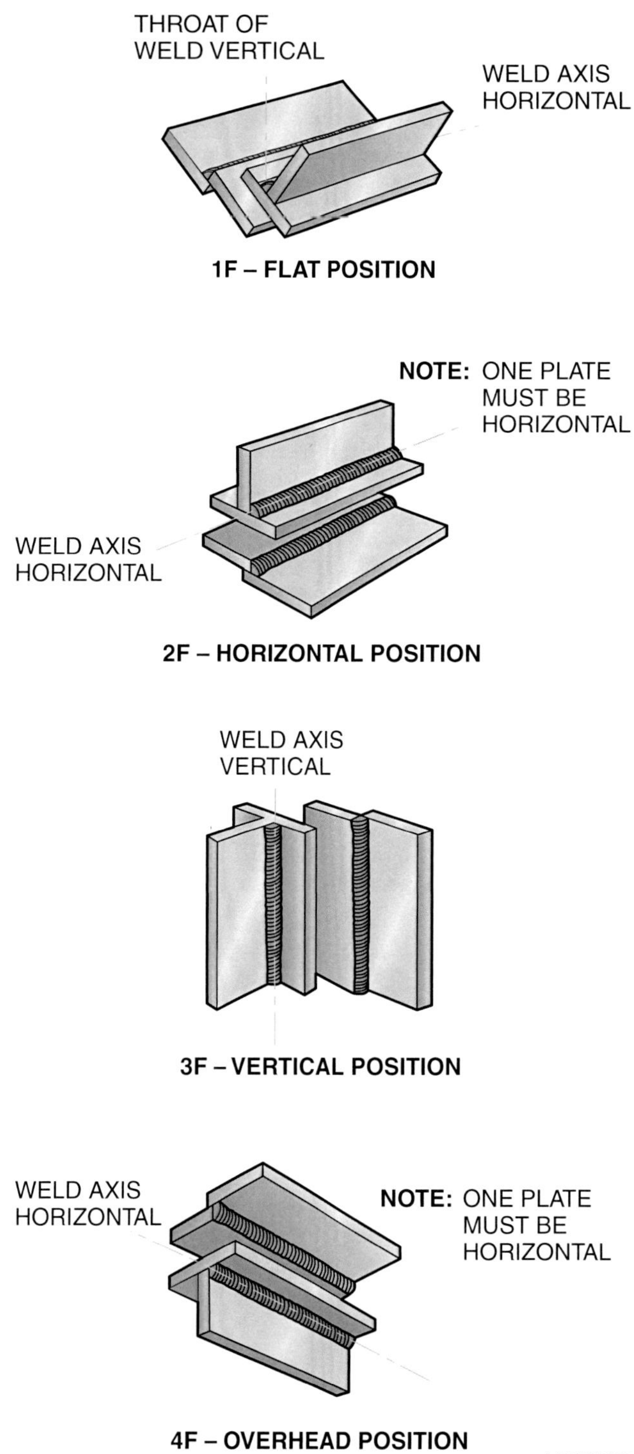

Figure 25 Fillet weld positions.

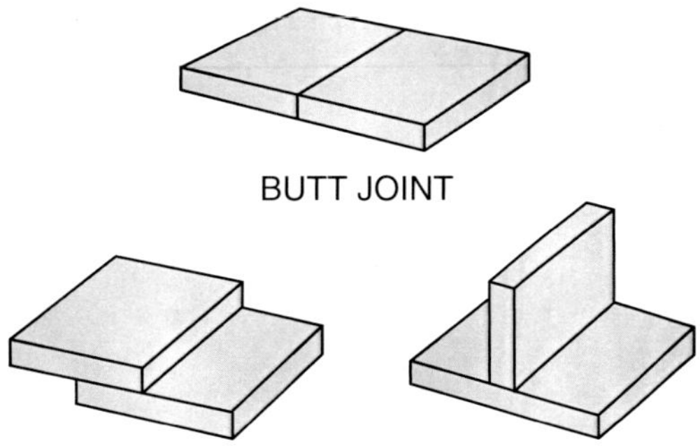

Figure 26 Butt, T- and lap joints.

these defects, such as underbead cracking, may not be visible to the naked eye.

Low-hydrogen electrodes absorb moisture faster than any other type of electrode. Also, a smaller amount of moisture causes defects. This is because of the nature of the base metal welded with low-hydrogen electrodes. For these reasons, special care must be used when handling and storing low-hydrogen electrodes.

6.1.0 Code Requirements

All welding codes require electrodes with a low-hydrogen coating to receive special handling. They must be shipped in hermetically sealed containers. Immediately after opening the sealed containers, low-hydrogen electrodes must be stored in special ovens at a temperature of 250°F (120°C). When low-hydrogen electrodes are removed from the oven, they can only be exposed to the air for a specified period of time, which varies by the classification of the electrode. It generally ranges from one-half hour to ten hours. *AWS D.1.1, Table 5.1 Allowable Atmospheric Exposure of Low-Hydrogen Electrodes* gives the allowable exposure times. *Table 5* shows the *Table 5.1* from *AWS D.1.1.* After the specified time, the electrodes must then be returned to the oven. They will generally require redrying at higher temperatures. The exact procedures for handling filler metal at your site are typically written out in a quality standard. These procedures must be followed. If you are unsure of the requirements at your site, check with your supervisor.

CAUTION

The exposure times shown are general guidelines only. Excessive exposure times cause weld defects. Check your site quality standards for the actual exposure times for your location.

CAUTION

The consequences of not following your site's standards on electrode handling and storage may be severe. Be sure you understand your site's standards before handling electrodes. Improper handling of electrodes can cause weld defects. The information provided in this module is of a general nature. Your site's specific quality standards must be followed.

6.2.0 Receiving Filler Metal

Inspect all filler metal as it is received. All low-hydrogen electrodes (EXX15, 16, or 18), stainless steel electrodes, and nickel alloy coated electrodes must be supplied in hermetically sealed containers. If the seal has been broken, reject the electrodes. Mark them Not to Be Used and notify your supervisor.

NOTE

Your company insurance inspectors may require material test reports (MTRs) for welding materials.

Table 5 Allowable Atmospheric Exposure of Low-Hydrogen Electrodes

Electrode	Column A (hours)	Column B (hours)
A5.1		
E70XX	4 max	Over 4 to 10 max
E70XXR	9 max	
E70XXHXR	9 max	
E7018M	9 max	
A5.5		
E70XX-X	4 max	Over 4 to 10 max
E80XX-X	2 max	Over 2 to 10 max
E90XX-X	1 max	Over 1 to 5 max
E100XX-X	½ max	Over ½ to 4 max
E110XX-X	½ max	Over ½ to 4 max

Notes:

1. Column A: Electrodes exposed to atmosphere for longer periods than shown shall be redried before use.
2. Column B: Electrodes exposed to atmosphere for longer periods than those established by testing shall be redried before use.
3. Electrodes shall be issued and held in quivers, or other small open containers. Heated containers are not mandatory.
4. The optional supplemental designator, R, designates a low-hydrogen electrode which has been tested for covering moisture content after exposure to a moist environment for 9 hours and has met the maximum level allowed in *AWS A5.1, Specification for Carbon Steel Electrodes for Shielded Metal Arc Welding.*

112T05.EPS

All non-low-hydrogen electrodes must be shipped in moisture-resistant containers. Inspect these containers for damage. The electrodes cannot be used if their coatings are damaged. They also cannot be used if they have come into direct contact with water or oil. Mark electrodes that cannot be used and notify your supervisor.

Each electrode container must be marked with at least the following information:

- ASME, AWS, or other electrode specification number and electrode classification designation
- Electrode size
- Weight

If the containers are not properly marked, mark them Not to Be Used and notify your supervisor.

6.3.0 Storing Filler Metal

All filler metal must be stored in a warm (40°F minimum), dry storage area. Filler metals are not to be stored directly on the floor. Electrodes that are stored on the floor form **condensation**. Use pallets or shelves to store electrodes off the floor. They must also be stored in such a manner that they will not be damaged. Once low-hydrogen, stainless steel, or nickel alloy electrode containers have been opened, the electrodes must be stored in an oven immediately.

6.4.0 Storage Ovens

An electrode storage oven is typically an electrical oven. Most are controlled by thermostats. Storage ovens for large shops can hold several hundred pounds of electrodes. The ovens can reach maximum temperatures of more than 800°F, which is too high for most electrodes. Always follow the electrode manufacturer's guidelines for storing electrodes in ovens. Smaller portable ovens are designed to be used on the job. These may hold only 25 or 30 pounds of electrodes. Depending on the WPS or job quality standards, portable ovens may be required to hold electrodes at the point of use. The electrodes remaining in the portable ovens after performing the task must be returned to the holding oven. *Figure 27* shows electrode storage ovens.

Storage ovens should be set between 250°F and 300°F. The oven must remain on at all times. Never unplug an oven. Never plug it into an outlet that could accidentally be shut off.

NOTE

Do not mix electrode types in an oven. If electrodes of various types must be stored in the same oven, keep them organized by placing them on separate shelves.

6.4.1 Exposure Times

The time that a low-hydrogen, stainless steel, or nickel alloy electrode can be exposed to the air is limited and depends on the electrode. It is the welder's responsibility to take out of the oven only the number of electrodes that can be used before the maximum exposure time is reached. The exposure time is defined as the period of time that the electrode is not in a heated holding or portable oven.

Samples of electrodes being used at a site are often collected by the quality control group for moisture testing. Depending on the results of this testing, the maximum exposure times can be adjusted. In addition, if high moisture content is found in the electrodes, the welds made with the electrodes may have to be replaced.

Depending on the site quality standards, electrodes that exceed the maximum exposure time must either be destroyed or dried. Check your site quality standards to see if you can dry the electrodes or if they must be destroyed. Low-hydrogen electrodes that are exposed to water must be destroyed.

6.4.2 Drying Electrodes

Electrodes that exceed the maximum exposure time can be dried once. They are dried by placing them in a holding oven at a higher temperature for a period of time. The amount of time and temperature depend on the nature of the moisture pickup and the electrode. The rebake procedure must be approved by a metallurgist.

CAUTION

Do not attempt to dry electrodes before determining whether your site quality standards allow drying. Improperly drying electrodes causes weld defects. Check with your supervisor.

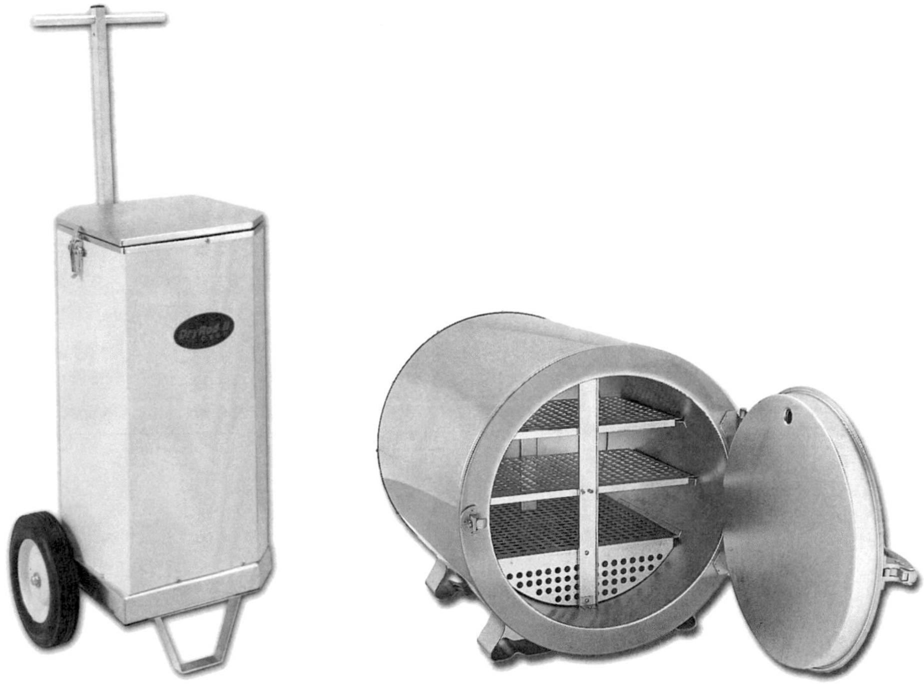

Figure 27 Electrode storage ovens.

7.0.0 Filler Metal Traceability Requirements

Welding codes and the WPS on a project or job site normally specify the type of electrode to be used. Once a weld is made, there must be a means to trace and document the type of electrode that was used to make the weld or welds. This is necessary to confirm that the correct welding codes and WPS were followed.

Traceability requirements vary according to client requirements and specifications. On many job sites, these traceability requirements may be found in the quality control documentation mandating the job. One of the more common requirements followed may be found in one of the *ASME Boiler and Pressure Vessel Code* sections.

> **NOTE**
> The traceability requirements in site quality control manuals must be followed. The penalties for violation of the requirements can be severe. Understand traceability requirements before proceeding with a welding job. Confirm this information with your supervisor, project engineer, or welding inspector.

8.0.0 Joint Design

Welded joints are chosen mainly for the safety and strength required for the conditions that will be encountered. When selecting the joint for a particular application, many factors must be taken into account. These include load considerations, the environment in which it will be used, materials, processes, and cost.

8.1.0 Load Considerations

The loads in a welded steel structure or part are transferred through the welded joint. The welded joint must withstand the stresses caused by the loads. *Figure 28* shows how force creates stress on welded joints.

Any or all of these forces may act on welded joints. When two parts are welded together, in whatever manner, they are supposed to become one. The surface areas of the two parts being welded must be properly prepared. Otherwise the welds may fail if one or more of these forces are applied. The following are brief explanations of load terms as they relate to welds:

- *Tensile* – Capable of stretching or being stretched. When metals are pulled in opposite directions by forces (F_1 vs. F_2), they are expected to stretch some, but if stretched too far, they eventually break. Their strength to remain one is referred to as being their tensile strength. When metals are heated, as when they are welded, the molecules in the metals change. When the metals are cooled, their molecules change again. These heating and cooling cycles can weaken the tensile qualities of metals. When two pieces of metal are welded end-to-end, they become one piece. When they are pulled at each end, the tensile strength of the two pieces, and the weld, are tested. If the weld is a good weld, one of the welded pieces will usually break before the weld breaks.

- *Compression* – The reduction of volume or mass. Metals can be compressed. If you strike a steel nail or bolt with a hammer, you can eventually compress the nail or bolt into a different shape. When the two pieces of metal are welded end-to-end and forces (F_1 and F_2) are applied inwardly on each end, the pieces and the weld both feel compression.

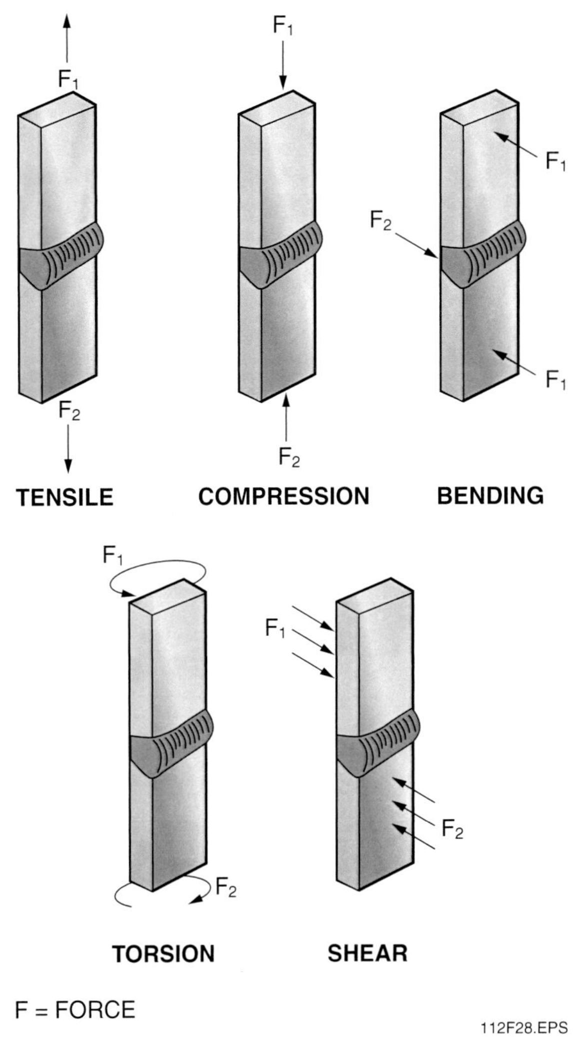

Figure 28 Forces that act on welded joints.

- *Bending* – To make curved. For example, you may be able to manually bend a section of straight lightweight flat bar around a 6" pipe. The pipe in this case is a force (F_2) pushing against the center of the bar as you pull or apply force (F_1) in the opposite direction at each end of the bar. The result of such action is that the flat bar becomes curved. When the two pieces of metal are welded end-to-end, they may be tested in what is known as a bend test. The purpose of a bend test is to see if the weld holds when force is applied against the ends of the pieces.
- *Torsion* – Twisting. The amount of stress the welded bars can withstand as a counterclockwise force (F_1) is applied at one end and a clockwise force (F_2) is applied at the other end. The driveshaft of a truck is twisted as it transmits the rotational movements of the drive motor and transmission to the drive axle.
- *Shear* – To cut off or trim. When a piece of sheet metal is placed into a press brake machine, one side of the sheet metal rests against a stationary surface of the machine. That stationary surface applies some force to that side of the sheet metal. In *Figure 28*, that stationary force is shown as F_1. When the brake is activated, a moveable part of the machine applies force to the opposite side of the sheet metal. F_2 represents the force created by that moveable part. When the moveable force (F_2) is increased, it eventually shears or cuts the metal because the stationary surface (F_1) is applying force to the opposite side of the sheet metal.

8.2.0 Types of Joints

There are five basic types of joints: the butt joint, lap joint, corner joint (inside and outside), T-joint, and edge joint. *Figure 29* shows illustrations of these joint types. Each of these joints may be made using various types of welds.

8.3.0 Types of Welds

Most types of welds require some degree of base metal edge preparation. The preparation depends on the type of joint being welded. Common types of welds include the following: **surfacing**, plug or slot, fillet, square groove, bevel groove, V-groove, J-groove, and U-groove. If a weld requires a specified **root opening**, it is typically shown on the welding drawing or WPS. Base metals may require that some type of groove is cut on one or both edges. Edge preparations are discussed for each type of weld as it is covered in this section.

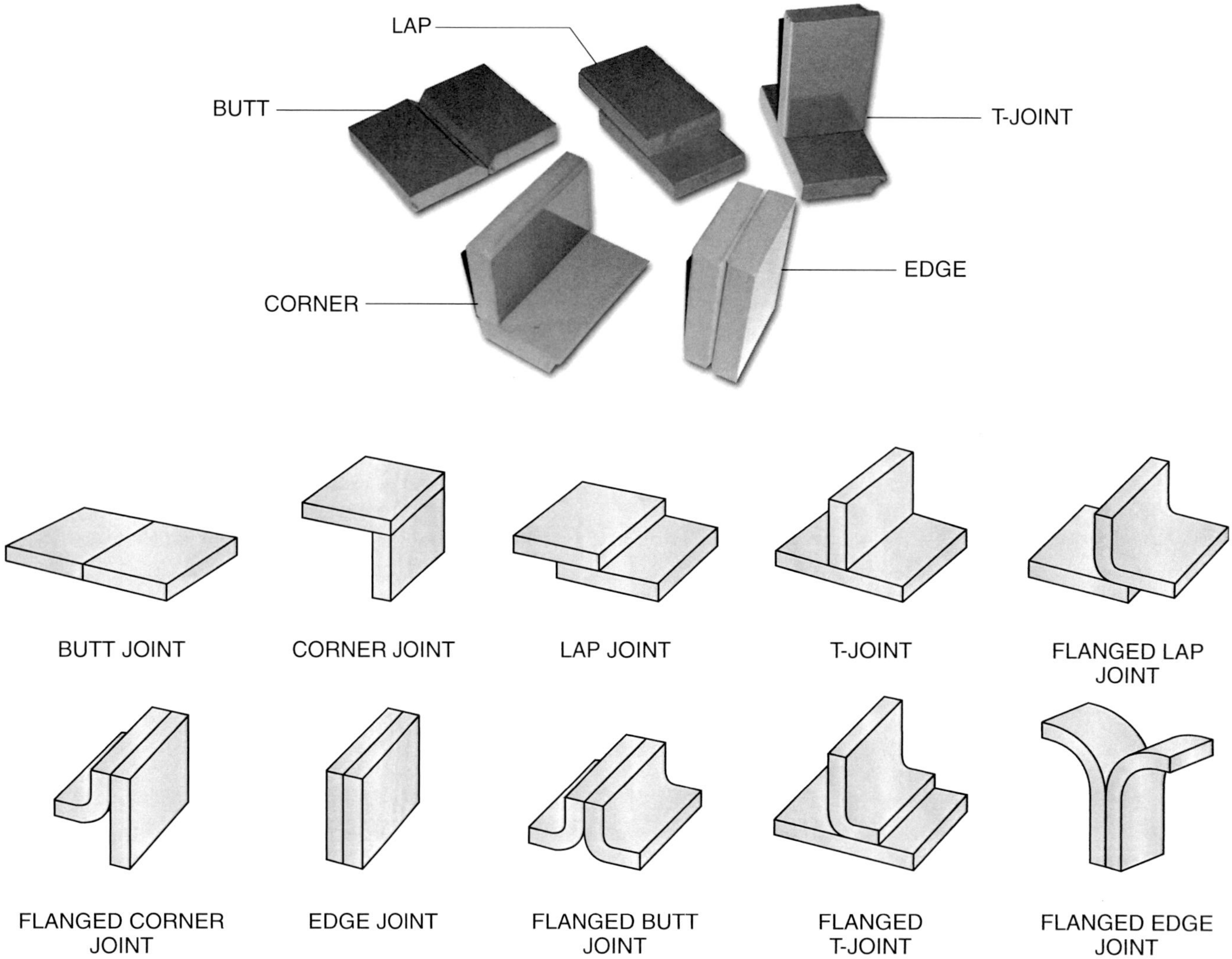

Figure 29 Five basic joint types with variations.

8.3.1 Surfacing Welds

Surfacing is the only type of weld that is applied to just one base metal surface. It is used to build up a base surface that has become worn below the desired thickness or dimension. It is also used to add material to a base to increase its strength and stability. An example of surfacing is shown in *Figure 30.*

Before applying the first layer, the base metal surface must be cleaned to remove contaminants such as oxidation and dirt. Manual or power cleaning tools are used. When applying the surfacing weld using an oxyfuel process, the surface should be evenly preheated before surfacing. Preheating reduces warping of the base material.

For proper penetration and layering to take place, each layer of surfacing must be properly chipped, scraped, and cleaned before applying additional layers of surfacing.

Surfacing is often applied in the tool and die industry to rebuild relatively expensive tools and

Figure 30 Surface welding on an inclined vertical surface.

dies that have worn down. It is also used in the repair of heavy equipment to build up areas that are exposed to excessive stress factors such as shear, tensile stress, and torsion.

8.3.2 Plug and Slot Welds

Plug and slot welds are used to join metal pieces when the edges cannot be welded. The plug or slot weld may be applied on lap, corner, and T-joints. The hole or slot can either be completely or partially filled with weld material when joining it to the other base metal. If the hole is round, the weld is referred to as a plug weld. If the weld is slotted or elongated, it is called a slot weld. *Figure 31* shows how a plug or slot weld may be used on a lap joint. A typical application is to overfill the hole or slot, and then grind the excess material to a level, finished surface.

Preparation of the base metals for this type of weld requires proper cleaning and removal of any oxides or dirt from both pieces. Then the hole or slot is drilled to receive the weld. The walls of the hole or slot must also be cleaned of any oxides, dirt, and oils prior to welding.

Plug and slot welds are commonly used where a finished surface is required. They are also used where overall dimensions cannot exceed those of the plate thickness.

Figure 31 Plug and slot welds.

8.3.3 Fillet Welds

Fillet welds may be applied to lap joints, T-joints, or corner joints. The fillet weld does not require any base metal edge preparation other than removing contaminants. This is done by cleaning with a wire brush or other appropriate cleaning tools. The two base metals are positioned together with or without a root opening. One or more passes (welding beads) are applied where the two base metal edges meet. Whenever more than one pass is made on a fillet weld, or on any weld, the previous pass must be cleaned of all contaminants such as dross, dirt, oxides, and porosity before the next pass is applied. *Figure 32* shows fillet welds and includes the drawing symbols for the welds.

To make a fillet weld on outside corner welds, a corner-to-corner or a half-lap fit can be used, as illustrated in *Figure 33*. The corner-to-corner joint is difficult to assemble because neither plate can support the other. Care must also be used when welding to prevent burning through the corner. The half-lap joint is easier to assemble, requires less welding material, and has less chance of burning through the corner. The half-lap does require a second weld on the inside of the corner. If a half-lap fit is used, allowances in the plate dimensions must be made for the lap.

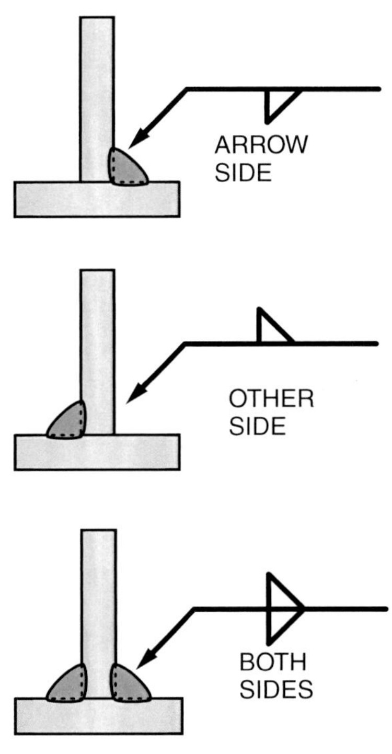

Figure 32 Fillet welds and drawing symbols.

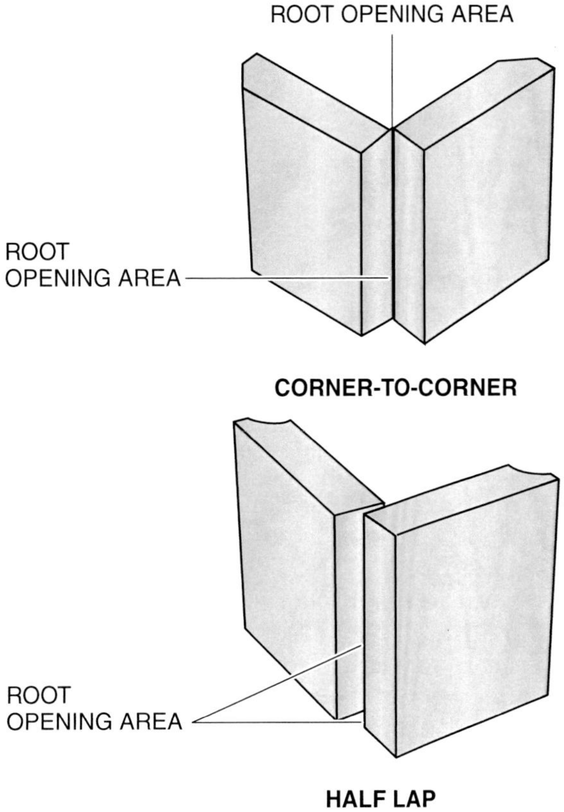

112F33.EPS

Figure 33 Fit-up of corner-to-corner and half-lap fillet welds.

8.3.4 Square-Groove Welds

Square-groove welds can be used with butt joints, corner joints (inside and outside), T-joints, and edge joints. There is one main difference between the fillet and the square groove. A square groove requires a root opening to be set or prepared between the two base pieces prior to welding, as illustrated in *Figure 34.* This allows penetration into a greater portion of the surface area of the two base metals. A partial joint-penetration weld has much less strength than a complete joint-penetration weld.

Welding codes impose restrictions for complete joint-penetration welds. With SMAW, the maximum base metal thickness is ¼". Welding from both sides is required. In addition, a root opening of half the thickness of the base metal is required. The root of the first weld must be gouged before the second weld is made. For GMAW and FCAW, the maximum base metal thickness is ⅜". These processes share the requirements for welding from both sides and for back gouging with SMAW. Welding codes and specifications are discussed later in this module.

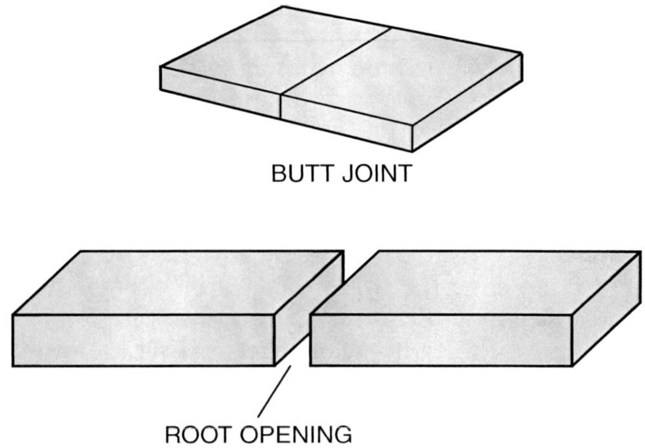

112F34.EPS

Figure 34 Preparation for a square groove on a butt weld.

8.3.5 Bevel-Groove Welds

Bevel-groove welds are common in the welding industry. The base metal preparation is relatively simple and provides greater surface penetration than the square groove. The bevel-groove weld can be applied to all five of the basic weld joints. Typically, a single bevel is cut on the edge of one of the base metal pieces. Preparations for a bevel-groove weld are shown in *Figure 35.* The preparation may include a specified root opening along with the beveled cut, as shown in the butt weld arrangement in the figure. The bevel may be cut by various methods.

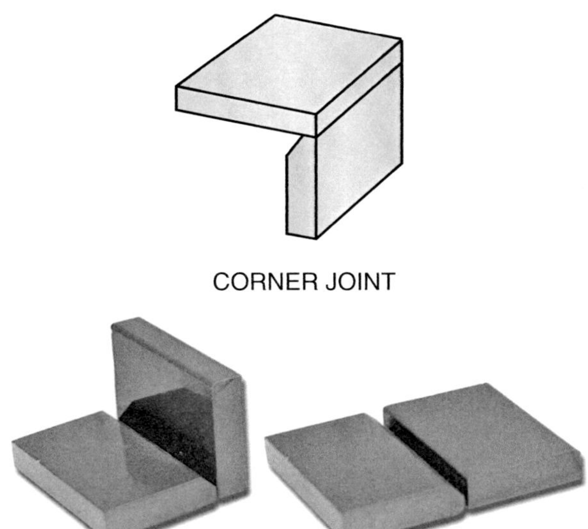

112F35.EPS

Figure 35 Preparations for corner and butt bevel-groove welds.

8.3.6 V-Groove Welds

V-groove welds are generally applied to inside or outside corner joints, butt joints, or edge joints. *Figure 36* shows an illustration of a prepared V-groove on a butt joint to be welded. V-grooves and bevel grooves are more economical to prepare than other types of welds. Typically, a 45-degree angle must be cut on both edges to prepare the base metals to receive the weld. In preparing inside or outside corner joints or butt joints for V-groove welds, the angle may not extend from the top of the bevel edge to the bottom of the bevel edge on each piece. It may be cut short of the bottom corner if a **root face** is required. In this edge preparation, the flat surface from the point where the angle stops to the bottom of the piece is referred to as the root face.

8.3.7 Single Versus Double V-Groove Welds

When possible, a double V-groove weld should be used in place of a single V-groove. The double V-groove requires half the weld metal compared to the single V-groove. Welding from both sides also reduces distortion, because the forces of distortion work against each other. *Figure 37* compares a double V-groove and a single V-groove preparation.

8.3.8 J- and U-Groove Welds

J- and U-groove welds are similar (*Figure 38*). The main difference is that a J-groove weld requires only one base metal to have its edge grooved in the form of a J. A U-groove is formed by preparing two matching J-grooves, one on each base metal edge. Even though both types require much less weld material than bevel or V-groove welds, they require much more preparation time because of the shape of the grooves. A J-groove weld can be applied to all five types of weld joints. The U-groove can only be used on butt, inside and outside corner, and edge joints.

8.3.9 Combination Fillet and Groove Welds

A combination groove and fillet weld is sometimes used in place of a fillet weld alone. A groove weld requires more preparation time than a fillet weld. A fillet weld, in contrast, requires more welding time and material than a groove weld. The combination weld requires more preparation time than a fillet weld alone, but helps to reduce distortion.

8.3.10 Groove Angles and Root Openings

The purpose of the groove angle is to allow access to the root of the weld. The root preparation is sized to control **melt-through**. Increasing or decreasing the root preparation (root opening

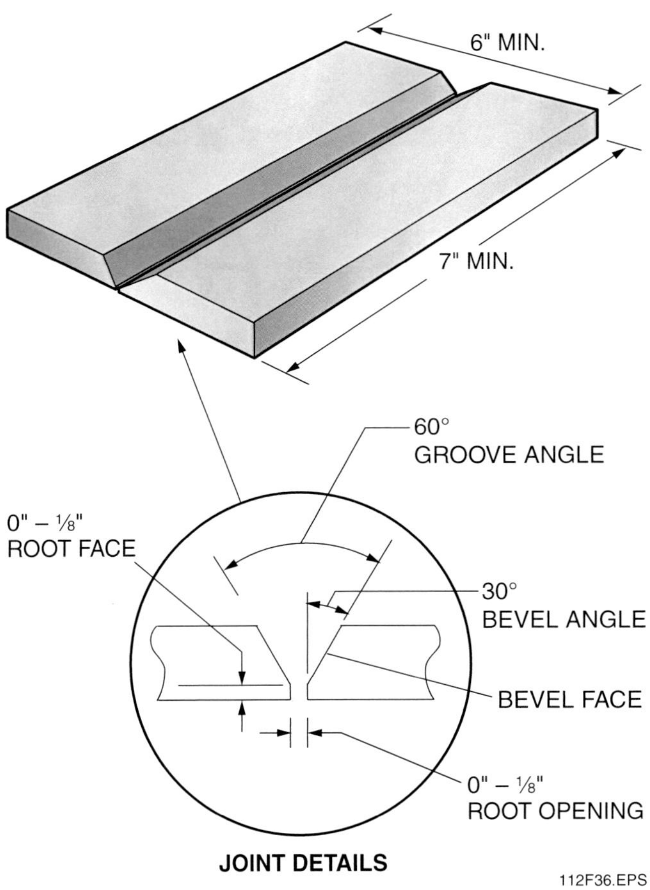

Figure 36 Open-root V-groove weld.

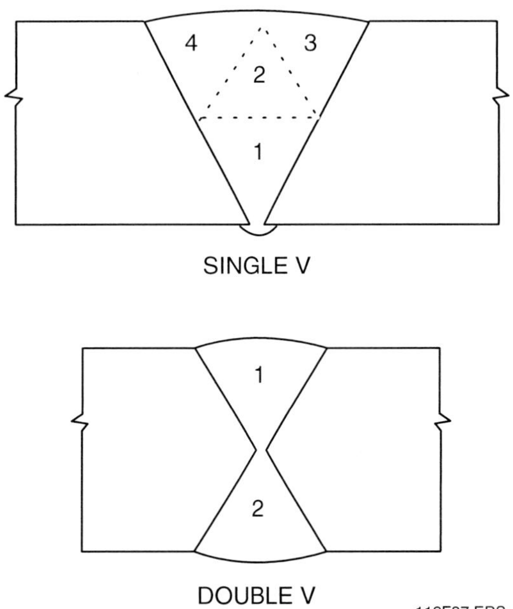

Figure 37 Comparison of double and single V-grooves.

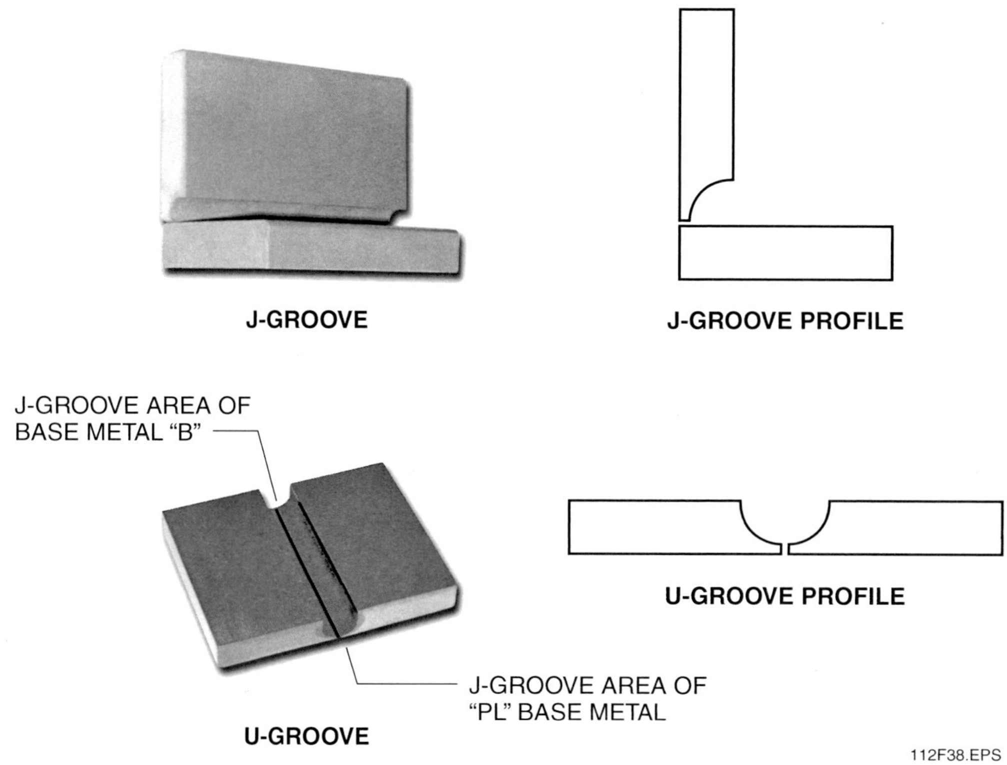

Figure 38 J- and U-groove welds.

and root face) results in excess melt-through or insufficient root penetration. As a general rule, as the groove angle decreases, the root opening increases to compensate (*Figure 39*). Root faces are used with open root joints, but not when metal backing strips are used.

CAUTION

If the groove angle is larger than necessary, it requires additional weld metal to fill. This increases the time and cost to complete the weld. It also increases distortion. If the groove angle is too small, it may result in weld defects.

8.3.11 Open Root Welds

For open root welds on plate, the groove angle should be 60 degrees. The maximum size of the root opening and root face is ⅛". For open root welds on pipe, the groove angle is 60 or 75 degrees, depending on the code or specifications used. The maximum size of the root opening and root face is ⅛". *Figure 40* shows illustrations of open root joint preparation.

8.3.12 Welds with Backing on Plate

Backing for plate can be a gas, flux-coated tape, fiberglass-coated tape, ceramic tape, or strips made from the same material as the base metal (*Figure 41*). When gas is used as a backing, the same groove angles and root preparation used for open root welds are used. When flux-coated tape, fiberglass tape, or ceramic tape is used as a backing, the groove angle should be 60 degrees, the maximum root opening should be $^3/_{16}$", and the maximum root face should be $^3/_{16}$". When applying tape, be sure to clean the surface to ensure the tape tightly adheres. Also, make sure that the tape is centered on the weld joint.

NOTE

Check the WPS or manufacturer's recommendations for joint preparation before using flux-coated, fiberglass, or ceramic tape.

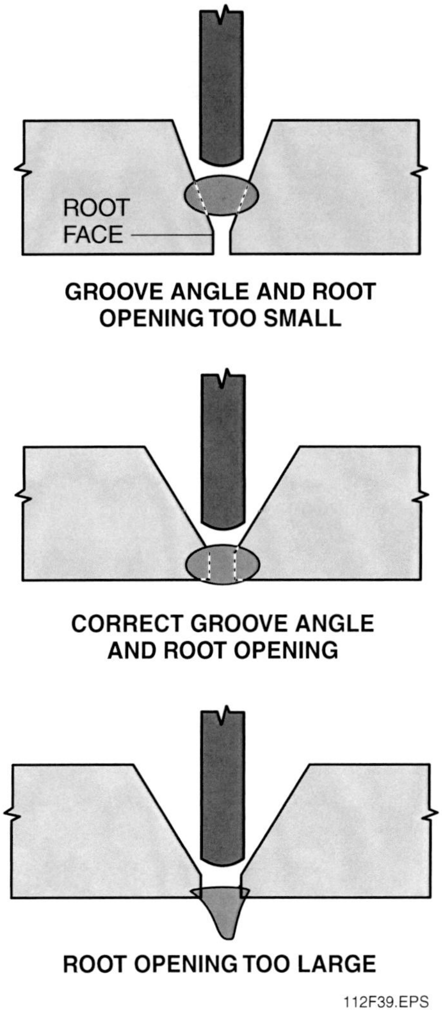

Figure 39 Groove angles and root openings.

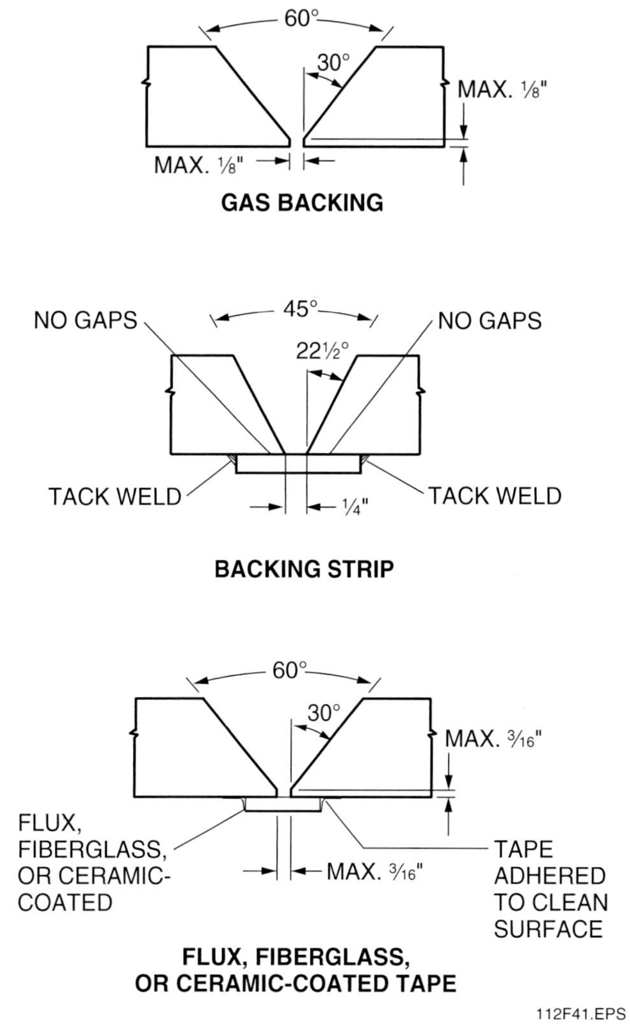

Figure 41 Plate joint preparation for backing.

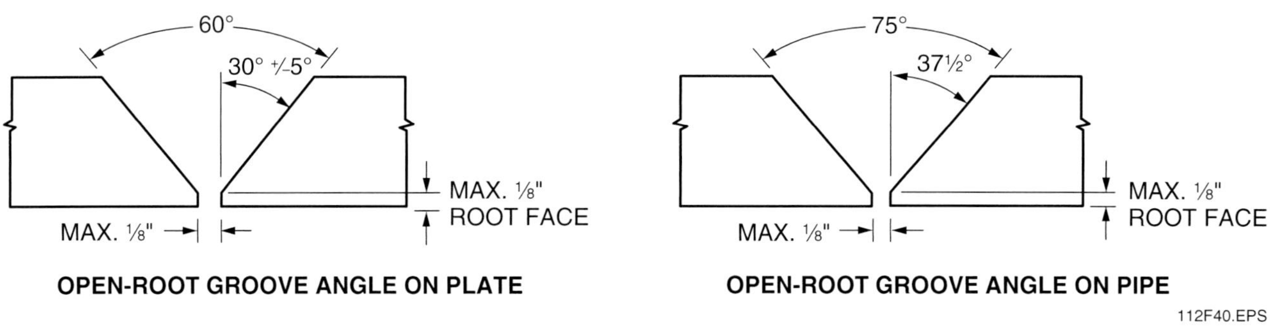

Figure 40 Open root joint preparation.

WARNING

Nitrogen and argon are used as backing and purge gases for some welding applications. These gases displace all the oxygen in the space where they are used. At a refinery in Corpus Christi, Texas, while the welder was at lunch, a helper entered a pipe in which argon had been used as a backing gas. When the welder returned, he found the helper dead. Nitrogen is lighter than air and, if allowed, floats upward. Argon, on the other hand, is heavier than air and settles in low-lying areas. When removing nitrogen from a confined space, open the top to allow it to float out. When removing argon from a confined space, provide some way for it to drain out the bottom.

When backing is made from strips of the same material as the base metal, a groove angle of 45 degrees, a root opening of ¼", and a feather edge (no root face) are used. The backing strip must be thick enough and wide enough to prevent burn-through at the root. It must also be large enough to absorb and dissipate the heat of the root pass. The recommended size for a backing strip for welding mild steel up to ¾" thick is ⅜" thick by 3" wide.

NOTE

Check the WPS or site procedures for the backing size requirements for your site.

The backing strip must be tack welded for it to be held securely in place during welding. It must also be centered on the groove and be in close contact (no gaps) with the back of the plate being welded.

8.4.0 Welding Position

It is easier and faster to weld groove welds in the flat position and fillets in the flat or horizontal position than it is to weld out of position. Always try to position the weldment so that welding is performed in the flat position. If a weldment is an assembly of parts, look for welds that can be made on the bench before tacking the assembly together. When welding pipe, weld as many fittings as possible before tacking the pipe in position, and try to leave welds that will be the easiest to reach and perform. When preparing weld joints, try to prepare the joint so that welding takes place from the most accessible position. *Figure 42* shows the positioning of weld joints for welding.

8.5.0 Codes and Welding Procedure Specifications

A welding code is a detailed listing of the rules and principles that apply to specific welded products. Codes ensure that safe and reliable welded products are produced. They also ensure that persons associated with the welding operation are safe. Clients may specify which welding codes should be followed when they place their orders or award contracts.

If codes are mandated, a WPS must be written for each weld. A WPS (*Figure 43*) is a written set of instructions for producing reliable welds. It includes the type of joint to be used and the type of weld and any groove preparation that may be required for that particular weld. Each WPS is written and tested according to a particular welding code or specification. All welding requires that acceptable industry standards be followed, but not all welds are mandated by codes. The requirement for use of a WPS is often listed on drawings as a note or in the tail of the welding symbol.

If you are unsure whether or not the welding being performed requires a WPS, you should not proceed until you confirm this information.

8.6.0 Weld Symbols and Drawings

Like drivers who must know how to read road signs and navigate by using road maps, ironworkers and welders must learn to read drawings to construct a building or a vessel. To be a welder, one must learn the symbols of the welding trade and understand how to read the design drawings produced by engineers and architects. The engineers and architects have calculated all the stress and load points of a structure or vessel and have

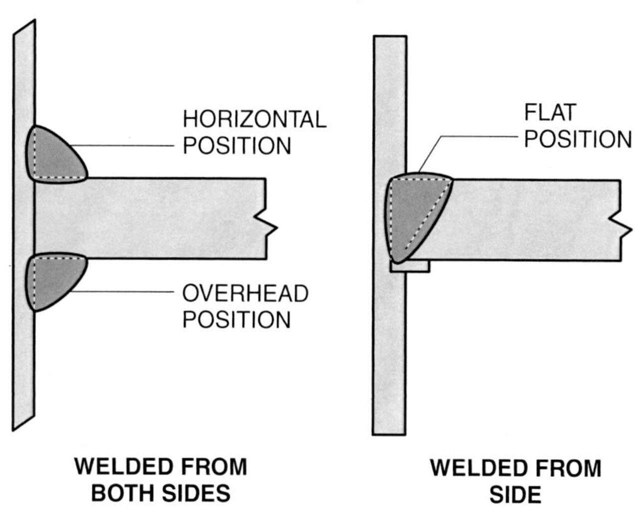

Figure 42 Positioning welds.

Weld Procedure Number	30 P1 TIG 01 Issue A
Qualifying Welding Procedure (WPAR)	WP T17/A

Manufacturer:	National Fabs Ltd 25 Lane End Birkenshaw Leeds	Method Of Preparation and Cleaning:	Machine and Degrease
		Parent Metal Specification:	Grade 304L Stainless Steel
		Parent Metal Thickness	3 to 8mm Wall
		Pipe Outside Diameter	25 to 100mm
Location:	Workshop	Welding Position:	All Positions
Welding Process:	Manual TIG	Welding Progression:	Upwards
Joint Type:	Single Sided Butt Weld		

Joint Design

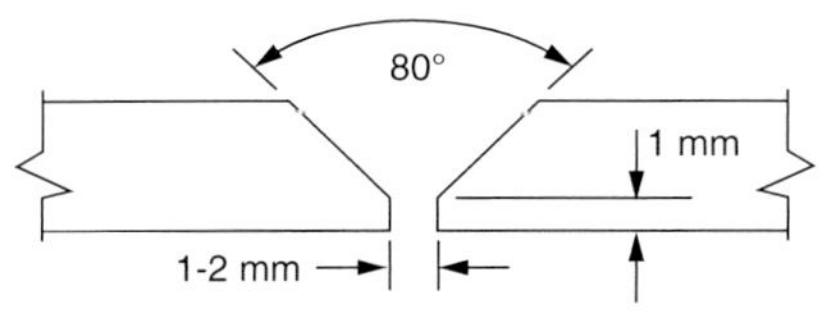

Welding Sequences

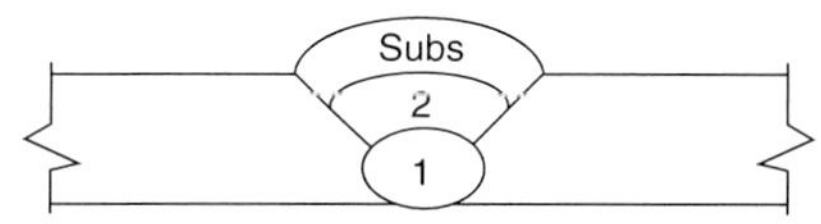

Run	Process	Size of Filler Metal	Current A	Voltage V	Type of Current/Polarity	Wire Feed Speed	Travel Speed	Heat Input
1 2 And Subs	TIG TIG	1.2 mm 1.6 mm	70 - 90 80 - 140	N/A	DC– DC–	N/A	N/A	N/A

Welding Consumables:-	
Type, Designation Trade Name:	BS 2901 Part 2 : 308S92
Any Special Baking or Drying:	No
Gas Flux:	Argon 99.99% Purity
Gas Flow Rate - Shield:	8 - 12 LPM
- Backing:	5 LPM
Tungsten Electrode Type/ Size:	2% Thoriated 2.4mm Dia
Details of Back Gouging/Backing:	Gas Backing
Preheat Temperature:	5°C Min
Interpass temperature:	200°C Max
Post Weld Heat Treatment Time, temperature, method: Heating and Cooling Rates*:	Not Required

Production Sequence

1. Clean weld and 25mm borders to bright metal using approved solvent.
2. Position items to be welded ensuring good fit up and apply purge.
3. Tack weld parts together using TIG, tacks to at least 5mm min length.
4. Deposit root run using 1.2mm dia. wire.
5. Inspect root run internally.
6. Complete weld using 1.6mm dia wire using stringer beads as required.
7. 100% Visual inspection of completed weld

Revision History

Date	Issue	Changes	Authorization
26/11/2000	A	First changes	Jack Straw *Jack Straw*

112F43.EPS

Figure 43 Example of a WPS.

indicated the materials to be used on the project. The design drawings they produce for welding work use a set of standardized symbols accepted in the welding industry. After the design drawings have been completed and signed, copies are given to the craftworkers who will be working on the structure or vessel. The craftworkers must be able to properly read the drawings in order to successfully do the work and complete the project. This section introduces you to the symbols and drawings used in the welding trade.

Most companies use the symbols that have been standardized by the American Welding Society (AWS) in *AWS A2.4:2007, Standard Symbols for Welding, Brazing, and Nondestructive Examination*. Most companies also have a site quality standard that provides guidelines and examples of how welding symbols are to be used. Always refer to your site quality standard when interpreting welding symbols.

Having a general understanding of fillet and groove joints will help you learn the official symbols used on drawings to indicate the kind of weld the engineer or architect wants.

8.6.1 Weld Symbol Basics

The base for all welding symbols is the horizontal **reference line** with an arrow at one end. The **arrow line**, which can be on either side of the reference line, points to the location to which the welding symbol applies *(Figure 44)*. The welding symbol describes the type of weld, its size, and its surface finish.

The opposite end of the reference line is used for information that aids in making the weld (*Figure 45*). For instance, it is used to indicate the welding and cutting processes, the reference to a note, or the welding procedure or electrode type to be used. The tail is omitted when a reference is not required.

In the case of a T-joint, there are two sides to which the arrow points, and welding may be accomplished on either side. The welding symbol distinguishes between the two sides of a joint by using the arrow and the spaces above and below the reference line. The side of the joint to which the arrow points is called the arrow side, and its weld is made in accordance with the instructions below the reference line. The opposite side of the joint is referred to as the other side, and its weld is made in accordance with the instructions found above the reference line. *Figure 46* shows examples of these symbols as applied to fillet and groove welds.

Regardless of which end of the reference line the arrow line is on, information on the reference line is always read from left to right. *Figure 47* shows a welding symbol base and the numerous elements of a welding symbol.

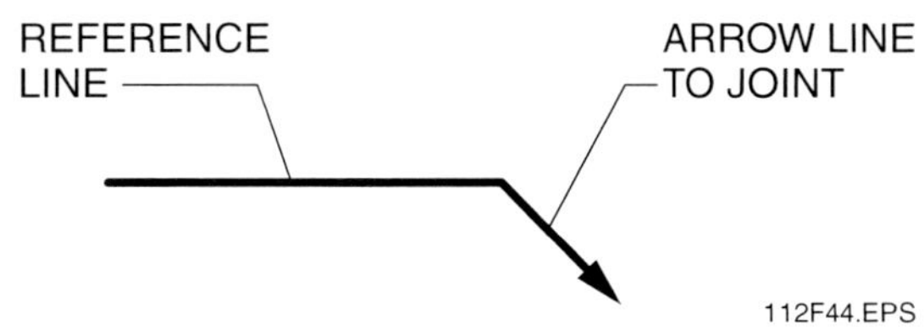

Figure 44 Horizontal reference line.

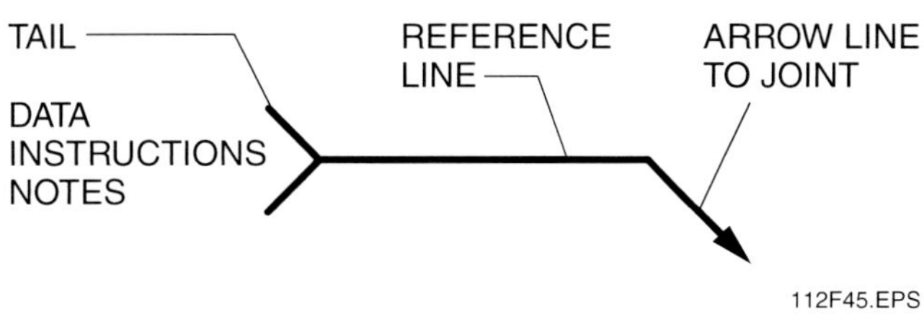

Figure 45 Tail side information.

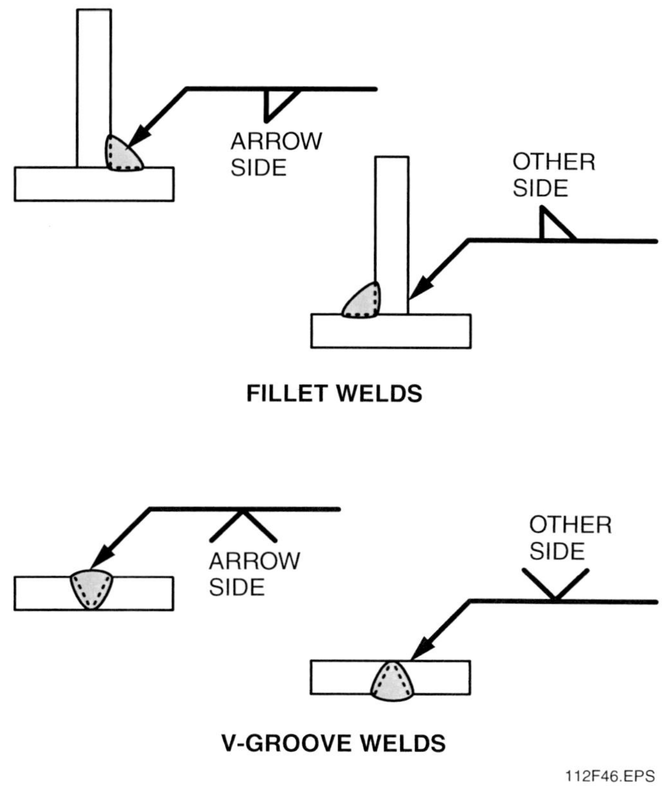

Figure 46 Arrow side and other side significance.

The **weld symbols** used to identify the type of weld to be made signify a basic type of weld or joint preparation (*Figure 48*).

The type of weld used is determined by the way the members to be welded are positioned to form the joint. The basic types of weld joints are the butt, lap, corner, T-, edge, flanged butt, flanged corner, flanged T-, flanged lap, and flanged edge joints (*Figure 49*).

When the weld symbol is on both the top and bottom of the reference line, the information applies to both sides of the joint or surface to which the arrow is pointing. Symbols appear on both sides of the reference line when a double-fillet or double-groove weld is specified. *Figure 50* shows symbols for arrow side, other side, and both sides for fillet and V-groove welds.

8.6.2 Weld Symbol Measurements for Fillet Welds

Unless defined in a drawing note, the size data for a weld is always shown to the left of the symbol to which it applies. For example, in *Figure 51*, the size of the fillet weld shown is ¼". The length of the weld is always shown to the right of the symbol to which it applies. In this drawing, the length is 6".

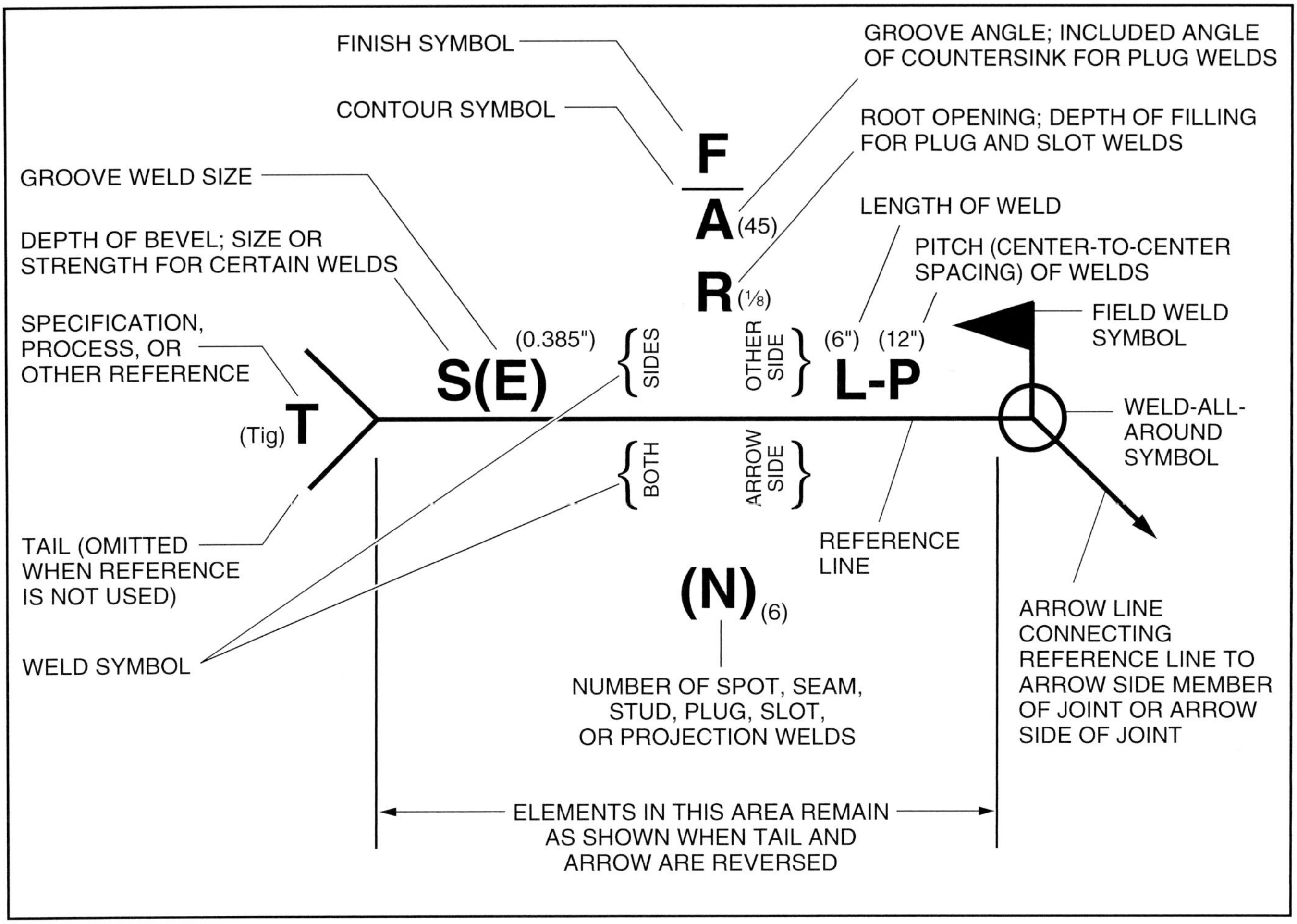

Figure 47 Standard location of elements of a welding symbol.

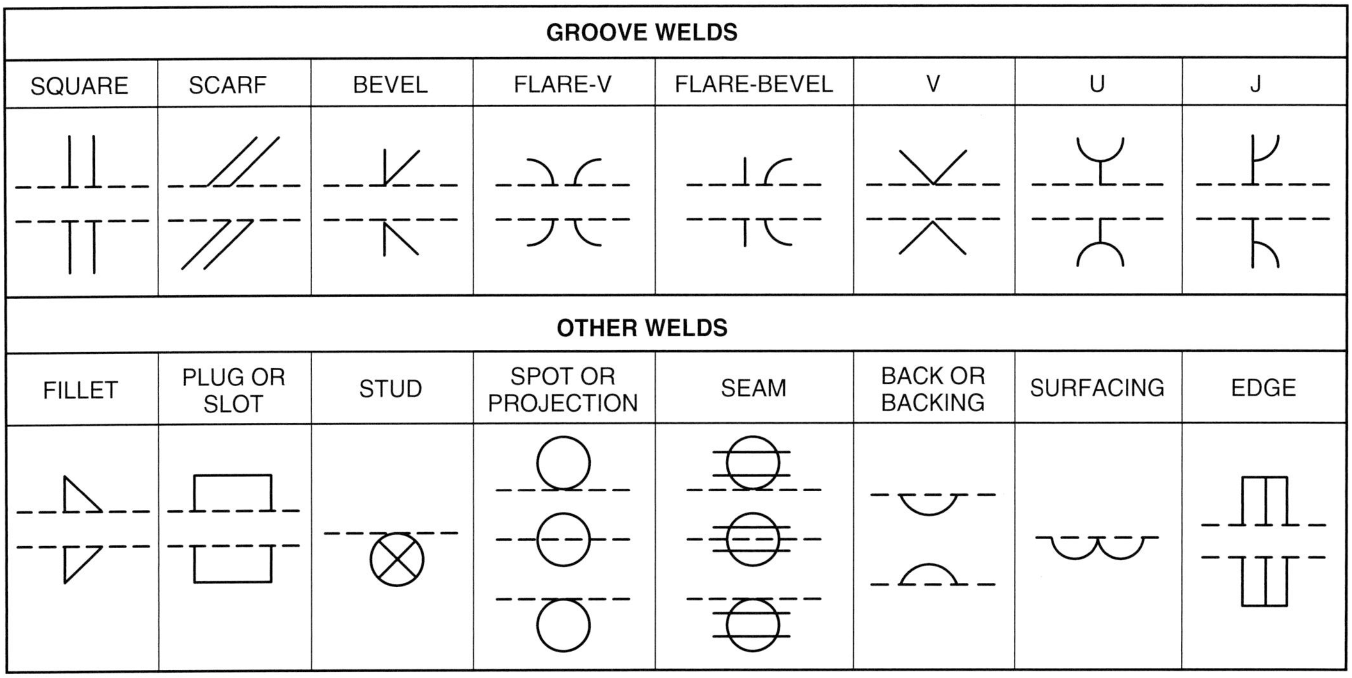

Figure 48 Basic weld symbols.

Dimensions can be given as fractions, decimals, or metric measurements.

A fillet weld is a weld made in a corner. It is roughly triangular in shape and joins two perpendicular welding surfaces, such as in a lap, T-, or corner joint. The size of a fillet weld is the length of the legs (sides) of the largest triangle that can be inscribed within the fillet weld cross section (*Figure 52*). The lengths of the legs on a typical fillet weld are equal, so only one dimension is

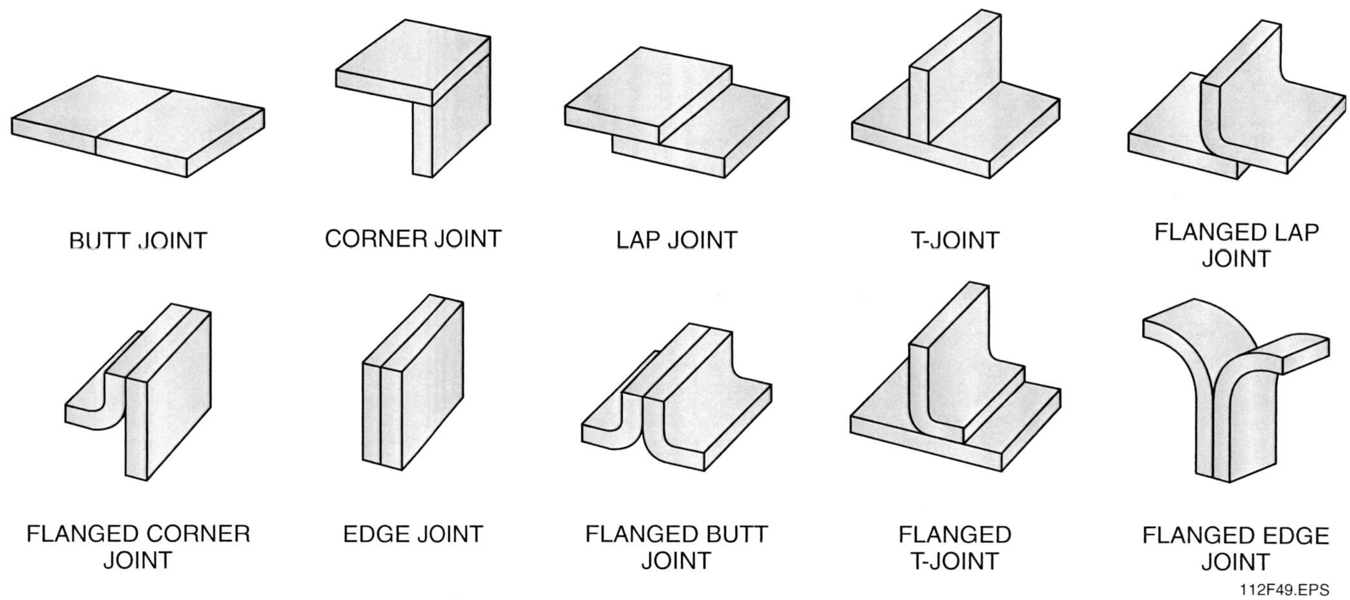

Figure 49 Basic types of weld joints.

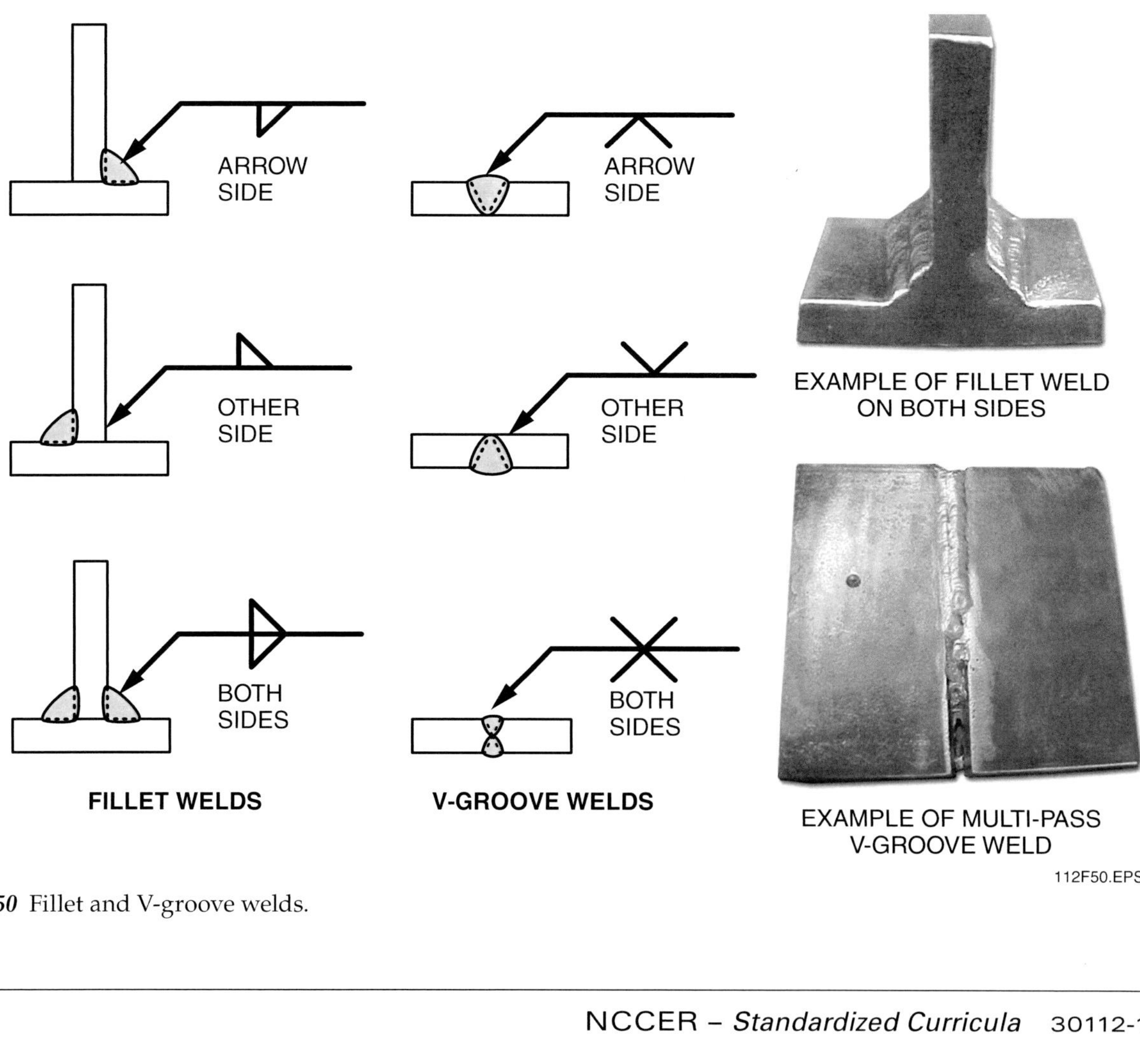

Figure 50 Fillet and V-groove welds.

given. Equal-leg fillet welds are sized by placing the size to the left of the fillet weld symbol or as defined in a drawing note. If the fillet weld has unequal legs, the size of each leg is given in parentheses or brackets to the left of the fillet weld symbol with a drawing or explanation of leg orientation. When there is no dimension to the right of the fillet weld symbol, the weld extends the full length of the joint.

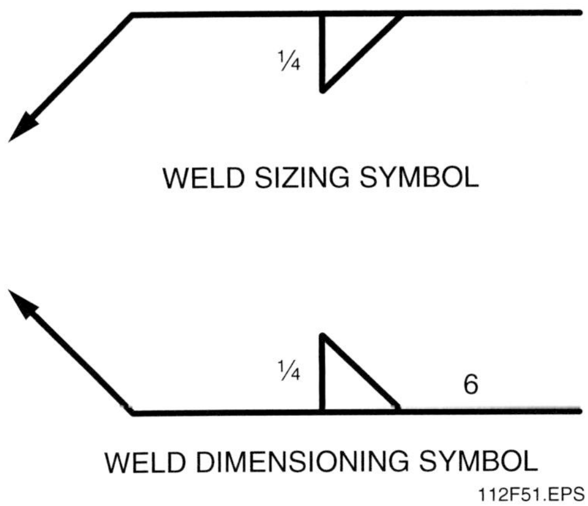

Figure 51 Sizing and dimensioning welds.

8.6.3 Weld Symbol Measurements for Groove Welds

A groove weld is made in the groove formed by two beveled pieces of metal placed edge to edge (*Figure 53*). Several factors must be considered when sizing all types of groove welds.

Groove preparation is the depth to which the groove penetrates into the base metal. When the groove preparation extends all the way through the joint, as shown in *Figures 54(A)* and *54(B)*, no preparation size needs to be shown. When the groove preparation extends only part of the way through the joint, as shown in *Figures 54(C)* and *54(D)*, the depth of the groove is shown to the left

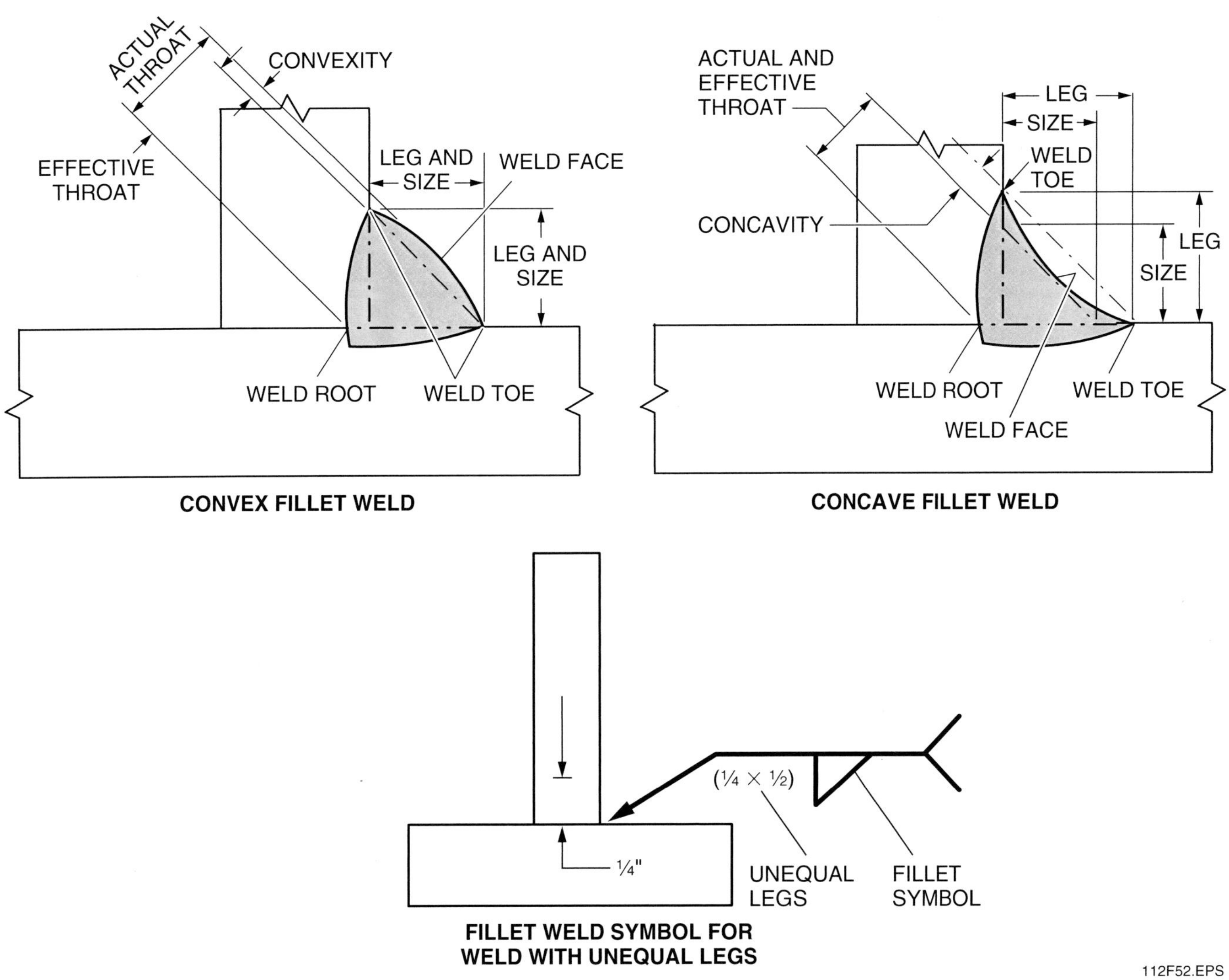

Figure 52 Fillet weld profiles and weld symbol.

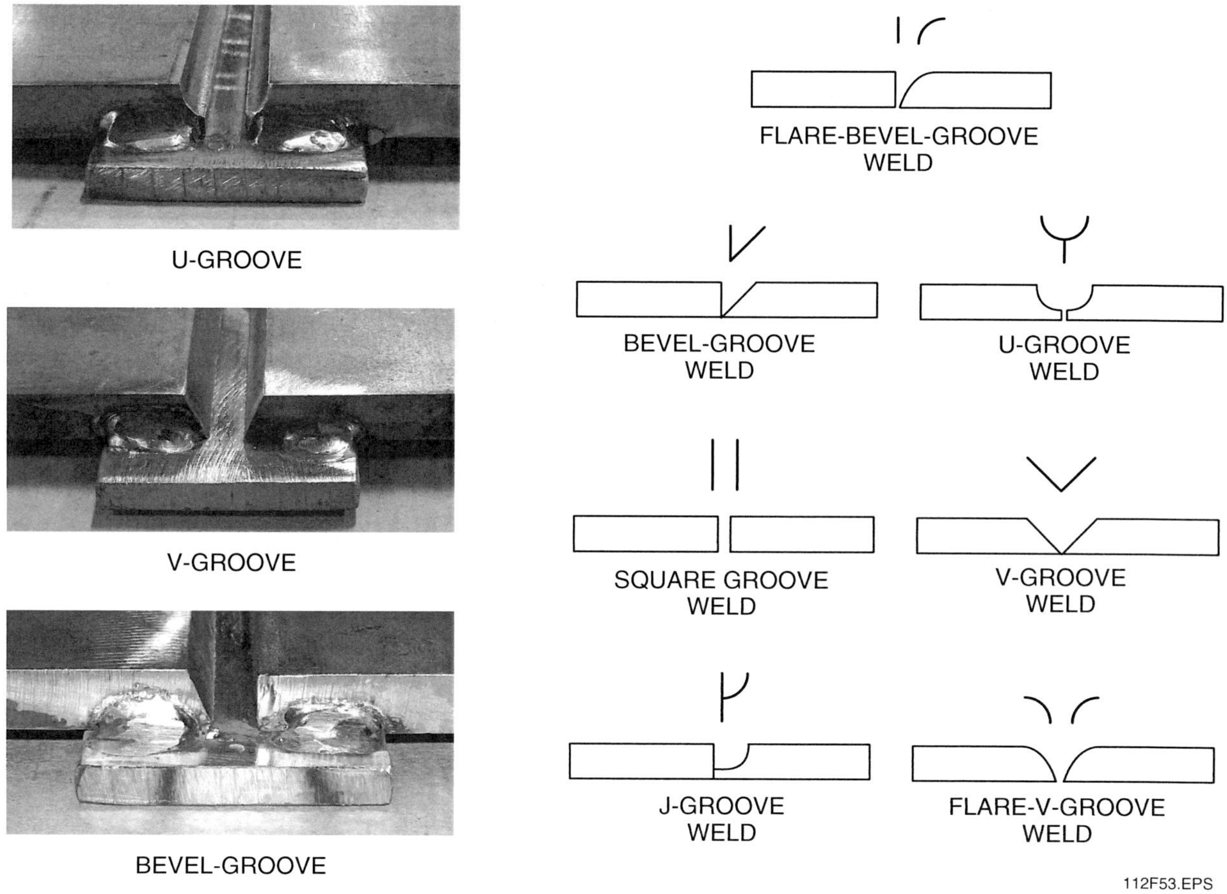

Figure 53 Types of groove welds.

of the symbol. *Figures 54(C)* and *54(D)* show sizing a groove with a depth of ¼".

For partial joint penetration (PJP) groove welds, both the depth of groove penetration and weld size (joint penetration) are shown, as in *Figures 55(A)* and *55(B)*. The depth of the groove is shown first and the weld size is shown next in parentheses. In this example the groove depth is 0.25" and the weld size is 0.385". For complete joint penetration (CJP) groove welds, the weld size (joint penetration) is the thickness of the base metal, not including face and root reinforcement.

If a root opening is required, the root opening dimension of a groove weld is shown inside the symbol. In *Figure 56*, the root opening is ⅛" and is shown on only the arrow-side symbol because it is common to both grooves.

The angle for the type of groove is shown above or below the welding symbol, depending upon whether the symbol is above or below the reference line. In *Figure 57*, the groove angle is shown as 45 degrees and is shown only by the arrow-side symbol because the groove is the same on both sides. The bevel angle for groove preparation on each workpiece is 22½ degrees.

8.6.4 Weld Drawings

Weld drawings come in a variety of styles. Official weld drawings are stamped, dated, and signed by the engineer or architect. Unofficial weld drawings may be scribbled on a piece of paper or drawn in chalk on a piece of scrap steel. Whatever method is used, the person doing the welding has to interpret the provided drawing and then perform the indicated task(s). *Figure 58* shows a weld drawing from the American Welding Society. Such drawings are used to actually certify welders.

Other drawings are not as complex, but just as important. *Figure 59* shows a weld drawing with different views of a mounting bracket that needs to be fabricated. The bracket is shown to the right of the drawing. The weld symbol is shown with the Top View drawing.

The information provided in this section in no way makes you a certified welder, but it does give you an overview of the symbols used on weld drawings.

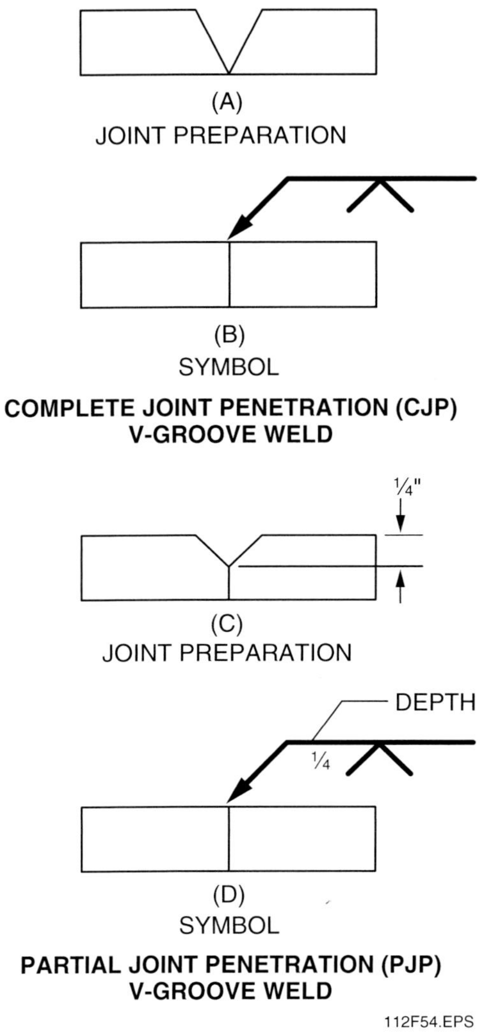

Figure 54 Sizing groove welds.

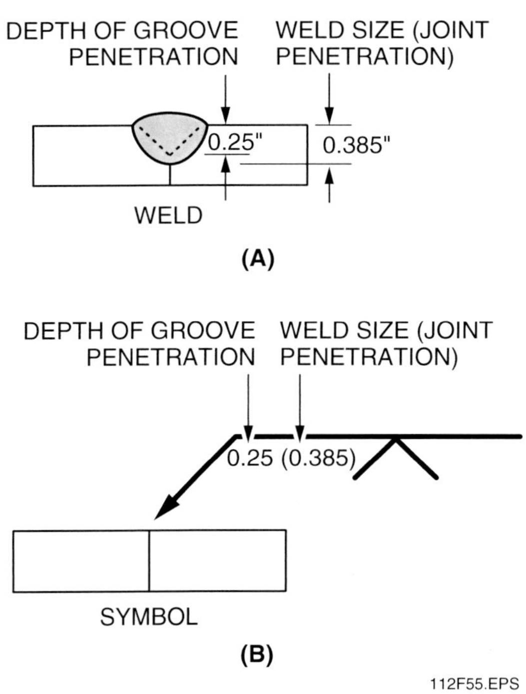

Figure 55 Sizing partial joint preparation (PJP) welds.

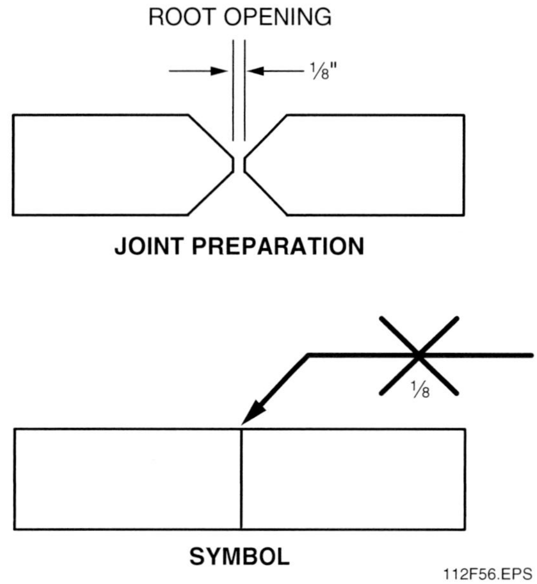

Figure 56 Sizing the root opening.

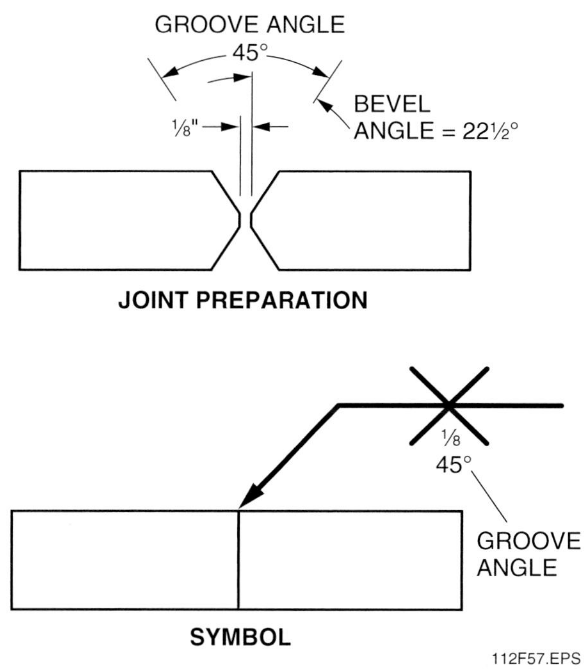

Figure 57 Specifying the groove angle.

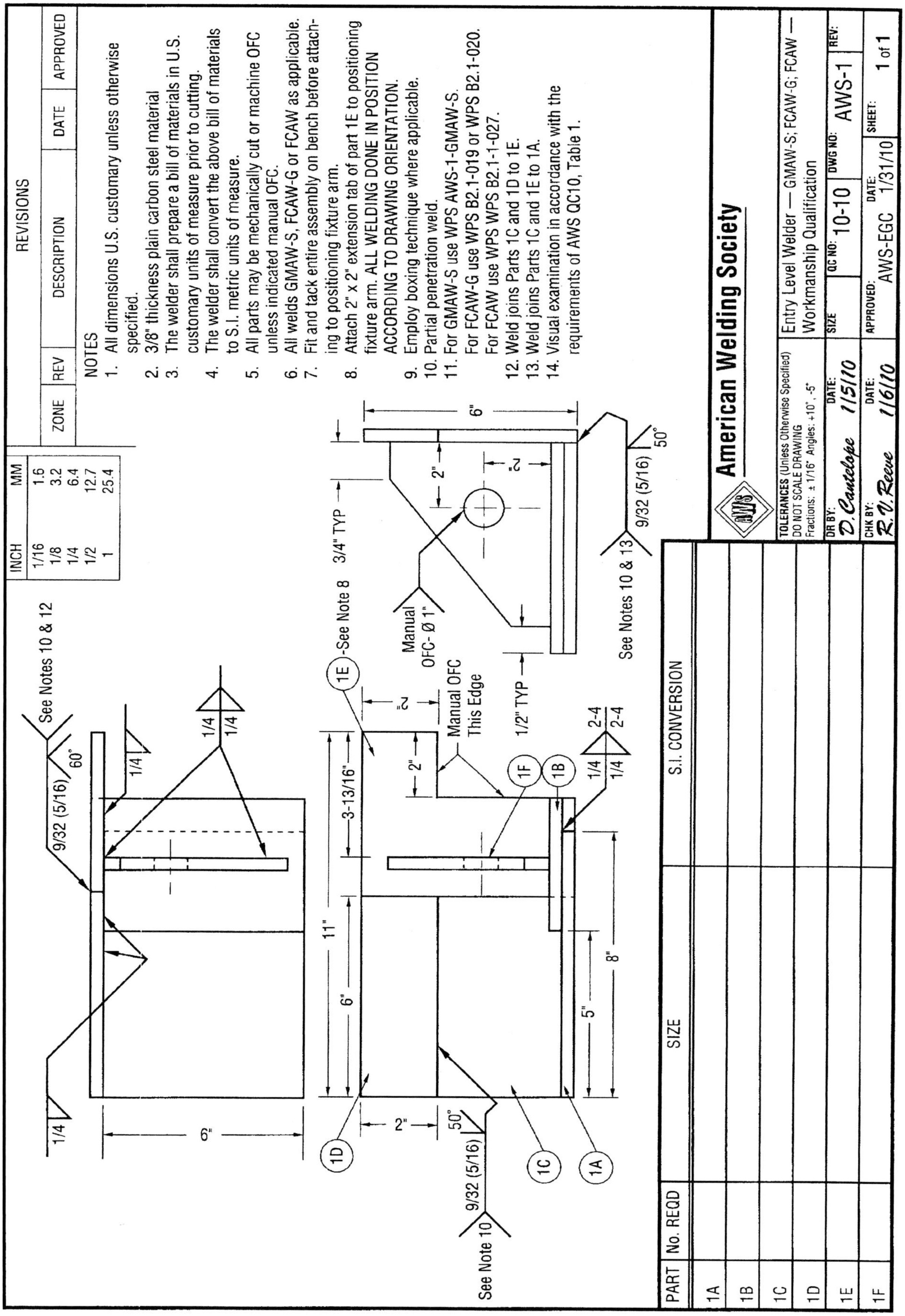

Figure 58 AWS weld drawing.

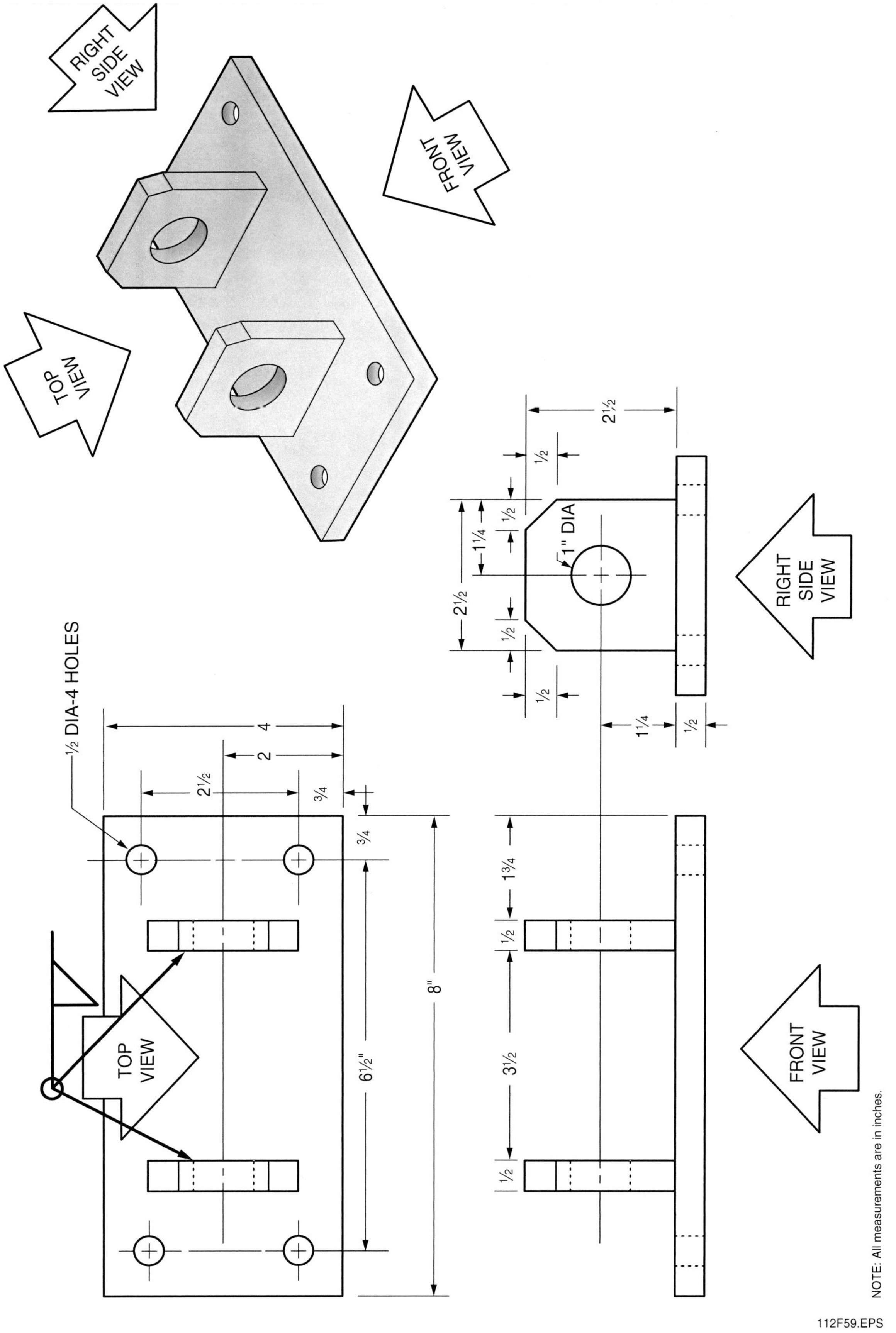

Figure 59 Weld drawing with different views.

9.0.0 Arc Welding Machines

Several different types of SMAW welding machines are available to produce the current necessary for welding. They include the following:

- Transformer welding machines
- Transformer-rectifier welding machines
- Engine-driven generators and alternators

Both transformer and transformer-rectifier welding machines require electrical power from commercial power lines to operate. The current from commercial power lines is high-voltage, low-amperage current. When it comes into the welding machine, it is called the primary current. The primary current flows first from the power lines to an electrical circuit breaker or disconnecting fuse box. From the disconnecting device, the current flows to a receptacle. The disconnecting device can be used to shut off the flow of current to the receptacle. The welding machine is plugged into the receptacle. The primary current required for welding machines can be 110V or 220V single-phase or 480V three-phase.

WARNING

Coming into contact with the primary circuit of a welding machine can cause electrocution. Ensure that proper safety precautions are taken to avoid contact.

The operator controls in the front of the welding machine vary from one welding machine to the next as you can see in *Figure 60*. The unit at the top of the figure has a built-in gasoline engine and thus has an Engine Start switch, for example. The unit at the bottom of the figure is powered by electricity, so it does not require an engine control.

10.0.0 Arc Welding With SMAW Equipment

Before welding can take place, the area has to be made ready, the welding equipment must be set up, and the metal to be welded must be prepared. The following sections explain how to set up equipment for welding and how to run a bead, using SMAW equipment and electrodes

10.1.0 Preparing the Welding Area

To practice welding, a welding table, bench, or stand is needed (*Figure 61*). The welding surface must be steel, and provisions must be made for placing **weld coupons** out of position.

To set up the area for welding, follow these steps:

Step 1 Ensure that the area is properly ventilated. Make use of doors, windows, and fans.

Step 2 Check the area for fire hazards. Remove any flammable materials before proceeding.

Step 3 Check the location of the nearest fire extinguisher. Do not proceed unless the extinguisher is charged and you know how to use it.

Step 4 Set up flash shields around the welding area.

10.2.0 Preparing the Weld Coupons

Welding activities are practiced on small sections of steel called coupons. For this module, the weld coupons should be carbon steel plate, ¼" to ¾" thick. Use a wire brush or grinder to remove heavy mill scale or corrosion. Prepare weld coupons to practice the welds indicated as follows:

- *Striking an arc* – The coupons can be any size or shape that is easily handled.
- *Running beads* – The coupons can be any size or shape that is easily handled.
- *Overlapping beads* – The coupons can be any size or shape that is easily handled.
- *Fillet welds* – Cut the metal into 4" × 6" rectangles for the base and 3" × 6" rectangles for the web.

Figure 62 shows the weld coupons for fillet welding.

Steel for practice welding is expensive and difficult to obtain. Every effort should be made to conserve and not waste the material that is available. Reuse weld coupons until all surfaces have been welded upon. Weld on both sides of the joint, and then cut the weld coupon apart and reuse the pieces. Use material that cannot be cut into weld coupons to practice striking an arc and running beads.

10.3.0 Preparing the Welding Machine

Ironworkers can expect to find different types and makes of welding machines in welding shops and on jobs in the field. Most machines will have no operator manuals with them. As a trainee, and even as a trained ironworker, you must be able to recognize the different types of welding machines and figure out how to safely set up and operate them. Select a welding machine to use and then follow these steps to set it up for welding:

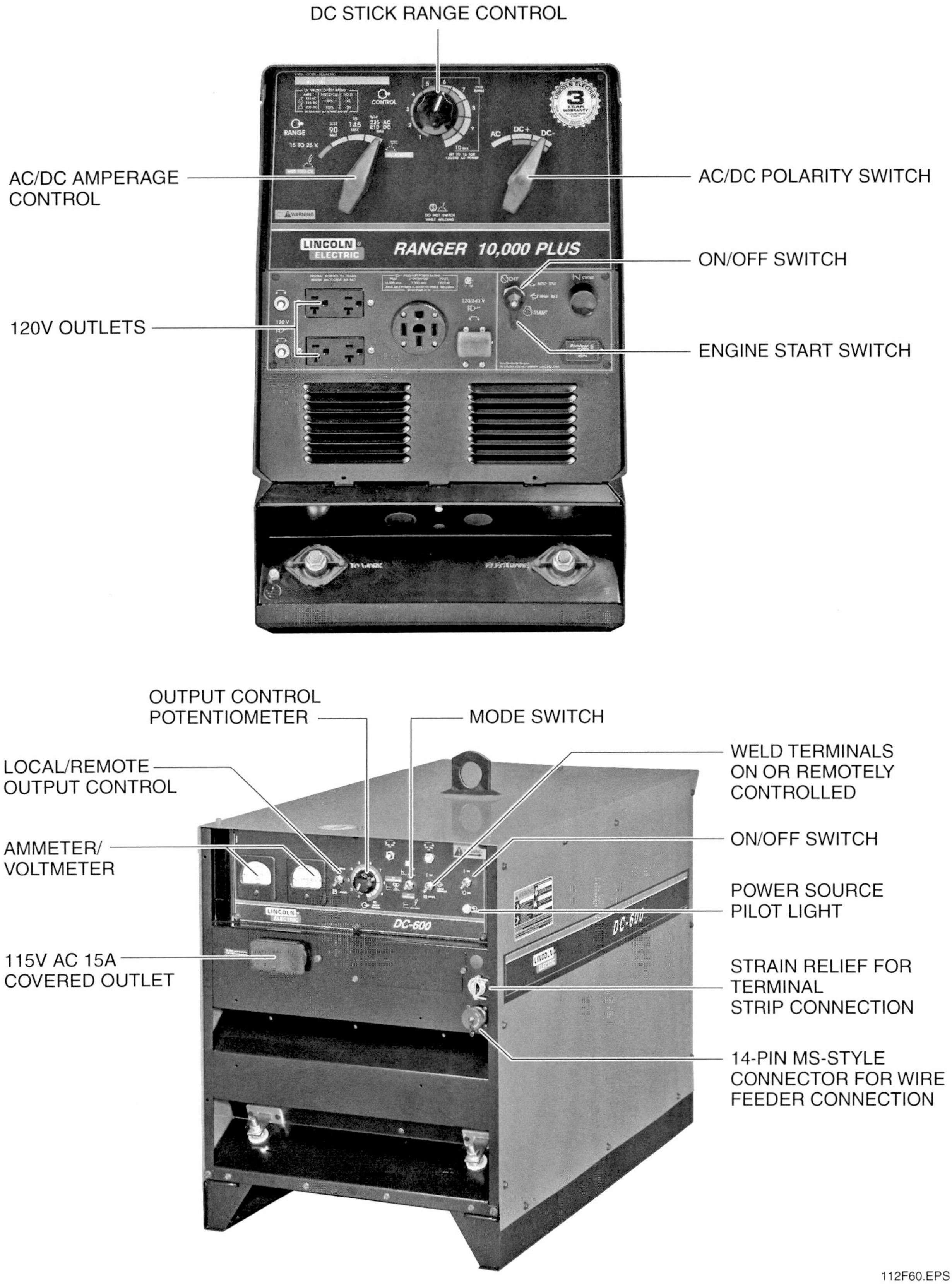

Figure 60 Typical AC/DC and DC welding machines.

Figure 61 Welding station.

Step 1 Verify that the welding machine can be used for DC welding.

NOTE

An AC welding machine can be used if a DC machine is not available. If an AC welding machine is used, use E6011 and E7018 electrodes.

Step 2 Verify the location of the primary current disconnect to the machine.

Step 3 Check the area for proper ventilation.

Step 4 Set the polarity to direct current electrode positive (DCEP).

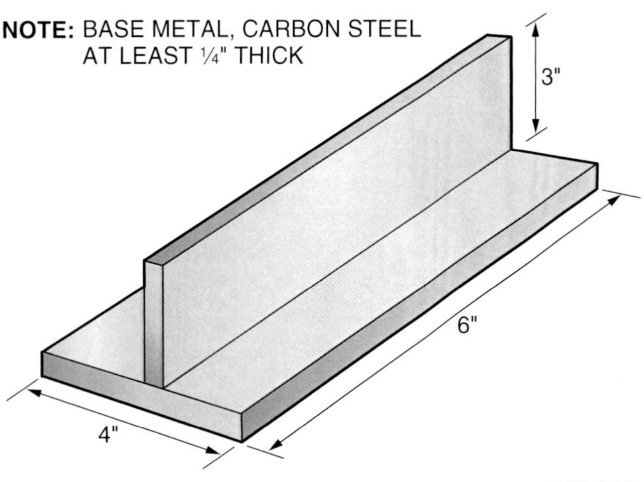

Figure 62 Fillet weld coupons.

Step 5 Connect the clamp of the workpiece lead to the workpiece.

WARNING

Even though workpiece leads and clamps are sometimes called ground leads or clamps, they may not actually be grounded. In this case, the full open-circuit voltage of the welding machine may exist between the workpiece lead or clamp and any grounded object.

Step 6 Set the amperage for the electrode type and size to be used. Typical settings are shown in *Table 6*.

NOTE

Amperage recommendations vary by manufacturer, position, current type, and electrode brand. For specific recommendations, refer to the manufacturer's literature for the electrode being used.

Step 7 Ensure that the electrode holder is not touching the workpiece lead clamp, the workpiece, or a grounded object.

Step 8 Turn on the welding machine.

10.4.0 Striking an Arc

The first step in starting a weld is striking the arc. To strike an arc, touch the end of the electrode to the base metal and then quickly raise it to the correct arc length. The general rule is that the arc length should be the diameter of the electrode being used. If a ⅛" electrode is being used, the arc length should be ⅛". The arc length is measured from the end of the electrode core to the base metal. If the electrode has a heavier flux coating, such as E7018 or E6013, the end of the electrode core will

GOING GREEN

Recycling Metal

Salvage any coupons or scrap metal that is no longer usable for welding practice and place it into a recycling bin. Almost all metals are recyclable and valuable. One technical school took in donated steel, used it for cutting and welding practice, and then sold all their recyclable metals to fund their school's annual picnic. The bottom line is that metal that is no longer usable for cutting or welding practice can be recycled and kept out of the landfills. Recycled metals are one of America's largest exports.

Table 6 Amperage Settings for Electrodes

Electrode	Size	Amperage
E6010	⅛"	75A to 130A
E6010	5⁄32"	90A to 175A
E6011	⅛"	75A to 120A
E6011	5⁄32"	90A to 160A
E7018	⅛"	90A to 150A
E7018	5⁄32"	120A to 190A
E6013	⅛"	110A to 150A
E6013	5⁄32"	150A to 200A

be recessed in the coating. For these electrodes, the visible arc length (from the end of the coating to the base metal) should be slightly less to compensate for the part of the arc that is recessed. This will ensure that the actual arc length is correct. *Figure 63* shows arc lengths for a ⅛" rod.

There are two ways to strike an arc: the scratching method and the tapping method.

10.4.1 Scratching Method

The scratching method (*Figure 64*) is the easiest way to strike an arc, and it is done by trainees who are learning to weld. It is similar to striking a match. The end of the electrode is simply scratched along the base metal to establish an arc. When the arc has been established, the electrode is raised to establish the correct arc length.

When attempting to strike an arc using a cold, partially consumed electrode, the cup formed by the hardened flux sheath extending beyond the metal core (*Figure 65*) may not allow the core to contact the workpiece to establish an arc. Scratch striking or hard tapping can be used to break off the sheath. Unfortunately, this sometimes causes large pieces of the flux coating to break off the sides, exposing part of the core. Electrodes with missing flux are difficult to strike. If an arc is struck, it will be unstable and will blow or wander until the core is consumed up to where the entire flux sheath cup is reestablished. During this time, poor welds will occur.

One method of removing the flux sheath without chipping flux away from the sides of the rod is to draw the end of the rod across the face of a medium-coarse file. If code requirement welding is being performed, electrodes with chipped, missing, or cracked flux coatings must be discarded.

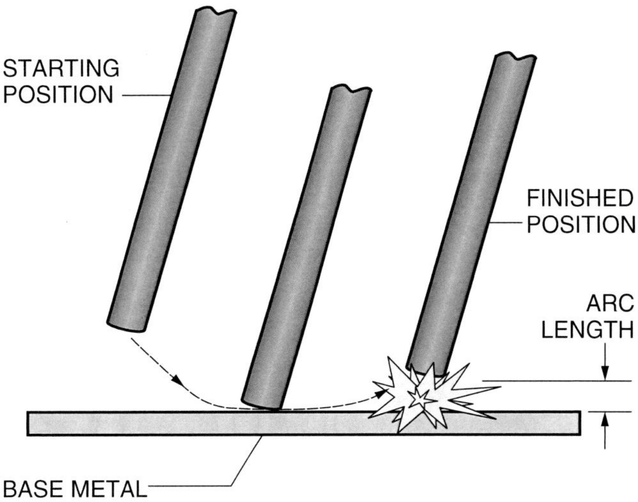

Figure 64 Scratching method of striking an arc.

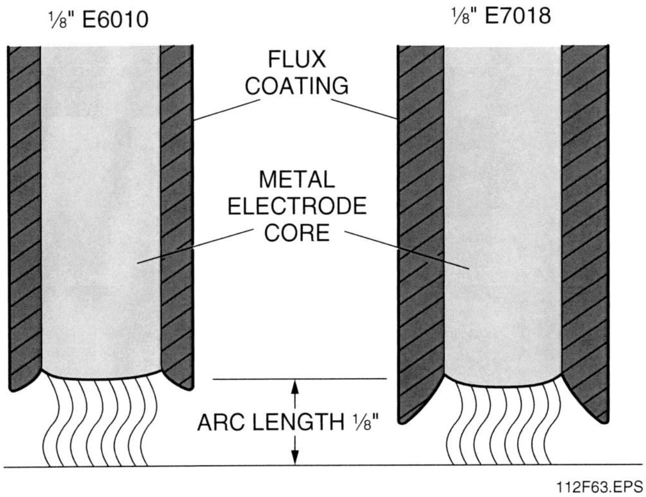

Figure 63 Arc lengths.

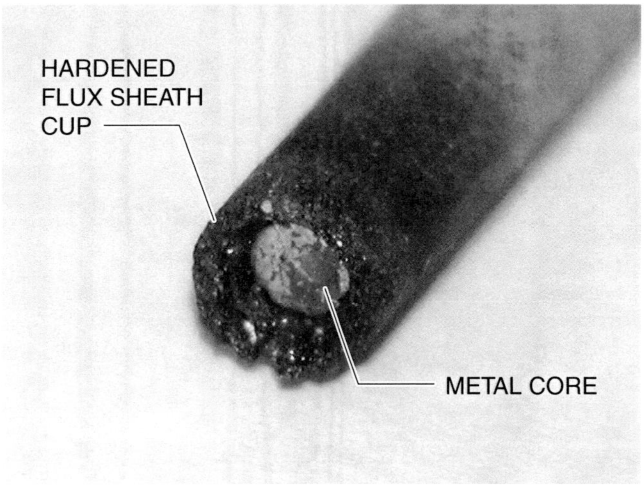

Figure 65 Cold electrode.

10.4.2 Tapping Method

The tapping method (*Figure 66*) is the best method for establishing an arc when using a transformer DC welding machine. It is more difficult and takes more practice to perfect. But as ironworkers develop their skills, it is the method that should be used. The scratching method leaves **arc strikes** on the base metal that are not allowed by the welding codes unless they occur within the welded area. With the tapping method, the arc is usually established just after the point where the welding should begin (¼" to ½"). The arc is then moved back to the correct starting point as it stabilizes. To strike an arc with the tapping method, move the electrode quickly to the base metal, lightly tap it, and then raise the electrode to establish the correct arc length.

> **NOTE**
> Electrodes do get stuck (attached) to the base metal or weld and the welder operator must break them loose. When an electrode gets stuck, break it loose, and carefully inspect its contact end. If any of the flux coating is missing, cracked, or badly burned, discard the electrode.

10.4.3 Practicing Striking and Extinguishing an Arc

Striking the arc correctly is an important step in learning to weld. Practice both methods using 1/8" E6010 or E6011 electrodes until you can strike and maintain an arc consistently. When striking an arc, place the base metal flat on the welding table. To extinguish the arc at the end of a weld, quickly lift the electrode away from the work.

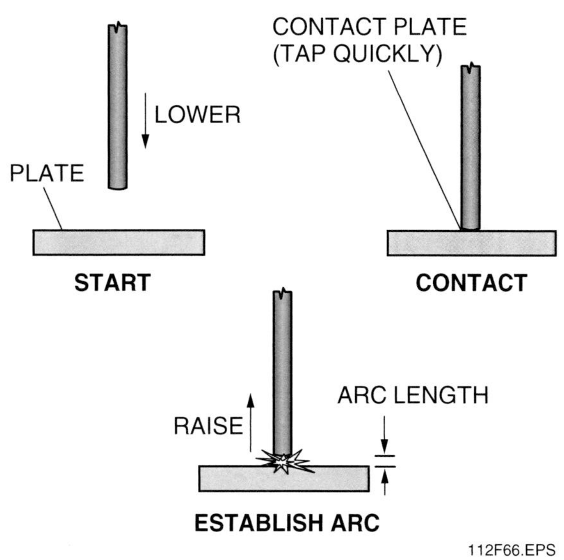

Figure 66 Tapping method of striking an arc.

> **NOTE**
> Striking an arc with E7018 (low-hydrogen) electrodes is more difficult than with most other electrodes because they tend to stick to the base metal more easily. Practice striking an arc with E7018 after you become more skilled at running beads with E6010 or E6011.

When practicing striking an arc, most trainees generally experience these two problems:

- The end of the electrode welds itself to the base metal. This is caused by keeping the electrode in contact with the base metal too long or by trying to maintain too short an arc length. If this condition occurs, free the electrode by quickly moving the electrode holder from side to side. If this fails, pull the holder from the electrode and remove the electrode from the base metal with pliers. When the electrode is released from the holder, arcing generally occurs between the holder jaws and the electrode. If arcing has occurred, inspect the holder jaws for any heavy pitting or gouging damage, and replace them if necessary.
- The arc breaks (goes out). This is caused by raising the end of the electrode too far above the base metal.

These problems will be eliminated as you gain experience in striking and controlling the arc.

10.5.0 Dealing with Arc Blow

When current flows, especially DC current, strong magnetic fields may be created. The magnetic fields tend to concentrate in corners, in deep grooves, or in the ends of the base metal. When the arc approaches these concentrated magnetic fields, it is deflected. This phenomenon is known as **arc blow**. In some cases, ferrous metal welding fixtures or jigs that are part of a DC welding current path and subjected to repeated current flow can become magnetized and contribute to arc blow. In AC welding, arc blow is rarely a problem because the magnetic field is constantly reversing at twice the frequency of the primary power source. This effectively cancels any strong magnetic field effects. DC arc blow can cause defects such as excessive weld spatter and porosity.

If arc blow occurs, try one or more of these methods to control it:

- Change the position of the workpiece lead clamp. This changes the flow of welding current, affecting the way the magnetic fields are created.

- Shorten the arc length. With a shorter arc length, the magnetic field has less effect on the arc.
- Change the angle of the electrode (*Figure 67*). The normal electrode angle is 10 degrees to 15 degrees in the direction of travel. Raising the electrode angle toward 90 degrees or, in extreme cases, up to 20 degrees in the opposite direction of travel will compensate for the arc blow.
- If possible, tack-weld the workpieces together at the ends and in the middle.
- Periodically degauss welding fixtures or jigs with an AC-powered degaussing coil or probe.

10.6.0 Making Stringer and Weave Beads

A **stringer bead** is a weld bead that is made with very little or no side-to-side motion of the electrode. The width of a stringer bead will vary with the electrode type. A **weave bead** is a weld bead that is made with a side-to-side motion of the electrode. Outside of welding joints, stringer and weave beads (*Figure 68*) are used for resurfacing or hardsurfacing.

NOTE

The width of stringer and weave beads is usually specified in the drawing welding symbols, welding code, or welding procedure specification being used for the work at your site. Do not exceed the widths specified for the work.

10.6.1 Practicing Stringer Beads

Practice running stringer beads in the flat position using ⅛" E6010 or E6011 electrodes. After striking the arc, the electrode axis angle should be a 10- to 15-degree **drag angle** in the direction of travel for the weld and at a zero **work angle** that is perpendicular (90 degrees) to the base metal across the weld. *Figure 69* shows electrode axis angles.

Travel angles used for SMAW and other types of welding are categorized as either drag or **push angles**. An electrode drag angle is a travel angle in which the electrode axis points at the weld bead during the running of a weld bead. A push angle is the opposite condition; it occurs when the electrode axis points away from the weld bead and toward the direction of travel when running a weld bead. Except for vertical welds, SMAW is usually accomplished with an electrode drag angle. Electrode work angle is the side-to-side tilt of a plane containing the electrode axis from a perpendicular line to the major work surface. The plane is formed by the intersection of the **weld**

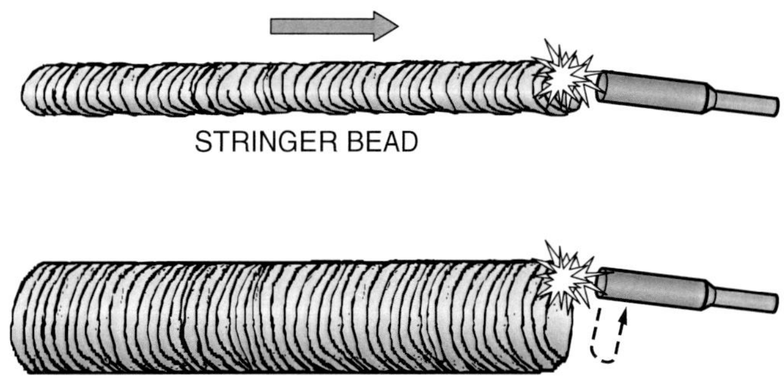

Figure 68 Stringer and weave beads.

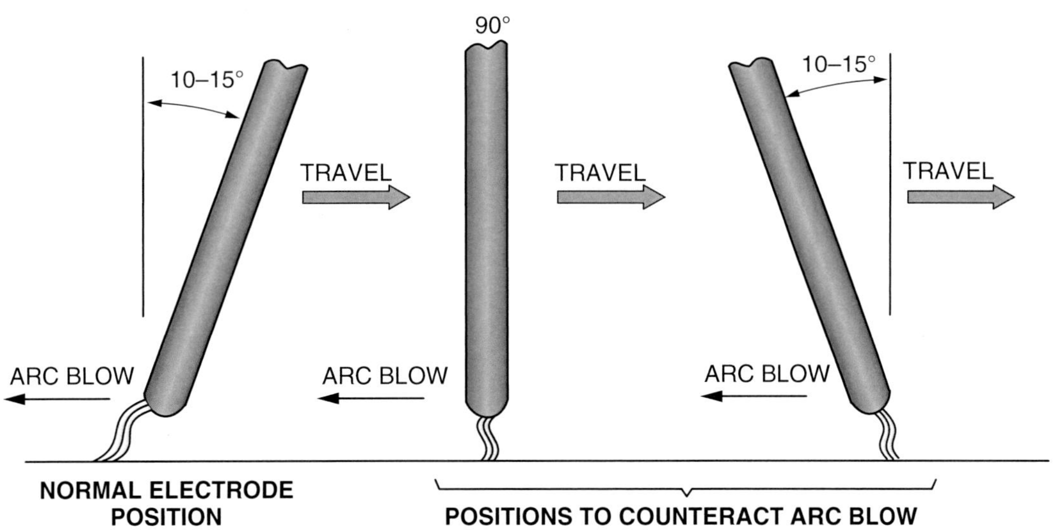

Figure 67 Controlling arc blow.

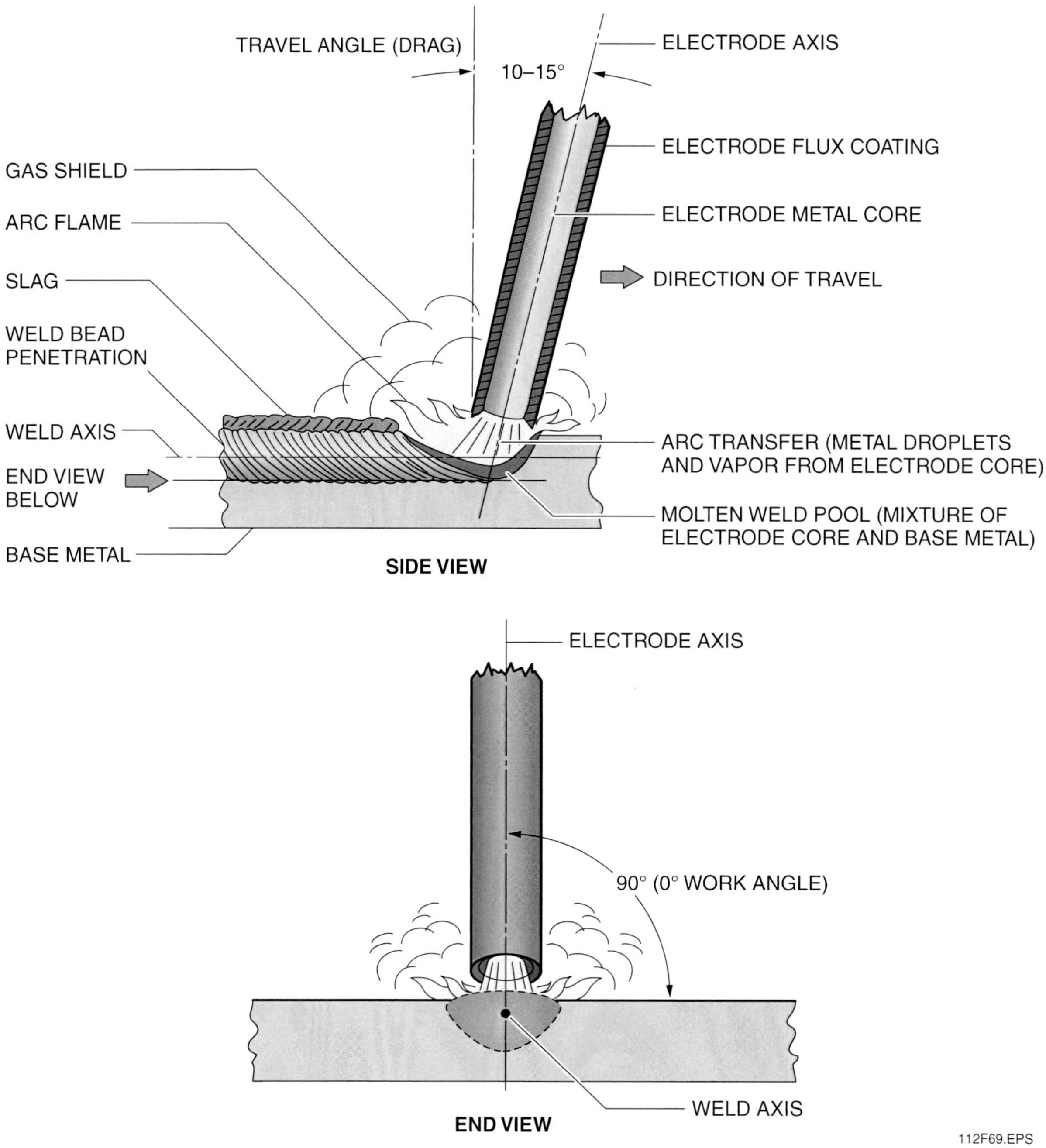

Figure 69 SMAW characteristics with proper travel and work angle.

axis and the electrode axis. Work angles typically range from 0 to 25 degrees. In the case of a single bead being run on a flat plate, the work angle is zero. The work angle for a fillet weld in a T- or corner joint is always taken from a perpendicular line to the non-butting member.

The whipping motion (*Figure 70*), also called a stepping motion, can be used when depositing the stringer bead to control the weld puddle. It is performed by moving the electrode up and forward about ¼", then down and backward about 3⁄16". A short pause at the end of the backward travel deposits the weld metal. The exact length of travel can vary. By momentarily lengthening and advancing the arc, the molten weld pool is allowed to cool. Lengthening the arc lowers the arc temperature (amperage) and reduces metal transfer from the rod core. Advancing the arc preheats the base metal ahead of the weld and burns off contaminants.

You should be in a relaxed, comfortable position when welding. Because of the limited view of the welding area through a dark helmet lens, a beginner may sway due to a loss of sense of balance. To counteract this, sit or lean against something to achieve and maintain a stable and relaxed position. This reduces fatigue and ensures personal safety.

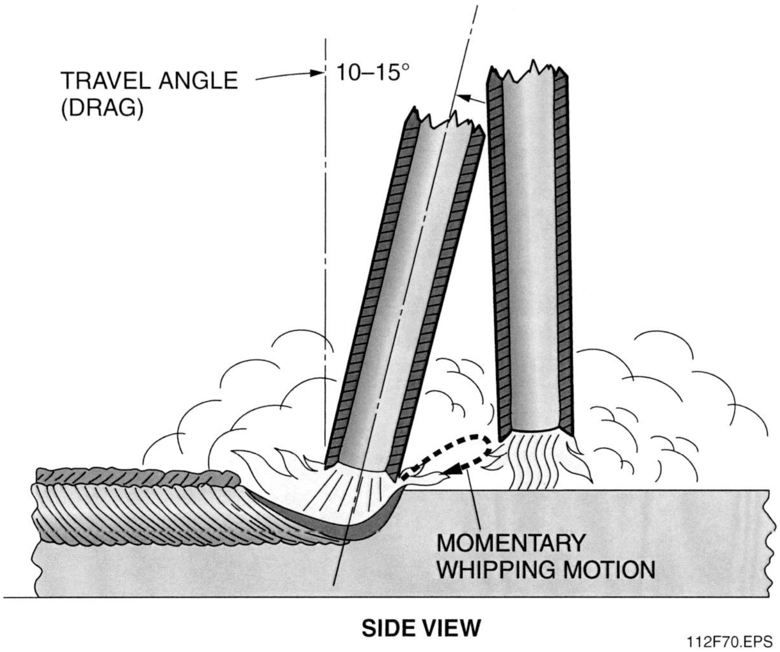

Figure 70 Whipping motion.

Beginners must learn to view the entire electric arc welding work area through the shaded lens of a helmet and to listen to the sound of the arc. At first, a beginner's focus is normally concentrated only on the electric arc, because of the difficulty of striking and keeping an arc established. Once you have managed to strike and maintain an arc, however, the tremendous light and heat energy generated within the arc are usually fascinating and distracting for a time. With practice, you will gradually be able to shift focus and see all sides of a molten weld pool, the weld buildup at the trailing edge of the pool, the cooling slag cover over the weld buildup, and the adjacent welding area on the workpiece. You will then become attuned to the correct sound of the arc for the rod and work being welded, and the arc itself may not even be noticed anymore. Once you are able to view the entire weld area and listen to the arc, running straight and correct welds is relatively easy.

Carefully observe and listen to the weld as it is being made. If the arc length, speed, angles, and motions are correct, a distinctive frying sound can be heard. Experiment by making minor changes in these factors and notice the effects.

Follow these steps to run stringer beads:

Step 1 Shield your eyes, strike the arc, and move the electrode back slightly to the correct starting point. Then position the electrode for the travel angle and perpendicular to the base metal.

Step 2 Hold the arc in place until the weld puddle widens to about two times the diameter of the electrode.

Step 3 Slowly move the arc forward, keeping a constant arc length. As necessary, use the whipping motion and adjust the forward speed to control buildup and preheat the weld zone.

> **WARNING**
>
> Always wear face protection to prevent hot slag from striking your face.

Step 4 Continue to weld until a bead about 2" to 3" long is formed and then break the arc by quickly lifting the electrode straight up. A crater will be left at the point where the arc was broken.

Step 5 Chip the weld slag with a chipping hammer.

Step 6 Clean the bead with a wire brush.

Step 7 Have your instructor inspect the bead for the following characteristics:

- Straightness
- Uniform appearance on the bead face
- Smooth, flat transition with complete fusion at the toes (edges) of the weld
- No porosity

- No undercut at the toes
- No inclusions
- No cracks
- No overlap

Step 8 Continue welding beads until you can make acceptable welds every time. *Figure 71* shows proper and improper weld beads.

The welding craft has a slogan called LASH that is used to ensure good welding results. Here is an explanation of LASH:

L (*length of arc*) – The distance between the electrode and the base metal (usually the electrode diameter).

A (*angle*) – Two angles are critical:

- *Travel angle* – The longitudinal angle of the electrode in relation to the axis of the weld joint
- *Work angle* – The traverse angle of the electrode in relation to the axis of the weld joint

S (*speed*) – Travel speed is measured in inches per minute (IPM). The width of the weld will determine if the travel speed is correct.

H (*heat*) – Controlled by the amperage setting and dependent upon the electrode diameter,

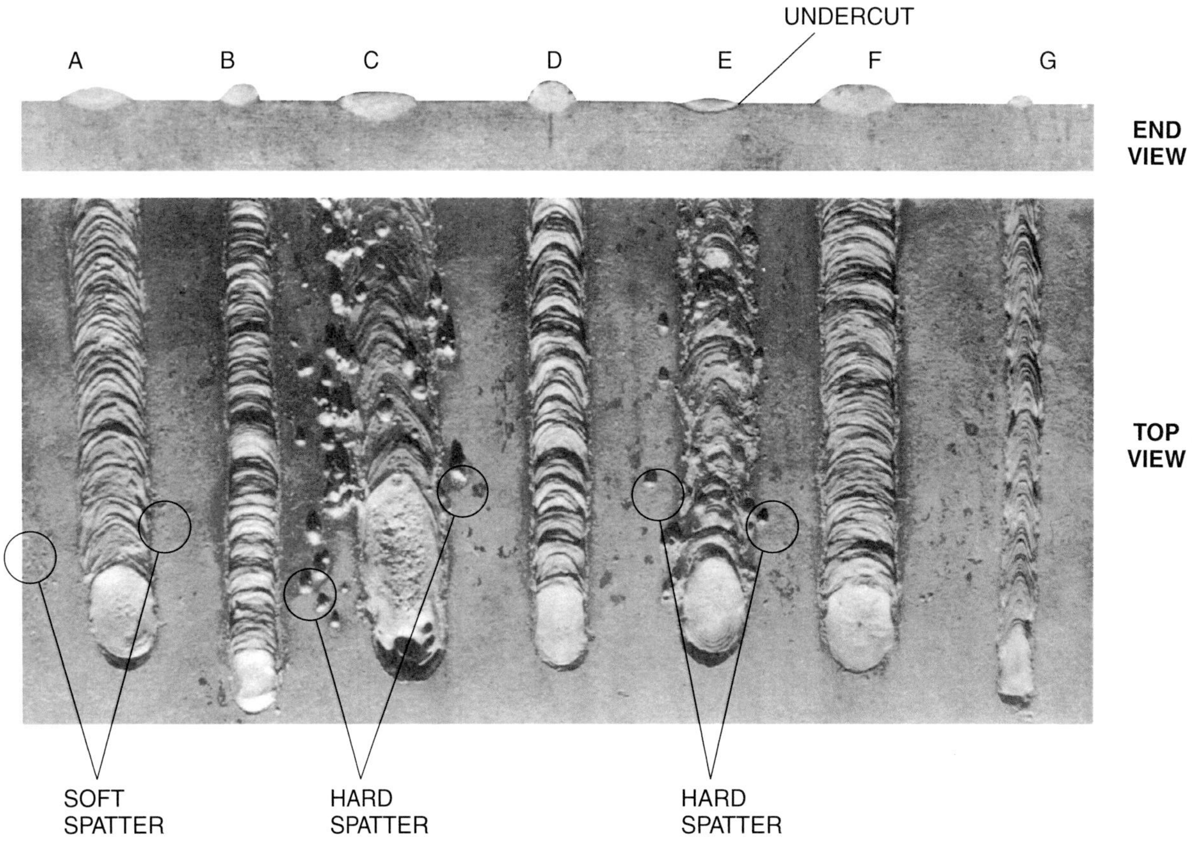

A = Correct current, arc length, and travel speed. Note the easily removed spatter (soft spatter).

B = Current set too low.

C = Current set too high. Note the hard spatter (spatter firmly bonded to base material that must be ground or chiseled off). Note the pointed ends of the bead indicating the weld pool was too hot and cooled too slowly. Impurities are usually trapped in the weld due to the slow cooling.

D = Arc length too short (narrow, high bead caused by arc pressure).

E = Arc length too long. Note the hard spatter and bead undercut of the edges.

F = Travel speed too slow (wide high bead).

G = Travel speed too fast. Note the pointed ends of bead.

112F71.EPS

Figure 71 Effect of current, arc length, and travel speed on SMAW beads.

base metal type, base metal thickness, and the welding position.

10.6.2 Making Restarts

A **restart**, sometimes referred to as a tie-in, is the point where one weld bead stops and another starts (*Figure 72*). Restarts are important because most SMAW welds cannot be made without at least one restart, and an improperly made restart creates a weld defect. A restart must be made so that it blends smoothly into the rest of the weld and does not stand out. The technique for making a restart is the same as for stringer and weave beads. Follow these steps to make a restart:

Step 1 Just before the previous electrode is used up, quickly increase the weld speed to taper the weld for ¼" to ⅜", and then break the arc.

Step 2 Chip any slag from the tapered section and crater.

Step 3 With a new electrode, restrike the arc in front of the crater and in line with the weld. (The welding codes do not allow arc strikes outside of the area to be welded.)

Step 4 Move the electrode back onto the tapered weld section and develop the weld puddle with a slight circular motion, maintaining the correct arc length and electrode angles.

Step 5 Start the forward motion as the bead achieves the proper width, and continue to weld.

Step 6 Inspect the restart. A properly made restart will blend into the bead, making it hard to detect.

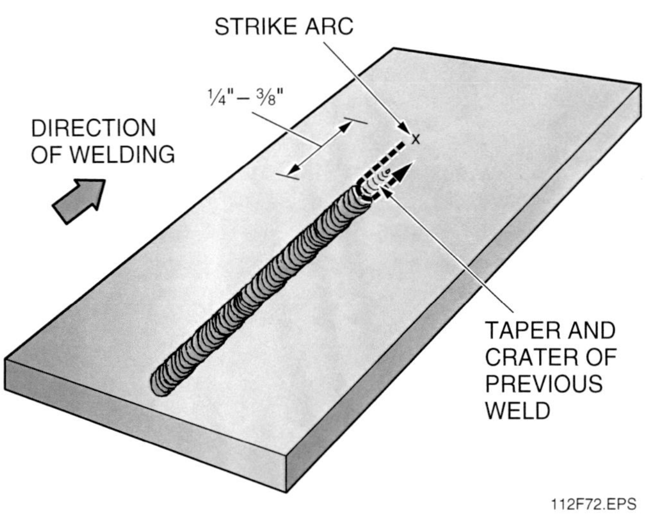

Figure 72 Making a restart.

If the restart has undercut, not enough time was spent on the taper or the crater to fill it. If the undercut is on one side or the other, use more of a side-to-side motion as you move into the taper. If the restart has a hump, it was overfilled; too much time was spent before resuming the forward motion.

Continue to practice restarts until they are correct. Use the same techniques for making restarts whenever performing SMAW for other types of welds.

10.6.3 Terminating a Weld

Terminations are made at the end of a weld. A termination leaves a crater. When making a termination, the welding codes require that the crater be filled to the full cross section of the weld. This can be difficult because most terminations are at the edge of a plate where welding heat tends to build up, making filling the crater more difficult.

The technique for making a termination is basically the same for all SMAW (*Figure 73*):

Step 1 As you approach the end of the weld, start to stand the electrode up toward 90 degrees, and slow the forward travel.

Step 2 Stop forward movement about ⅛" from the end of the plate and slowly angle the electrode to about 10 degrees toward the start of the weld.

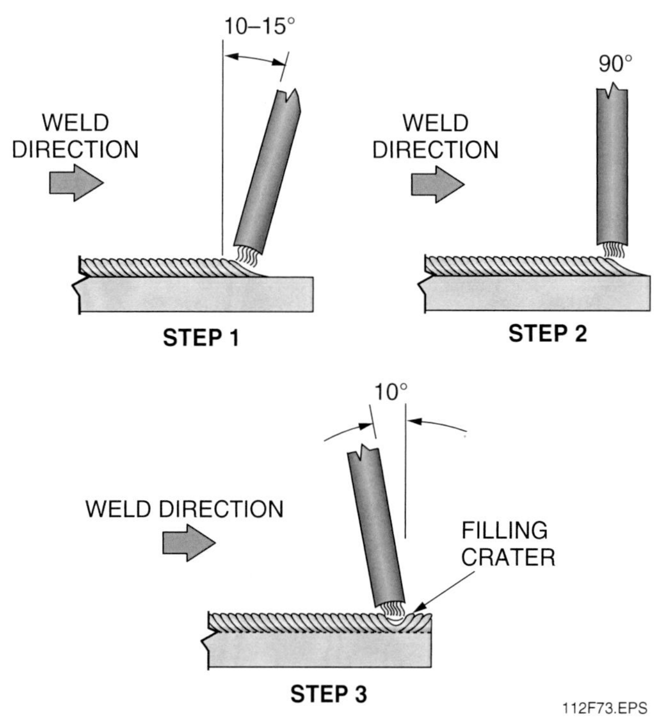

Figure 73 Weld termination.

Step 3 Move about ⅛" toward the weld, and break the arc when the crater has been filled.

Step 4 Inspect the termination. The crater should be filled to the full cross section of the weld.

10.6.4 Practicing Weave Beads

Practice running weave beads in the flat position using ⅛" E6010 or E6011 electrodes. After striking the arc, the electrode angle should be 10 to 15 degrees in the direction of travel and perpendicular (90 degrees) to the base metal.

The weave bead is made by moving the electrode back and forth. Many different patterns can be used to make a weave bead, including zigzags, Js, crescents, squares, circles, and figure 8s (*Figure 74*). When making a weave bead, use care at the toes to ensure proper tie-in to the base metal. To ensure proper tie-in at the toes, slow down or pause slightly at the edges. Pause longer when using the zigzag motion. The pause at the edges will also flatten out the weld, giving it the proper profile.

Carefully observe and listen to the weld as it is being made. If the arc length, speed, angles, and motions are correct, a distinctive frying sound will be heard. Experiment by making minor changes in these factors and notice the effects. *Figure 75* shows an end view of how stringer and weave beads should be made to fill in the groove between two plates. Also, notice the weld symbols above the grooves.

Follow these steps to run weave beads:

Step 1 Strike the arc and position the electrode for a drag angle. Make sure the electrode is perpendicular to the base metal (0-degree work angle).

Step 2 Hold the arc in place until the weld puddle widens to about two times the diameter of the electrode.

Step 3 Slowly move the arc forward in a weaving motion, keeping a constant arc length.

Step 4 Continue to weld until a weave bead about 2" to 3" long is formed, and then break the arc by lifting the electrode straight up. A crater will be left at the point where the arc was broken. Chip any slag away from around the crater.

Step 5 Make a restart, and continue welding to the end of the plate.

Step 6 Chip the weld slag with a chipping hammer.

Step 7 Clean the bead with a wire brush.

Step 8 Have the instructor inspect the bead, which should have the following characteristics:

- Weld bead straight to within ⅛"
- Uniform appearance on the bead face
- Smooth, flat transition with complete fusion at the toes of the weld

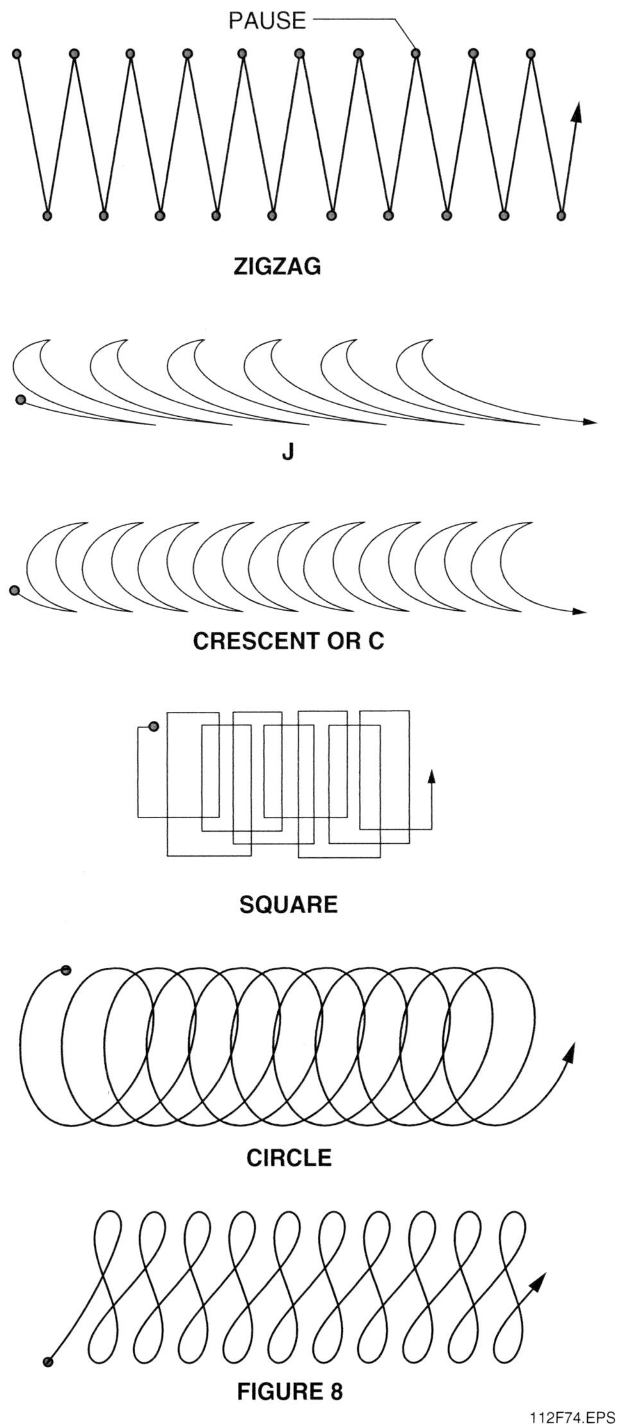

Figure 74 Weave motions.

- Crater and restarts filled to the full cross section of the weld
- No porosity
- No excessive undercut at the toes
- No inclusions
- No cracks
- No overlap

Continue welding weave beads until you can make acceptable welds every time. Although the usual width of a straight bead is two to three times the electrode diameter, the bead width for a weave bead can usually be up to, but should not exceed, eight times the electrode diameter.

11.0.0 Codes Governing Welding

A code is a set of requirements covering permissible materials, service limitations, fabrication, inspection, testing procedures, and qualifications of welders. Welding codes ensure that safe and reliable welded products are produced. They also ensure that persons associated with the welding operation are safe. Clients specify in the con-

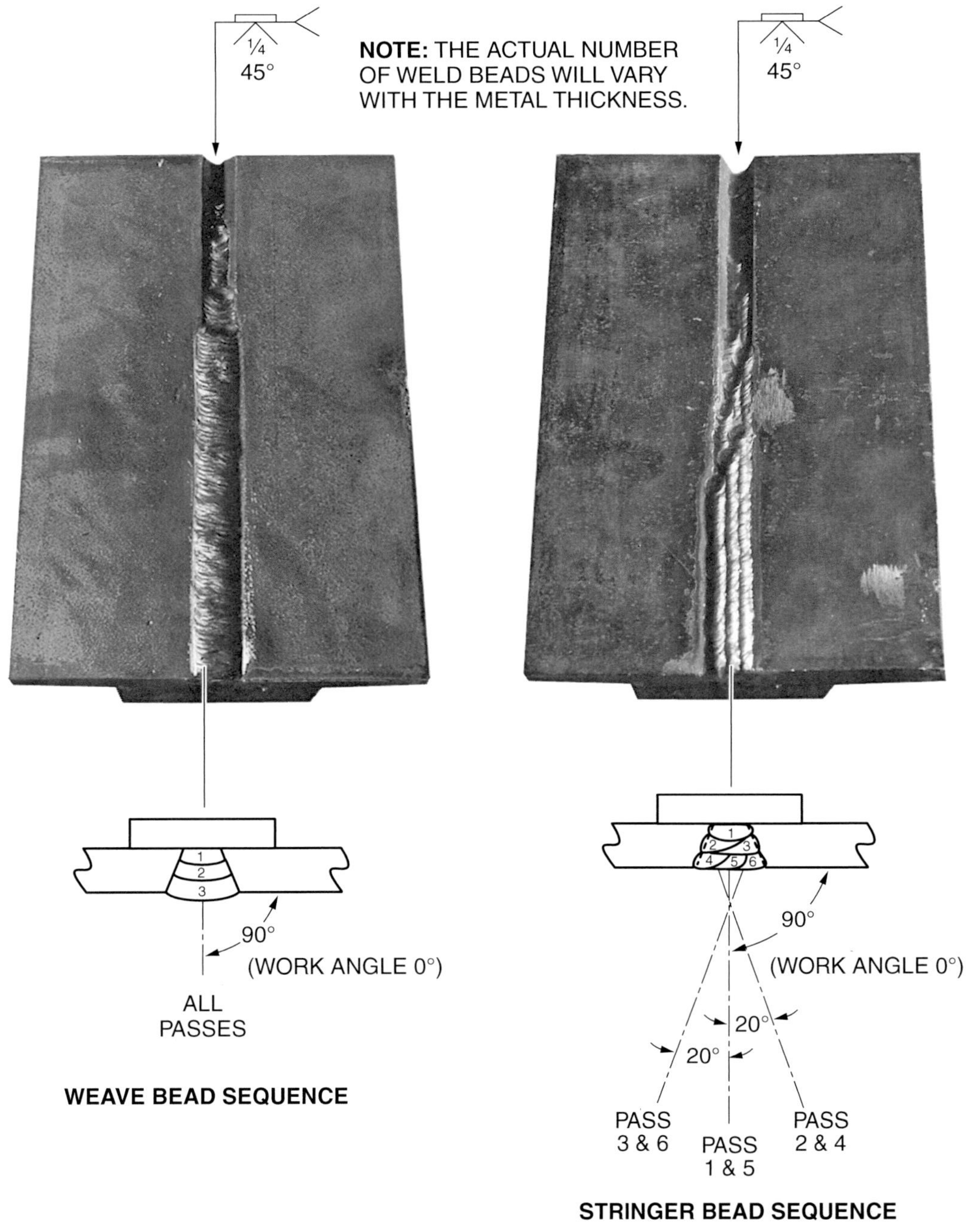

Figure 75 Weave bead.

tract the codes that will be used on the project. All welding must then be performed according to the guidelines and specifications outlined in that code. There are a number of codes. Each is updated periodically. You must know which welding codes and code year apply to your job.

Several nationally recognized agencies have developed and published codes and specifications that apply to welding safety and quality, and to the erection of structural metals. To eliminate the need to write a code for each new job, sections of these existing codes are referenced by the project contract. Agencies and societies that have established codes include the following:

- American Society of Mechanical Engineers (ASME)
- American Institute of Steel Construction (AISC)
- American Welding Society
- American National Standards Institute (ANSI)

This module does not cover all the detailed requirements of each individual code or code section. Instead, general considerations that apply to weld quality are covered.

11.1.0 American Society of Mechanical Engineers (ASME)

The American Society of Mechanical Engineers (ASME) deals more with piping and vessels, but ironworkers may be involved in building the structures supporting the piping and vessels. Check with your supervisors or engineers when welding on or near piping or vessels.

11.2.0 American Institute of Steel Construction (AISC)

The structural members provided for any job are selected for specific reasons. They are to be part of the overall structure. Any alteration, by grinding, cutting, or welding, on any structural member must be approved by a person designated to speak on behalf of the customer (owner). Situations such as the temporary installation of bracing or backing of structural members must be approved prior to their installation. Their removal also must be approved prior to their actual removal. The American Institute of Steel Construction (AISC) produces a document called the *Code of Standard Practice for Steel Buildings and Bridges.* All ironworkers must understand the guidelines within this code, which covers the fabrication and erection of structural steel. There are sections within this code that address when the cutting or welding on structural members is authorized.

11.3.0 American Welding Society (AWS)

AWS publishes numerous documents covering welding. These documents include codes, standards, specifications, recommended practices, and guides. *AWS D1.1, Structural Welding Code – Steel,* is the code most often cited. It covers welding and qualification requirements for welded structures of carbon and low-alloy steels. It is not intended to apply to pressure vessels, pressure piping, or base metals less than ⅛" thick. When aluminum is involved, the standard that applies to it is *AWS D1.2, Structural Welding Code – Aluminum.*

Among the many documents and standards published by the American Welding Society are documents covering weld inspection. These include the following:

- *AWS B1.10, Guide for the Nondestructive Inspection of Welds*
- *AWS B1.11, Guide for the Visual Inspection of Welds*
- *Certification Manual for Welding Inspectors*

11.4.0 American National Standards Institute (ANSI)

ANSI is a private organization that does not actually prepare standards. Instead, it adopts standards that it feels are of value to the public interest. ANSI standards deal with dimensions, ratings, terminology and symbols, test methods, performance, and safety specifications for materials, equipment, components, and products in many fields, including construction. Many codes used today have been adopted as ANSI standards.

11.5.0 Basic Elements of Welding Codes

A major function of welding codes is the qualification of operators to perform a particular operation. The operational qualification usually entails specific aspects of the procedure in question.

All welding codes provide detailed information about qualification in the following general areas:

- Welding procedure qualification
- Welding performance qualification
- Welding operator qualification

Machine welding is covered in some codes but is not common to all codes. Each type of qualification mentioned is different and subject to different requirements.

> **NOTE**
> The information in this module is provided as a general guideline only. Check with your supervisor if you are unsure of the codes and specification requirements for your project.

11.5.1 Welding Procedure Qualification

A welding procedure is a written document that contains materials, methods, processes, electrode types, techniques, and all other important information about the weldment. Welding procedures must be qualified before they can be used. **Procedure qualification** has nothing to do with the skills of the individual ironworker.

Welding procedure qualifications are limiting instructions. They are written to explain how welding should be done. These limiting instructions are listed in a document known as a welding procedure specification (*Figure 76*). The purpose of the WPS is to define and document in detail the variables involved in welding a certain base metal. The WPS lists the following information in detail:

- Various base metals to be joined by welding
- Filler metal to be used
- Range of preheat and postheat treatment
- Thickness and other variables described for each welding process

WPS variables are identified either as essential or nonessential variables. Essential variables are items in the WPS that cannot be changed without requalifying the welding procedure. Nonessential variables are items in the WPS that may be changed within a range identified by the code, but that do not affect the qualification status. The following can be considered essential variables in a welding procedure:

- Filler metal classification
- Material thickness
- Joint design
- Type of base metal
- Welding process

Examples of nonessential variables that may be changed without having to requalify the procedure include:

- Amperage
- Travel speed
- Shielding gas flow (if applicable)
- Electrode and filler wire size

> **NOTE**
> Do not change any variable (essential or nonessential) without discussing it with your supervisor.

The WPS is qualified for use by welding test coupons (samples of the welded material) and by testing the coupons in accordance with the applicable code. A test weld is made and test coupons are cut from it. The test coupons are used to make tensile tests, root bends, and face bends as required by the code. The test results are then recorded on a document known as a **procedure qualification record (PQR)**. If the weldment produced by the particular procedure meets the standards of the code, the procedure becomes qualified. Under most codes, each WPS must have a PQR to document the quality of the weld produced. *Figure 77* shows a sample PQR.

The methods used to qualify procedures are more detailed and thorough than those used to qualify either ironworkers or welding operators. This is because procedures must qualify physical and metallurgical properties.

11.5.2 Welding Performance Qualification

Once a procedure has been qualified, the welder using it must be qualified to use that procedure by passing a welding performance qualification test. No single performance test can qualify welders for all the different types of welding that must be done. A welder may be required to pass several performance qualification tests.

The welder becomes qualified by successfully completing a weld made in accordance with the WPS. It is general practice for code welding to qualify welders on the groove weld tests. Passing these tests also permits the welder to weld fillet welds.

11.5.3 Welding Operator Qualification

When fully automatic welding equipment is used, the operators of the equipment must demonstrate their ability to set up and monitor the equipment so that it will produce acceptable welds. The codes also contain qualification tests for these operators.

ANNEX B—SAMPLE FORMS

WELDING PROCEDURE SPECIFICATION (WPS) Yes ☐
PREQUALIFIED ________ QUALIFIED BY TESTING X
or PROCEDURE QUALIFICATION RECORDS (PQR) Yes ☐

Identification # PQR 231
Revision 1 Date 12-1-07 By W. Lye
Authorized by J. Jones Date 1-18-08
Type—Manual ☐ Semi-Automatic ☐
Machine ☐ Automatic ☐

Company Name Red Inc.
Welding Process(es) FCAW
Supporting PQR No.(s) –

JOINT DESIGN USED
Type: Butt
Single ☒ Double Weld ☐
Backing: Yes ☒ No ☐
Backing Material: ASTM A131A
Root Opening 1/4" Root Face Dimension –
Groove Angle: 35° Radius (J–U) –
Back Gouging: Yes ☐ No ☒ Method –

BASE METALS
Material Spec. ASTM A131
Type or (Grade) A
Thickness: Groove 1" Fillet –
Diameter (Pipe) –

FILLER METALS
AWS Specification A5.20
AWS Classification E71T-1

SHIELDING
Flux – Gas CO_2
Composition
Electrode-Flux (Class) Flow Rate 30-40 cfh
Gas Cup Size 5/8" or 3/4"

PREHEAT
Preheat Temp., Min 75° Ambient
Interpass Temp., Min 75°F Max 350° F

POSITION
Position of Groove: 4G Fillet: –
Vertical Progression: Up ☐ Down ☐

ELECTRICAL CHARACTERISTICS
Transfer Mode (GMAW) Short-Circuiting ☐
Globular ☒ Spray ☐
Current: AC ☐ DCEP ☒ DCEN ☐ Pulsed ☐
Other
Tungsten Electrode (GTAW)
Size:
Type:

TECHNIQUE
Stringer or Weave Bead: Stringer
Multi-pass or Single Pass (per side) Multipass
Number of Electrodes 1
Electrode Spacing Longitudinal –
Lateral –
Angle –
Contact Tube to Work Distance 3/4-1"
Peening None
Interpass Cleaning: Wire Brush

POSTWELD HEAT TREATMENT
Temp. N.A.
Time

WELDING PROCEDURE

Pass or Weld Layer(s)	Process	Filler Metals: Class	Filler Metals: Diam.	Current: Type & Polarity	Current: Amps or Wire Feed Speed	Volts	Travel Speed	Joint Details
1	FCAW	E71T-1	.045"	DC+	180	26	8	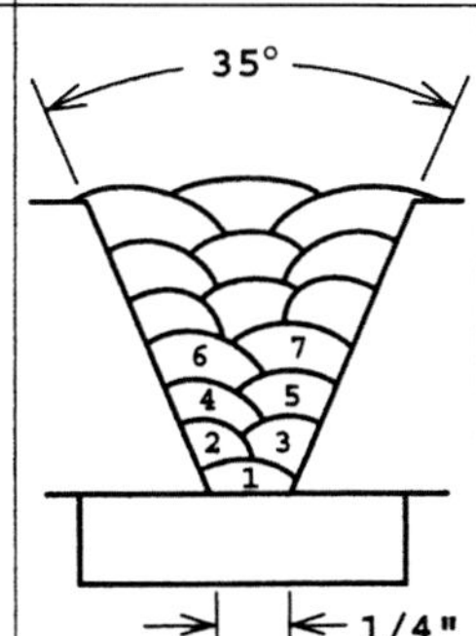
2-8	"	"	"	"	200	27	10	
9-11	"	"	"	"	200	27	11	
12-15	"	"	"	"	200	27	9	
16	"	"	"	"	200	27	11	

Form E-1 (Front)

112F76.EPS

Figure 76 Typical WPS.

Procedure Qualification Record (PQR) # 231
Test Results

TENSILE TEST

Specimen No.	Width	Thickness	Area	Ultimate Tensile Load, lb	Ultimate Unit Stress, psi	Character of Failure and Location
231-1	.75"	1.00"	.75"	52 500	70 000	Ductile
231-3	.75"	1.00"	.75"	52 275	69 700	Ductile

GUIDED BEND TEST

Specimen No.	Type of Bend	Result	Remarks
231-2	Side	Pass	
231-4	Side	Pass	Small (< 1/16") opening acceptable
231-6	Side	Pass	
231-5	Side	Pass	

VISUAL INSPECTION
Appearance acceptable
Undercut acceptable
Piping porosity none
Convexity none
Test date 12-3-2002
Witnessed by D. Davis

Radiographic-ultrasonic examination
RT report no.: D231 Result passed
UT report no.: ______ Result ______

FILLET WELD TEST RESULTS
Minimum size multiple pass Macroetch
1. ______ 3. ______
2. ______

Maximum size single pass Macroetch
1. ______ 3. ______
2. ______

Other Tests

All-weld-metal tension test
Tensile strength, psi 83,100
Yield point/strength, psi 72,600
Elongation in 2 in., % 28
Laboratory test no. PW 231

Welder's name W. T. Williams Clock no. 261 Stamp no. ______

Tests conducted by RED Inc. & ABC Testing Laboratory

Test number PQR 231

Per D. Miller

We, the undersigned, certify that the statements in this record are correct and that the test welds were prepared, welded, and tested in conformance with the requirements of Section 4 of AWS D1.1/D1.1M, 2002 (year) Structural Welding Code—Steel.

Signed RED Inc.
Manufacturer or Contractor
By R. M. Boncrack
Title Q.C. Mgr.
Date 12-15-2002

112F77.EPS

Figure 77 An example procedure qualification record (PQR).

SUMMARY

Welding is a process in which intense heat is used to raise the temperature of metals above their melting point, creating a pool of molten metal. When the heat source is removed, the pool solidifies, causing the metals to fuse. SMAW, also called stick welding, is the most common welding method. Other methods include GMAW and FCAW. These two methods are similar in that both use a consumable wire electrode fed from a spool as a filler metal. The main difference is that FCAW uses a wire with a flux core, while GMAW uses a solid wire. GTAW, unlike the other methods, uses a nonconsumable tungsten electrode. GTAW also uses rods of consumable filler metals that melt into the weld when heated by the tungsten electrode.

Welding involves more than just melting base metal and electrodes. There are many types of electrodes. Selection of electrodes for a given type of weld depends on the base metal being welded, the welding method to be used, and the weld position. Ironworkers must be able to interpret the coded identifiers on the electrodes in order to select the correct electrode for a given situation. Learning how to properly handle and store electrodes is also very important.

Ironworkers must be familiar with the various types of welds and the circumstances in which they are used. Welding is governed by published codes and is performed in accordance with certified welding procedures. Most piping and structural welding must be performed by ironworkers who have been certified to perform the specific procedure. Ironworkers must pass performance tests in order to obtain these certifications. All ironworkers start by running stringer and weave weld beads as learned in this module.

Review Questions

1. Which of these welding processes is referred to as stick welding?
   a. GMAW
   b. FCAW
   c. SMAW
   d. GTAW

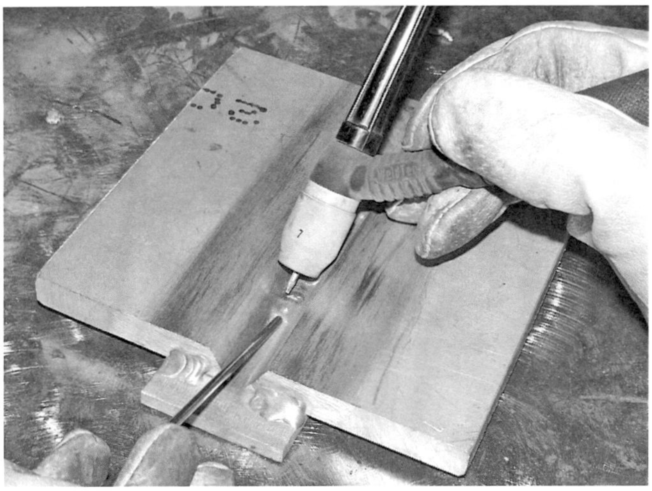

108RQ01.EPS

Figure 1

2. The welding process pictured in *Figure 1* is _____.
   a. GMAW
   b. FCAW
   c. SMAW
   d. GTAW
3. In which of these processes is gas passed through an electric arc inside a torch?
   a. GMAW welding
   b. CAC-A cutting
   c. PAC cutting
   d. Oxyfuel cutting
4. Inert gases such as argon and helium are used in welding to _____.
   a. increase the intensity of the arc
   b. provide a protective barrier around the welding
   c. blow slag and dross away from the weld pool
   d. extinguish hot particles created by the welding process
5. Polyester and synthetic fabrics should *not* be worn when welding because _____.
   a. these fabrics break down when exposed to ultraviolet light
   b. sparks will cause them to melt, inflicting serious burns
   c. they will burst into flames in the presence of pure oxygen
   d. the will give off toxic fumes if exposed to acetylene
6. Proper ventilation is important when welding because _____.
   a. it keeps the operator from getting overheated
   b. it keeps smoke away from the work
   c. the fumes and smoke given off by the process can be toxic
   d. the airflow keeps sparks away from the operator
7. In SMAW, welding current is transferred from the electrode holder and into the work, through the ____.
   a. base metal
   b. flux coating
   c. welding ground
   d. wire core
8. Electrodes designed for welding high-sulfur, phosphorus, and medium- to high-carbon steels belong to which group?
   a. Fast-fill
   b. Fill-freeze
   c. Fast-freeze
   d. Low-hydrogen
9. The type of electrode that absorbs moisture the fastest is the _____.
   a. fast-fill electrode
   b. fill-freeze electrode
   c. fast-freeze electrode
   d. low-hydrogen electrode

10. All filler metals must be stored in a dry storage area at a minimum temperature of _____.
    a. 30°F
    b. 40°F
    c. 50°F
    d. 60°F

11. The five basic types of welding joints are butt, lap, corner, T-joint, and _____.
    a. chamfer
    b. edge
    c. bevel
    d. square

12. Lap joints, T-joints, and corner joints are welded using _____.
    a. surface welds
    b. butt welds
    c. fillet welds
    d. spot welds

13. What is the maximum size allowed for the root opening when making SMAW groove welds on plate?
    a. ⅛"
    b. 3⁄16"
    c. ¼"
    d. 5⁄16"

14. The written set of instructions for producing reliable welds is the _____.
    a. welding procedure specification
    b. welding standard
    c. welding code
    d. site quality standard

15. Which welding code covers structural steel?
    a. *AWS D1.1*
    b. *AWS D1.2*
    c. *AISC B32.2*
    d. *AISC B36.7*

Cornerstone of Craftsmanship

Nelson Plumb

Crossland Construction Co.
Steel Superintendent

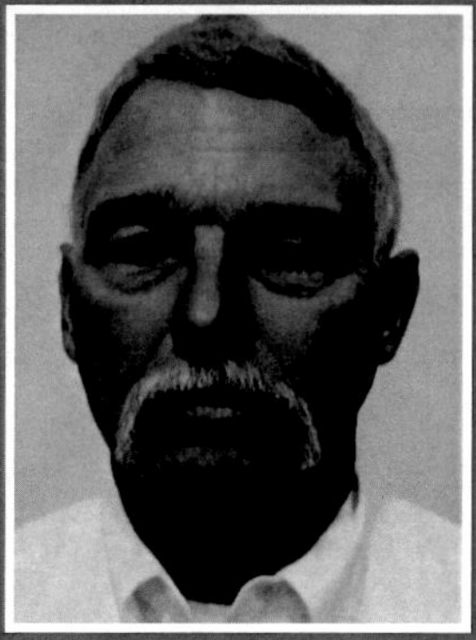

After serving time in the Army, Nelson was discharged and in need of a job. His wife was pregnant so he wasn't going to be able to go back to school and he went looking for a way to support his family. Nelson learned from his best friend's dad that Tulsa Ironworker Local Union #584 was accepting applications for apprentices. Nelson jumped on the chance to get the apprenticeship that would develop his skills in the construction industry and support his growing family.

Nelson feels the construction industry has been a very rewarding career choice for him. He has received several certifications that have enabled him to become a marketable commodity within the industry. He highly recommends a construction career if a person has the temperament and personality needed to work hard and proactively strive for training and education within the field. Nelson says the best part about his job is that at the end of the day, he can see the accomplishments that have been made and enjoys the satisfaction of knowing he has done something well.

Trade Terms Introduced in This Module

Alloy: A metal that has had other elements added to it that substantially change its mechanical properties.

Alternating current (AC): Electrical current that reverses its flow at set intervals.

Amperage: A measurement of the rate of flow of electric current.

Arc: Flow of electricity through an air gap that melts the electrode.

Arc blow: The deflection of the arc from its intended path by magnetic forces.

Arc strike: A discontinuity consisting of localized melting of the base metal or finished weld caused by the arc. It occurs as a result of striking the arc outside the area to be welded.

Arrow line: The line drawn at an angle from the reference line (either end or both ends) to an arrow head at the location of the weld.

Backing: A weldable or nonweldable material used behind a root opening to allow defect-free welding at the open root of a joint.

Condensation: The process in which water from the atmosphere forms on a cool surface.

Direct current (DC): Electrical current that flows in one direction only.

Distortion: The expansion and contraction of welded parts caused by the heating and subsequent cooling of the weld joint.

Drag angle: The travel angle when the electrode is pointing in a direction opposite to the welding progression.

Ductile: The mechanical property of a material that allows it to be bent or shaped without breaking.

Electrode: Point from which the arc is produced to perform welding.

Flux: Material used to prevent, dissolve, or facilitate the removal of oxides and other undesirable substances on a weld.

Ground: An object that makes an electrical connection with the earth.

Heat-affected zone: The part of the base metal that has been altered but not melted by the heat.

Hermetically sealed: Having an airtight seal.

Hot work permit: Official authorization from a site manager to perform work that may pose a fire hazard.

Low-hydrogen electrode: An electrode specially manufactured to contain little or no moisture.

Melt-through: Complete joint penetration.

Notch toughness: The ability of a material to resist breaking at points where stress is concentrated.

Porosity: Gas pockets, inclusions, or voids in the weld metal.

Procedure qualification: The demonstration that welds made following a specific process can meet prescribed standards.

Procedure qualification record (PQR): The document containing the results of the nondestructive and destructive testing required to qualify a welding procedure specification (WPS).

Push angle: The travel angle when the electrode is pointing in the same direction as the welding progression.

Reference line: The horizontal line in the center of the welding symbol from which all elements of the welding symbol are referenced. The reference line is one of the most important elements of the welding symbol.

Restart: The point in the weld where one weld bead stops and the continuing bead is started.

Root face: A small facing on the end of a bevel for a groove weld.

Root opening: The space between the base metal pieces at the bottom or root of the joint.

Root pass: The initial pass of a weld that requires multiple passes.

Spatter: Metal particles discharged into the air during welding.

Stringer bead: A type of weld bead made without appreciable weaving motion. With shielded metal arc welding (SMAW), stringer beads are not more than three times the electrode diameter.

Surfacing: The application by welding, brazing, or thermal spraying of a layer of material to a surface to obtain desired properties or dimensions.

Traceability: The ability to verify that a procedure has been followed by reviewing the documentation step by step.

Vertical welding: Welding with a downward progression from top to bottom.

Voltage: A measurement of electromotive force or pressure that causes current to flow in a circuit.

Weave bead: A type of weld bead made by transverse oscillation of the electrode.

Weld axis: A straight line drawn through the center of a weld along the length of the weld.

Weld coupon: The metal pieces to be welded together as a test or practice.

Weld symbol: A graphic character connected to the reference line of a welding symbol specifying the weld type.

Welding symbol: A graphical representation of the specifications for producing a welded joint; includes a reference line and arrow and many also include a weld symbol.

Work angle: An angle less than 90 degrees between a line perpendicular to the major workpiece surface and a plane determined by the electrode axis and the weld axis. In a T-joint or corner joint, the line is perpendicular to the non-butting member. The definition of work angle for a pipe weld is covered in a later module.

Workpiece: The part being welded, brazed, soldered, or cut.

Additional Resources

This module presents thorough resources for task training. The following resource material is suggested for further study.

Welding Handbook, Volume 5, 2001. Miami, FL: The American Welding Society.

The Procedure Handbook of Arc Welding, 2000. Cleveland, OH: The Lincoln Electric Company.

OSHA Standard 1926.351, Arc Welding and Cutting, Current edition. Washington, DC: Occupational Safety and Health Administration.

Stick Electrode Welding Guide, 2004. Cleveland, OH: The Lincoln Electric Company.

Stick Electrode Product Catalog, 2008. Cleveland, OH: The Lincoln Electric Company. www.lincolnelectric.com.

Lincoln Electric Web site: http://www.lincolnelectric.com offers sources for products and training.

Figure Credits

Lincoln Electric Company, Figures 1, 2, (photo), 4, 10, 11 (photo), 12 (photo), 16, 17, 22, 23, and 60

Terry Lowe, Figure 8

Bryan McClure, Figure 13

Topaz Publications, Inc, Figures 24, 29 (photo), 30, 31, 35 (photos), 38 (photos), 50 (photos), 61, 65, 69, and 75 (photo)

AWS D1.1/D1.1M:2010, Table 5.1, reproduced with permission from the American Welding Society (AWS), Miami, FL, Table 5

Phoenix International, Inc., Figure 27

AWS A2.4:2007, Figure 2, reproduced with permission of the American Welding Society (AWS), Miami, FL, Figure 47

Specialty Welding & Fabricating of New York, Inc., Figure 53 (photos)

AWS EG2.0-95, Figure 2, reproduced with permission of the American Welding Society (AWS), Miami, FL, Figure 58

©American Welding Society (AWS) 2001, *Welding Handbook, Vol. 2*, Welding Processes, Miami: American Welding Society, Figure 71

American Welding Society, Figure 76

AWS D1.1/D1.1M:2010, Procedure Qualification Record, reproduced with permission of the American Welding Society (AWS), Miami, FL, Figure 77

NCCER CURRICULA — USER UPDATE

NCCER makes every effort to keep its textbooks up-to-date and free of technical errors. We appreciate your help in this process. If you find an error, a typographical mistake, or an inaccuracy in NCCER's curricula, please fill out this form (or a photocopy), or complete the online form at **www.nccer.org/olf**. Be sure to include the exact module ID number, page number, a detailed description, and your recommended correction. Your input will be brought to the attention of the Authoring Team. Thank you for your assistance.

Instructors – If you have an idea for improving this textbook, or have found that additional materials were necessary to teach this module effectively, please let us know so that we may present your suggestions to the Authoring Team.

NCCER Product Development and Revision

13614 Progress Blvd., Alachua, FL 32615

Email: curriculum@nccer.org
Online: www.nccer.org/olf

❏ Trainee Guide ❏ Lesson Plans ❏ Exam ❏ PowerPoints Other ______________________

Craft / Level: Copyright Date:

Module ID Number / Title:

Section Number(s):

Description:

Recommended Correction:

Your Name:

Address:

Email: Phone:

Bar Joists and Girders

30113-11

Trainees with successful module completions may be eligible for credentialing through NCCER's National Registry. To learn more, go to **www.nccer.org** or contact us at **1.888.622.3720**. Our website has information on the latest product releases and training, as well as online versions of our *Cornerstone* newsletter and Pearson's Contren® product catalog.

Your feedback is welcome. You may email your comments to **curriculum@nccer.org,** send general comments and inquiries to **info@nccer.org**, or use the User Update form at the back of this module.

V.1 1/11

30113-11
Bar Joists and Girders

Objectives

When you have completed this module, you will be able to do the following:

1. Recognize the various types of bar joists.
2. Explain how bar joists are designated.
3. Describe the proper procedures for rigging and storing steel joists.
4. Describe the proper erection procedures for bar joists.
5. Explain the use of joist girders in steel joist construction systems.

Performance Tasks

Under the supervision of the instructor, the trainee should be able to do the following:

1. Identify selected types, shapes, and grades of bar joists and joist girders.
2. Interpret connection details for bar joists and girders.
3. Demonstrate the ability to handle, store, and rig different types of bar joists and girders.
4. Identify different bridging and mounting devices used with bar joists and joist girders.

Trade Terms

Bridging
Chord
Dunnage
End bearing
Joist girder
Open-web steel joist (bar joist)
Project engineer of record
Tie joists (OSHA joists)

Contents

Topics to be presented in this module include:

Figures and Tables

1.0.0 INTRODUCTION

Ironworkers erect systems that utilize bar joists. The purpose of this module is to introduce the basics of bar joists. This module covers the yarding, handling, rigging, and erection of bar joists.

2.0.0 OPEN-WEB BAR JOISTS

Bar joists are horizontal or sloping structural members with open webs. They are used to support floors and roof decks in buildings. Bar joists derive their structural strength in the same way as a lattice boom. The alternating bracing between the top and bottom **chords** forms strong triangular shapes. Bar joists, however, have little lateral strength.

The use of bar joists allows economical and efficient designs. Their light weight puts less dead load on beams, columns, and foundations. The open-web system allows for easy installation of piping and duct systems, increasing the maximum ceiling height.

2.1.0 Types and Designations

There are six standard types of **open-web steel joists (bar joist)** (*Figure 1*). They are differentiated by the pitch of the top chord and the end configuration.

The bar joists are identified as belonging to six different series. The joists are identified on the plans by a three-part code. The first section of the code is a number identifying the depth of the joist. The second part is a letter identifying the series of the joist. The last part is a number called the chord section. This number shows the relative strength of the joist, and the higher the number, the greater strength the joist has. An example of a joist number might be 22K9.

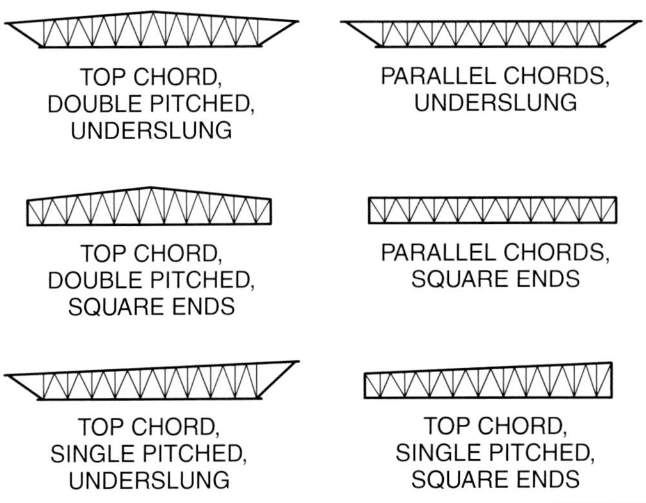

Figure 1 Open-web steel joist types.

The common series of joists are the following:

- K series are common open-web joists. These are standard sized joists. The spans may be from 8 feet to 60 feet, and the depths may run from 8 inches to 30 inches. K series chord section numbers run from 1 to 12.
- LH series are also called longspan joists. Their spans run from 25 feet to 96 feet. Depths may be from 18 inches up to 48 inches, and chord sections range from 2 to 17.
- DLH series, also called deep longspan, can be from 89 to 144 feet long, and from 52 inches to 72 inches deep. The chord section numbers are from 10 to 19.
- SLH series, called super longspans, are from 111 feet to 240 feet long, from 80 inches to 10 feet deep, and the chord sections are from 15 to 25.
- CS series, or constant shear joists, are for loads that may vary from point to point. The depths run from 10 to 30 inches, and chord section numbers are from 1 to 5.
- CJ series, called composite joists, are in the same size range as K series.

There are also special varieties, such as bowstring joists, gable joists, scissor joists, and barrel joists.

Joist girders are structural trusses that are designed to support joists. Joist girders usually bear on columns, and the joists bear on the joist girders. A typical joist girder designation might be 52G10N9F. The 52 is the nominal joist depth at midspan in inches; the G means this is a joist girder; the 10N shows that there are 10 joist spaces; and the 9F means the concentrated load at each joist location is 9 kips from a load and resistance factor design. (A *kip* is a unit of force that equals 1,000 pounds-force.)

2.2.0 Handling, Storing, and Rigging

Proper handling of open-web steel joists is crucial. Bar joists are light and can be seriously damaged by improper handling. When unloading bar joists, never throw them off the truck.

Joists are usually shipped resting on their top chords. They should be stored in the same position. Stacking bar joists on their bottom chords or sides can cause them to bend.

Stack bar joists on **dunnage** to keep them above damp ground. Place the dunnage under the panel points. Place the dunnage for the second tier of joists as well as the second tier itself directly over the first. Never stack joists so high that the bottom pile bends. The proper height is determined by the size and weight per foot of the particular joists. Joists are shipped in bundles weighing no more than 1,000 pounds per bundle. *Figure 2* shows the proper way to stack bar joists using dunnage.

Always stack bar joists with the sequence of their erection in mind. All shipping tags should be placed at the same end.

Joists are best lifted in bundles and placed at the intended storage location. Use two basket hitches placed at the ⅓ points of the joists where the panels vee at the bottom chord. The lift should be made smoothly and evenly. Excessive snatching of bar joists can bend them. If joists are being moved with a forklift, be sure the forks are placed so that the members of the joist are not bent or jammed against the forks, and so that the joist or bundle is well balanced on the lift. Pressure on the chords directly can bend the chords and weaken the joist. *Figure 3* shows the proper rigging for bundles of joists or trusses.

When joists are landed on a structure, they must be secured until they can be placed and fastened to prevent them from falling. Do not climb onto any joist until both sides of one end are fastened, and any required bridging has been installed. Where joists are used to stabilize columns, both ends must be field-bolted before welding and before releasing the cables on them. The bottom chord of the joist must also be stabilized before it is released to prevent it from rolling. No steel joist may be placed on any support structure that has not been stabilized.

When opening bundles of joists, keep hands and feet clear when the strapping is being cut. Do not stand in line with the strapping as it is cut, as the tension may make it lash out.

2.3.0 Bridging

Joists are supported laterally by **bridging** (*Figure 4*). Various kinds of bridging can be used, depending on the design. The manufacturers of open-web steel joists often ship bridging of their own design along with the joists.

Erection bridging is required for particular spans of specific joist series. *OSHA 1926.757(c) and (d)(1). Table A, Erection Bridging for Short Span Joists* and *Table B, Erection Bridging for Long Span Joists* give the requirements for bridging. If erection bridging is required, the hoisting cables must

113F03.EPS

Figure 3 Proper rigging of joists or trusses.

113F02.EPS

Figure 2 Proper way to stack bar joists using dunnage.

not be released until one end of the joist has been fastened on both sides and some of the bridging has been installed. The location of the bridging depends on the length of the span. Be careful to keep control of the cables until they are clear of the joist and of personnel, so that they will not hook the joist or strike anyone.

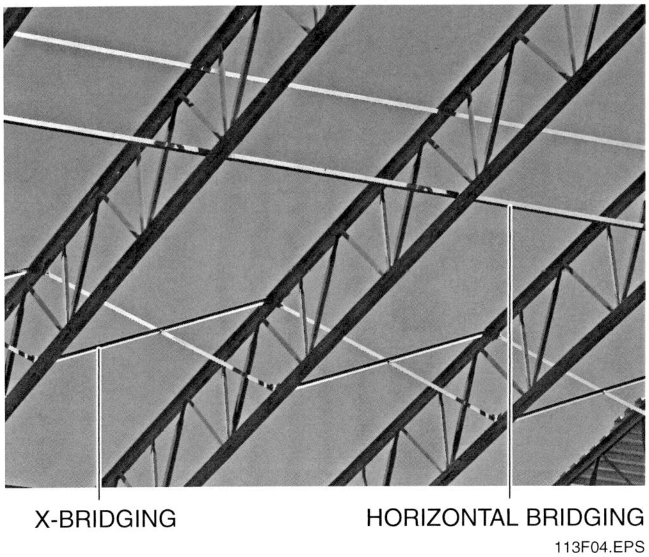

Figure 4 Types of bridging.

If the span requires bridging, and is between 60 and 100 feet in length, the bridging must be placed nearest to the ⅓ points of the joist. If the span is from 100 to 144 feet, all the rows of bridging must be installed before releasing the hoisting cables. If any of the joists is bottom chord bearing, a row of bolted bridging must be installed and anchored before the hoisting cable or cables are released. The two common types of bridging are horizontal bridging and X-bridging.

Angle iron is used for bridging, and is usually placed horizontally (or as specified on the drawings) and welded to the top and bottom chords. *Figure 5* shows an example of horizontal bridging.

The second type of bridging, X-bridging, can be fabricated on the job according to the design specifications, or may be supplied by the manufacturer. It is either bolted or welded to the top chord of one joist and the bottom chord of the adjacent joist. There are specifications regarding the spacing of bridging within spans of bar joists. Always consult the erection drawings to obtain this information.

When bridging is not required at installation, joists for a particular bay are often lifted into place as a bundle, and slid into position by workers. The joists are sorted on the ground and made

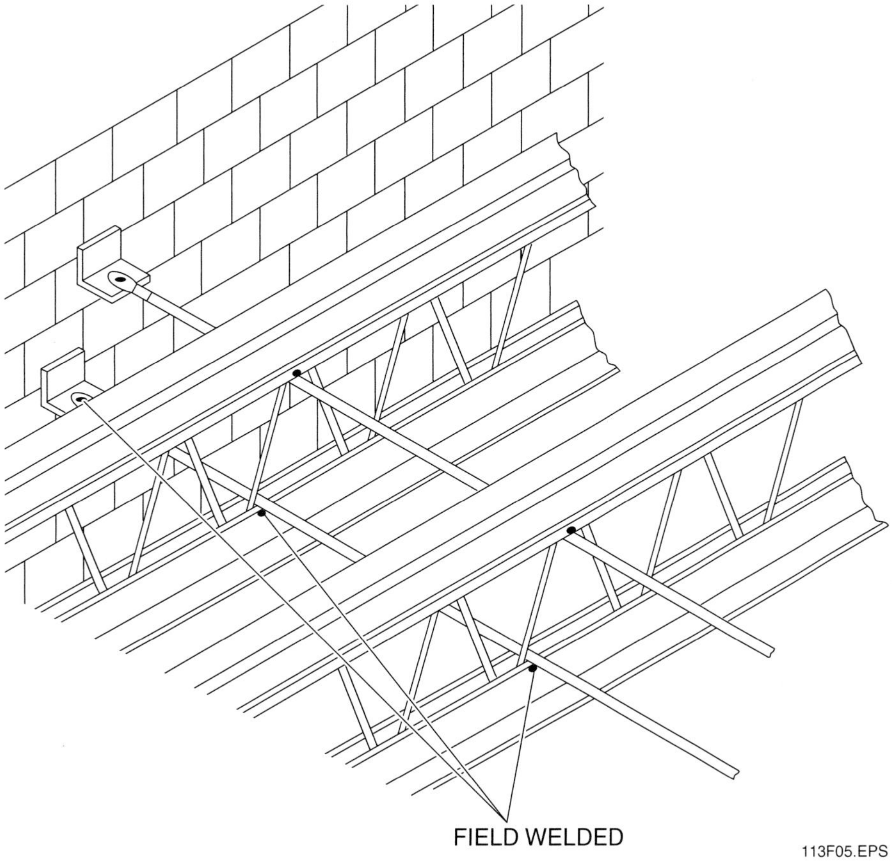

Figure 5 Cutaway of horizontal bridging.

into a bundle in the order they are to be installed, with all of the tags turned the same way, the ends aligned correctly, and the joists right side up. When bundles of joists are landed on a structure, the bundle should be secured until the individual joists can be placed and fastened, to prevent them from falling. Where joists are used to stabilize columns, both ends must be field bolted before releasing the cables on them. The bottom chord of the joist must also be stabilized before it is released to prevent it from rolling. No steel joist may be placed on any support structure that has not been stabilized.

WARNING

Do not climb onto any joist until both sides of each end of the joist have been field bolted, and the required bridging has been installed.

Bridging is frequently attached to joists before lifting the joist into place. The joist is usually placed on its side and the X-bridging fastened to the side (*Figure 6*). Where the X-bridging crosses, a bolt is inserted and finger tightened. The joist is raised and fastened into place. The workers then connect the bridging. This reduces the time spent working above the ground, contributing to the safety of the workers.

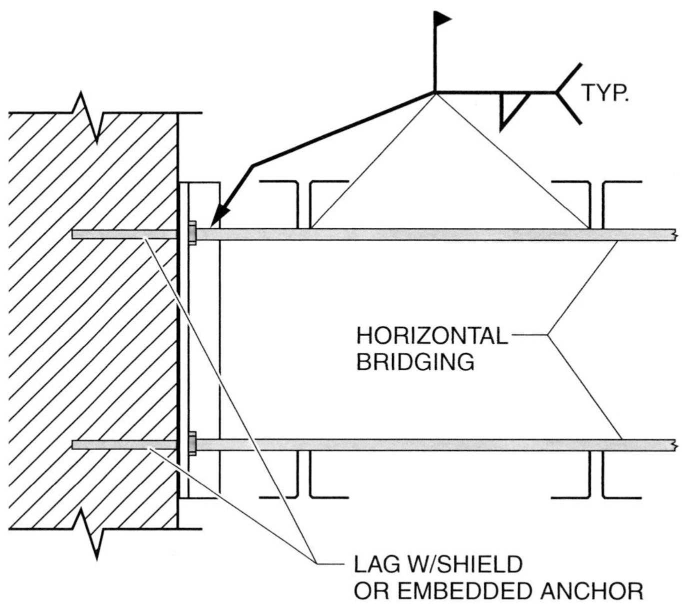

113F06.EPS

Figure 6 X-bridging.

2.4.0 End Bearing

The term **end bearing** refers to that part of the structure the joist rests on. Joists may rest on steel or masonry, depending upon the design.

Joists that rest on steel must have at least 2½ inches resting on the beam and can be either bolted or welded to the steel. Joist girders that rest on steel must have at least 4 inches resting on the beam, and 6 inches resting on a bearing place on masonry. The specifications on the erection drawings usually contain other information such as the size of bolts to use, the type of electrode to use, and the length and size of the weld. *Figure 7* shows sample detail drawings of each type of connection for anchorage to steel.

When joists rest on masonry, a variety of anchors may be used (*Figure 8*). Some designs require that steel plates or special anchors be embedded in the masonry as it is being erected. The joists can then be welded or bolted to the anchors. The specifications and codes contain the necessary information regarding the types of anchors to use, the required number of joists that must be anchored, the bolts to use, and the size of the anchor. At least 2½ inches of the joist must rest on the steel-bearing plate. It must be either fitted with embedded anchors or have 4 inches resting on the masonry.

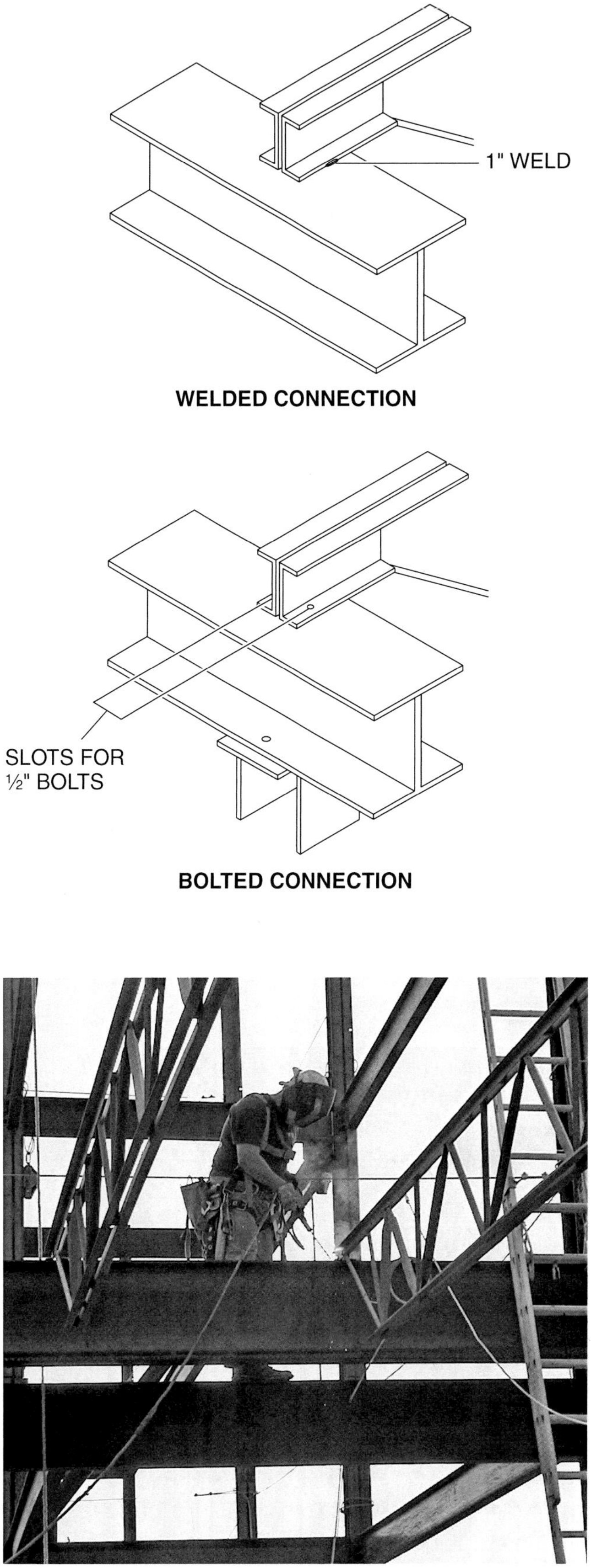

113F07.EPS

Figure 7 Steel anchorage detail drawings.

3.0.0 Erecting Steel Joists

Bundles of joists are broken apart and the number needed for the bay are selected. They should be rigged with a basket hitch so that the top chord is up.

The joists are then installed in their proper location within the structure. **Tie joists (OSHA joists)** are usually set first, depending upon their length. Generally, joists should be installed so that the tagged ends bear on the same wall or as designed on the drawings.

Joist assemblies are often panelized. Panelizing is the assembling of a bay section or portion on the ground and lifting the assembly onto the structure. This limits the amount of work that must be performed off the ground, reducing the fall hazards. *Figure 9* shows a panelized structure being lifted into place. This lines up the panel points better and makes it easier to install ductwork, conduit, or other piping in the joists.

> **WARNING**
>
> Before making any attempt to walk out on a bar joist, make sure that both ends are secured and any required bridging is installed. Unsecured joists can fall, causing injury or death.

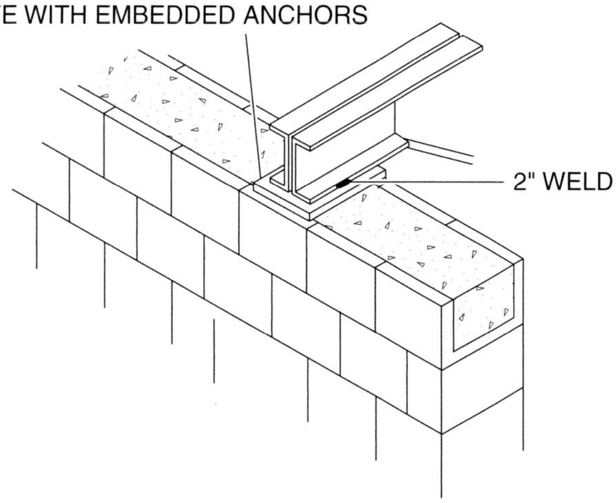

Figure 8 Masonry anchorage detail drawings.

Figure 9 Panelized structure.

Joists tend to have a certain amount of curvature. To eliminate this condition, one end of the joist should be tack welded to the bearing surface during the erection process. The welder would then lean to one side to eliminate the curve before welding.

All bridging should be installed before the joist is permanently fastened. Before any construction loads may be applied to the joists, all joists must be permanently fastened. This is very important for preventing safety hazards.

4.0.0 JOIST GIRDERS

Many times, structures using open-web bar joists also use **joist girders**. Joist girders resemble bar joists. They are used to span between columns the same way structural steel girders are used. The bar joists are attached to the steel girders. *Figure 10* shows a joist arrangement.

The same recommendations that apply to bar joists concerning handling, storing, rigging, and erecting also apply to joist girders. Note that no modifications may be made to any bar joist or girder without the approval of the **project engineer of record** or the joist manufacturer's engineer.

The standard connection for joist girders to columns is 13⁄16-inch holes for 3⁄4-inch bolts. Normally, A-325 bolts are used for erection. Stabilizer plates must be used on columns to keep joists or joist girders from rotating. The plates must be at least 6" × 6", and must extend at least 3 inches below the bottom chord of the joist girder. If, for some reason, stabilizer plates are not an option, the engineer will have designed an alternative stabilization system. For joists over 60 feet, both ends of the joist must be attached before the hoisting cables are released. Also, joists over 60 feet long and near a column must be installed in tandem, with the bridging installed. If a joist does not require bridging according to the OSHA tables, only one worker is allowed on the joist until all bridging has been secured. Unless a qualified person says to do so, don't anchor fall protection on a steel joist or joist girder. Braces used for the bottom chord may be either bolted or welded to the girder, but must be welded to the bar joist. Additional holes are added for guying.

Figure 10 Joist girders attached to column.

SUMMARY

Open web bar joists and joist girders are used frequently in construction. Ironworkers must know how to handle, store, and erect bar joists. All of the necessary information concerning the erection of bar joists is found on the erection drawings.

Review Questions

1. To form strong triangular shapes, joists have alternating bracing between the top and bottom _____.
   a. chords
   b. stringers
   c. rafters
   d. posts

2. Each joist is assigned a designation that consists of _____.
   a. numbers and letters
   b. letters and symbols
   c. numbers and symbols
   d. shapes and symbols

3. The depth of a joist with a designation of 24LH10 is _____.
   a. 2 inches
   b. 4 inches
   c. 10 inches
   d. 24 inches

4. The CS series is used for applications where _____ may vary from point to point.
   a. anchors
   b. loads
   c. corners
   d. angles

5. Joists should be stored resting on their ______.
   a. bottom chords
   b. panel points
   c. top chords
   d. webbing

6. Joists are supported laterally by _____.
   a. bars
   b. bundles
   c. ties
   d. bridging

7. The two common types of bridging are X-bridging and _____.
   a. V-bridging
   b. horizontal bridging
   c. Y-bridging
   d. vertical bridging

8. Joists that rest on steel must have at least _____ inches resting on the beam and can be either bolted or welded to the steel.
   a. 1½
   b. 2½
   c. 3½
   d. 4½

9. Generally, joists should be installed so that the tagged ends _____.
   a. bear on opposite walls
   b. bear on alternating walls
   c. bear on the same wall
   d. never bear on the same wall

10. Joist girders are usually erected using _____ bolts.
    a. A-325
    b. A-454
    c. B-315
    d. B-410

Trade Terms Introduced in This Module

Bridging: Lateral supports for joists.

Chord: The top and bottom of bar joists.

Dunnage: Lumber or other material that supports and protects cargo from damage during transport or storage.

End bearing: The part of the structure that supports the joists.

Joist girder: Girders manufactured in the same manner as open-web joists.

Open-web steel joist (bar joist): Joists made of chords and bars rather than solid web.

Project engineer of record: Registered licensed professional responsible for the design of structural steel framing and whose seal appears on structural documents.

Tie joists (OSHA joists): Joists located at or near column lines.

Additional Resources

This module presents thorough resources for task training. The following resource material is suggested for further study.

OSHA Standards for the Construction Industry (29 CFR Part 1926). Latest edition. Washington, DC: Occupational Safety & Health Administration.

Steel Construction Manual. 13th edition. Chicago, IL: American Institute of Steel Construction.

Figure Credits

Topaz Publications, Inc., Module opener

LPR Construction, Figures 2–4, 6 (photo), 7 (photo), 9, and 10

U.S. Department of Labor, Figure 6 (illustration)

NCCER CURRICULA — USER UPDATE

NCCER makes every effort to keep its textbooks up-to-date and free of technical errors. We appreciate your help in this process. If you find an error, a typographical mistake, or an inaccuracy in NCCER's curricula, please fill out this form (or a photocopy), or complete the online form at **www.nccer.org/olf**. Be sure to include the exact module ID number, page number, a detailed description, and your recommended correction. Your input will be brought to the attention of the Authoring Team. Thank you for your assistance.

Instructors – If you have an idea for improving this textbook, or have found that additional materials were necessary to teach this module effectively, please let us know so that we may present your suggestions to the Authoring Team.

NCCER Product Development and Revision

13614 Progress Blvd., Alachua, FL 32615

Email: curriculum@nccer.org
Online: www.nccer.org/olf

❑ Trainee Guide ❑ Lesson Plans ❑ Exam ❑ PowerPoints Other ____________________

Craft / Level: Copyright Date:

Module ID Number / Title:

Section Number(s):

Description:

Recommended Correction:

Your Name:

Address:

Email: Phone:

Metal Decking

30114-11

Trainees with successful module completions may be eligible for credentialing through NCCER's National Registry. To learn more, go to **www.nccer.org** or contact us at **1.888.622.3720**. Our website has information on the latest product releases and training, as well as online versions of our *Cornerstone* newsletter and Pearson's Contren® product catalog.

Your feedback is welcome. You may email your comments to **curriculum@nccer.org,** send general comments and inquiries to **info@nccer.org**, or use the User Update form at the back of this module.

V.1 11/10

30114-11
METAL DECKING

Objectives

When you have completed this module, you will be able to do the following:

1. Identify and explain types of decking and deck profiles.
2. Describe how decking is packaged, shipped, and stored.
3. Erect decking and observe job-site safety.
4. Explain the effects of deck penetrations and damage.
5. Demonstrate how to place concrete.

Performance Tasks

Under the supervision of the instructor, the trainee should be able to do the following:

1. Demonstrate safe lifting methods.
2. Properly place decking.
3. Identify safety precautions for decking operations.
4. Demonstrate proper decking layout.
5. Identify types of decking.

Trade Terms

Button punch
Cellular decking
Fuse
Galvanize
Phosphatized
Profile
Rolled-in embossment
Shear stud
Substrate
Weld washer

Contents

Topics to be presented in this module include:

Figures and Tables

1.0.0 INTRODUCTION

Steel decking is cold-formed, structural grade sheet steel. It is formed into a repeating pattern of parallel ribs. The shape of the ribs and the material properties of the steel give the panels their strength and stiffness. Decking is made in various lengths to fill different needs. However, it is usually less than 40 feet long, for shipping reasons. Although standard decking widths vary, they normally fall within the range from 12 inches to 36 inches, in 6-inch increments. The sheets are square on the ends. Any openings or angle cuts are done at the site. The decking is normally attached using puddle welds, powder- or pneumatic-driven pins, or self-drilling screws. Sheets are fastened together with welds, screws, or button punching. Metal decking is classified as follows:

- Composite floor deck
- Roof deck
- Cellular deck
- Form deck

1.1.0 Composite Floor Deck

Installed and fastened floor decking serves a number of different purposes. It stabilizes the frame, becomes a work platform, reinforces the slab so it will carry the expected design loads, and acts as a form for the slab. On floor decking, shear connectors are part of the deck. They lock the deck and concrete together mechanically, enabling them to act as a unit to carry floor loads. Shear connectors may be lugs, wires, **rolled-in embossments**, or holes. Concrete and steel may also be interlocked by the deck **profile** configuration (*Figure 1*). The profile of a piece of decking is a combination of the height of the protrusion and the end-to-end length of a single protrusion. The coverage is the length from overlap to overlap of the edges of a single unit. Putting it another way, coverage is where the edges of two side-by-side pieces would overlap (*Figure 2*).

Composite decks may be coated with zinc (**galvanized**) or painted and **phosphatized**. On a phosphatized deck, the bare or phosphatized surface is placed on top to contact the concrete. The bottom side has a coat of primer. The top surface will start to rust before the concrete has been poured. This can be attributed to the acidic properties of rainwater.

Galvanized decking does not rust quickly because of the zinc coating on both sides. Zinc helps to prevent corrosion, and it can be painted with the proper paints.

Because of the shear connectors or profile shapes, composite floor decking is usually not installed with the ends overlapping on the support beams. Overlapped sheet ends create an air gap that will prevent shear stud welds from developing the proper fusion. Decking must have the end bearing amount shown on the erection drawings. If the bearing is less than the drawings show, check with the design engineer.

1.2.0 Roof Deck

Roof decking is not composite; it does not rely on other materials to transfer vertical and horizontal loads onto the frame of the building. When compared to floor decking, roof decking has narrower rib openings that provide better support for rigid thermal insulation board.

Roof decking (*Figure 3*) normally end laps 2 inches or more over support members, although it may be butted instead for field-fitting. Roof decking is frequently designed as part of the horizontal support for the structure. This being the case, any change of fastenings must be checked with the engineer. The outside edge of the deck must be supported all the way around or else the edge could sag or bend on the finished roof. There is also the possibility that strains may not be transferred from the roof to the structure or from the structure to the roof.

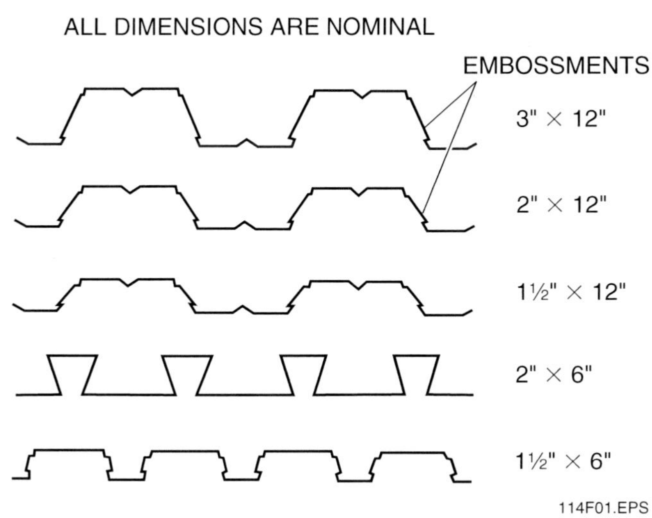

Figure 1 Typical composite floor decking profiles.

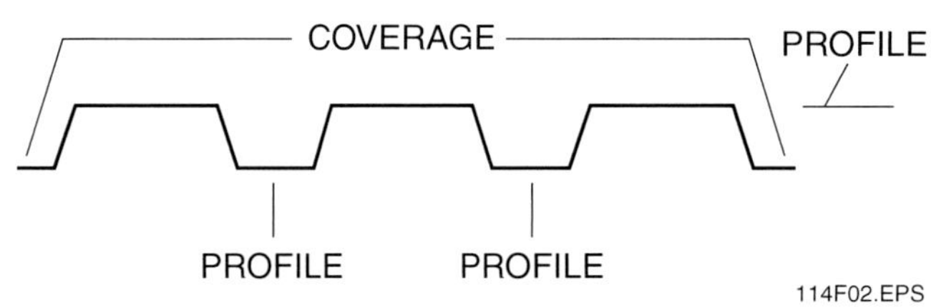

Figure 2 Coverage and profile.

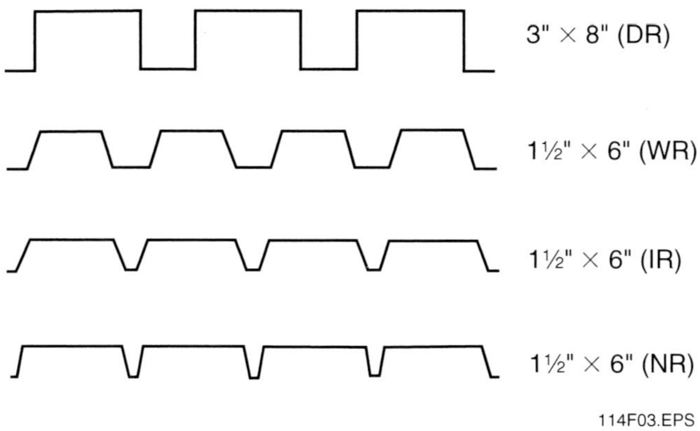

Figure 3 Typical roof deck profiles.

Roof decking is generally either galvanized or primed with paint. If the decking was primed, the primers applied at the factory are not designed to provide long-term weather resistance. As a result, the decking may need to be painted, unless it will be concrete-coated. Field painting and repairs to the primer coat are the responsibility of the buyer of the decking, as is the choice of the paints used both for priming and field painting.

1.3.0 Cellular Deck

Cellular decking (*Figure 4*) consists of a normal composite floor or roof decking profile that has a steel sheet attached to the bottom. Cellular decking can be used similar to the way that floor decking is used. Wiring can be passed through the tubes (chases) created between the profile of the deck and the flat sheet. For wiring to pass through the decking, the ends of the ribs must be lined up to form a smooth tube. The joint at the ends must be protected by tape or some other means that will prevent concrete from entering the cell. In addition, sharp intrusions from welding or cutting must not be allowed in the cells, as these could damage the wires.

In cases where cellular decking is used for roof decking, the cellular decking usually remains exposed and visible. Great care must be taken to achieve the proper alignment. It is also very important to install the decking carefully to prevent damage. Cellular decking is always either galvanized, or galvanized and then painted. This protects the surface from damage.

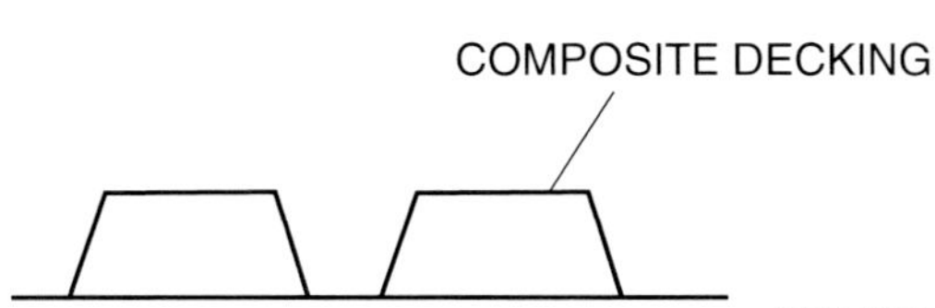

Figure 4 Cellular decking.

Sometimes the flat plate of cellular decking is placed on the top, as a walking surface. This surface is slippery and must be treated to make it safe. Installing the decking upside down in this way requires special fastening methods. This is because the plate will prevent workers from reaching the bottom of the groove where it rests on the supports.

1.4.0 Form Deck

Floor or roof decking products may be used as concrete forms. The attachment methods for form decking are basically the same as those used for roof and floor decking systems. Welding washers should be used for metal that is less than 22 gauge or 0.028 inch.

Form decking (*Figure 5*) may arrive uncoated, or it may be primed or galvanized. Form decking must be galvanized when it is used as a roof deck to carry lightweight insulating concrete fill. Proprietary variations of roof decking are available that use various types of insulation on form decking.

2.0.0 PACKAGING

Decking is transported in bundles of various weights, up to several thousand pounds. Bundle sizes and weights are arranged by the supplier and various individuals in positions of responsibility on the project. The bundles arrive with an attached tag (*Figure 6*) that contains a portion or all of the following information:

- Bundle weight (maximum 4,000 lbs)
- Manufacturer's contract identification number
- Customer and job information
- Product description details
- Number of pieces, including their lengths
- Bundle identification number
- Special information

Certification tags (Underwriters Laboratories, etc.) are also applied to the bundle.

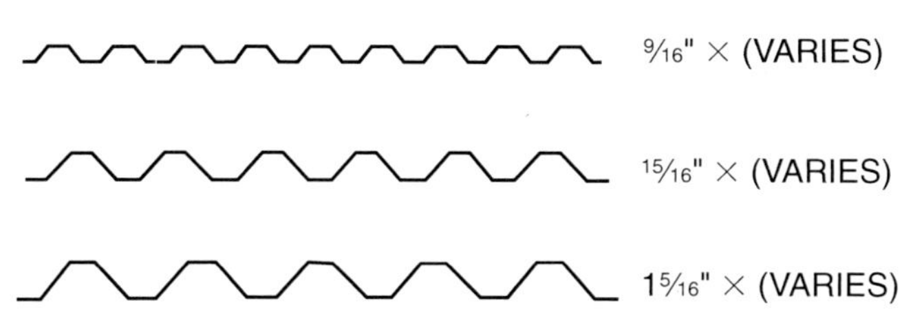

Figure 5 Typical form deck profiles.

STEEL DECK FABRICATOR

CONTRACT NO.

CUSTOMER NAME
PROJECT NAME
PROJECT ADDRESS

DECK TYPE, GAUGE, FINISH

NUMBER OF PIECES	LENGTH FT.	IN.	INCHES	MARK
20	27	10.00	334.00	
20	25	11.00	311.00	
5	25	7.00	307.00	
45				

NOTES:

BUNDLE NO. X OF X TRUCK: XX
CONTROL NO. XXXX-XXX-XXXXX
BOM BUNDLE NO. XX

114F06.EPS

Figure 6 Sample bundle tag.

2.1.0 Loading and Shipping

Bundle tags indicate where the bundle should be placed at the site. Decking should be delivered where and when it needs to be used so the operation sequence can be met. The decking manufacturer should be informed of the sequence of delivery when the order is placed, or as soon as possible afterward. Weight limits, special blocking, and other relevant job conditions must also be passed on to the shipper.

When trucks are being loaded with decking, the following guidelines apply:

- Deck bundles must rest on and be separated with dunnage at least 1½ inches thick (the thickness of a nominal 2 × 4, unless more is required). This will allow slings and forklift forks to go between the bundles as they are unloaded.
- To make sure that the load is balanced, place the longest bundles closest to the bed.
- Deck bundles may be turned in alternating directions (for example, from side-to-side and lengthwise) or in any other configuration that will make the load as stable as possible.
- Deck bundles may be positioned against the bulkhead of the trailer or truck so that they cannot slide forward when there are sudden stops.
- Bundles must be securely strapped down to ensure that none of the sheets blow off or shift during transport.

When in transit, loads often compress from the vibration and shocks of the highway. As a result, the straps and chains used to secure decking bundles may loosen. Overtightening the tie-downs damages the decking. To avoid this, shippers must periodically retighten the tie-downs. When they perform this procedure, shippers also adjust the protective coverings as needed in order to keep the product dry and free from water damage.

Full-truckload shipments are standard, unless other arrangements have been made. When less-than-truckload shipments are transported, damages may occur from transfers or handling by the shipping company. Such damages would not be considered the responsibility of the manufacturer.

2.2.0 Receiving and Unloading

Decking must be received close to the structure, using equipment that can safely lift the bundles into the proper place on the structure. As material comes in, it must be checked. The bundles must be counted and compared against the bill of lading. Any damaged or missing material must be noted on the bill of lading before signing for the product. In addition, a supervisor should be notified and the shipper told at once. Otherwise, the material may not be available when it is needed on the job.

2.3.0 Storage and Protection

When decking is to be stored on the ground, raise the bundles up off the ground on some type of dunnage that has been sloped to ensure proper drainage. Stack the bundles carefully so that they are stable during storage. Check bundles for secure strapping that will protect against sheets becoming loosened in the wind. Check the security of the bundles periodically, and strap or band them as needed. Do not use strapping to move or lift bundles, and always keep bundles tag-side-up. In addition, make sure that the structural frame is strong enough to hold the weight of the bundles.

WARNING

Never place decking bundles on unbolted frames or unbridged joists. Always place the bundles near a major supporting beam and column or wall. Falling bundles can cause personal injury or death.

3.0.0 Erecting Decking and Observing Job-Site Safety

During the erection of steel decking, all federal, state, and local safety regulations must be observed and the general safety guidelines in this section followed. Failure to understand and comply with these regulations and guidelines can lead to serious injury or death.

A site-specific erection plan must be developed for each building. The erection plan must cover unloading, hoisting, landing, or installing deck. Refer to the latest *OSHA Standard 29 CFR, 1926 Subpart R* for guidance.

Deck erectors build their working platform as they go. They must often work on open steel frames or on ladders (*Figure 7*) or scaffolding to reach the work areas. Many of these areas lack protected edges or protected openings. Because of this, it is important that the deck erector remain constantly aware of the danger from falling.

When decking is being laid on a level, the area may be a controlled decking zone (CDZ), according to *OSHA Standard 29 CFR, 1926 Subpart R*. A CDZ is an area that is more than 15 feet above, and not more than 30 feet above, a lower level where metal decking is being installed. This is a case in which decking forms the leading edge of the work area. The CDZ is marked with control lines, caution tapes, or comparable markings that are not less than 6 feet and not more than 90 feet from the leading edge. The control lines must meet certain standards, including the ability to resist weight. Netting or other fall arrest systems must be in place to protect workers from injury, meeting the specific requirements of *OSHA Standard 29 CFR, 1926 Subpart R*.

At least two safety deck attachments must be installed to each sheet of decking throughout the CDZ, up to the leading edge. Only those personnel working on the leading edge of the area are allowed into the CDZ. Workers must have been specifically trained in the safe working methods

114F07.EPS

Figure 7 Unsafe ladder.

for installing decking. They must also understand safe fall arrest techniques and use them at all times when working on the leading edge of the decking.

No more than 3,000 square feet of unsecured decking must be in a CDZ. Final deck attachments and shear connectors are not installed in the CDZ. The CDZ must not be more than 90 feet wide and 90 feet long from the leading edge.

3.1.0 General Safety

Unless a deck is being installed on the first floor of a building without a basement, the decking work will be done on an elevated structure. There is always some danger of falling from an elevated structure. Falls can happen at any time or place. Ladders must be used correctly and must be secured to a safe structure or scaffolding. If stairs will be used, they must be firmly attached to and supported by the building's frame. To prevent tripping hazards, all areas providing access to the work area must be kept clear of equipment, material, and debris.

Always protect yourself from the sharp edges of decking material. Sharp, projecting corners are also hazardous and can endanger you as well as your fall-protection strapping.

Before placing decking on the structure, be sure to check frame plumbness and connections. Also check any bracing to ensure that the structure will be held in plumb until the decking is placed and secured. Make sure that the steel joists are fastened securely at their bearing ends and that their bridging is properly and completely attached.

3.2.0 Lifting Safety

Bundles must be rigged so that they will not be subjected to excessive shifting or tipping (*Figure 8*). In addition, make sure that the lifting equipment will not damage the decking and that it is capable of handling the lifting requirements of the job. All rigging and hoisting must be performed by properly trained personnel. Tag lines must be fastened to the bundles to properly position the material. Never pull on the banding to move the material. As a precaution, it is a good idea to have workers secure the bundles until they have been placed.

When possible, spread the bundles out near several columns (*Figure 9*) instead of creating one large stack. Make sure that both ends of the bundles rest on a safe bearing structure. Bundles must be placed so that the sheets do not require turning before being spread into position. Once the bundles have been unbanded, they must be properly secured to prevent them from being blown by the wind.

3.3.0 Safe Working Platform

For safety reasons, and to keep the decking from being damaged, it should be spread and attached as soon as possible. If decking sheets are being used as a temporary platform to spread and fasten the decking from the bundles, the sheets must be end-bearing and securely fastened to the frame to prevent them from slipping. A working area that is used to move sheets off the bundle should be at least 12 feet wide. This area should be located next to each bundle and must be attached to the frame. The working platform can then extend in whatever way is most useful for the job. Specific job requirements must be considered in order to determine the starting points and the sequence for erecting the deck. As the platform is extended, at least one worker will be required to work from the structure itself, rather than from the platform. All workers must follow OSHA regulations at all times, including using fall protection equipment properly while erecting deck. Keep in mind that decking sheets are slippery when wet, so be sure to follow all applicable safety precautions. *Figure 10* shows the correct technique to use for load placement.

3.4.0 Placing Decking

One edge of the deck is always open, or leading, as the sheets are placed. Only go to the leading edge to place the next sheet. Stay at a safe distance from the end of the section. When aligning the side edge or lap of a sheet, always kneel to lower your center of gravity and to create a wider base, which helps to lessen the chance of falling.

As soon as possible after erecting the deck, the edges of the building and of all the openings must be protected with cables, fencing, or other OSHA-required warning and protection devices. All floor holes must be protected with a cover, which is secured against accidental displacement, and must meet OSHA strength and construction standards. When a cover is not in place, the hole should be protected by a railing that meets OSHA standards.

3.5.0 Decking Capacities

Personnel not involved in the decking erection process should not be on the working platform or immediately under it when work is being performed unless designated to do so. Workers should be trained in all aspects of deck safety be-

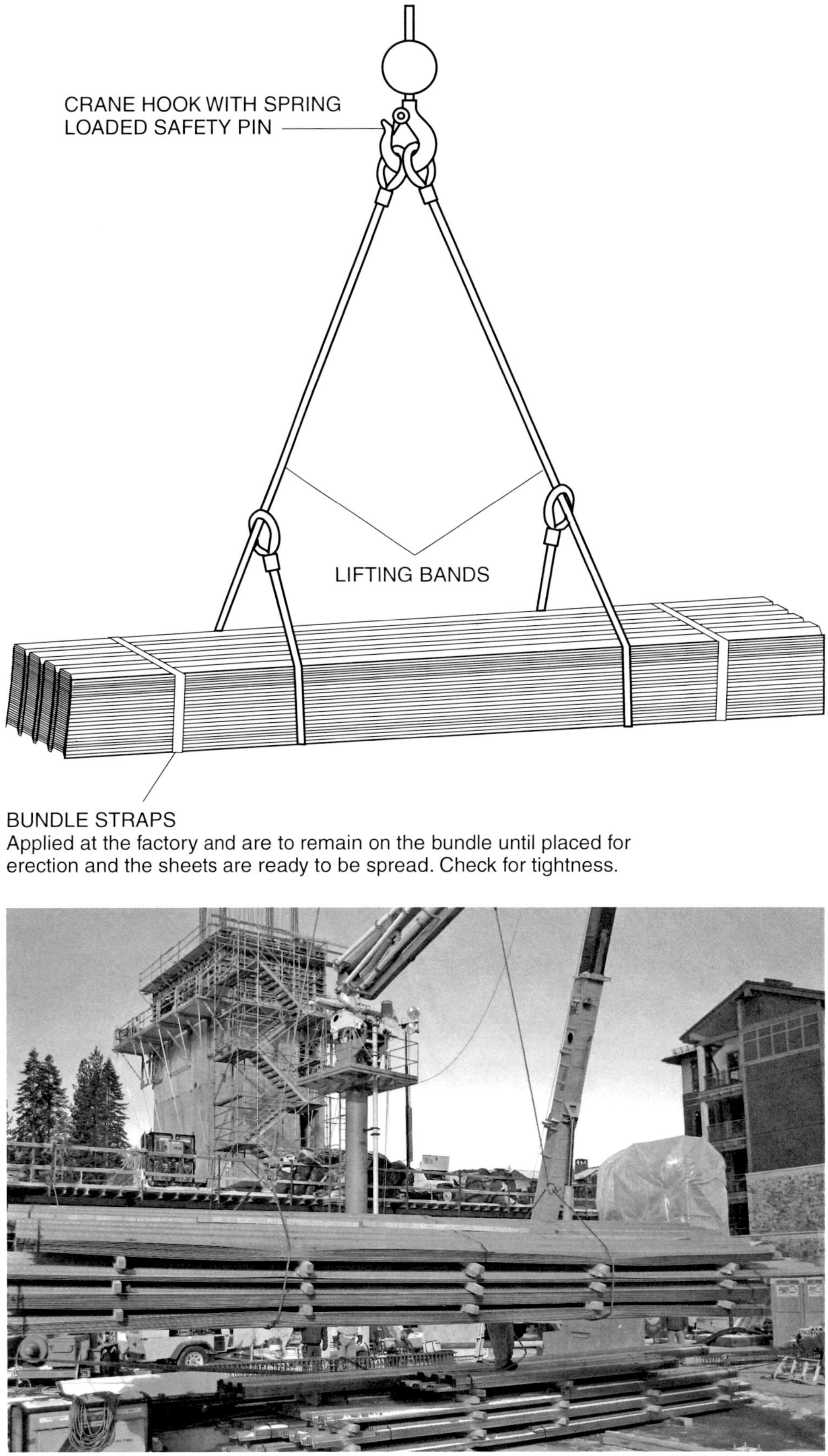

Figure 8 Example of proper lifting.

114F09.EPS

Figure 9 Spreading bundles.

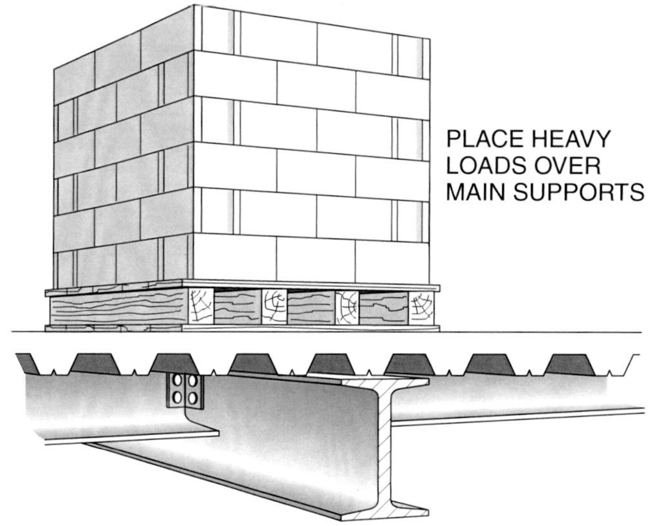

114F10.EPS

Figure 10 Proper load placement.

fore any installation takes place. Check for dunnage under or on the bundles before cutting the strapping. Also, take the necessary precautions to ensure that neither banding nor dunnage falls on personnel or equipment below.

If floor decking is used as a working platform, it must be able to support at least 50 pounds per square foot (psf). If shoring is needed to reach this capacity, it must be installed properly before the erection process starts. Concrete weight is not included in the calculation of the 50 psf.

Roof decking must support at least 30 psf. If the decking does not meet this requirement, it must be supported during the erection work. Any locations that will have to support weight, such as areas exposed to material staging or traffic, must be planked or otherwise supported. Never exceed the maximum recommended roof deck spans. The following safety rules apply when erecting decking:

- Adjust rigging carefully to balance loads before they are lifted.
- Never stand under raised loads unless designated to do so.
- Always keep an eye on raised loads.
- Be sure that loads are placed and positioned correctly to prevent having to move them twice. Review erection drawings as needed.
- Use the correct crane hand signals.
- Never cut the banding until the bundle is safe and stable.
- When cutting banding, do not stand in line with the banding, and avoid getting too close to the bundle.
- Always wear eye protection and gloves when cutting banding.
- Be careful when working with single span bundles because they are especially prone to instability.
- Take special care when working on short units and single span units, and make sure that platform units are secure before attempting to work from them.
- Properly support and guard all cutouts and openings.
- Mark the location of supporting steel with carefully placed chalklines.
- Be aware of slipping hazards on wet decking, and take the necessary precautions.
- Wear leather work gloves to protect hands from sharp edges.
- Always wear the proper eye protection when performing welding work.
- Be sure to wear sunglasses and apply sunscreen when working on galvanized decking on sunny days.
- Avoid creating tripping or slipping hazards by keeping the work area litter free.

3.6.0 Fastening and Installing Decking

It is very important to have consistent rib alignment in the decking as it crosses the structure. Misalignment is not only unsightly, but it can lead to other problems. In the case of floor decking, if the ribs are not continuously aligned, the concrete may leak more at the laps. Roof panels also look unattractive if they are not oriented accurately, parallel with the girders. Errors in lap cover width and in the alignment of panels cannot be corrected at the end of a run.

The deck starting point must be carefully chosen at the beginning of a level. The starting point is determined by the desired orientation of the deck or roof. Once the first point has been established, use a chalkline to mark the line for the edge of the decking. Use chalklines to align the rest of the deck as it is laid and fastened. Similar to shingle roofing, it is very easy for the alignment to slowly shift away from the original orientation. The more carefully the initial alignment is adhered to, the better. Keep in mind that each deck panel must be adjusted as it is laid.

Decking must be properly attached. Avoid substituting fasteners for the ones specified unless the substitution has been approved by the designer. Decking often plays a major part in the design strength of the structure. The use of incorrect fasteners or the wrong fastening pattern can result in structural problems.

CAUTION

Never change or substitute the fastener type or pattern without the designer's permission.

Deck can be attached to the structure with air-driven or powder-driven fasteners, welding, or with screws, which are usually self-drilling (*Figure 11*). Powder- and air-actuated tools may only be used by qualified operators. If decking is to end lap, the end lap location should be placed in the center of the support. When the support is a bar joist, the center should rest on one of the top chords.

3.6.1 Attaching Decking

Self-drilling screws are usually #12 or ¼-inch diameter. The screws have a distinctive drill point, and their selection is based on the thickness of the metal being joined (the decking and the frame). The screws are driven by an electric screwgun that has a clutch and a depth-limiting chuck so that the torque can be correctly set.

Air-driven fasteners (*Figure 12*) have flat heads and bullet-like points. Sizes vary based on the penetration requirements of the steel **substrate**. The fasteners are driven by tools that are set at a pressure level that meets the needs of the deck attachment. The tools are driven by a compressor that can be controlled to deliver the correct pressure, which prevents the fastener from being over driven or under driven.

Another system of fastening decking uses powder-actuated tools. These are tools that use gunpowder to fire a fastener through the deck. While these tools have all appropriate safety interlock mechanisms to prevent them from discharging without being pressed against the deck, it is always possible to make a mistake. Be careful about where the tool is pointing at all times.

3.6.2 Welding Decking

All welding performed on decking must meet American Welding Society (AWS) regulations. Using shear lugs on lapped composite decking is not advisable because welding requires tight metal-to-metal contact. Built-in hanger tags in

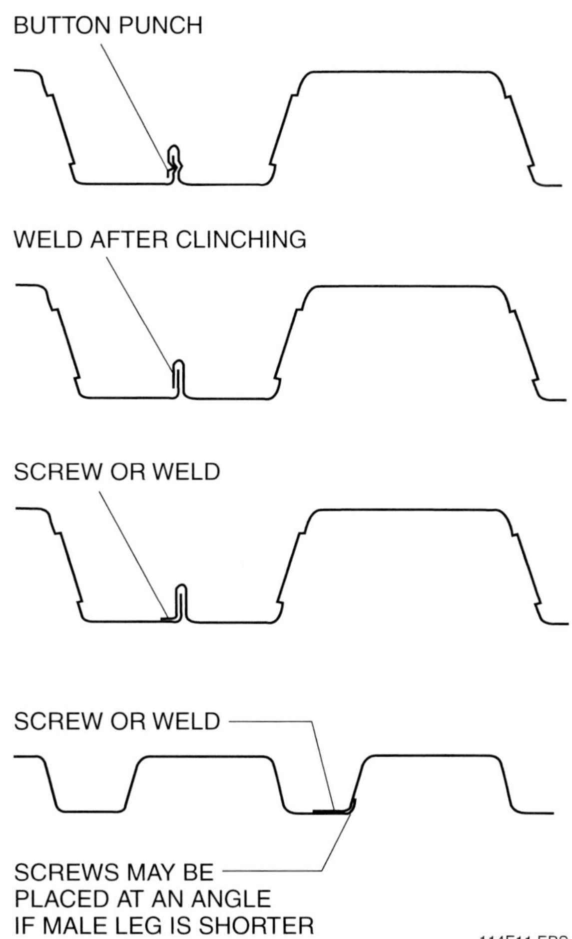

Figure 11 Side lap connections.

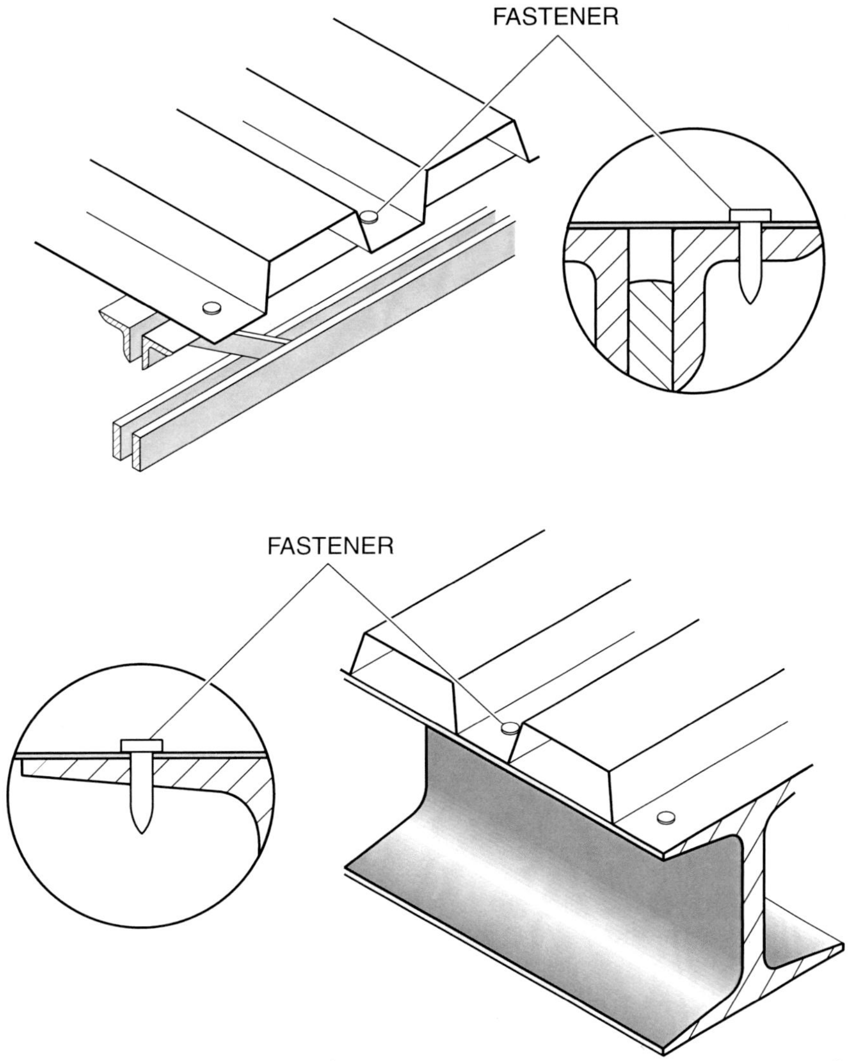

Figure 12 Air-driven fasteners.

floor decking must be flattened or removed. Do not use weld washers for metal thicker than 0.028 inches (about 22 gauge). Welding requires dry weather and may only be done by qualified welders. *Figure 13* shows welding patterns for decking.

> **WARNING**
>
> Never perform any type of heating, cutting, or welding until a hot work permit is obtained and a fire watch is established. If unsure of the procedure, check with a supervisor. Violation of the hot work permit and fire watch can result in serious injury or death.

A weld bead can be either a stringer bead or a weave bead, based on the technique used to produce it (*Figure 14*). If the bead is formed along the joint with little or no side-to-side motion, it is referred to as a stringer bead. A weave bead is made by moving the electrode laterally, from side-to-side, as the bead is deposited along the joint. A weave bead is typically wider than a stringer bead. The travel speed of a weave bead is less than that of a stringer bead because of its side-to-side motion, as measured along the longitudinal axis of the weld.

3.6.3 Shear Studs

Shear studs (*Figure 15*) that are welded in place with special equipment, in accordance with AWS regulations, can be used to hold the deck to the frame. In general, the studs are installed after the decking has been spread and attached to the structure as a working platform. Shear studs can be installed on cellular deck and are capable of penetrating both metal thicknesses. All fasteners must be installed according to the design drawings.

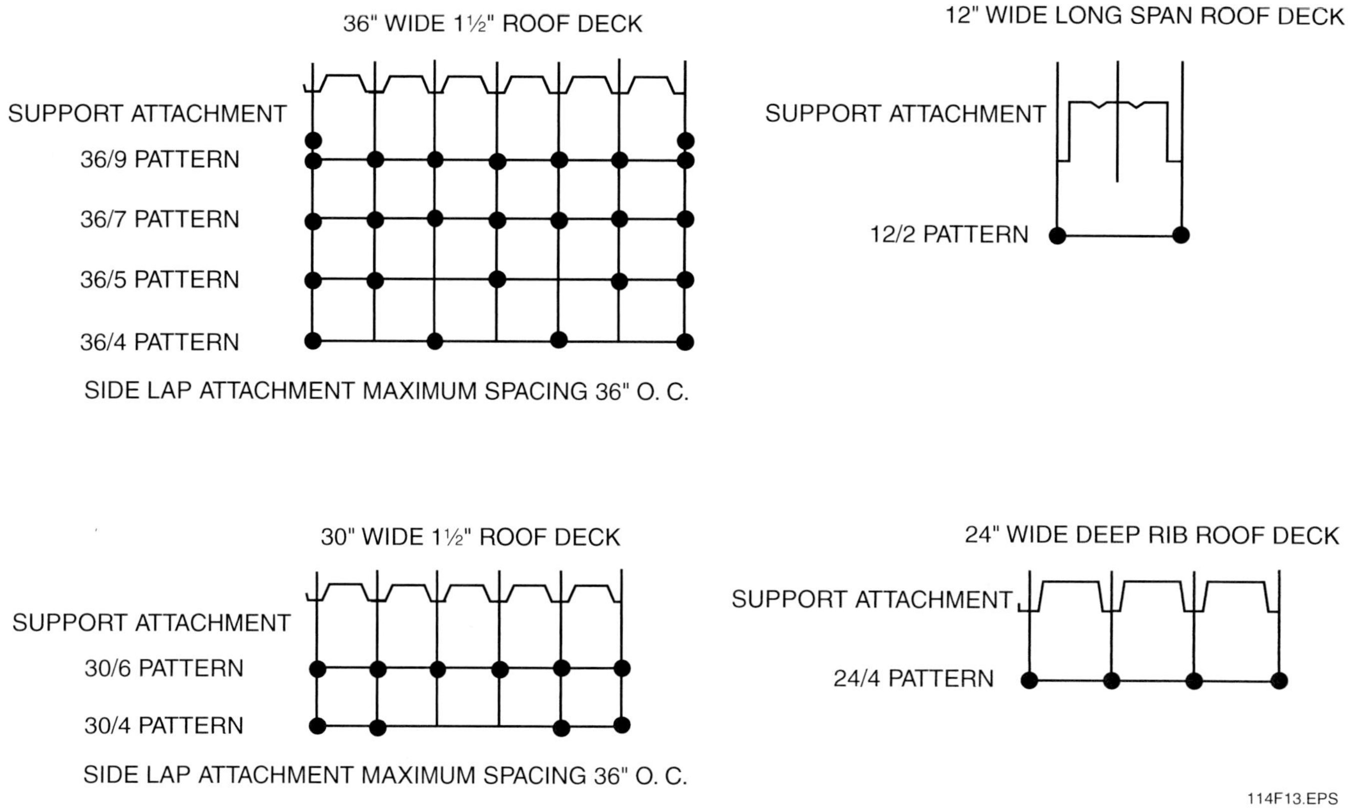

Figure 13 Typical fastening patterns.

However, if the decking is thicker than 16 gauge, be sure to check with the stud manufacturer for any special instructions. Personnel involved in or working around the welding operations must wear full ultraviolet-light eye protection, even if they are not performing the welding. Eye burns from welding flash can be very painful.

Ventilation must be provided unless welding is being done in open areas. Ventilation becomes even more critical when the metal being welded is galvanized decking. The fumes from this process can result in serious illness or death. When exposure has been minor, milk can alleviate some of the symptoms. When ventilation is inadequate, respirator equipment may be necessary.

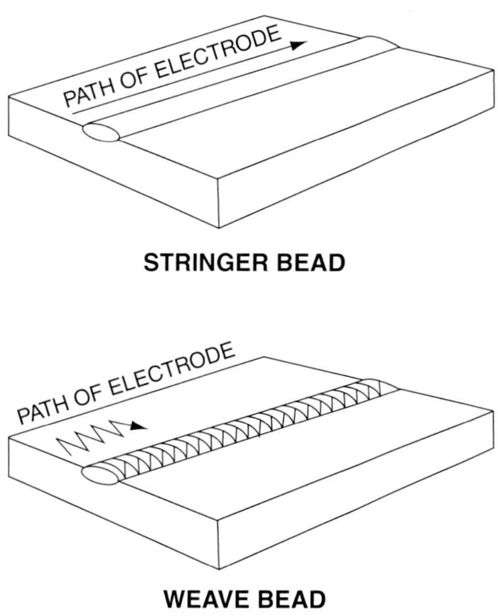

Figure 14 Examples of stringer and weave beads.

Figure 15 Shear studs.

WARNING

Torching or welding galvanized metals produces a very toxic smoke. The fumes can cause serious illness or death. Always follow all appropriate safety precautions when working with heat applied to galvanized metals.

3.6.4 Side Lap Connections

Connections must frequently be made between the deck sheets, in addition to those made when fastening the decking to the frame. These sheet-to-sheet connections, also called stitch connections, may be **button punched**, fastened with self-drilling screws, or welded. The screws used for stitch connections are usually larger than #10 and normally range from #8 through ¼-inch diameter. Make sure that the sheets are in close contact. Button punching or screws are the preferred methods, but welding may be specified.

Button punching requires workers to use their body weight to hold the tops of the two sheets level across the joint. The manual button punch is occasionally replaced with an automatic crimping device. The quality of the button punch connection depends on the skill and experience of the worker and on the consistency of the procedure.

Good metal-to-metal contact is needed for acceptable side lap welds (*Figure 16*). To be strong, the weld must have three-quarters or more of its perimeter **fused.** Welding side laps is not generally used for decks thinner than 22 gauge. Weld washers should not be used for welding side laps between supports. As shown in *Figure 17*, welds may also be used at supports.

3.6.5 Housekeeping

All debris, including dunnage, banding, deck cutoffs, and welding rod stubs, as well as any flammable materials, should be removed from the

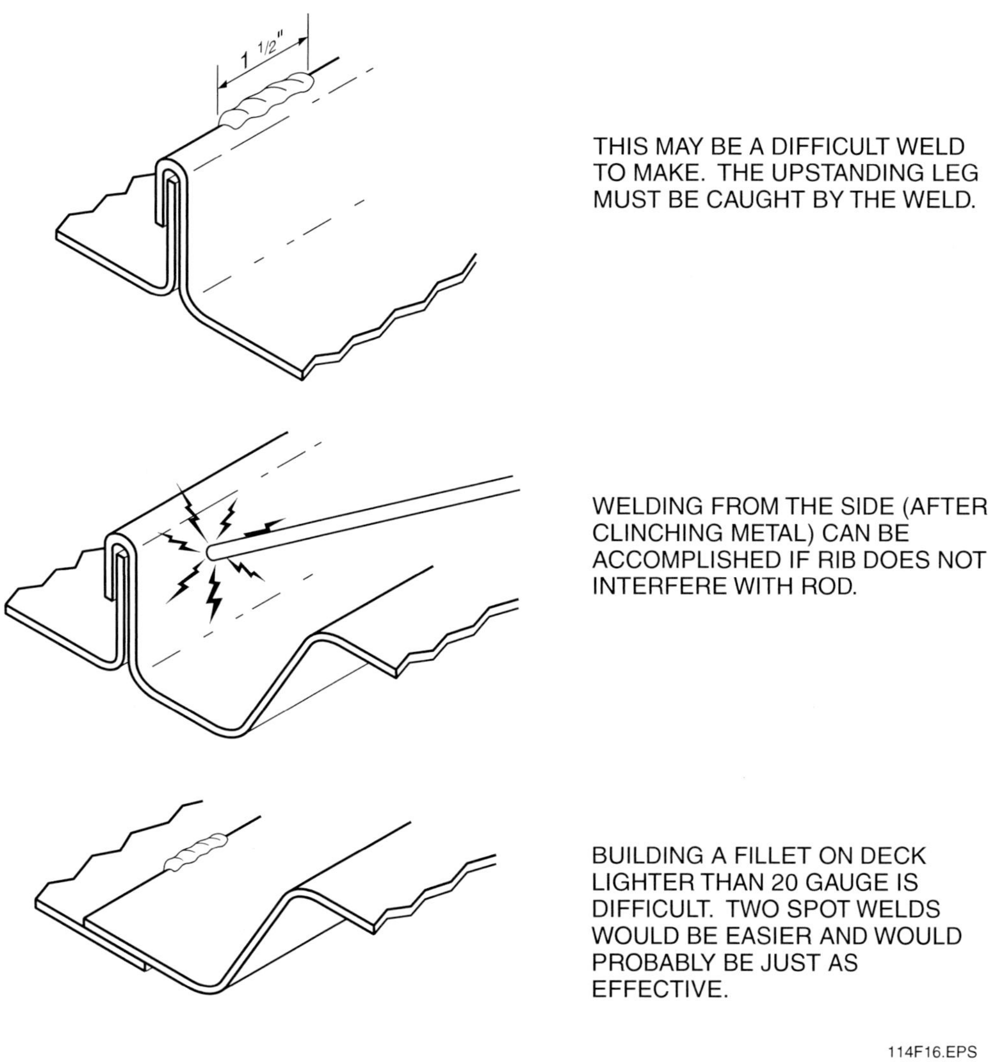

Figure 16 Side lap welds between supports.

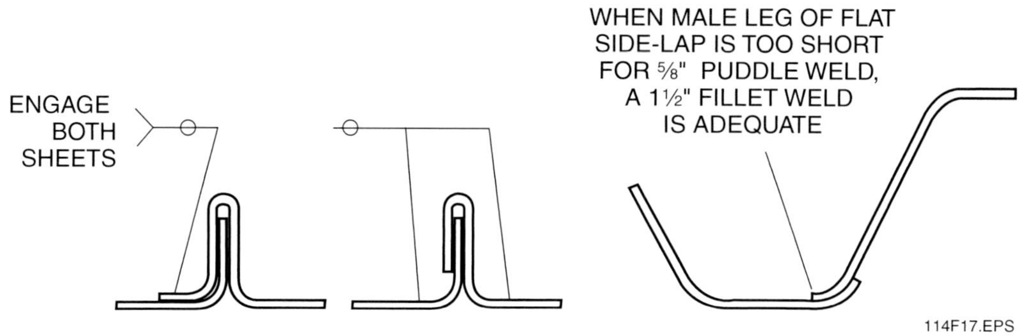

Figure 17 Side lap welds at supports.

working platform every day as a safety practice. Tools must be stored to prevent tripping hazards and to help ensure overhead safety for workers below. Stud welding ferrules and all debris must be removed before concrete is poured. Loose deck sheets or partial bundles must be secured at the end of every working day. In addition, all combustible material must be properly stored or discarded.

4.0.0 Deck Damage and Penetrations

The sizes and exact locations of penetrations, such as vent stacks or conduit, are often uncertain at the time of the design process. This means that the reinforcement and damage repair requirements will not be known until the placement becomes certain at the time of installation. The Steel Deck Institute (SDI) specification states that "Openings not shown on the erection diagram, such as those required for stacks, conduits, plumbing, vents, etc., shall be cut (and reinforced, if necessary) by the trades requiring the openings." When reinforcements are required, details are frequently included in the drawings. However, if the holes are small and the details are not included in the drawings, no deck reinforcement may be required.

WARNING

If the area of the hole cannot be guardrailed or covered, workers must be provided with personal fall protection systems.

The decision as to what defines damaged decking is made by the designer. This decision is based on architectural and structural design issues, visual appearance, and possible delay of the job. If the decking will be publicly shown, little or no visible damage may be accepted. When the decking is hidden by a dropped ceiling or other material, the structural function is the sole concern.

4.1.0 Roof Deck

Toleration of damage in roof decking caused by denting or penetration varies based on the number of webs removed and the size of the damaged area. If the roof deck's depth is the common 1½ inches, then one rib per sheet can be lost to denting or penetration without an unacceptable loss of strength. It is not necessary to reinforce a 6-inch hole unless more than two ribs have been removed. The diaphragm strength of the deck should be more than sufficient to handle a 6-inch hole per sheet. A dent can be larger than 6 inches without affecting the strength of the sheet. However, the dent may have to be covered with a plate to allow insulation board to go over it. If the dent or hole is less than 8 inches, the plate can be made of 0.045-inch material, and it must extend to neighboring ribs. If the hole or dent is larger than 8 inches, use at least 0.057-inch material. If the holes are close enough to supports, it may not be necessary to take any action. On the other hand, there may be a group of holes so close together that they must be supported with a structural frame. Several examples of penetrations are shown in *Figure 18*.

When laying down roof deck, use chalklines or paint markers to identify the location of frame members so that welders do not burn holes in the roofing when performing side lap welds. Burn holes are not necessarily a problem if they are small and close to a structural member. However, they can cause problems when they are not supported and are too closely spaced.

Foot traffic may create small dents over fairly large areas. While these are not necessarily a structural problem, they may cause difficulties when attaching insulation board. In such cases, insulation board is often attached with mechanical fasteners rather than with adhesives, if this is permitted. The worker should notice and repair any soft spots in roofs that could result in insulation cracking from foot traffic.

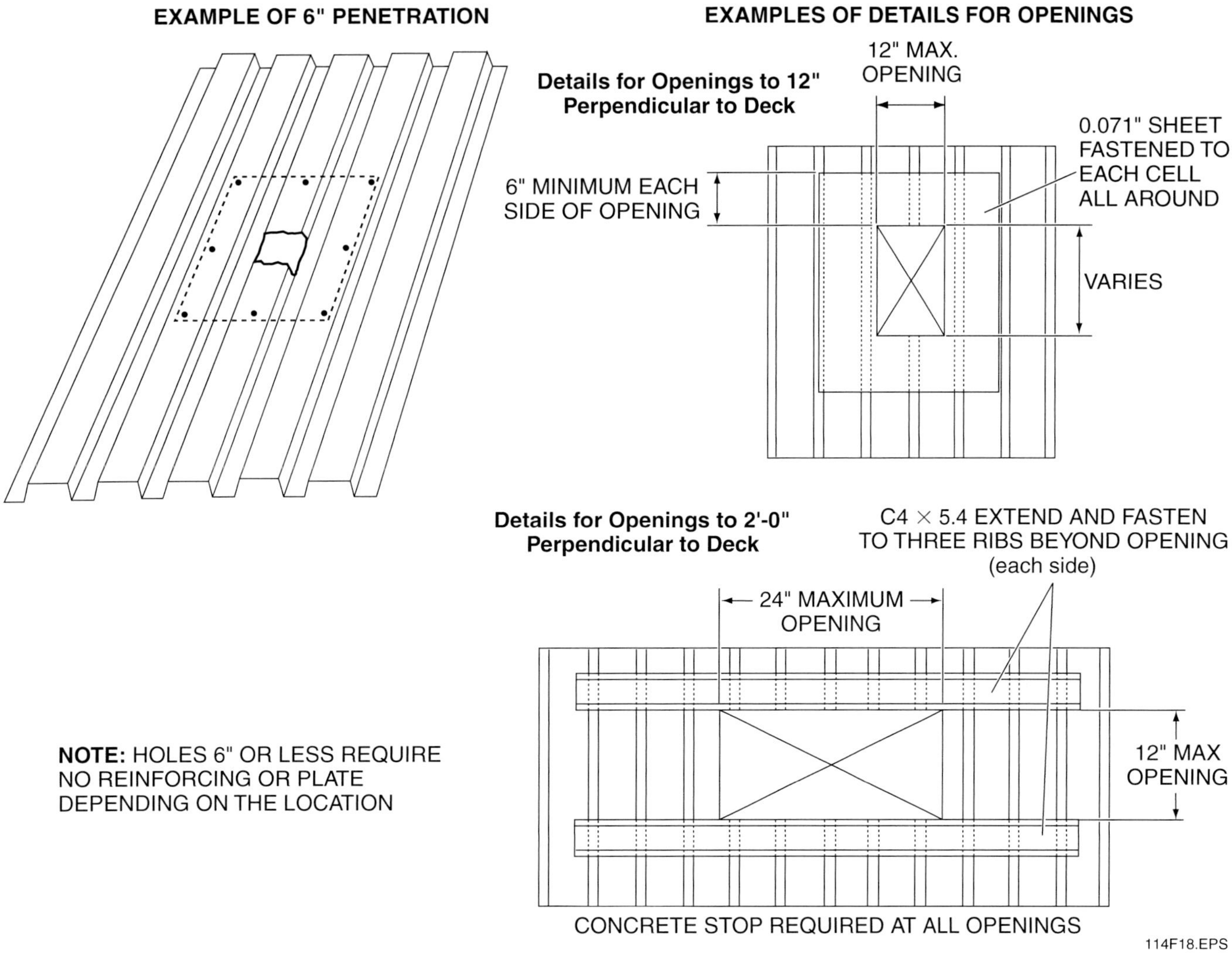

Figure 18 Penetration examples.

4.2.0 Sump Pans

A sump pan is a section of a roof that is designed to hold water for a roof drain or scupper. It can carry the load of the decking it replaced if it is attached and reinforced correctly (*Figure 19*). The sump pan's reinforcing channels help to transfer load to other uncut sheets.

4.3.0 Floor Deck

Floor deck damage or penetrations can affect the capacity of the deck to carry concrete. Any damage or penetration must be examined and possibly supported before the pour. Side laps must be checked to ensure good attachment at supports, as well as at the laps themselves. The connections must be tight to prevent them from opening during the concrete pour.

Areas that buckle during the pour are most likely caused by allowing the concrete to build up, over-spanning the deck, or previous damage. The buckled areas do not generally affect the live load capacity.

Floor decking may be designed as a cantilever (*Figure 20*).

Concrete discourages corrosion. Composite deck is usually located in dry interior areas. Because of this, field painting of burned, cut, or abraded areas is seldom necessary. If there are any touch-up requirements, they will be found in the job specifications.

5.0.0 Placing Concrete

When the floor or form deck has been completed and fastened, it is used as a working platform for a number of trades. The minimum capacity of the deck is 50 psf. If the load is likely to be more than 50 psf, the deck must be supported to meet the demands of the greater load. Before concrete is poured, the ferrules must be removed from the studs, dirt and debris should be disposed of, and

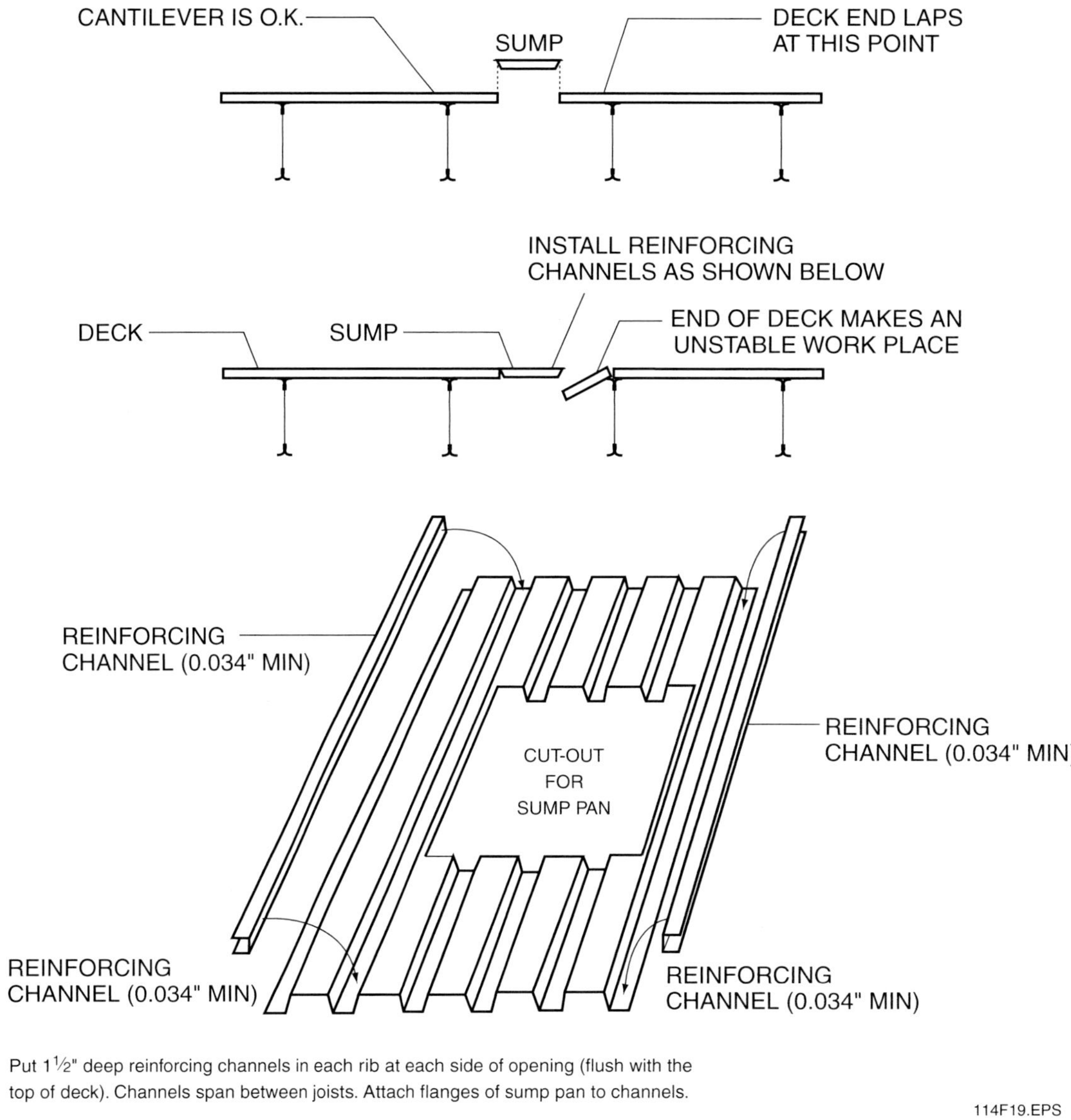

Figure 19 Sump reinforcing at the end of a deck.

all reinforcement material must be in place. In addition, any damaged areas must be approved or repaired.

Proper concrete pumping technique calls for the concrete to be released from as low a level as is reasonable so that the impact on the decking will be limited. Concrete should be poured directly over the structure and then spread onto the decking. The pour should proceed so that load is first placed on the top sheet of the laps in order to close the openings between the sheets as the concrete approaches. Take care to prevent excessive buildup during the pour.

If the concrete will be delivered on buggies, they must be run on planking that is stiff enough to transfer the load without damage to the deck. Whether the concrete is pumped or poured from buggies, workers not engaged in spreading the material should move out of the pour area to avoid increasing the load in that location.

To some extent, both the deck and the frame are deflected by the weight of the concrete. The amount of deflection is taken into consideration by the design engineers, and the amounts of concrete are calculated based on the strength limitations of the frame and deck.

In cases where a single span or short span is installed, the fastenings are even more critical. A common placement for single spans is between elevator shafts. The gauge of the fastener is selected on the assumption that the weight of concrete is greater than in cases involving standard or multiple spans. Assume a weight of concrete 50 percent greater than under normal circumstances, and perform a very close inspection of all fastenings.

If the deck is acting as a diaphragm in shear (if the deck is holding the frame in its horizontal

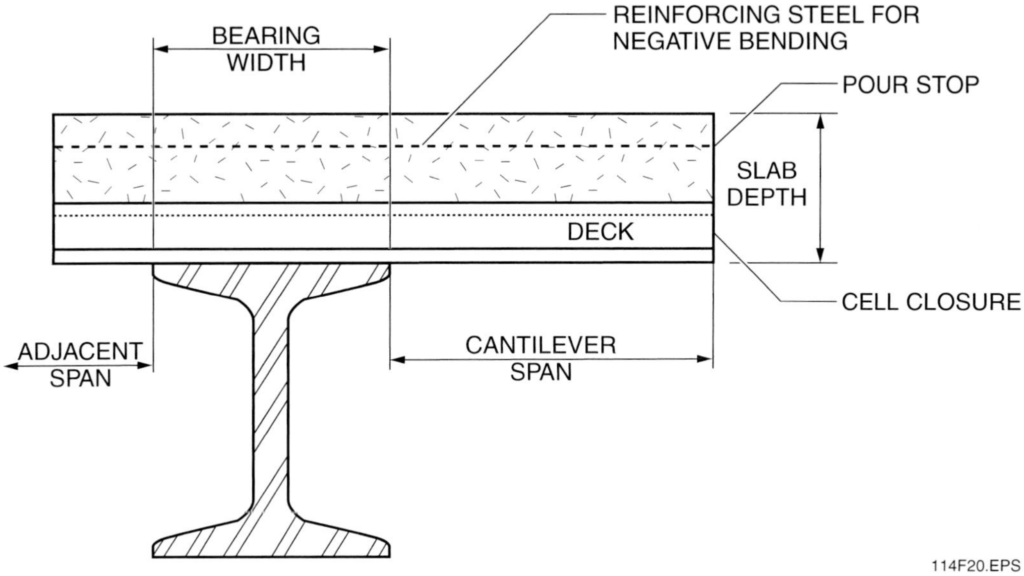

Figure 20 Cantilever span.

shape), the workers must be informed of this. The notes should warn that the decking is a part of the structural support of the building and that none of the panels may be removed unless replacement bracing is supplied at that time. When framed openings are present, the deck fasteners should be spaced and designed to match the spacing used in attaching the deck to the frame. When the deck is acting as a diaphragm in shear, keep in mind that the diaphragm is not fully working until it has been completed. Incomplete areas must be braced during any delays.

SUMMARY

Steel decking is divided into four classifications: composite floor, roof, cellular, and form. Composite decking consists of metal decking that is attached to a layer of concrete by shear connectors. To prevent corrosion, the surface of the metal is treated by galvanizing or phosphatizing.

Roof decking is used independently, rather than as part of the structural system. Although it may be used as a diaphragm, to reduce shear stress, or as part of the horizontal bracing system.

Cellular decking is made by attaching a bottom steel sheet to roof or composite floor decking. The chase between the deck panel and the bottom sheet can be used to house wiring or part of the electrical distribution system if the ribs are aligned and any intrusions removed.

Form decking is any floor or roof deck used as a concrete form. Decking is packed in standard bundles of 4,000 pounds, although it may be ordered in other weights, as required by the job. Decking is shipped and arranged on site in the sequence in which it will be erected. Once on site, be sure to store decking with the identification tag facing up. Also ensure that decking is properly protected against moisture and severe weather.

Before installing decking, workers must be familiar with the assembly methods, load capacities, welding requirements, and all applicable safety measures and precautions. When decking is damaged or penetrated, careful consideration must be given to decisions about replacement or additional support.

Review Questions

1. Floor decking can be identified by the _____.
   a. profile
   b. zinc coating
   c. shear connectors
   d. phosphatized surface

2. Roof deck does *not* rely on other materials to transfer horizontal and vertical loads onto the _____.
   a. diaphragm
   b. building frame
   c. bottom steel plate
   d. walking surface

3. By attaching a steel sheet to the bottom of a normal composite floor or roof decking profile, manufacturers produce _____.
   a. temporary shoring
   b. thermal insulation
   c. cellular decking
   d. form decking

4. Form decking may be furnished primed, galvanized, or _____.
   a. concreted
   b. plastic coated
   c. beaded
   d. uncoated

5. Tags on bundles of decking include the _____.
   a. sequence of erection
   b. job conditions
   c. delivery schedule
   d. area that is to receive the bundle

6. A working area that is used to move sheets off the bundle should be at least _____.
   a. 8 feet wide
   b. 10 feet wide
   c. 12 feet wide
   d. 16 feet wide

7. Floor deck that is used as a working platform should provide a capacity of at least _____.
   a. 30 psf
   b. 50 psf
   c. 75 psf
   d. 90 psf

8. Roof deck rib alignment must be parallel to the girders at all girder lines to prevent _____.
   a. an unsightly appearance
   b. excessive spans
   c. a cantilevered condition
   d. end lap conditions

9. A weld bead can be either a stringer bead or a _____.
   a. spinner bead
   b. wave bead
   c. weave bead
   d. pattern bead

10. Sheet-to-sheet connections at side laps are also called _____.
    a. knit connections
    b. stitch connections
    c. tongue connections
    d. mortise connections

11. The decision as to what defines damaged deck depends on _____.
    a. the manufacturer's considerations
    b. architectural and structural considerations
    c. length of span considerations
    d. alignment considerations

12. When properly attached and reinforced, the section of roof that can carry the load of the deck that it replaces is a _____.
    a. sump pan
    b. floor hole
    c. cross brace
    d. wooden cover

13. When pouring concrete on floor deck, buckled areas do *not* generally affect the _____.

  a. floor appearance
  b. smoothness of the floor
  c. live load capacity
  d. positioning of equipment

14. To avoid impacting the deck, concrete should be poured from a _____.

  a. wheelbarrow
  b. low level
  c. high angle
  d. low angle

15. As concrete is placed, the entire frame, as well as the deck, will to some extent be _____.

  a. rotated
  b. deflected
  c. adsorbed
  d. raised

Trade Terms Introduced in This Module

Button punch: A tool used to pinch a round protrusion in a decking lap, to hold the two pieces in relation to each other.

Cellular decking: A normal composite floor or roof decking profile that has a steel sheet attached to the bottom.

Fuse: To weld and join together base metal components.

Galvanized: To apply a zinc coating to a base metal as an effective barrier of protection. The zinc coating isolates the metal from the environment, adheres to it, and provides relatively superior abrasion resistance.

Phosphatized: A metal that has been treated with a thin coating of phosphoric acid or a phosphate compound. The phosphate causes a chemical reaction on the surface of the metal that allows powder coatings to adhere effectively. Phosphatizing also inhibits corrosion from forming underneath the powder coating.

Profile: The ribbed pattern shown in a cross section of decking.

Rolled-in embossment: Angular indentions and patterns in the ribs of decking that are formed when the steel is rolled. They are designed to interlock with the concrete slab.

Shear stud: A steel rod with an enlarged top welded to the deck and frame to lock the concrete in place.

Substrate: The underlying layer, or the material onto which another layer is applied or fabricated.

Weld washers: Ring-shaped metal pieces, used to add strength to the weld between light gauge sheet metal and either frame members or other sheet metal.

Additional Resources

This module presents thorough resources for task training. The following resource material is suggested for further study.

SDI Manual of Construction With Steel Deck. Second ed. Fox River Grove, IL: Steel Deck Institute, 2006.

Figure Credits

Topaz Publications, Inc., Module opener

Steel Deck Institute, Figures 1, 3, 5, 6 (illustration), 8 (illustration), 10 (illustration), 11-13, and 16–20

LPR Construction, Figures 6, (photo), 7, 8 (photo), 9, 10 (photo), and 15

NCCER CURRICULA — USER UPDATE

NCCER makes every effort to keep its textbooks up-to-date and free of technical errors. We appreciate your help in this process. If you find an error, a typographical mistake, or an inaccuracy in NCCER's curricula, please fill out this form (or a photocopy), or complete the online form at **www.nccer.org/olf**. Be sure to include the exact module ID number, page number, a detailed description, and your recommended correction. Your input will be brought to the attention of the Authoring Team. Thank you for your assistance.

Instructors – If you have an idea for improving this textbook, or have found that additional materials were necessary to teach this module effectively, please let us know so that we may present your suggestions to the Authoring Team.

NCCER Product Development and Revision

13614 Progress Blvd., Alachua, FL 32615

Email: curriculum@nccer.org
Online: www.nccer.org/olf

❑ Trainee Guide ❑ Lesson Plans ❑ Exam ❑ PowerPoints Other ________________________

Craft / Level: Copyright Date:

Module ID Number / Title:

Section Number(s):

Description:

Recommended Correction:

Your Name:

Address:

Email: Phone:

Field Fabrication

30115-11

Trainees with successful module completions may be eligible for credentialing through NCCER's National Registry. To learn more, go to **www.nccer.org** or contact us at **1.888.622.3720**. Our website has information on the latest product releases and training, as well as online versions of the *Cornerstone* newsletter and Pearson's Contren® product catalog.

Your feedback is welcome. You may email your comments to **curriculum@nccer.org,** send general comments and inquiries to **info@nccer.org**, or send us the User Update form at the back of this module.

V.1 1/11

30115-11
FIELD FABRICATION

Objectives

When you have completed this module, you will be able to do the following:

1. Identify safety hazards associated with ironworking fabrication.
2. Use common layout tools.
3. Fabricate angle iron to given dimensions.
4. Fabricate channel iron to given dimensions.
5. Fabricate T-shapes to given dimensions.
6. Fabricate W-shapes to given dimensions.

Performance Tasks

Under the supervision of the instructor, the trainee should be able to do the following:

1. Use common layout tools.
2. Fabricate angle iron to given dimensions.
3. Fabricate channel iron to given dimensions.
4. Fabricate T-shapes to given dimensions.
5. Fabricate W-shapes to given dimensions.

Trade Terms

Angle-iron clips
Circumference
Control line
Cope
Edge distance
Heel
Intersection
Magnetic drill
Morse taper
Perpendicular line
Radius
Tangent
Toe

Contents

Topics to be presented in this module include:

Figures and Tables

Figures and Tables (continued)

1.0.0 INTRODUCTION

In the ironworking trade, field fabrication involves a variety of tasks. Walkways, access platforms, ladders, conveyor supports, and columns all could be included under the broad heading of field fabrication. The purpose of this module is to introduce the trainee to some of the basic skills needed for field fabrication. These skills include working from shop drawings, handling and cutting various structural shapes, drilling, and layout work. Once the basic skills have been mastered, they can be applied easily to the different forms of field fabrication required on the job site.

2.0.0 SAFETY

If a job cannot be done safely, it should not be done. Fabrication presents its own types of safety hazards. The ironworker's best defenses are knowledge and common sense.

Adhere to the following safety guidelines when performing fabrication work:

- Clean the oil from all fabricated materials with an approved solvent or degreaser.
- A **magnetic drill** (Hogan drill) must always be tied off. Tag the plug. Do not disconnect.
- When working above ground, be aware of those underneath the work area.
- Remember that water and electric tools are a potentially deadly combination.
- Wear an approved face shield when doing any grinding, even if only touching up an edge. A grinder throws off bits of hot, sharp metal that can damage eyes.
- Always shut off the cutting torch when finished with a cut.
- Do not set a grinder down while the disk is still turning.
- Do not leave hot work unattended. Mark the piece HOT.
- When moving workpieces to another location, always check that the path to the new location is clear.
- Never leave the track torch unattended while it is lit.
- Make sure that the workbench support will support the weight of the workpiece.
- Make sure the sparks from grinding do not contact others working in the area.
- Always shield others from the arc when welding and cutting.
- Make sure that there are no flammable materials nearby while grinding.

3.0.0 LAYOUT TOOLS

Many different tools are used for measuring and laying out parts for fabrication. In addition to common hand tools, ironworkers may also use the following:

- Scribers
- Steel rules
- Combination set
- Tri-square
- Protractors
- Dividers
- Trammel points
- Prick punches
- Center punches
- Straightedges

3.1.0 Scribers

Scribers, which are used to mark lines on metal parts, are made of tooled steel and have a handle or holder (*Figure 1*). The scriber is ground to a needle point. The point of some scribers is bent at an angle to mark the inside of cylindrical objects. The point of the scriber must be kept sharp so that it makes a distinct mark.

To use a scriber, hold it at an angle to and tight against the pattern or straightedge and pull it toward you.

3.2.0 Steel Rules

A steel rule is a flat measuring tool made of tempered steel that has graduations down to one sixty-fourth of an inch. Some steel rules are marked on both sides. One side may be marked in eighths and sixteenths of an inch, and the other side in thirty-seconds and sixty-fourths of an inch. They are available in lengths from 1 to 144 inches. The steel rule is used to measure distances and to mark lines. *Figure 2* shows two types of steel rules.

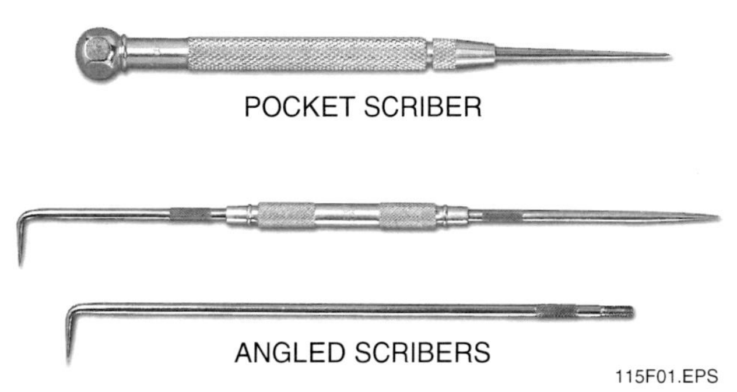

Figure 1 Scribers.

3.3.0 Steel Squares

The steel square (*Figure 3*), also called a framing square, is used to square up large stock and patterns and to lay out and check right angles. It is also used as a rule or straightedge. The steel square has a 24-inch blade and a 16-inch tongue. The blade and the tongue are marked in graduations down to one sixty-fourth of an inch.

3.4.0 Combination Set

The combination set (*Figure 4*) can be used as a combination square to lay out 45-degree angles, 90-degree angles, and parallel lines, and to measure heights. It can also be used as a protractor to measure and lay out angles to within 1 degree of accuracy. In addition, it can be used to locate centers of round or square bars and serve as a steel rule, level, or marking gauge.

The combination set consists of a steel rule or blade, a square head, a center head, and a protractor head. The square head contains a spirit level, and sometimes the protractor head contains a spirit level.

Figure 2 Steel rules.

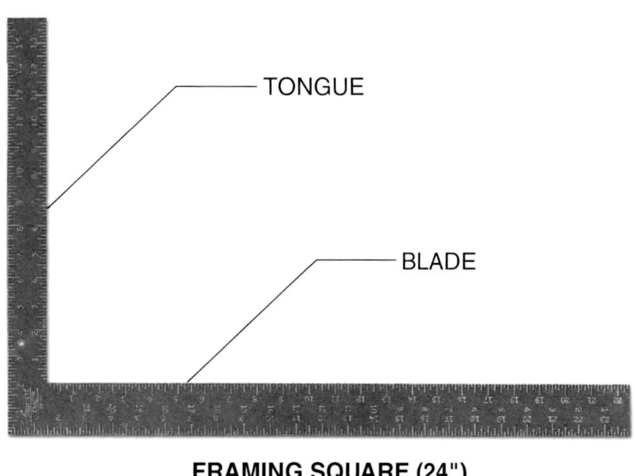

Figure 3 Steel square.

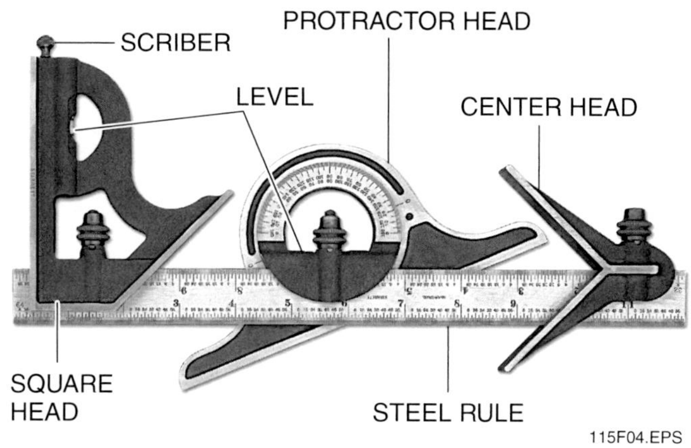

Figure 4 Combination set.

3.5.0 Protractors

A protractor (*Figure 5*) is an instrument used for measuring or laying out angles. It is made of plastic or light-gauge metal.

The protractor is shaped like a half circle. The curved part is divided into 180 degrees. The straight bottom has a center mark and is usually marked in sixteenths of an inch. The protractor should be kept clean and straight. It should be stored in a protective envelope when not in use.

3.6.0 Dividers

Dividers (*Figure 6*) are used to transfer measurements, to compare distances, and to scribe circles and arcs. Dividers have hardened steel points attached to two legs. These points should be kept sharp. The distance between the legs can be adjusted by a screw. Dividers should be kept clean and stored in a container to prevent damage. A drop of oil should be placed on the threads of the adjusting screw for lubrication and rust prevention. The points should be sharpened on an oil stone when necessary.

3.7.0 Trammel Points

Trammel points (*Figure 7*) are used to scribe large circles and arcs. Trammel points have two legs with hardened steel points. The legs are attached to a blade and can be adjusted along the blade. Trammel points should be handled carefully to prevent damage. The points should be sharpened with an oil stone.

3.8.0 Prick Punches

A prick punch (*Figure 8*) is used with a light hammer to mark centers and to make small indentations along layout lines. It is made of steel and is about 4 to 6 inches long. The point is ground at an angle from 30 to 60 degrees. The point of the

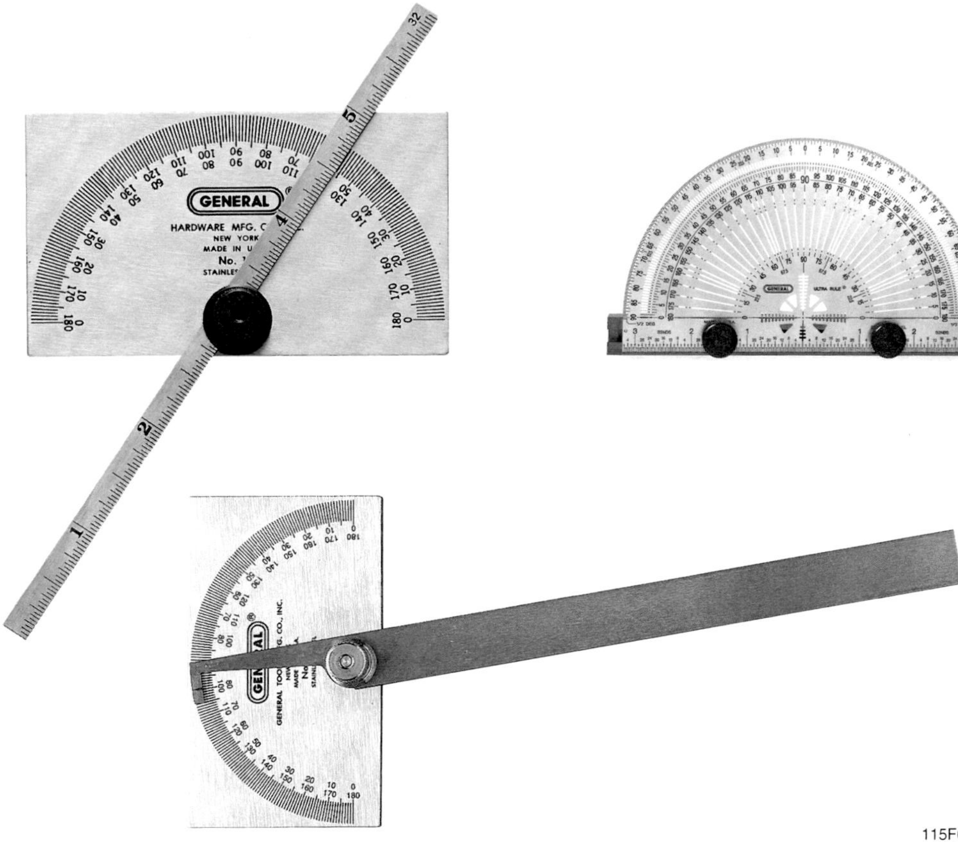

115F05.EPS

Figure 5 Protractors.

115F06.EPS

Figure 6 Divider.

prick should be kept sharp and at the proper angle. The struck end should be ground flat and the edge beveled to prevent mushrooming.

3.9.0 Center Punches

The center punch (*Figure 9*) is used to enlarge center marks made with a prick punch. This is done so a drill will start easily when drilling holes. The center punch is similar to a prick punch but has a more blunt point that is ground at a 90-degree angle. The point of the center punch should be kept sharpened to the proper angle and the struck end beveled to prevent mushrooming.

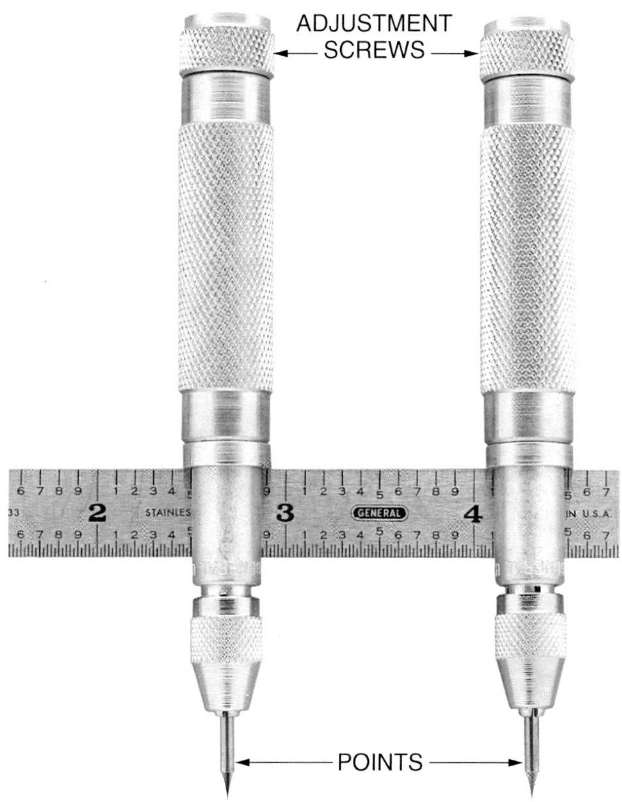

115F07.EPS

Figure 7 Trammel points.

115F08.EPS

Figure 8 Prick punch.

115F09.EPS

Figure 9 Center punch.

3.10.0 Straightedges

A straightedge (*Figure 10*) is a piece of flat metal with the edges machined perfectly straight. The straightedge is used to check machined surfaces for flatness or straightness. It is also used to clamp along the edge of a piece of metal as a reference line from which other lines are scribed. The straightedge should be cleaned with solvent and a cloth. A thin coat of lightweight machine oil should be applied to prevent rusting. The straightedge should be stored in a case when not in use.

3.11.0 Magnetic Drills

Cutting holes in steel members is often required. This can be performed with a torch or drill. However, a torch cannot be used to cut a hole in a beam.

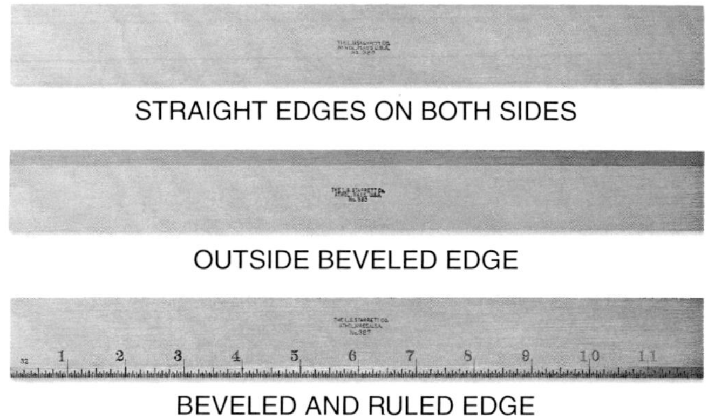

115F10.EPS

Figure 10 Straightedge.

One of the most versatile tools used for fabrication in the field is the magnetic drill (*Figure 11*). A magnetic drill, sometimes called a Hougen® drill, is a portable drill press that provides the accuracy and stability of a stationary drill press in a tool light enough to carry by hand. The design and operation of a magnetic drill are simple. A heavy-duty drill motor is mounted to a base that contains a powerful electromagnet. The entire unit is placed on a beam or plate near the spot that needs drilling. The operator throws a switch on the back of the base assembly that energizes the magnet. The magnet locks the base directly onto the beam or plate and holds the tool steady during the entire drilling operation.

WARNING

Always tie off or secure the magnetic drill from falling and tag the plug.

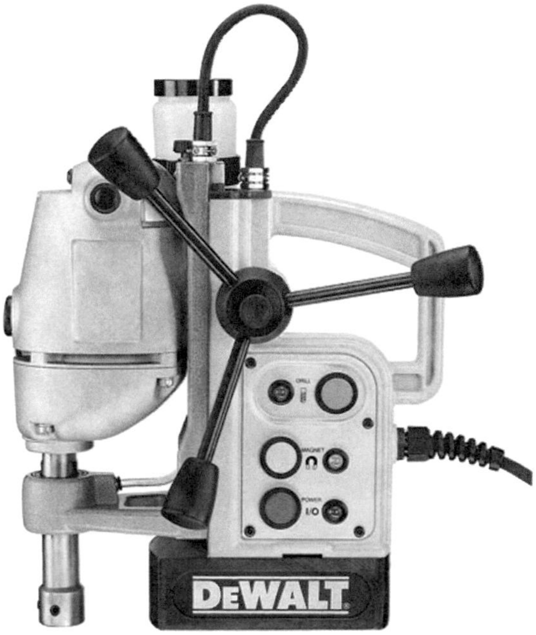

115F11.EPS

Figure 11 Magnetic drill.

The drill motor operates the same as a stationary drill press. When the handle is rotated, the drill motor moves up and down. To obtain the accuracy required for fabrication, an adjusting handle frees the stand from the base and allows it to turn to either side.

Magnetic drills operate on either 120V or 220V electric power. Be sure to know which type is being used. Always clamp the workpiece. Do not hold it by hand.

Heavy-duty magnetic drills usually have a number-3 **Morse taper** socket instead of a three-jaw chuck to hold the bits. This allows larger bits to be used. For some models, an adapter is available to convert the drill motor to a number-4 Morse taper socket, allowing even greater drill bit capacity.

Hawg bits (*Figure 12*) are bits that, rather than cutting out all the material in the hole, cut out a circle around a core. They are quicker than regular bits but have the limitation of being able to drill holes to a limited depth that is shorter than normal bits. They are faster than regular bits and do not require a pilot hole. They use a lubricant called Hawg Wash®, which is water soluble; therefore, if the holes are being drilled in painted, finished surfaces, the cleanup is much easier and faster.

To use the magnetic drill, obtain a piece of plate that is long enough and wide enough to accommodate the base of the magnetic drill. Install a drill bit. (For this exercise, it doesn't matter what size bit is used.) Then proceed as follows:

Step 1 Clean the area around the location of the hole to ensure a good bond.

Step 2 Centerpunch a mark on the plate.

Step 3 Move the magnetic drill up to the mark so that it centers on the mark.

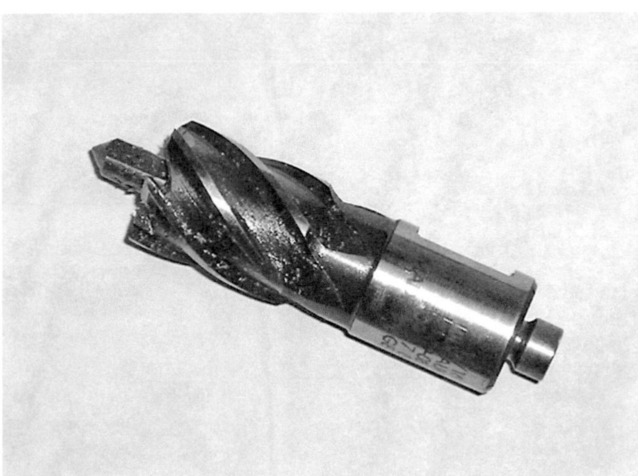

115F12.EPS

Figure 12 Hawg bit.

CAUTION

Ensure that the plate is securely clamped to prevent movement. If the magnetic drill moves, the workpiece will be damaged.

Step 4 Turn on the magnet.

Step 5 Recheck the alignment of the bit to the hole.

NOTE

If a fine adjustment is needed, unlock the housing from the base and readjust.

Step 6 Turn on the drill motor.

Step 7 Pull down on the spoke until the drill bit meets the hole.

Step 8 Squeeze oil on the area to keep the bit from overheating.

Step 9 Apply just enough pressure to produce a curled chip.

NOTE

Too little pressure produces small, broken chips and increases drilling time. Too much pressure causes the chips to turn blue or brown and may overheat the bit, drawing the temper from it and ruining it.

Step 10 Oil the bit periodically as you drill the hole.

4.0.0 Layout Tasks

The ironworker may be required to lay out holes to be drilled; circles, arcs, and shapes to be cut; and locations of equipment, stairs, and columns. Each of these tasks follows a specific layout procedure.

4.1.0 Base Line Layout

Two base lines at right angles (90 degrees) are required for most layout tasks. All dimensions and additional lines are taken from these base lines. Base lines may be used as center lines, or additional center lines (**control lines**) may be laid out parallel to the base line. *Figure 13* shows base lines and parallel lines.

The two most common methods used to lay out right angles in the field are the arc method and the 3-4-5 method. These methods are explained in the following sections.

4.1.1 Arc Method

When laying out **perpendicular lines** by the arc method, do not use string or line that can stretch. Instead, use a steel tape or metal rod. To use the arc method, follow these steps:

Step 1 Mark a base line by snapping a chalk line on the surface of the work.

Step 2 Locate two points, A and B, on the base line at equal distance from a given point. This is done by measuring or by swinging a steel tape from the given point. *Figure 14* shows how to locate points on a base line.

Step 3 Locate point C. This is done by swinging arcs from point A and point B on the base line. Use a **radius** length approximately 1½ times the distance from the given point to point A. Both arcs must have exactly the same radius. Point C is located where the arcs intersect. *Figure 15* shows how to locate point C.

Step 4 Draw a line from point C through the given point on the base line. The line should be at right angles to the base line. *Figure 16* shows this right-angle line.

4.1.2 3-4-5 Method

The 3-4-5 method is also called the triangle method. This is because when a triangle has sides in a 3-4-5 ratio, one angle is exactly a right angle (90 degrees).

Any multiple of 3-4-5, such as 6-8-10, 9-12-15, etc., may be used (*Figure 17*). When using the 3-4-5 method, use the largest possible measuring unit. The larger the measuring unit, the more accurate the layout. Follow these steps when laying out a 90-degree angle with the 3-4-5 method:

Step 1 Choose a measuring unit. Multiply each of the numbers in the 3-4-5 ratio by the measuring unit. The measuring unit will be limited by the work area space.

Step 2 Mark a base line by snapping a chalk line on the surface of the work. This line will form one leg of the triangle.

Step 3 Locate the given point on the base line. Use 5 feet as the measuring unit. Measure three units from the given point at approximately a right angle from the base line. Swing arc A. *Figure 18* shows an example of measuring at a right angle.

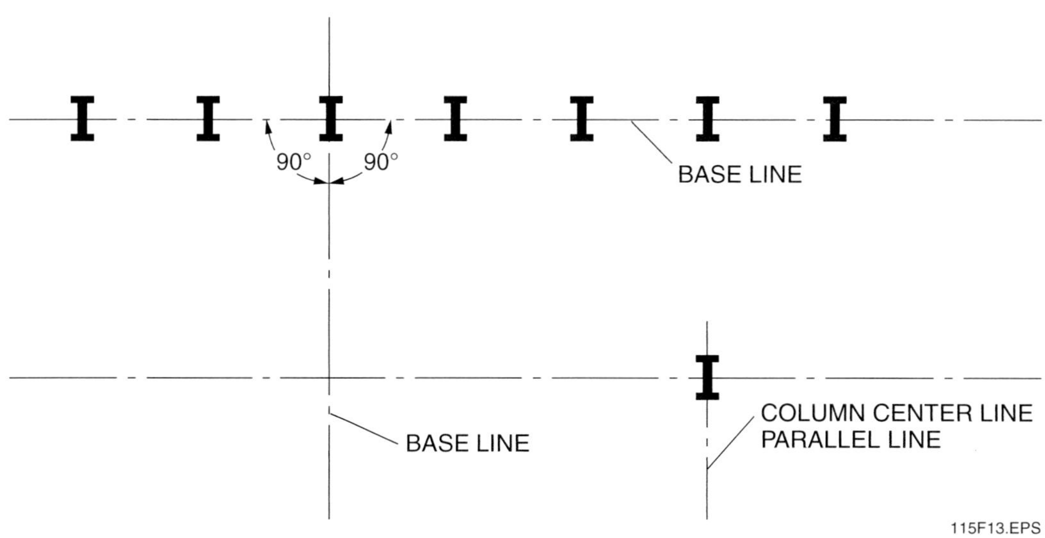

Figure 13 Base lines and parallel lines.

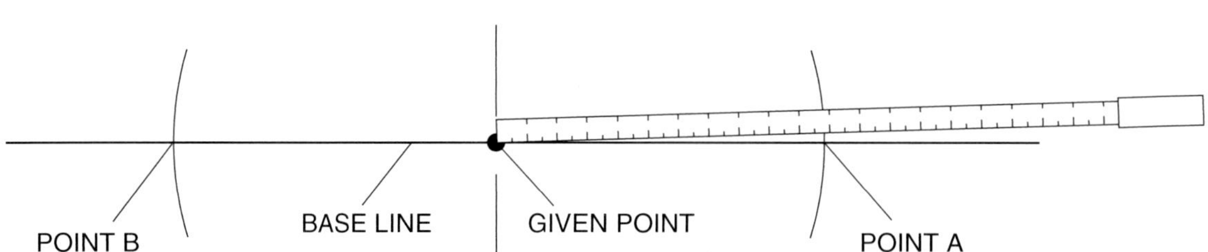

Figure 14 Locating points on a base line.

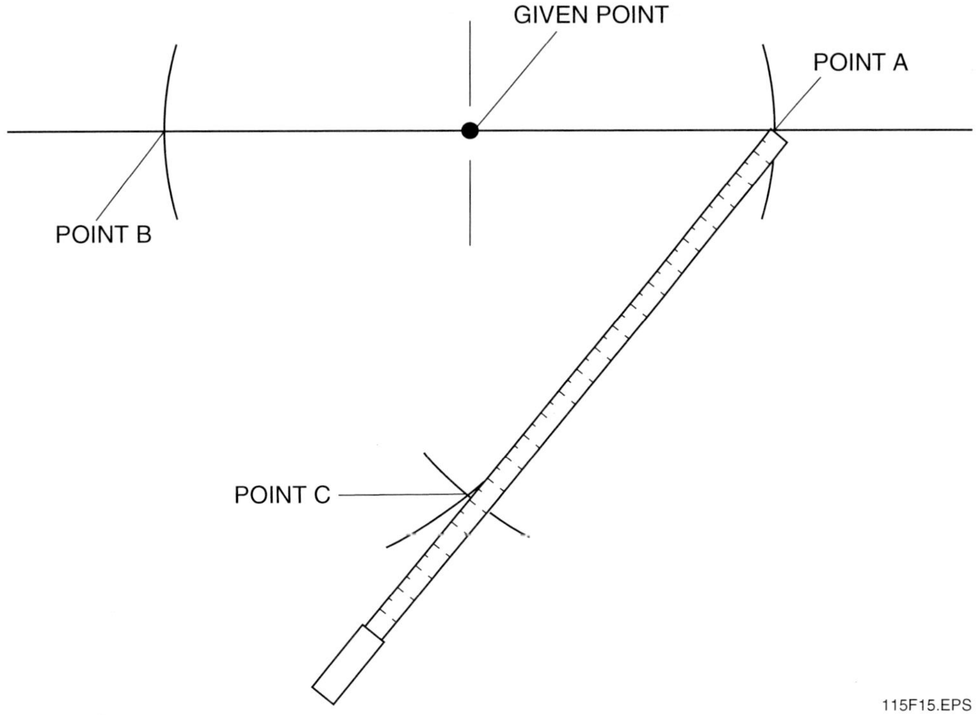

Figure 15 Locating Point C.

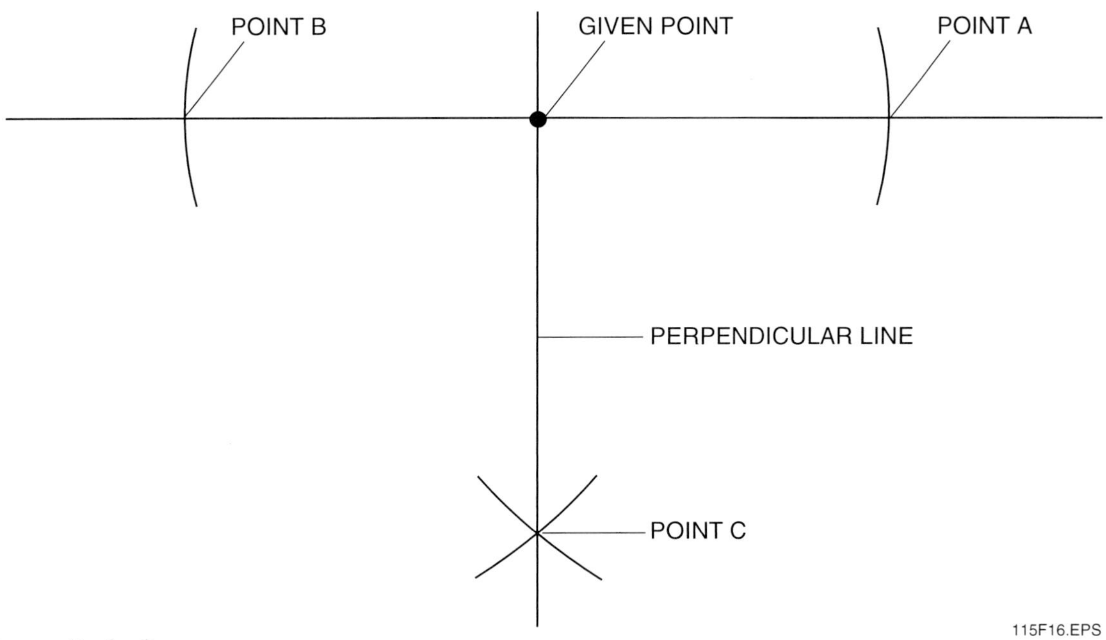

Figure 16 Perpendicular line.

Step 4 Measure 4 units from the given point along the base line and mark point B. *Figure 19* shows an example of locating point B.

Step 5 Measure 5 units from point B and mark point C on arc line A. *Figure 20* shows an example of locating point C.

Step 6 Mark a line from point C through the given point on the base line. This line will be perpendicular (90 degrees) to the base line. *Figure 21* shows an example of a perpendicular line.

When possible, another triangle of the same dimensions should be constructed on the opposite side of the base line to double-check the accuracy of the layout.

4.2.0 Scribing Lines

Accuracy is very important when scribing lines. Each layout should be planned to avoid material waste and loss of time. The following sections will explain how to scribe straight and angled lines.

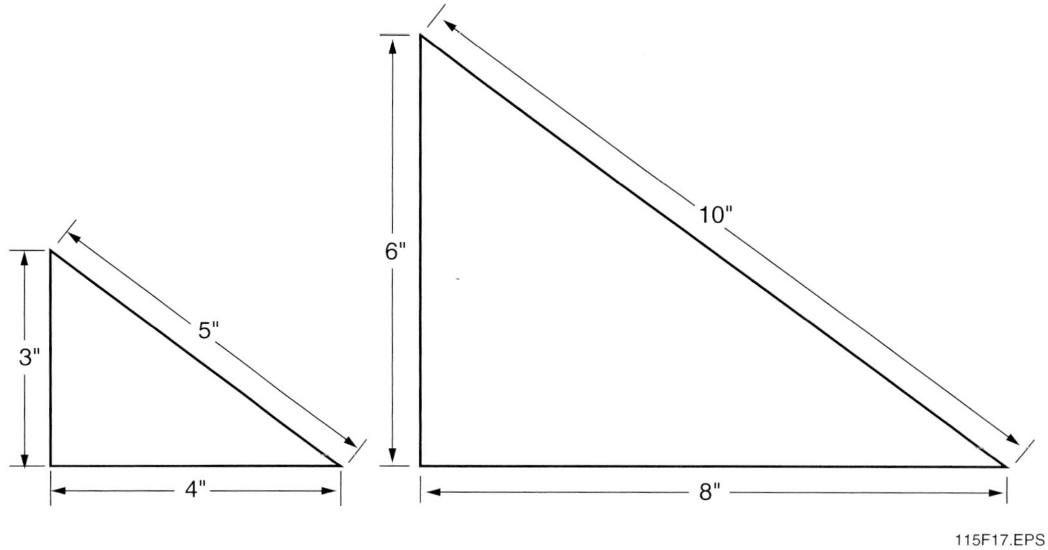

Figure 17 The 3-4-5 ratio.

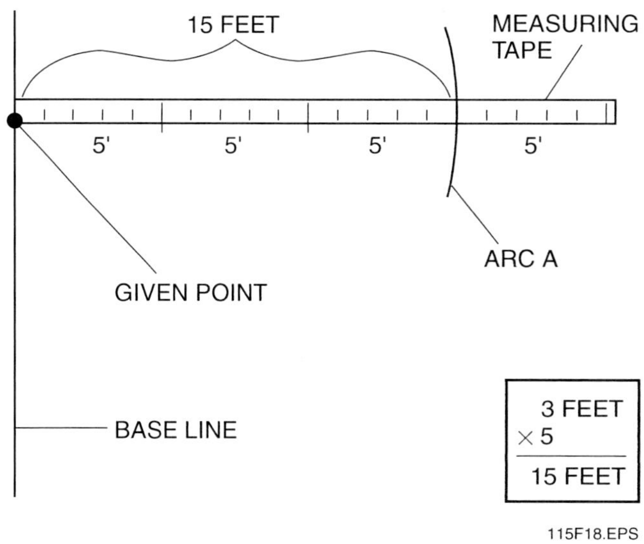

Figure 18 Measuring at a right angle.

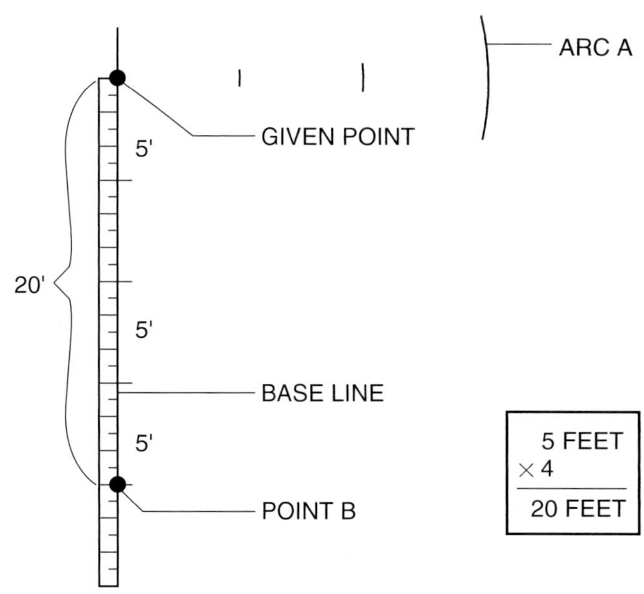

Figure 19 Locating point B.

4.2.1 Scribing Straight Lines

When scribing short lines, use a steel rule as a straightedge. Use a straightedge long enough to scribe the entire line without stopping. Do not retrace lines. Retracing lines causes two lines and results in inaccurate measurements. Follow these steps when scribing straight lines:

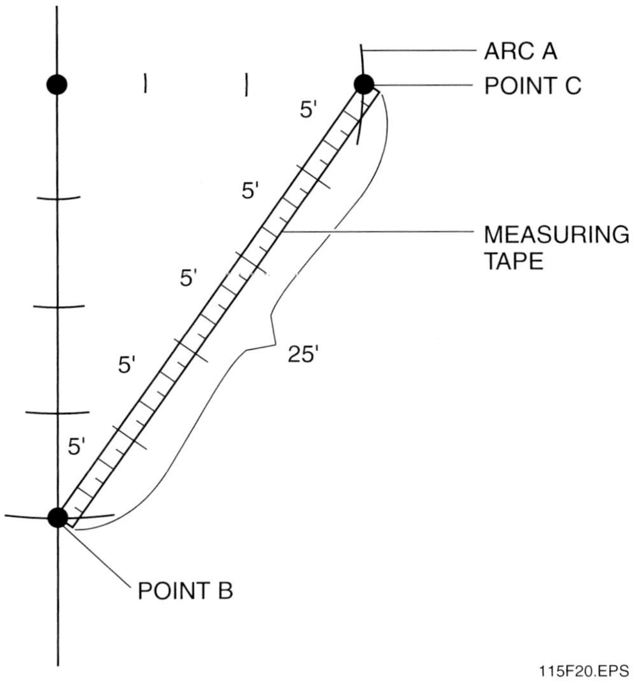

Figure 20 Locating point C.

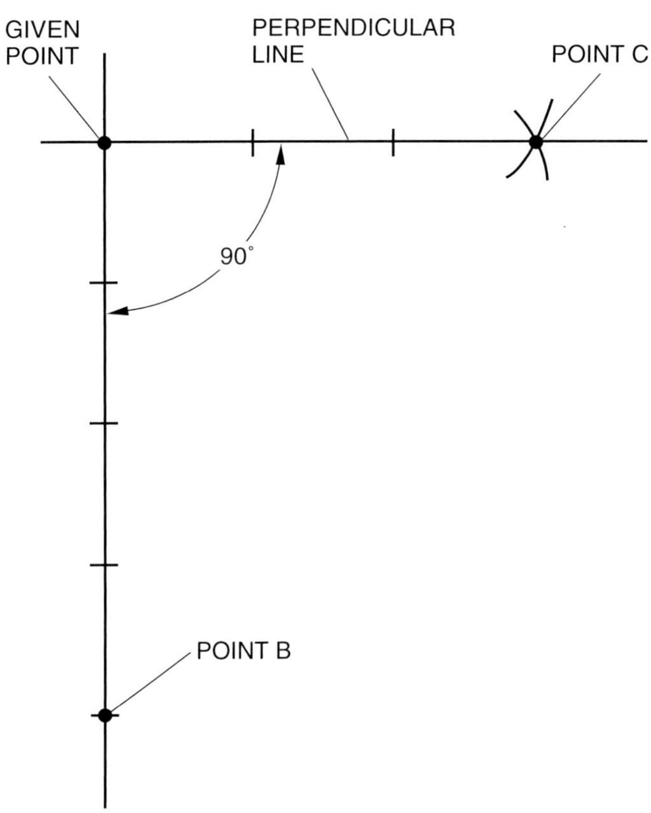

Figure 21 Perpendicular line.

Step 1 Locate and mark the starting point. This is done by marking a V on the material as the exact starting point, as shown in *Figure 22*.

Step 2 Locate and mark another point several inches from the starting point.

Step 3 Use a prick punch to make punch marks at the two Vs (*Figure 23*).

Step 4 Place the scriber point in one of the punch marks (point A). Move a straightedge in contact with the scriber. Align the other end of the straightedge with the punch mark (point B). *Figure 24* shows an example of aligning a straightedge.

Step 5 Hold the straightedge firmly to prevent movement. Starting at point A, place the scriber as close to the straightedge as possible, slightly angle the scriber, and scribe a line to point B (*Figure 25*).

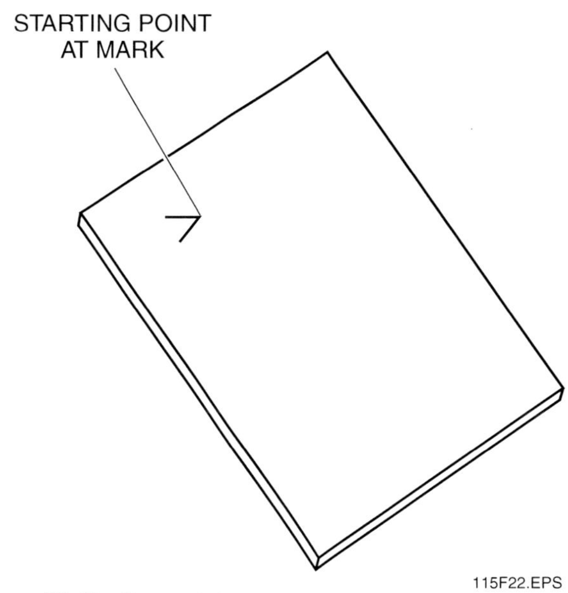

Figure 22 Starting point.

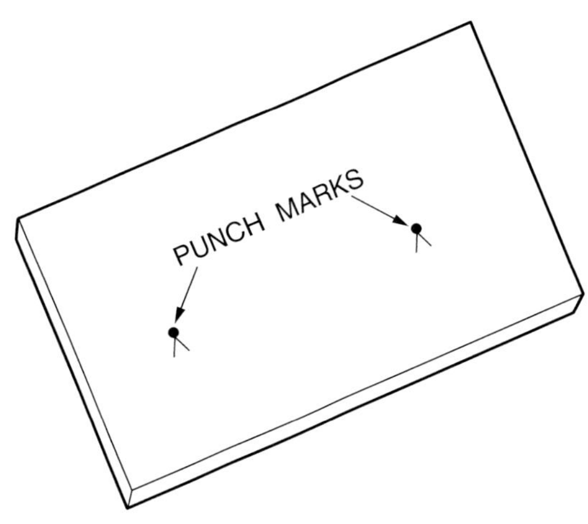

Figure 23 Punch marks.

At times, straight lines must be scribed that are parallel or perpendicular to a reference line on an edge. These lines are scribed with a framing square or combination square as a guide.

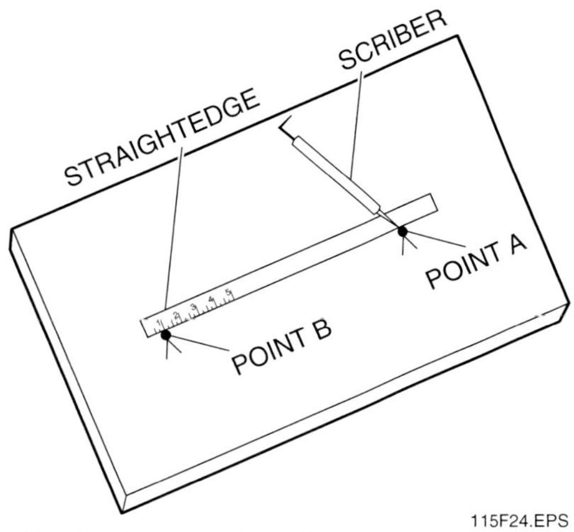

Figure 24 Aligning a straightedge.

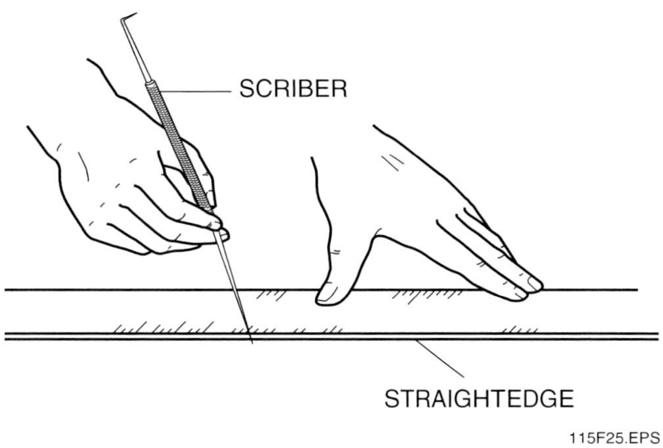

Figure 25 Scribing a line.

4.2.2 Scribing Perpendicular Lines to a Base Line Using the Framing Square

Follow these steps to scribe parallel lines that are perpendicular to a base line. Use a framing square and a straightedge.

Step 1 Locate and scribe a base line.

Step 2 Align the straightedge with the base line. Clamp the straightedge firmly to the material, as shown in *Figure 26*.

Step 3 Place the blade of the framing square against the straightedge. Scribe a line along the tongue of the square.

Step 4 Move the framing square a specified distance along the straightedge and scribe another line. This line will be parallel to the line scribed in Step 3. Both lines will be perpendicular to the base line. *Figure 27* shows an example of scribing parallel lines.

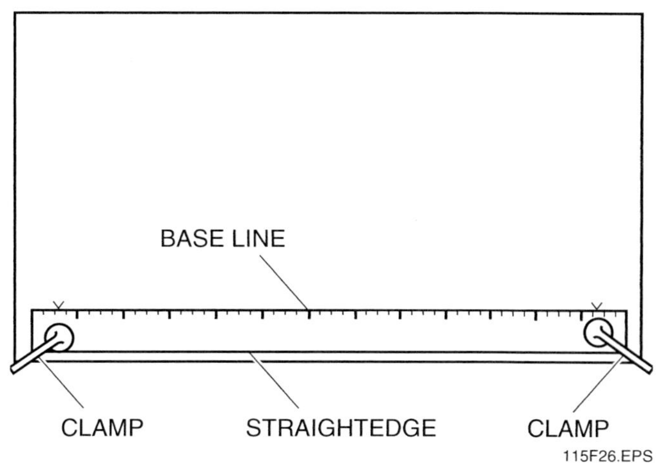

Figure 26 Straightedge clamped to material.

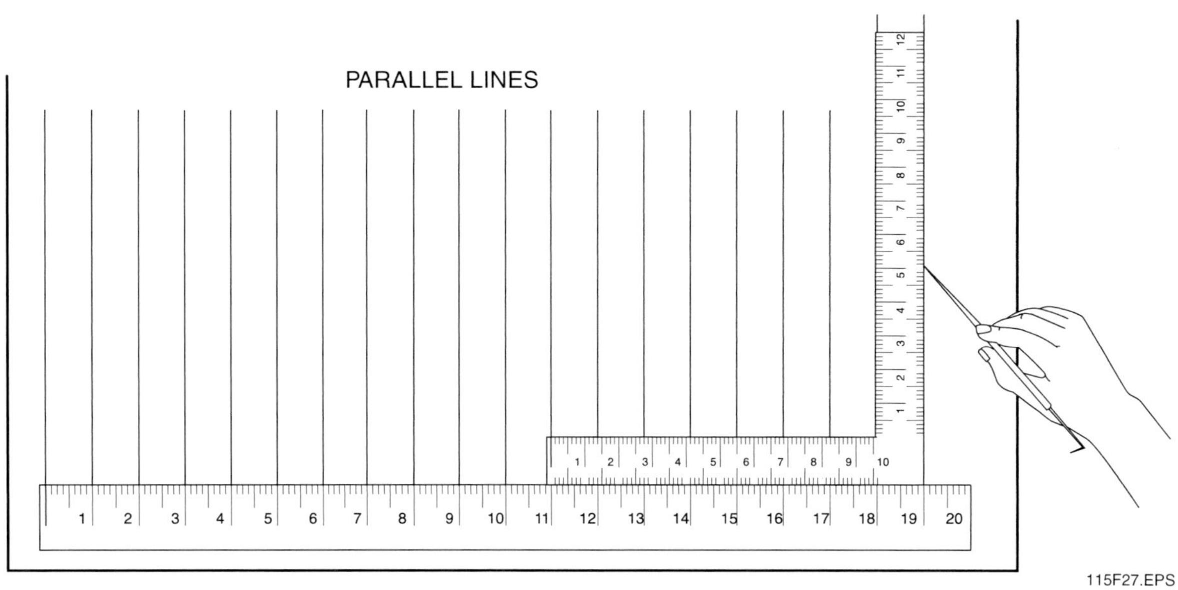

Figure 27 Scribing parallel lines.

4.2.3 Scribing Perpendicular Lines to an Edge Using the Combination Set

Follow these steps to scribe parallel lines that are perpendicular to an edge. Use a combination set.

Step 1 Select a smooth, straight edge on the material to use as a reference edge.

Step 2 Place the square head attachment of the combination set against this reference edge and scribe a line along the blade.

Step 3 Move the square head along the reference edge a specified distance and scribe another line. These lines will be parallel to each other and perpendicular to the reference edge (*Figure 28*).

4.2.4 Scribing Parallel Lines to an Edge Using the Combination Set

Follow these steps to scribe lines parallel to an edge. Use a combination set.

Step 1 Select a smooth, straight edge on the material to use as a reference edge.

Step 2 Adjust the blade in the square head to a specified distance. Tighten the locknut to prevent movement.

Step 3 Place the square head against the edge of the material. The blade should be perpendicular to the edge.

Step 4 Hold the point of the scriber at the end of the blade. Move the square and scriber across the material, scribing a line parallel to the edge (*Figure 29*).

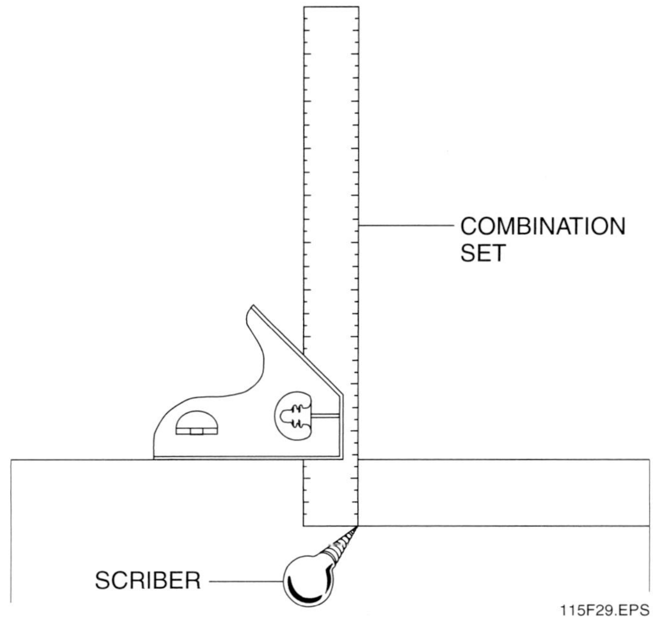

Figure 29 Scribing a line parallel to an edge.

4.2.5 Scribing Angled Lines Using a Combination Set

Follow these steps to scribe lines at a 45-degree angle to an edge using a combination set.

Step 1 Select a smooth, straight edge on the material to use as a reference edge.

Step 2 Place the 45-degree angle of the square head attachment against the reference edge. Hold the square head attachment firmly to prevent movement, and scribe a line along the blade.

Figure 28 Scribing a line perpendicular to a reference edge.

Step 3 Move the square head along the reference edge a specified distance and scribe the next line. These two lines will be parallel to each other and at a 45-degree angle to the edge. *Figure 30* shows an example of scribing lines at a 45-degree angle to a reference edge.

4.2.6 Scribing Angled Lines Using a Protractor

Follow these steps to scribe lines at different angles to a base line using a protractor and a straightedge:

Step 1 Locate and scribe a base line on the material.

Step 2 Punch mark a point on the base line.

Step 3 Align the straightedge of the protractor along the base line. Place the center mark of the protractor at the punch mark on the base line. Make another punch mark on the material at a specified degree on the protractor (*Figure 31*).

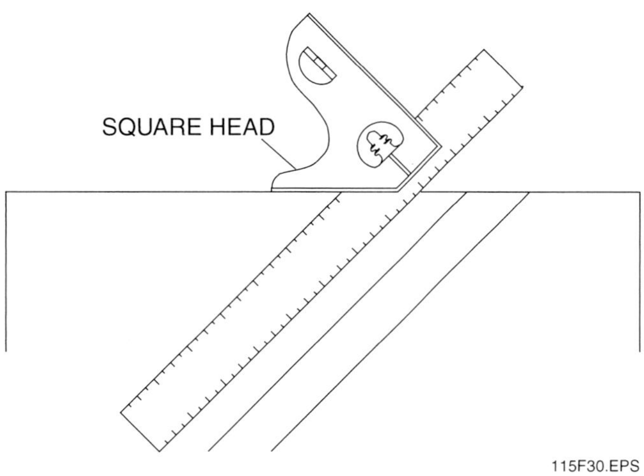

Figure 30 Scribing a line at a 45-degree angle.

Step 4 Scribe a line between the two punch marks. Use a straightedge to align the punch marks. *Figure 32* shows a 60-degree angle to the base line.

4.3.0 Divider and Trammel Point Uses

Dividers and trammel points are layout tools with many uses. Some of these uses are as follows:

- Scribing circles and arcs
- Scribing perpendicular lines
- Bisecting angles
- Dividing lines and circles

4.3.1 Scribing Circles and Arcs Using Dividers

Follow these steps to scribe circles and arcs with dividers:

Step 1 Punch mark a center point with a prick punch.

Step 2 Adjust the dividers to a specified radius. To adjust the dividers, place the point of one leg on the 1-inch mark of a steel rule.

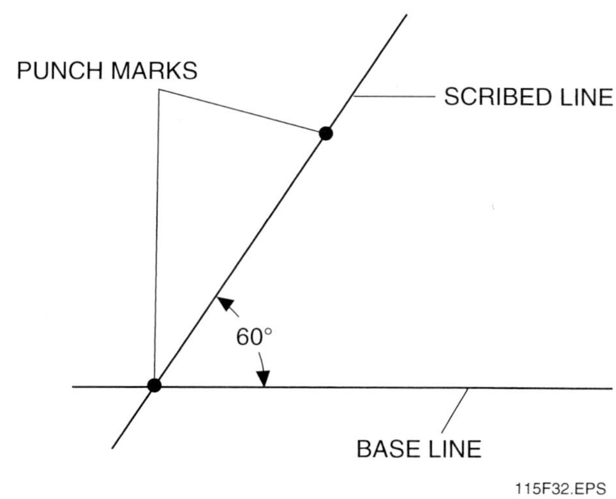

Figure 32 60-degree base line angle.

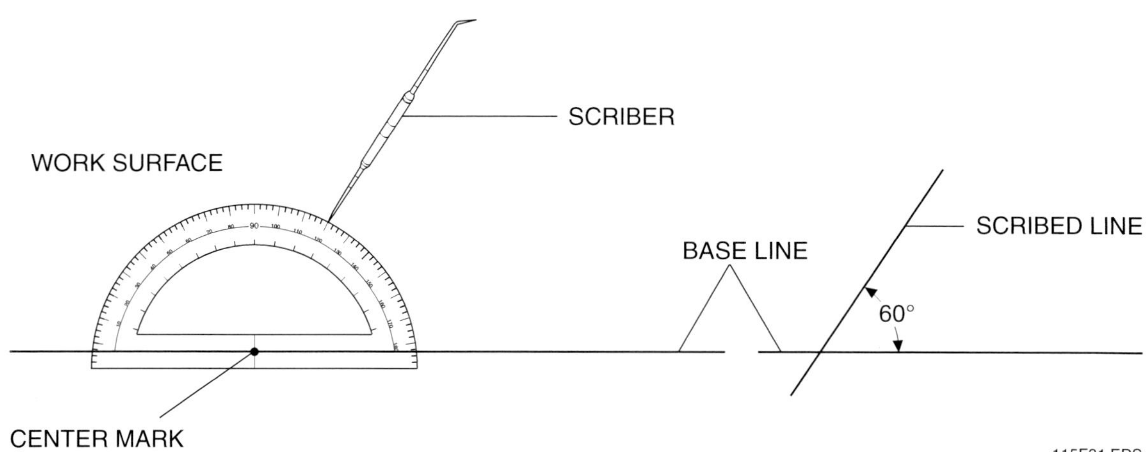

Figure 31 Using a protractor.

Rotate the adjusting nut until the point of the second leg reaches the specified dimension. *Figure 33* shows an example of adjusting dividers.

Step 3 Place the point of one divider leg into the punch mark at the center point. Tilt the dividers in the direction of travel and rotate 360 degrees with the thumb and index finger. This will scribe a complete circle. *Figure 34* shows an example of scribing a circle.

Step 4 Arcs are parts of a circle. The steps used to make an arc are the same ones used to make a circle, except the dividers are not rotated a full 360 degrees. *Figure 35* shows arcs scribed on a line.

Step 5 Check the circle diameter or the arc radius for the correct dimensions.

4.3.2 Scribing Circles and Arcs Using Trammel Points

Trammel points are used to scribe large circles and arcs. Follow these steps to scribe large circles and arcs with trammel points:

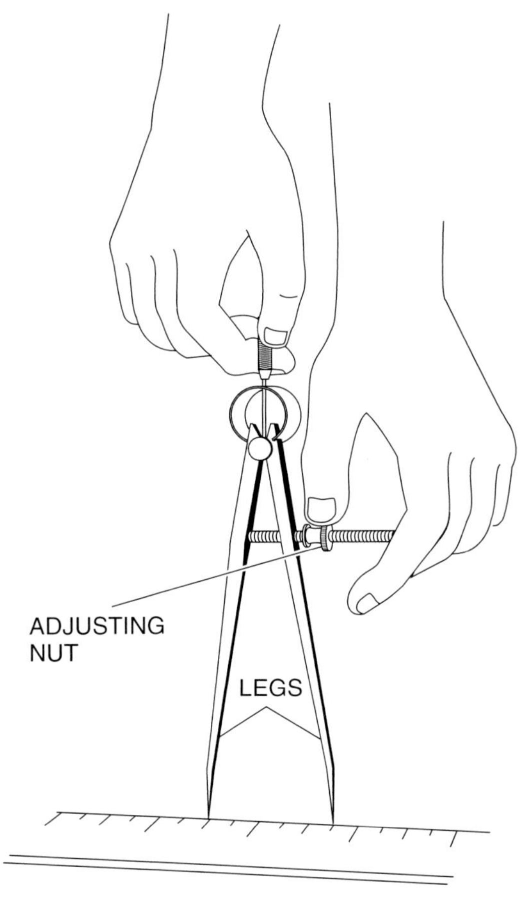

Figure 33 Adjusting dividers.

Step 1 Punch mark a center point with a prick punch.

Step 2 Adjust the trammel points to a specified radius. To adjust the trammel points, place one point on a 1" mark of a steel rule. Slide the other point to the proper radius and tighten the thumbscrew. *Figure 36* shows an example of adjusting trammel points.

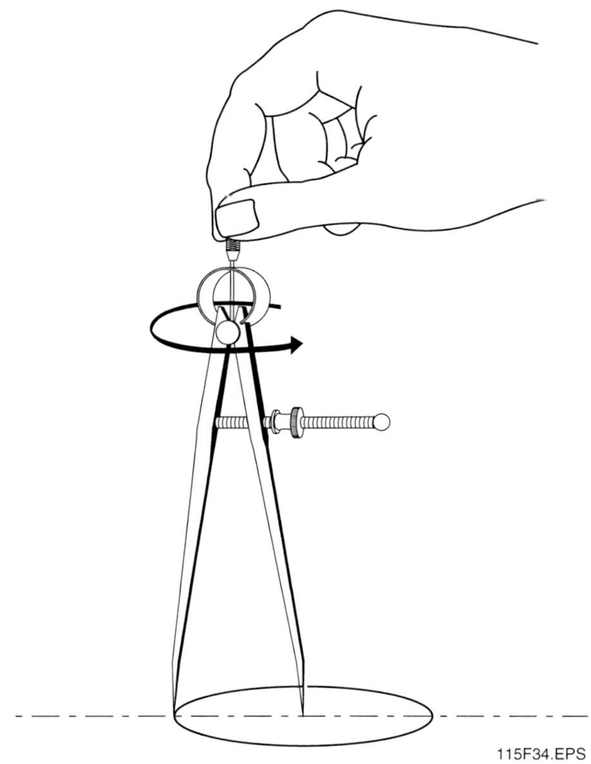

Figure 34 Scribing a circle with dividers.

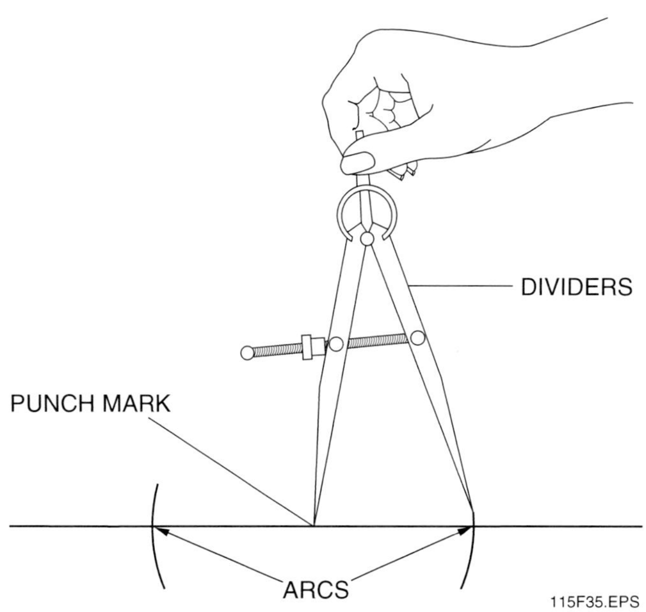

Figure 35 Scribing arcs.

Step 3 Place one trammel point in the punch mark at the center of the arc or circle and hold it firmly in place. Tilt the tool slightly in the direction of travel and rotate by pulling the trammel point toward you. This will scribe a circle. *Figure 37* shows an example of scribing a circle.

Step 4 Check the correct arc radius or circle diameter.

4.3.3 Scribing Perpendicular Lines From a Base Line Using Dividers

Follow these steps to scribe arcs for constructing lines that are perpendicular to a base line. Use dividers and a straightedge.

Step 1 Scribe a base line using a straightedge and a scriber.

Step 2 With a prick punch, punch mark the center point of the base line.

Step 3 Adjust dividers to a specified radius.

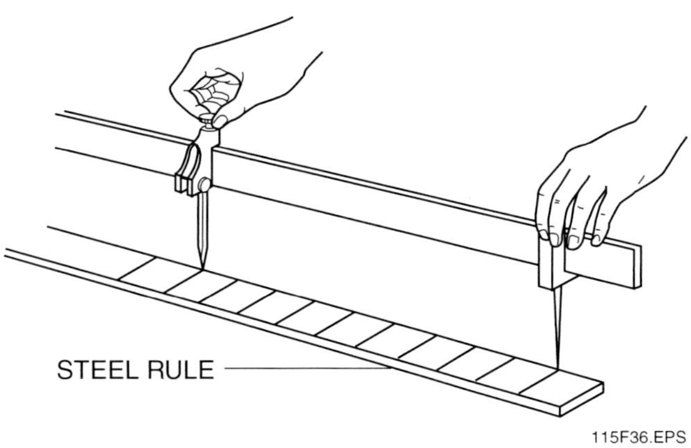

Figure 36 Adjusting trammel points.

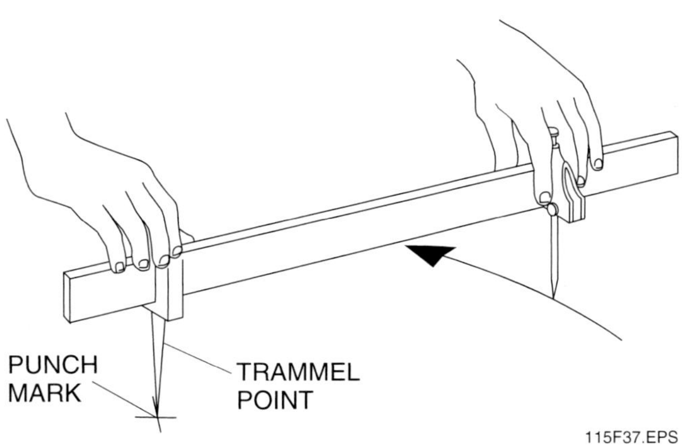

Figure 37 Scribing a circle using trammel points.

Step 4 Place the point of one divider leg into the punch mark at the center point of the base line.

Step 5 Scribe arcs intersecting the base line on both sides of the center point. *Figure 38* shows an example of the two arcs.

Step 6 Adjust the dividers to a radius larger than the one used in Step 5.

Step 7 Place the point of one divider leg at the exact point where one of the arcs intersects the base line. Scribe an arc above the center point of the base line. *Figure 39* shows an arc above the center point.

Step 8 Place the point of one divider leg at the other intersecting arc and scribe another arc above the center point. This arc should intersect the arc scribed in Step 7. *Figure 40* shows these intersecting arcs.

Step 9 Using a straightedge and scriber, scribe a straight line from the intersecting arcs through the center point of the base line. This line should be perpendicular to the base line. Check the angle with a protractor. *Figure 41* shows the perpendicular line constructed by intersecting arcs.

4.3.4 Scribing Perpendicular Lines From a Reference Point Using Dividers

Using dividers and a straightedge, follow these steps to lay out a perpendicular line from a reference point to a base line:

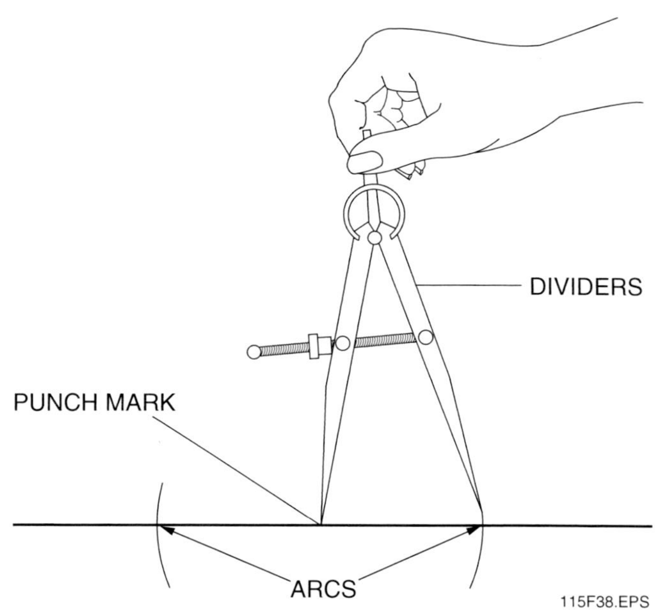

Figure 38 Arcs along a base line.

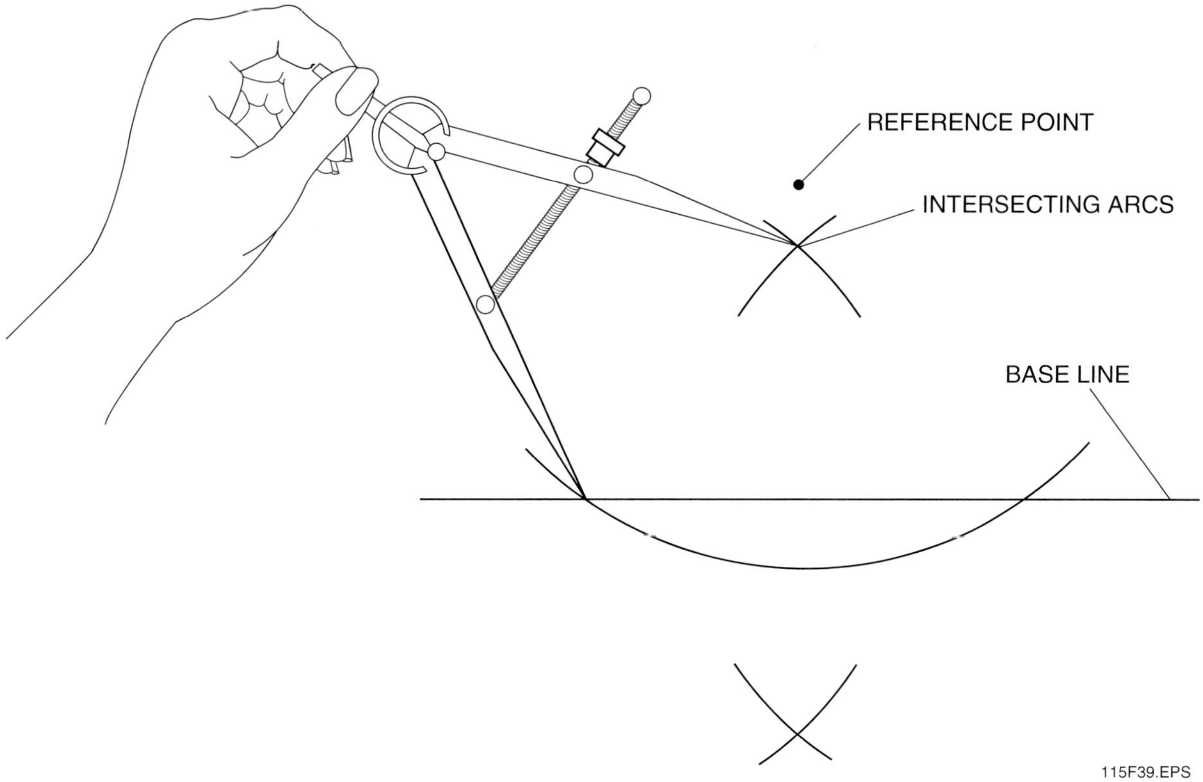

Figure 39 Arc above center point.

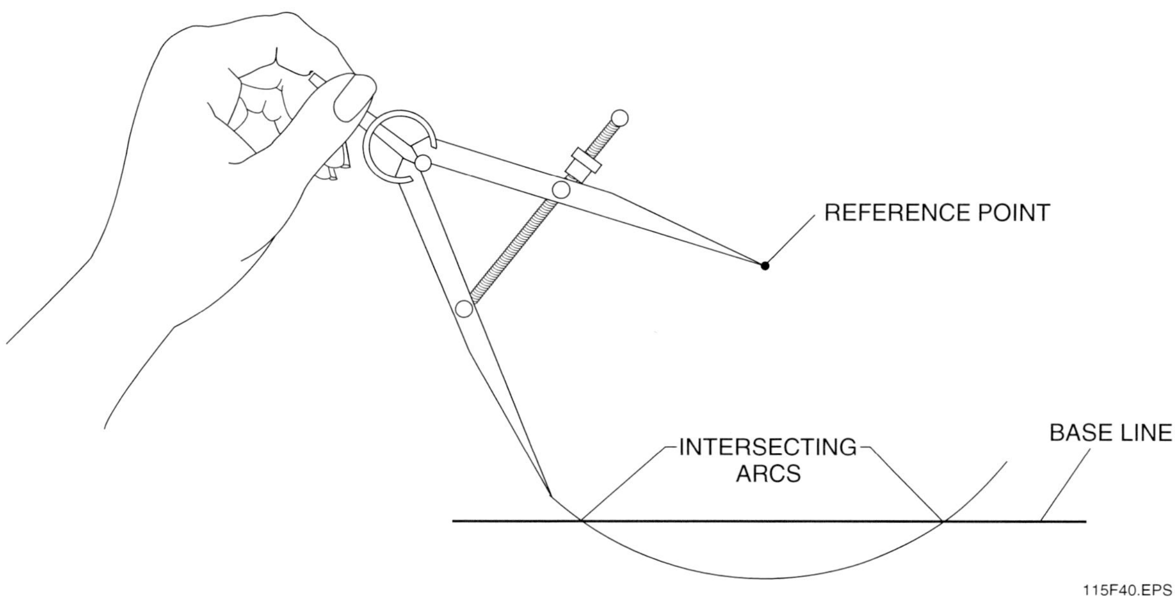

Figure 40 Intersecting arcs.

Step 1 Locate the reference point. With a straightedge, scribe a straight line 3 to 4 inches from the reference point.

Step 2 Place one divider leg on the reference point and adjust the other leg until the arc formed will intersect the base line in two places.

Step 3 Set one leg of the dividers at the point where one arc intersects the base line. Scribe an arc below the reference point. Repeat this step from the point where the other arc intersects the base line. These arcs should intersect below the reference point.

Step 4 Scribe a line from the reference point, through the intersecting arcs, to the base line. The line should be perpendicular to the base line.

Step 5 Using a protractor, check the angle formed by the two lines. The angle should be 90 degrees.

4.3.5 Bisecting Angles Using Dividers

Bisecting an angle means dividing it into two equal parts. Using dividers and a straightedge, follow these steps to bisect an angle:

Step 1 Using a protractor, construct a 60-degree angle by scribing two intersecting lines. The point of **intersection** will be point A.

Step 2 Place the point of one divider leg where the two lines intersect. Scribe an arc that intersects both sides of the angle. *Figure 42* shows an example of an arc.

Step 3 From the intersection points on each side of the angle, scribe two intersecting arcs (*Figure 43*).

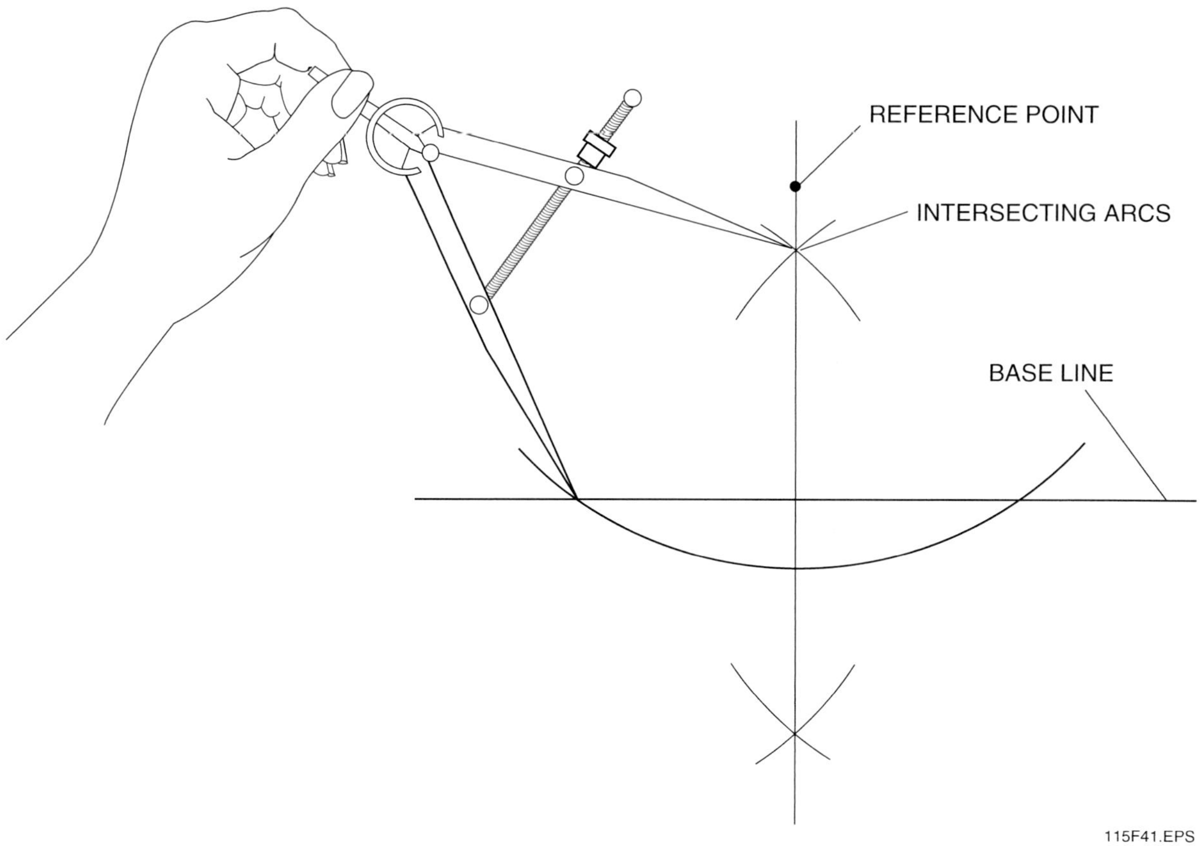

Figure 41 Perpendicular line through arcs.

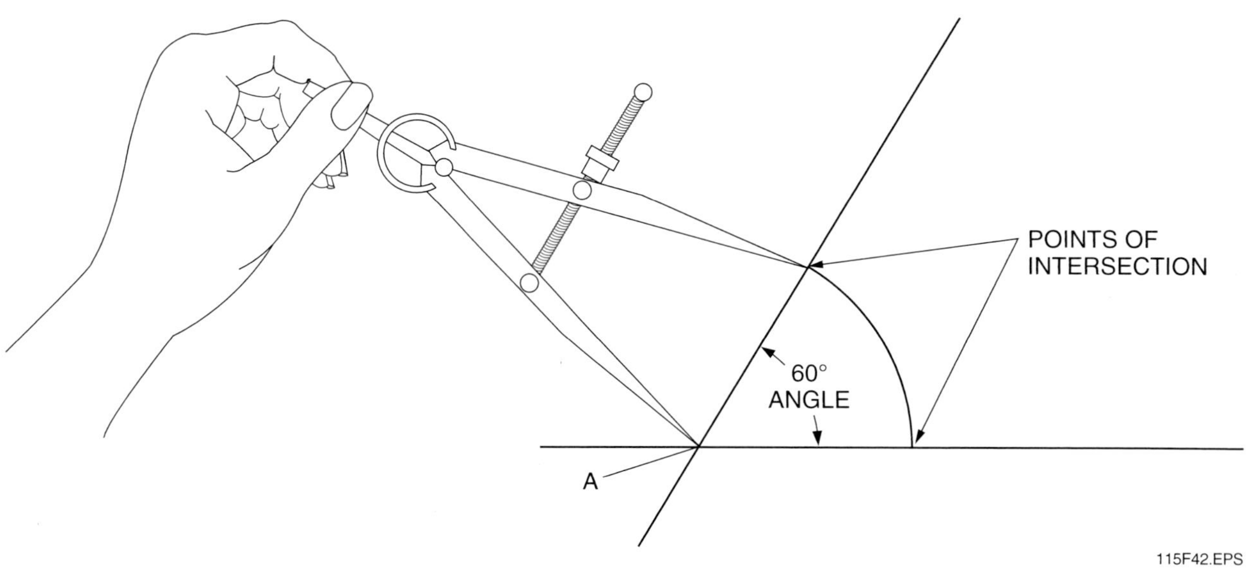

Figure 42 Arc intersecting legs of an angle.

Step 4 Using a straightedge, scribe a line from point A through point B. This line will form two equal angles. *Figure 44* shows an example of a line bisecting an angle.

Step 5 Check the accuracy of each angle with a protractor.

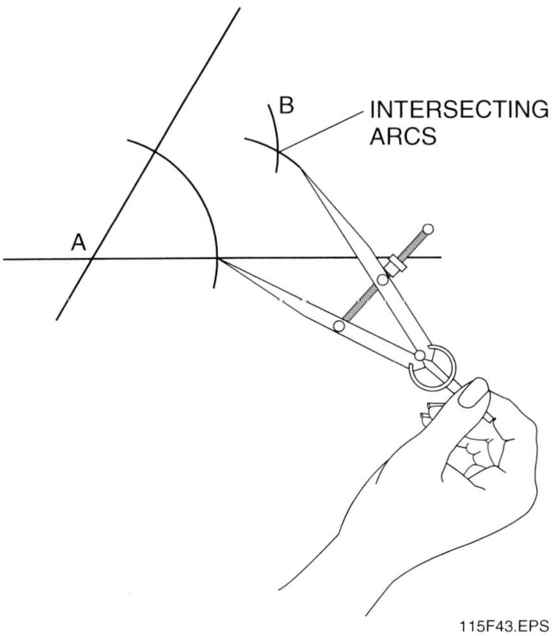

Figure 43 Intersecting arcs.

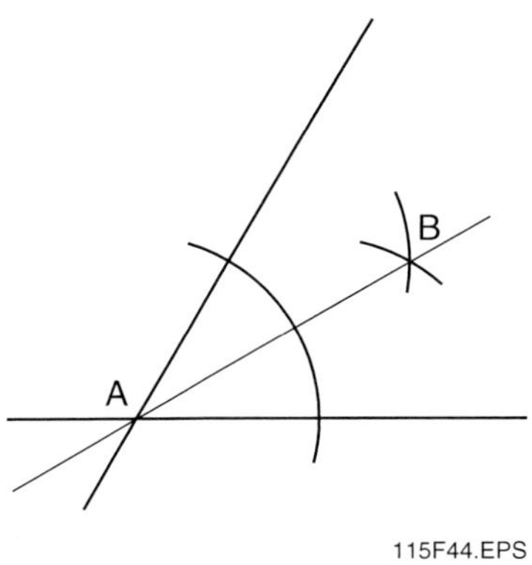

Figure 44 Line bisecting the angle.

4.3.6 Dividing a Line Into Equal Parts Using Dividers

Follow these steps to divide a line into equal parts, using dividers and a straightedge.

Step 1 Mark both ends of the line with a punch mark and label these marks point A and point B.

Step 2 Adjust the dividers to any radius and place the point of one leg at point A. Scribe an arc above the line. Using the same radius, scribe another arc below the line from point B. *Figure 45* shows arcs scribed above and below the line.

Step 3 From point B, scribe a line **tangent** to the arc above point A (*Figure 46*).

Step 4 To divide line A-B into five equal parts, adjust the dividers to any radius. Starting at point A, make five tick marks with that same radius along the line that is tangent to the arc below point B. Number these marks 1 through 5.

Step 5 Using the same radius, start at point B, and make five marks along the line that is tangent to the arc above point A. Number these marks 6 through 10 (*Figure 47*).

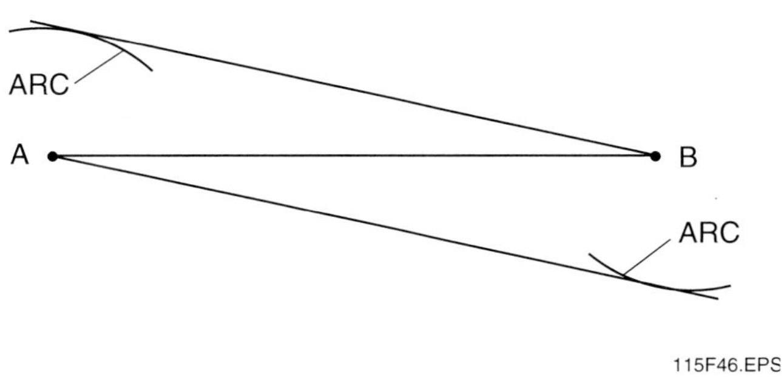

Figure 46 Lines tangent to arcs.

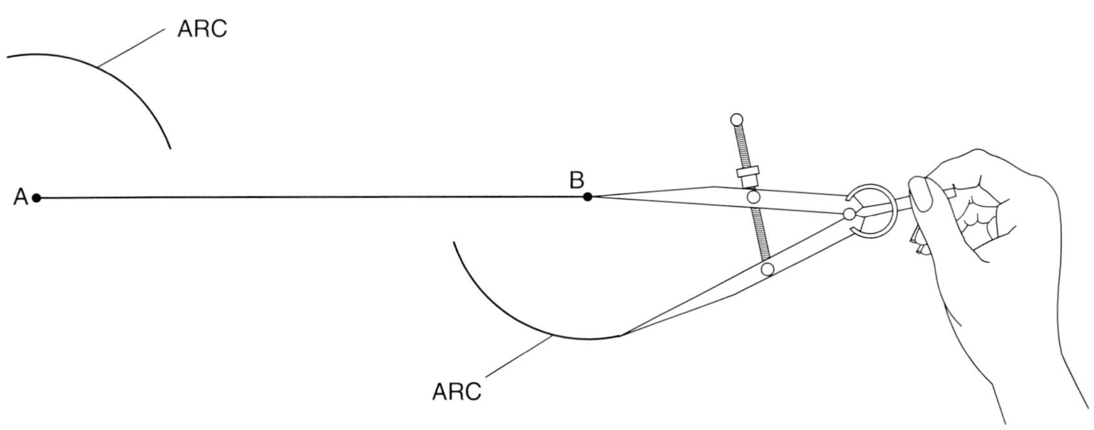

Figure 45 Arcs above and below the line.

Step 6 With a straightedge, scribe lines between points A-10, 1-9, 2-8, 3-7, 4-6, 5-B. These lines intersect A-B and divide it into five equal parts (*Figure 48*).

To divide any line into any number of equal parts, lay off an equal number of marks on each line.

4.3.7 Dividing A Circle Into Equal Parts Using Dividers

Follow these steps to divide a circle into six equal parts. Use dividers and a straightedge.

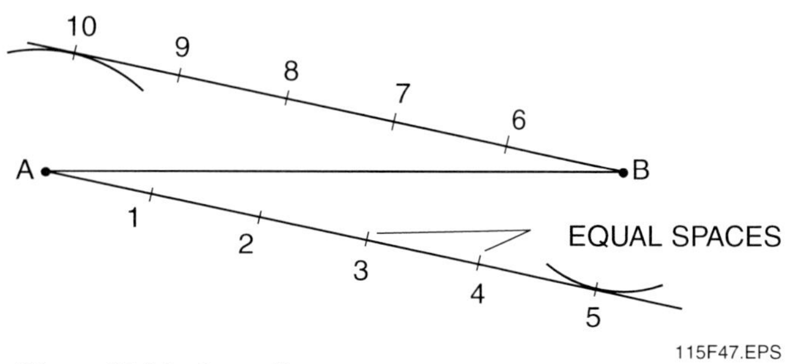

Figure 47 Marks on lines.

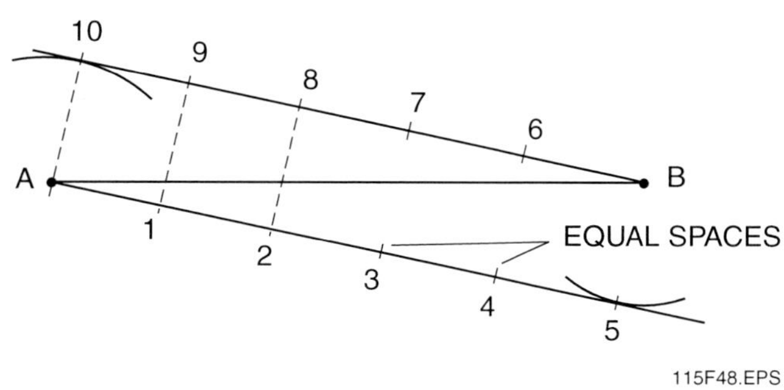

Figure 48 Lines scribed between marks on lines A and B.

Step 1 Scribe a circle to a specified diameter. Punch mark the center point. Do not change the radius of the dividers.

Step 2 Select any point on the **circumference** of the circle as a starting point. Mark the starting point number 1. Place the point of one divider leg at number 1 and scribe an arc on the edge of the circle. Place number 2 where this arc intersects the circle's edge (*Figure 49*).

Step 3 Place one divider leg at point number 2. Scribe another arc through the edge of the circle and label it number 3. Repeat this step around the circle, ending at point number 1. *Figure 50* shows a circle with six arcs.

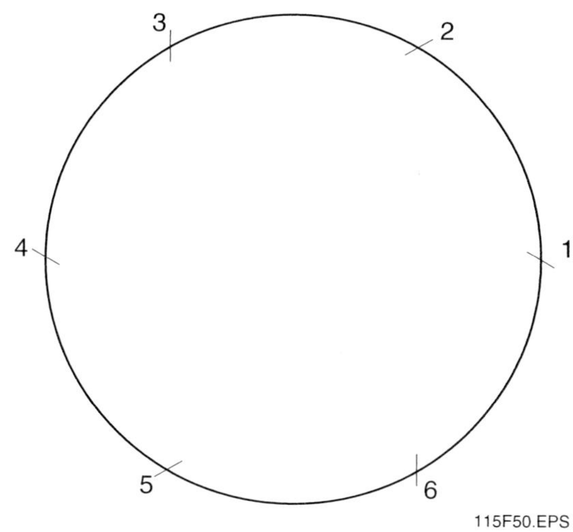

Figure 50 Lines scribed in a circle.

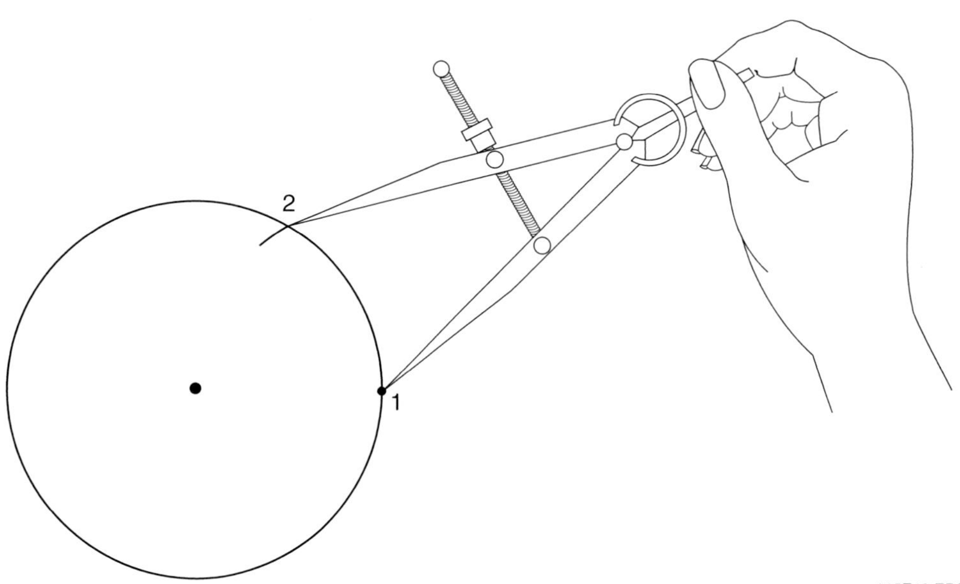

Figure 49 Circle and starting point.

Step 4 Using a straightedge, scribe lines connecting points 1 and 4, 2 and 5, and 3 and 6. All lines should intersect at the center point of the circle. The circle should now be divided into six equal parts, forming angles of 60 degrees each. *Figure 51* shows a circle divided into six equal parts.

Step 5 To divide the circle into 12 equal parts, bisect each of the 60-degree angles. *Figure 52* shows how to bisect one of the angles. *Figure 53* shows a circle divided into 12 equal parts.

4.4.0 Laying Out Bolt Pattern Templates

A template is a pattern used for marking holes and other cuts in metal parts. Templates are made from heavy paper or light gauge sheet metal. They are placed over the part to be cut as a guide for making punch marks.

Follow these steps to make a bolt pattern template. Use dividers and a straightedge.

Step 1 Review the necessary drawings and find the dimensions needed to lay out the template. *Figure 54* shows a drawing of a 6-inch flange.

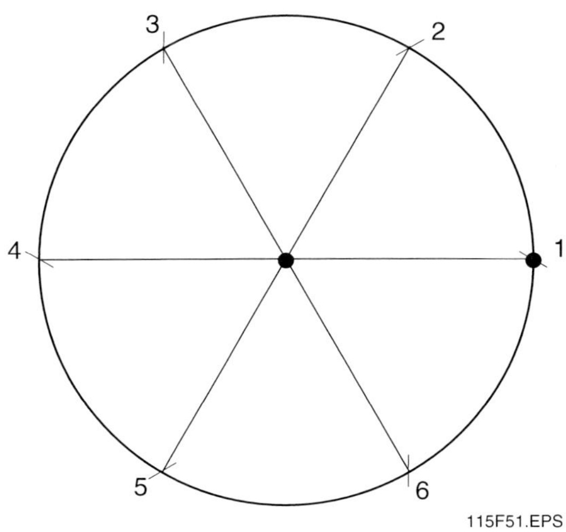

Figure 51 Six equal parts of a circle.

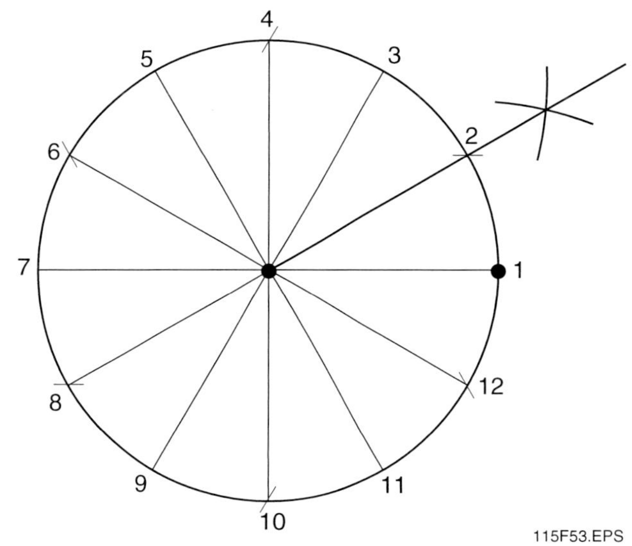

Figure 53 Twelve equal parts of a circle.

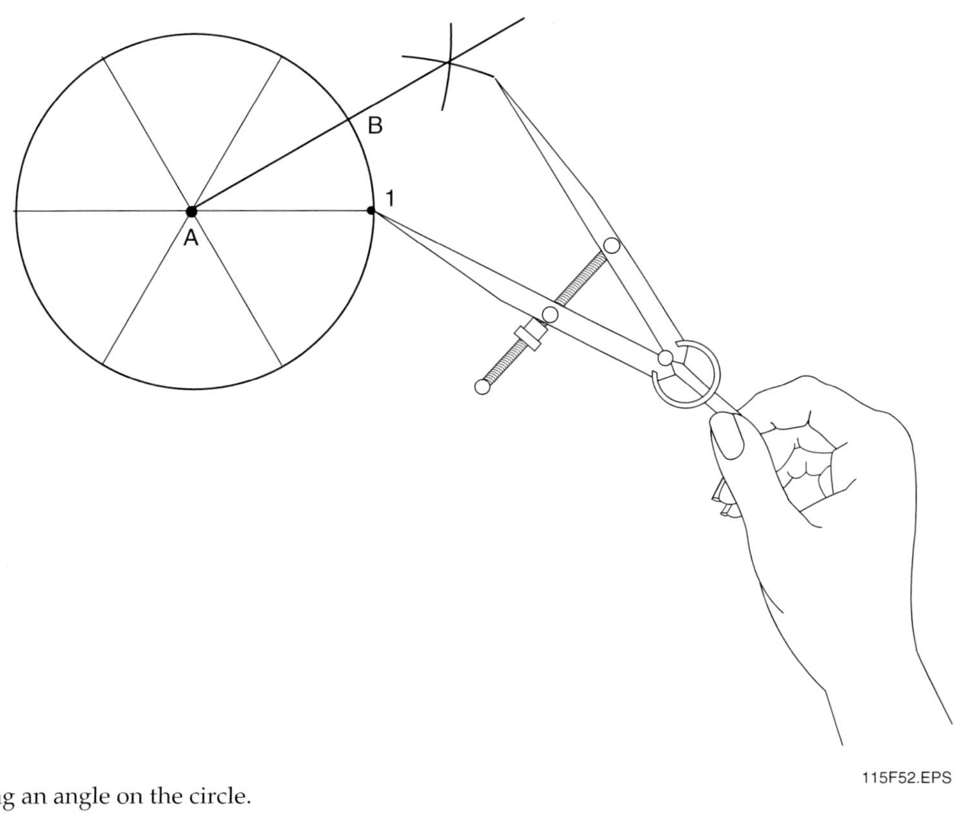

Figure 52 Bisecting an angle on the circle.

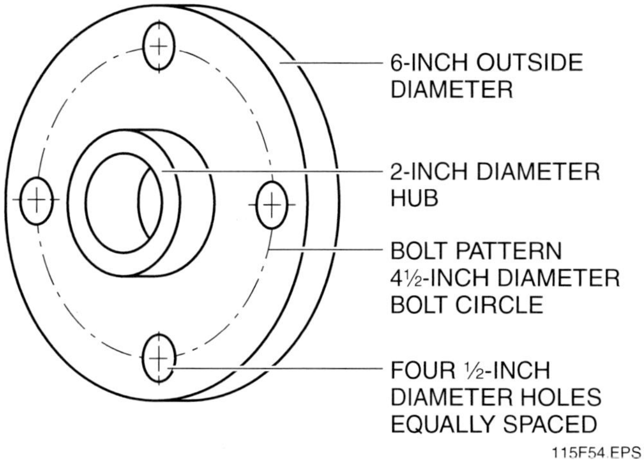

Figure 54 Six-inch flange.

Step 2 Lay out a horizontal reference line at least 8 inches long. Punch mark the midpoint of this line.

Step 3 Lay out a vertical reference line that intersects the horizontal line at a 90-degree angle through its midpoint. This line should extend at least 4 inches above and below the horizontal reference line. *Figure 55* shows horizontal and vertical intersecting reference lines.

Step 4 Set dividers for a 3-inch radius and scribe a circle from the midpoint of the reference line. The circle will have a diameter of 6 inches.

Step 5 Set dividers for a 2¼-inch radius and scribe a circle from the same midpoint. This circle will have a 4½-inch diameter. This smaller circle is called a bolt circle. The four points where the circle intersects the reference lines are the center points for the bolt holes. *Figure 56* shows this bolt circle.

Step 6 Prick punch the bolt hole centers. Set the dividers for a ¼-inch radius and scribe a ½-inch circle for each bolt hole.

5.0.0 Fabrication Tasks

The remainder of this module contains four separate tasks. Each is designed to provide hands-on experience in working with a specific structural shape. You will work with angle iron, channel iron, a T-shape, and a W-shape.

Figure 57 shows the parts of the various structural shapes. The terms are used in the fabrication tasks. Every ironworker should be familiar with these terms.

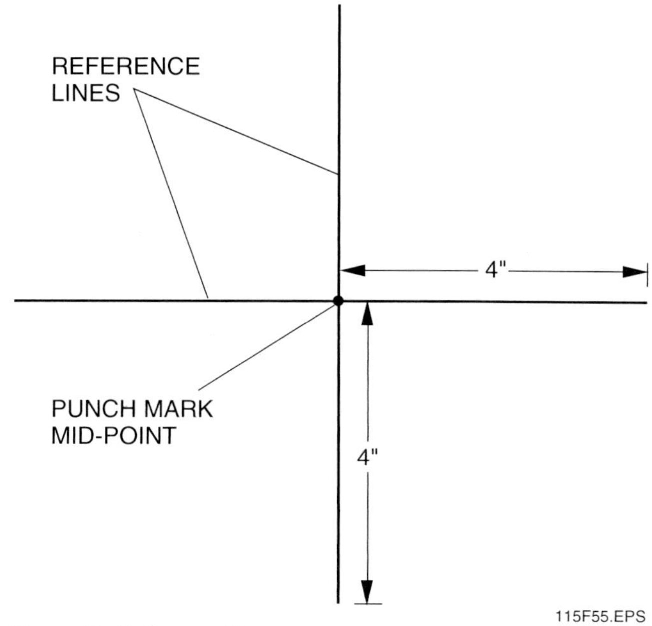

Figure 55 Reference lines.

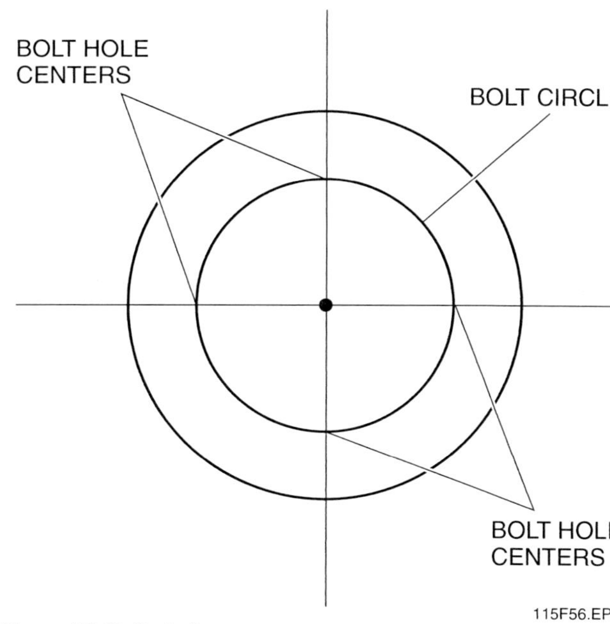

Figure 56 Bolt circle.

Some shop drawings contain a notation that reads GA and is followed by a dimension. This identifies the gauge.

Gauge is the spacing of holes about the center line of a structural member. Gauge is used on shop drawings when the thickness of a web must be taken into account. *Figure 58* shows two **angle-iron clips** placed on a W-shape. Notice how gauge is used to show the locations of the holes.

For angle iron, the gauge dimension may be given from the **heel** of the angle iron. *Figure 59* shows how far the holes are located from the angle iron.

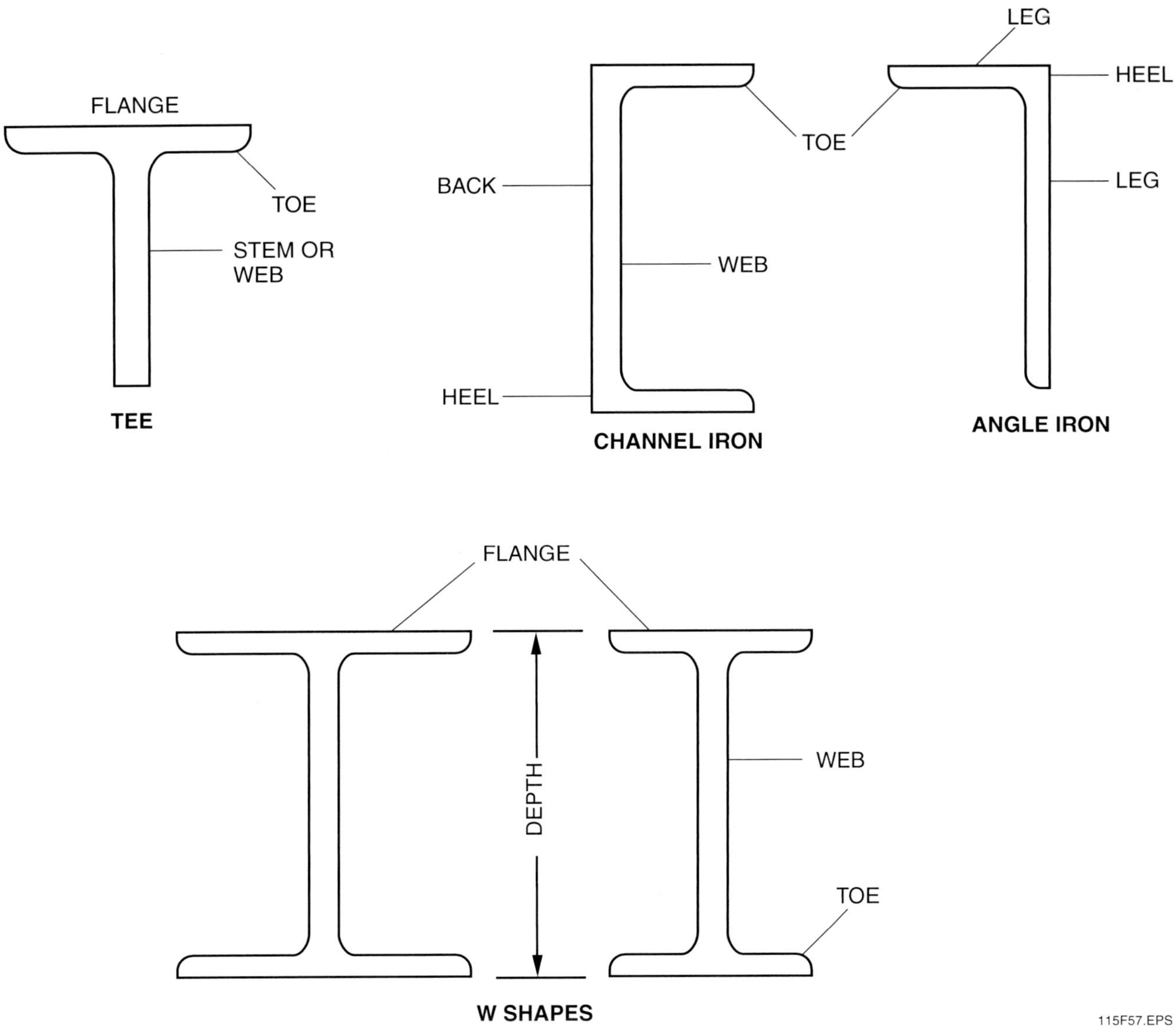

Figure 57 Structural shapes.

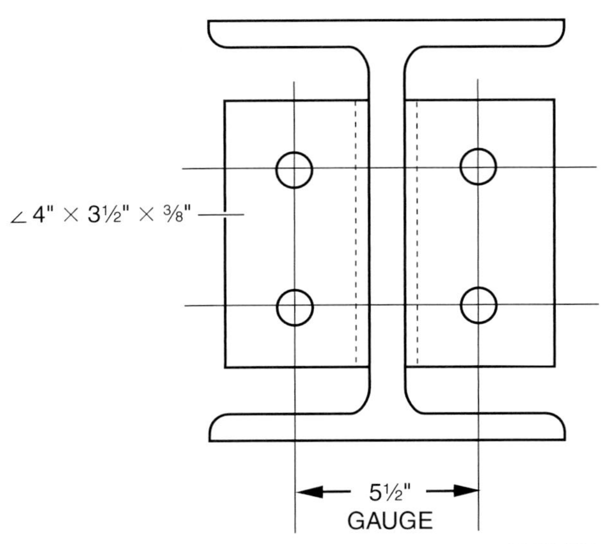

Figure 58 Two angle-iron clips on a W-shape.

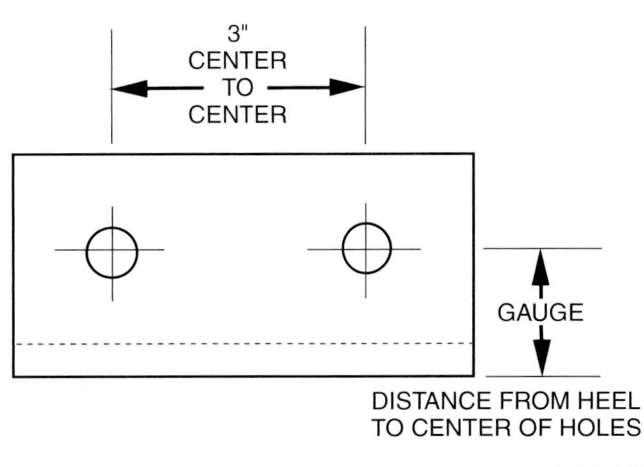

Figure 59 Locating holes on an angle iron.

In structural steel fabrication, the structural member is often trimmed so that it will fit the adjoining member. This trimming is called coping. In most cases, the drawing will detail the dimensions of the **cope**.

Unless otherwise noted, the cope should contain a radius. A sharp corner on a cope may lead to failure of the member. Follow the drawings carefully when coping.

Each of the following tasks contain a shop drawing and a list of materials and equipment you will need to complete the fabrication. Your instructor will set the standards to be used for evaluation.

Before doing any drilling, be sure to lay out the entire piece. All measurements in structural iron fabrication are taken from center lines or from an established working point. The tasks contain the necessary procedures.

5.1.0 Task 1: Fabricating Angle Iron

You will be fabricating 403DV6. The length is 6'-6¹¹⁄₁₆".

Figure 60, Shop Drawing I, shows lengths of 3½" × 3½" × ⁵⁄₁₆" angle iron that must be cut to specified lengths and drilled with six ¹³⁄₁₆" holes in specified locations.

Follow these steps to fabricate 403DV6:

Step 1 Obtain a piece of angle iron that is close to the required length.

Step 2 Draw a straight line on the angle iron from the heel to the **toe** using a tri-square and a carbide scratch awl.

> **NOTE**
> This line is the working point. All the holes are laid out from this line.

Step 3 Establish a center line by measuring 1¾" from the heel in two locations on the angle iron and connecting these points with a straight line.

> **NOTE**
> The holes are laid out on the center line of the angle iron. Notice that the center line of an angle iron is measured from the heel (outside edge) of the angle iron, not the toe.

Step 4 From the working point, measure off 1¼" along the center line.

> **NOTE**
> This is the location of the first hole. The remainder of the holes are laid out at 4¼", 7¼", 5'-11⁷⁄₁₆", 6'-2⁷⁄₁₆", and 6'-5⁷⁄₁₆". Notice also that the locations of the holes must be figured from the information given on the drawings.

Step 5 Centerpunch the location of each hole.

Step 6 Measure 6'-6¹¹⁄₁₆" from the working point and draw a straight line from the heel of the angle iron to the toe using a tri-square and a carbide scratch awl.

Step 7 Cut the angle iron to size with an oxy-acetylene torch.

Step 8 Dress the edges with a side grinder.

Step 9 Drill the ¹³⁄₁₆" holes, lubricating the bit as necessary.

> **NOTE**
> If a drill press is not available, clamp or tack the angle to a beam or other suitable support and use a magnetic drill.

5.2.0 Task 2: Fabricating Channel Iron

In this exercise, you will be fabricating channel 198G8, as shown on Shop Drawing II (*Figure 61*). The overall dimensions of this piece are listed in the Bill of Materials at the bottom of the figure. The designation C10 × 15.3 refers to a standard structural shape. (Look it up in the AISC Manual for an exact description.)

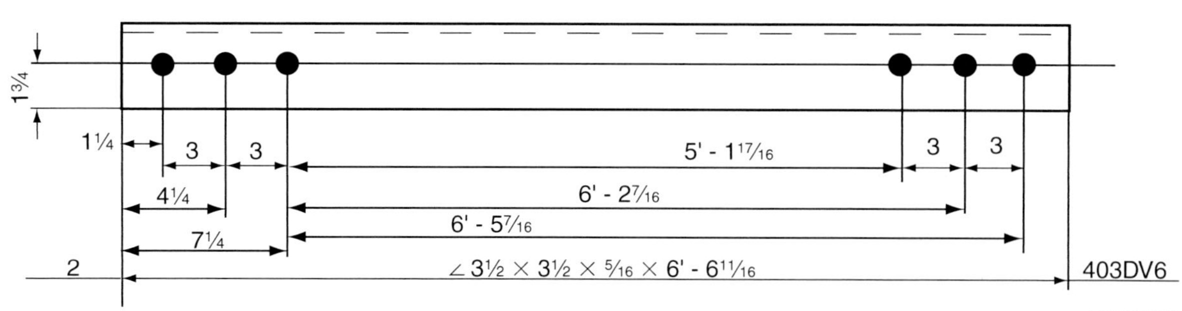

Figure 60 Shop drawing I.

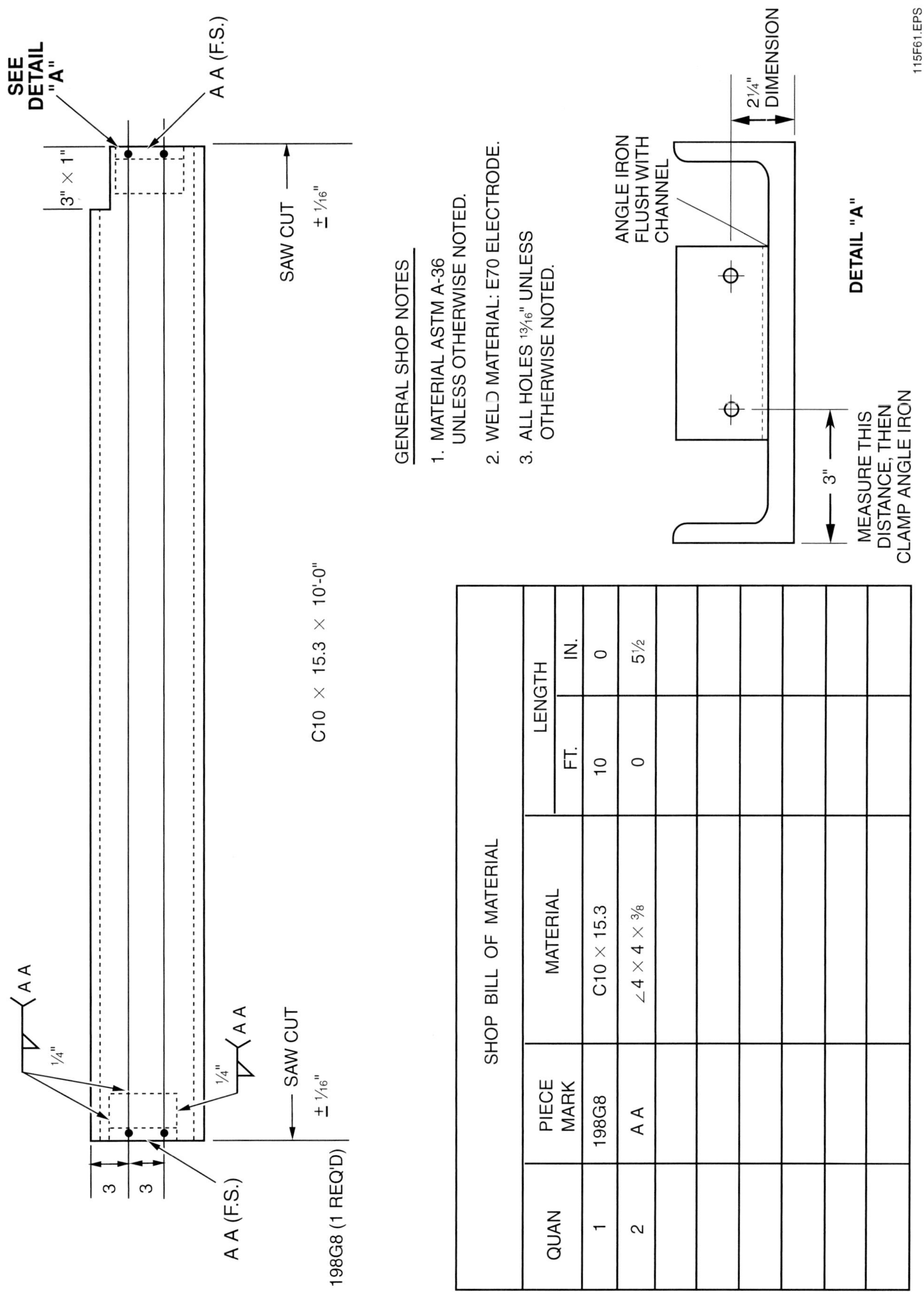

SHOP BILL OF MATERIAL

QUAN	PIECE MARK	MATERIAL	LENGTH FT.	LENGTH IN.
1	198G8	C10 × 15.3	10	0
2	A A	∠4 × 4 × 3/8	0	5 1/2

Figure 61 Shop drawing II.

The purpose of this exercise is to prepare a piece of channel to a given tolerance and to fabricate and install angle-iron clips.

Shop drawing II shows angle-iron clips welded on the ends of the channel. The welding symbol indicates that a ¼" fillet weld is to be placed on three sides of the angle iron. If you need to review welding symbols, see the *Trade Drawings* module.

Follow these steps to fabricate channel iron 198G8:

Step 1 Obtain a length of appropriate channel iron.

Step 2 Measure and mark 10' with the flanges facing down.

Step 3 Cut the channel with a bandsaw to within the tolerances given on the drawing.

> **NOTE** Be sure to keep the blade well oiled.

Step 4 Cut two lengths of angle iron near or slightly larger than 5½".

Step 5 Scribe a line near the end of each angle-iron clip with a tri-square and a carbide scratch awl. This is the working point from which all measurements are taken.

Step 6 Measure the web thickness of the channel iron and subtract the web thickness from the 2¼" dimension given in Detail A on Shop Drawing II. This establishes the center line of the holes as the center line is measured from the heel of the angle.

Step 7 Lay out the center line of the holes on each angle-iron clip.

Step 8 Lay out and centerpunch two holes on 3-inch centers on each angle-iron clip.

Step 9 Drill the 13⁄16" holes for the angle-iron clips using either a drill press or a magnetic drill.

Step 10 Place the clip so that there are 3 inches from the edge of the channel to the center line of the first hole.

> **NOTE** Be sure the flanges of the channel point in the right direction. The clip and channel must be flush at the edges.

Step 11 Clamp one end with a C-clamp.

Step 12 Recheck the measurements and adjust if needed.

Step 13 Clamp the other end.

Step 14 Repeat Steps 10 through 13 to fabricate the other clip.

> **NOTE** Be sure to take measurements from the same side of the channel.

Step 15 The clips are ready to be welded by a qualified welder.

> **NOTE** Be sure the clip is squared up to the channel.

Step 16 Mark the cope on the channel as shown on the drawing.

Step 17 Cut the cope with an oxyacetylene torch.

> **NOTE** Be sure to leave a radius.

Step 18 Dress the edges with a side grinder.

5.3.0 Task 3: T-Shape Fabrication

In this exercise, you will be fabricating T-shape 183DV2. *Figure 62*, Shop Drawing III, provides full details.

Follow these steps to fabricate T-shape 183DV2:

Step 1 Obtain a T-shape close to the size required.

Step 2 Establish and mark the center line of the T-shape from the web. This is done by measuring the thickness of the web, dividing it by 2, and marking the center line at the top of the flange at both ends of the T-shape.

Step 3 Mark a square line across the T-shape near the end with a tri-square and a carbide scratch awl. This is the working point from which all measurements must be taken.

Step 4 Measure 1½" from the working point along the center line of the T-shape and then 1⅜" at right angles on each side of the center line to lay out the first two holes. Since the gauge is given as 2¾", the center line of each hole straddles the center line of the T-shape by 1⅜". The distance from the edge of the T-shape to the center of the first set of holes is 1½". This distance is called the **edge distance**. Edge distance is a design standard that varies with the size of the hole and type of structural member.

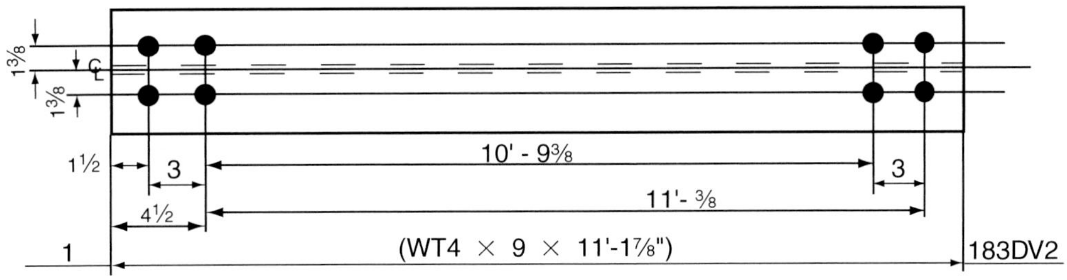

SHOP BILL OF MATERIAL				
QUAN	PIECE MARK	MATERIAL	LENGTH	
			FT.	IN.
1	183DV2	WT4 × 9	11	1⅞

GENERAL SHOP NOTES

1. MATERIAL ASTM A-36 UNLESS OTHERWISE NOTED.
2. WELD MATERIAL: E70 ELECTRODE.
3. ALL HOLES 13⁄16"" UNLESS OTHERWISE NOTED.

115F62.EPS

Figure 62 Shop drawing III.

Step 5 Measure 4½" from the working point along the center line of the T-shape and then 1⅜" at right angles on each side of the center line to lay out the second set of holes.

Step 6 Measure 10'-9⅜" from the working point along the center line of the T-shape and then 1⅜" at right angles on each side of the center line to lay out the third set of holes.

Step 7 Measure 11'-1⅜" from the working point along the center line of the T-shape and then 1⅜" at right angles on each side of the center line to lay out the fourth set of holes.

Step 8 Centerpunch the locations for the holes.

Step 9 Measure 11'-1⅞" from the working point along the center line of the T-shape and mark a square line with a tri-square and carbide scribing tool.

Step 10 Set up the magnetic drill and drill 13⁄16" holes.

> **NOTE**
> Use oil to protect the bit.

Step 11 Cut the T-shape to size with an oxyacetylene torch.

Step 12 Dress the edges with a side grinder.

5.4.0 Task 4: W-Shape Fabrication

In this exercise, you will be fabricating beam 303B2 from a standard W-shape. Angle-iron clips are to be welded to the ends. *Figure 63*, Shop Drawing IV, provides complete information, but certain dimensions must be calculated from the drawing.

Follow these steps to fabricate beam 303B2:

Step 1 Obtain a W-shape and cut it close to the size shown on the drawing.

Step 2 Establish the center line of the W-shape from the web.

Step 3 Cut the four angle-iron clips to the dimensions given in the Bill of Materials.

Step 4 Measure the thickness of the web, subtract the gauge, and divide by 2 to lay out the holes on "aa" and "ab."

Step 5 Centerpunch the locations of the holes on "aa" and "ab."

Step 6 Drill the holes with a magnetic drill or drill press.

Step 7 Fit clips "aa" and "ab" to the W-shape using C-clamps.

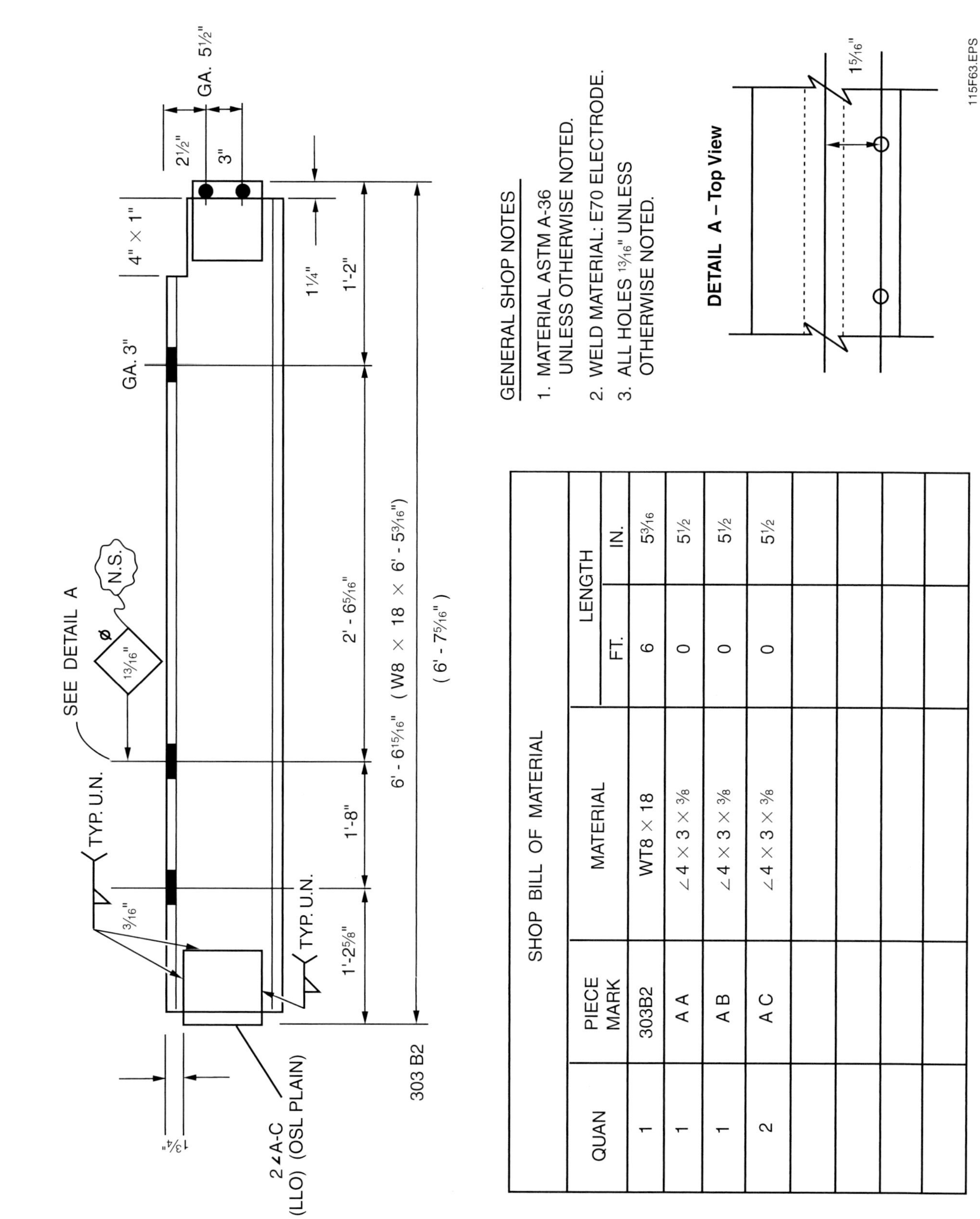

SHOP BILL OF MATERIAL

QUAN	PIECE MARK	MATERIAL	LENGTH FT.	LENGTH IN.
1	303B2	WT8 × 18	6	5 3/16
1	A A	∠4 × 3 × 3/8	0	5 1/2
1	A B	∠4 × 3 × 3/8	0	5 1/2
2	A C	∠4 × 3 × 3/8	0	5 1/2

Figure 63 Shop drawing IV.

> **NOTE**
> Follow the drawing carefully to obtain the correct placement. Notice that clips "aa" and "ab" extend 1¼" from the end of the W-shape. The clips fit LLO-long leg opposite. This means that the 3-inch side of each 4 × 3 × ⅜ angle-iron clip butts against the W-shape.

Step 8 Fit clips "ac" to the other end of the W-shape. These clips also fit LLO. Notice the dimensions given on the drawing.

Step 9 Measure the distance from one set of clips to the other. The overall distance should be 6'-7⁵⁄₁₆". Adjust the clips as needed to obtain this distance.

Step 10 Check the fitup of all the angle-iron clips against the dimensions given on the drawing. Readjust as needed.

Step 11 Have a qualified welder tack weld the clips to the W-shape.

Step 12 Lay out the two ¹³⁄₁₆" holes on the near side (NS) of the W-shape.

> **NOTE**
> Take all measurements from the same point, working from clips "aa" and "ab." See Detail A for the placement of the holes in relation to the center line of the W shape.

Step 13 Centerpunch the locations of the holes.

Step 14 Lay out the other set of holes.

> **NOTE**
> The gauge is 3 inches.

Step 15 Centerpunch the location of the holes.

Step 16 Drill the ¹³⁄₁₆" holes with a magnetic drill or a drill press. Oil the bit as you drill.

Step 17 Have a qualified welder weld the clips to the W-shape.

> **NOTE**
> The welding symbol gives details about the fillet.

Step 18 Lay out the 4" × 1" cope on the end of the W-shape.

Step 19 Cut out the cope with an oxyacetylene torch. Be sure to cut a radius.

Step 20 Dress the edge of the cope with a side grinder.

SUMMARY

Field fabrication is performed on every ironworking job. It may be simple or complex, depending upon the requirements of the particular project, but the basic skills needed for fabrication remain the same. Ironworkers must be able to work from shop drawings, cut steel to length, lay out holes, and drill them to specified tolerances. These are essential skills of the trade.

Review Questions

1. According to OSHA, when performing any grinding, even just a touch up, always wear _____.
   a. rubber gloves
   b. a welding mask
   c. a face shield
   d. a respirator

2. A steel rule may be marked in ⅛ and $\frac{1}{16}$ of an inch on one side and _____ on the other side.
   a. decimals
   b. whole numbers
   c. $\frac{1}{32}$ and $\frac{1}{64}$ of an inch
   d. tenths of an inch

3. A tool that can be used to lay out 45-degree angles, 90-degree angles, and parallel lines, as well as to measure the height of parts is called a _____.
   a. combination set
   b. protractor
   c. steel square
   d. divider

4. The magnetic drill is tied off and locked to the beam or plate by _____.
   a. bolting
   b. an electromagnet
   c. clamps
   d. welding

5. When using the 3-4-5 method to lay out a perpendicular line, always use _____.
   a. inches
   b. centimeters
   c. millimeters
   d. the largest unit possible

6. To lay out a number of parallel lines perpendicular to a reference edge, use a _____.
   a. combination square
   b. framing square
   c. straightedge
   d. tri-square

7. When scribing a circle using dividers, the dividers must be adjusted to the specified _____ of the circle.
   a. circumference
   b. diameter
   c. radius
   d. tangent

8. To divide a circle into 6 equal parts, scribe arcs on the circumference of the circle with the dividers set to _____ of the circle.
   a. the radius
   b. the diameter
   c. one-half the radius
   d. twice the radius

9. The notation GA followed by a number on a shop drawing indicates _____.
   a. that the material came from Georgia
   b. the thickness of the member
   c. the spacing of holes about a center line of the structural member
   d. the size of the hole to be drilled

10. The center line of a 3½ × 3½ × $\frac{5}{16}$ inch angle iron is _____ inches from the heel.
    a. $\frac{5}{16}$
    b. 1¾
    c. 1⅞
    d. 3½

Trade Terms Introduced in This Module

Angle-iron clips: Sections of angle iron welded to another shape for support or reinforcement.

Circumference: The outside edge of a circle.

Control lines: Additional center lines laid out parallel to the base line.

Cope: To trim a structural member so that it will fit to an adjoining member.

Edge distance: The distance from the edge of a shape to the center of specified holes.

Heel: The outside corner of an angle or channel.

Intersection: The point where two or more lines cross.

Magnetic drill: A portable drill press fitted with a magnetic base.

Morse taper: The chuck attached to the magnetic drill replacing the standard three-jaw chuck.

Perpendicular line: A line that is at a 90-degree angle to another line.

Radius: The distance from the center of a circle to the edge.

Tangent: A line that is touching but not intersecting a circle or arc.

Toe: The outside edge of a structural shape.

Additional Resources

This module presents thorough resources for task training. The following resource material is suggested for further study.

Steel Construction Manual, 13th edition. Chicago, IL: American Institute of Steel Construction.

Detailing for Steel Construction, 3rd edition. Chicago, IL: American Institute of Steel Construction.

Figure Credits

Topaz Publications, Inc., Module opener, Figures 6 and 12

L.S. Starrett Company, Figures 1–4 and 8-10

General Tools Manufacturing Co., LLC, Figures 5 and 7

DeWALT Industrial Tool Company, Figure 11

NCCER CURRICULA — USER UPDATE

NCCER makes every effort to keep its textbooks up-to-date and free of technical errors. We appreciate your help in this process. If you find an error, a typographical mistake, or an inaccuracy in NCCER's curricula, please fill out this form (or a photocopy), or complete the online form at **www.nccer.org/olf**. Be sure to include the exact module ID number, page number, a detailed description, and your recommended correction. Your input will be brought to the attention of the Authoring Team. Thank you for your assistance.

Instructors – If you have an idea for improving this textbook, or have found that additional materials were necessary to teach this module effectively, please let us know so that we may present your suggestions to the Authoring Team.

NCCER Product Development and Revision

13614 Progress Blvd., Alachua, FL 32615

Email: curriculum@nccer.org
Online: www.nccer.org/olf

❑ Trainee Guide ❑ Lesson Plans ❑ Exam ❑ PowerPoints Other ______________________

Craft / Level: Copyright Date:

Module ID Number / Title:

Section Number(s):

Description:

Recommended Correction:

Your Name:

Address:

Email: Phone:

Glossary

A-325 bolt: A high-strength bolt made of medium- or low-carbon steel that is used in bolting up steel.

A-490 bolt: A high-strength bolt made of low-alloy steel that is used in bolting up steel.

Aerial lift: A mobile work platform designed to transport and raise personnel, tools, and materials to overhead work areas.

Alloy: A metal that has had other elements added to it that substantially change its mechanical properties.

Alternating current (AC): Electrical current that reverses its flow at set intervals.

Amperage: A measurement of the rate of flow of electric current.

Anchor bolts (anchor rods): Threaded rods with a nut on the end used to transfer tension forces to the foundation. They are usually embedded into the concrete foundation.

Angle-iron clips: Sections of angle iron welded to another shape for support or reinforcement.

Anneal: To soften a metal by heat treatment.

Anti-two-blocking devices: Devices that provide warnings and prevent two-blocking from occurring. Two-blocking occurs when the lower load block or hook comes into contact with the upper load block, boom point, or boom point machinery. The likely result is failure of the rope or release of the load or hook block.

Apprentice: A worker who is learning a craft.

Arc: Flow of electricity through an air gap that melts the electrode.

Arc blow: The deflection of the arc from its intended path by magnetic forces.

Arc strike: A discontinuity consisting of localized melting of the base metal or finished weld caused by the arc. It occurs as a result of striking the arc outside the area to be welded.

Arrow line: The line drawn at an angle from the reference line (either end or both ends) to an arrow head at the location of the weld.

Backfire: A loud snap or pop as a torch flame is extinguished.

Backing: A weldable or nonweldable material used behind a root opening to allow defect-free welding at the open root of a joint.

Beam: A single piece that is a structural element mounted more or less horizontally.

Bent: All frames (whether rigid, truss-on-column, rafters-on-column) or other members used to span between columns in the principal plane.

Bevel washer: A washer that provides a square seat for a bolt or nut when it passes through a structural member that has a sloping inner surface.

Bird caging: A deformation of wire rope that causes the strands or lays to separate and balloon outward, similar to the vertical bars of a bird cage.

Blocking: Pieces of hardwood used to support and brace equipment.

Bracing: Diagonal or horizontal supports for joists or beams. Bracing helps to keep joists or beams from twisting or rolling.

Bridging: Lateral supports for joists.

Bull's tail (Christmas tree): A double strand of wire rope with a series of hooks, used to carry a series of beams or joists and lift them into place sequentially.

Button punch: A tool used to pinch a round protrusion in a decking lap, to hold the two pieces in relation to each other.

Calibrated wrench method: A technique of tightening a nut and bolt in which a specific amount of torque is applied to the connection.

Capacity: The total amount of weight the lift is capable of lifting. This includes personnel, tools and materials, and/or equipment.

Carbody: The part of a crawler crane base mounting that carries the rotating upper structure.

Carburizing flame: A flame burning with an excess amount of fuel; also called a reducing flame.

Cellular decking: A normal composite floor or roof decking profile that has a steel sheet attached to the bottom.

Center of gravity: The point at which an object is perpendicular to and balanced in relation to the earth's gravitational field.

Chord: The top and bottom of bar joists.

Circuit breaker: A device designed to protect circuits from overloads, which is able to be reset after tripping.

Circumference: The outside edge of a circle.

Column: A single vertical structural element.

Come-along: Portable hoist used with guys to pull frames into plumb.

Compliance: Acting in accordance with regulations.

Compressor: A mechanical device that generates compressed air to power a variety of tools.

Condensation: The process in which water from the atmosphere forms on a cool surface.

Connections: Devices used to attach units to each other or to the structure.

Control lines: Additional center lines laid out parallel to the base line.

Cope: To trim a structural member so that it will fit to an adjoining member.

Counterweights: Weights added to the opposite side of the crane boom to offset the weight of the load.

Cribbing: Timbers stacked in alternate tiers. Used to support heavy loads.

Deck: Material used as a horizontal surface on a building level. Decking may be used to hold a concrete floor, or may serve as a roof.

Direct current (DC): Electrical current that flows in one direction only.

Distortion: The expansion and contraction of welded parts caused by the heating and subsequent cooling of the weld joint.

Dog: A safety mechanism to prevent operation.

Drag angle: The travel angle when the electrode is pointing in a direction opposite to the welding progression.

Drag lines: The lines on the edge of the material that result from the travel of the cutting oxygen stream into, through, and out of the metal.

Drift pin: Pin used for aligning holes in structural steel members and for holding them in place prior to fastening.

Dross: The material (oxidized and molten metal) that is expelled from the kerf when cutting using a thermal process. It is sometimes called slag.

Drum: The part of a crane that holds the unused lifting line and that transmits the line pull to the lifting line.

Ductile: The mechanical property of a material that allows it to be bent or shaped without breaking.

Dunnage: Lumber or other material that supports and protects cargo from damage during transport or storage.

Edge distance: The distance from the edge of a shape to the center of specified holes.

Electrode: Point from which the arc is produced to perform welding.

End bearing: The part of the structure that supports the joists.

Equalizer beam: A beam used to distribute weight on multi-crane lifts.

Ferrous metals: Metals containing iron.

Fixed block: The upper block of a block and tackle; the block that is attached to the support.

Flashback: The flame burning back into the tip, torch, hose, or regulator, causing a high-pitched whistling or hissing sound.

Flux: Material used to prevent, dissolve, or facilitate the removal of oxides and other undesirable substances on a weld.

Forklift: A machine designed to move bulk items around the job site; also known as a powered industrial truck.

Frog: A device for carrying tools on a belt.

Fuse: To weld and join together base metal components.

Fuse holder: A device used to hold a fuse that may be part of a circuit or that may hold a spare fuse.

Galvanized: To apply a zinc coating to a base metal as an effective barrier of protection. The zinc coating isolates the metal from the environment, adheres to it, and provides relatively superior abrasion resistance.

Generator: A machine designed to generate electricity.

Grip: The thickness of the steel sections to be joined.

Ground: An object that makes an electrical connection with the earth.

Ground fault circuit interrupter (GFCI): A device that detects grounds on a power line and interrupts power flow to the attached outlet(s).

Guys: Cables and fixtures attached to a building frame used to rack the building into plumb.

Hauling line: The line of a lifting device that is pulled by hand to raise the load.

Heat-affected zone: The part of the base metal that has been altered but not melted by the heat.

Heel: The outside corner of an angle or channel.

Hermetically sealed: Having an airtight seal.

Hot work permit: Official authorization from a site manager to perform work that may pose a fire hazard.

Hydraulic: Operated by fluid pressure.

Interlocking systems: Control systems that are set to prevent accidents by keeping one system from overriding the other.

Intersection: The point where two or more lines cross.

Joist girder: Girders manufactured in the same manner as open-web joists.

Kerf: The gap produced by a cutting process.

Kinking: Bending a rope so severely that the bend is permanent and individual wires or fibers are damaged.

Leveling nuts: Leveling nuts are used to bear the weight of the frame until the bases are grouted.

Lifting lugs: Metal attachments welded to a metal item. The lug has a hole to which a shackle can be attached, allowing the piece to be lifted.

List angle: The tipping angle measured from the level of a floating crane.

Load: The total forces acting on a structural member.

Low-hydrogen electrode: An electrode specially manufactured to contain little or no moisture.

Luffing: Changing a boom angle by varying the length of the boom suspension ropes.

Magnetic drill: A portable drill press fitted with a magnetic base.

Melt-through: Complete joint penetration.

Moment: The rotation produced in a body when a force is applied; used to determine tipping load.

Morse taper: The chuck attached to the magnetic drill replacing the standard three-jaw chuck.

Mushrooming: The curling or flattening of the end of a tool from continual use.

National Center for Construction Education and Research (NCCER): An organization whose primary goal is to standardize construction craft training throughout the country so both employers and employees will benefit from the training, no matter their location.

Neutral flame: A flame burning with correct proportions of fuel gas and oxygen.

Notch toughness: The ability of a material to resist breaking at points where stress is concentrated.

Occupational Safety and Health Administration (OSHA): A division of the U.S. Department of Labor that is responsible for establishing and enforcing safety and health standards for all places of employment.

Open-web steel joist (bar joist): Joists made of chords and bars rather than solid web.

Outrigger: An extension that projects from the main body of a crane to add stability and support.

Oxidizing flame: A flame burning with an excess amount of oxygen.

Parts of line: The number of ropes between the upper and lower blocks of a block and tackle. Parts of line, also called falls, are the lines that carry the load.

Perpendicular line: A line that is at a 90-degree angle to another line.

Phosphatized: A metal that has been treated with a thin coating of phosphoric acid or a phosphate compound. The phosphate causes a chemical reaction on the surface of the metal that allows powder coatings to adhere effectively. Phosphatizing also inhibits corrosion from forming underneath the powder coating.

Pierce: To penetrate through metal plate with an oxyfuel cutting torch.

Plumb bob: A cone-shaped weight suspended at the end of a string or wire used to check the plumb of a vertical member.

Porosity: Gas pockets, inclusions, or voids in the weld metal.

Powered industrial trucks: See forklift.

Procedure qualification: The demonstration that welds made following a specific process can meet prescribed standards.

Procedure qualification record (PQR): The document containing the results of the nondestructive and destructive testing required to qualify a welding procedure specification (WPS).

Profile: The ribbed pattern shown in a cross section of decking.

Project engineer of record: Registered licensed professional responsible for the design of structural steel framing and whose seal appears on structural documents.

Proportional control: A control that increases speed in proportion to the movement of the control.

Purlin: The horizontal roof beam used to span the bay distance and on which the roof deck rests.

Push angle: The travel angle when the electrode is pointing in the same direction as the welding progression.

Push-pull jack: A fixture similar to the steamboat ratchet but capable of pulling or pushing.

Radius: The distance from the center of a circle to the edge.

Reaction: Forces acting on horizontal members of a structure that support the load.

Reciprocating saw: Saw used for cutting steel during the fabrication, demolition, and modification (reworking) of existing steel work.

Reference line: The horizontal line in the center of the welding symbol from which all elements of the welding symbol are referenced. The reference line is one of the most important elements of the welding symbol.

Related technical instruction: Classroom instruction that is designed to support on-the-job learning.

Restart: The point in the weld where one weld bead stops and the continuing bead is started.

Rolled-in embossment: Angular indentions and patterns in the ribs of decking that are formed when the steel is rolled. They are designed to interlock with the concrete slab.

Root face: A small facing on the end of a bevel for a groove weld.

Root opening: The space between the base metal pieces at the bottom or root of the joint.

Root pass: The initial pass of a weld that requires multiple passes.

Seize: To bind a rope in order to prevent it from untwisting after it has been cut.

Shackle: A piece of rigging equipment consisting of a horseshoe-shaped piece with a pin across the opening, used to connect slings, chains, and other lifting lines to loads, or to join slings together.

Shear stud: A steel rod with an enlarged top welded to the deck and frame to lock the concrete in place.

Shop drawings: All drawings prepared by the engineer, divided into erection drawings, anchor setting drawings, and production drawings.

Skidmore: A device that tests the tightness of a bolt.

Sleever bar: Bar used to provide extra leverage when positioning steel members and aligning bolt holes.

Sling angle: The angle formed by the legs of a sling with respect to the horizontal when tension is put upon the load.

Snug tight: As tight as a person can tighten a bolt without a tool.

Soapstone: Soft, white stone used to mark metal.

Softener: A piece of material that can be placed so as to prevent abrasion or cutting of slings or wire ropes by the sharp corners of loads.

Spandrel: The beams around the outside of the floor.

Spatter: Metal particles discharged into the air during welding.

Spools: Round fixtures attached to the end of a guy to reduce bending strain.

Spreaders: Beams used to distribute weight bearing along longer loads.

Spud wrench: A type of erection wrench designed to withstand high-leverage and heavy loads. It has a tapered handle that can be used to line up bolt holes.

Spur tracks: Special lines of railroad tracks that allow flat cars of material to be left at a site for unloading, so as not to interfere with railroad operations. Spur tracks may be single-ended or double-ended, depending on whether the spur returns to the main line or not.

Steamboat ratchet: A fixture used to pull two construction pieces together when more force is required than can be achieved with a turnbuckle.

Stickout: The amount of the threaded portion of the bolt that sticks out beyond the face of the nut after tightening.

Straight blade duplex connector: An electrical connector or style of outlet that pushes in to connect.

Stress: Forces acting on each structural member.

Stringer bead: A type of weld bead made without appreciable weaving motion. With shielded metal arc welding (SMAW), stringer beads are not more than three times the electrode diameter.

Substrate: The underlying layer, or the material onto which another layer is applied or fabricated.

Surfacing: The application by welding, brazing, or thermal spraying of a layer of material to a surface to obtain desired properties or dimensions.

Swing circle: The place where the upper mechanisms of a crane are joined to the base.

Tangent: A line that is touching but not intersecting a circle or arc.

Tension control (TC) bolt: A special bolt that has a snap-off spline.

Thread chaser: A die used for cleaning or repairing the threads on a rod or bolt. Also called a rethreader die.

Tie joists (OSHA joists): Joists located at or near column lines.

Toe: The outside edge of a structural shape.

Torque: A measure of turning force expressed in foot-pounds (ft lb) or Newton-meters (Nm).

Traceability: The ability to verify that a procedure has been followed by reviewing the documentation step by step.

Truss: Fabricated supports used to span between bents.

Turn-of-nut method: The technique of tightening a nut and bolt connection that requires a specific amount of turning of the nut after it has become snug.

Turnbuckle: A fixture used with guys to pull framing into plumb.

Twist-lock connector: A type of electrical connector that twists to connect.

Vertical welding: Welding with a downward progression from top to bottom.

Voltage: A measurement of electromotive force or pressure that causes current to flow in a circuit.

Weave bead: A type of weld bead made by transverse oscillation of the electrode.

Weld axis: A straight line drawn through the center of a weld along the length of the weld.

Weld coupon: The metal pieces to be welded together as a test or practice.

Weld symbol: A graphic character connected to the reference line of a welding symbol specifying the weld type.

Weld washers: Ring-shaped metal pieces, used to add strength to the weld between light gauge sheet metal and either frame members or other sheet metal.

Welding symbol: A graphical representation of the specifications for producing a welded joint; includes a reference line and arrow and many also include a weld symbol.

Wire rope: A rope made of steel wires, used for cranes and rigging.

Wire rope clips: U-bolt or clamp clips that clamp wire ropes to form loops for attachments.

Work angle: An angle less than 90 degrees between a line perpendicular to the major workpiece surface and a plane determined by the electrode axis and the weld axis. In a T-joint or corner joint, the line is perpendicular to the non-butting member. The definition of work angle for a pipe weld is covered in a later module.

Workpiece: The part being welded, brazed, soldered, or cut.

Yarding: The unloading and placement of material in the site area, including placement, storage and movement of material.

Index

A

C

F

M

N

O

S

T

U

V

W